STATISTICS OF MEDICAL IMAGING

CHAPMAN & HALL/CRC

Interdisciplinary Statistics Series

Series editors: N. Keiding, B.J.T. Morgan, C.K. Wikle, P. van der Heijden

Published titles

AN INVARIANT APPROACH TO STATISTICAL ANALYSIS OF SHAPES	S. Lele and J. Richtsmeier
ASTROSTATISTICS	G. Babu and E. Feigelson
BAYESIAN ANALYSIS FOR POPULATION ECOLOGY	Ruth King, Byron J.T. Morgan, Olivier Gimenez, and Stephen P. Brooks
BAYESIAN DISEASE MAPPING: HIERARCHICAL MODELING IN SPATIAL EPIDEMIOLOGY	Andrew B. Lawson
BIOEQUIVALENCE AND STATISTICS IN CLINICAL PHARMACOLOGY	S. Patterson and B. Jones
CLINICAL TRIALS IN ONCOLOGY, SECOND EDITION	J. Crowley, S. Green, and J. Benedetti
CLUSTER RANDOMISED TRIALS	R.J. Hayes and L.H. Moulton
CORRESPONDENCE ANALYSIS IN PRACTICE, SECOND EDITION	M. Greenacre
DESIGN AND ANALYSIS OF QUALITY OF LIFE STUDIES IN CLINICAL TRIALS, SECOND EDITION	D.L. Fairclough
DYNAMICAL SEARCH	L. Pronzato, H. Wynn, and A. Zhigljavsky
GENERALIZED LATENT VARIABLE MODELING: MULTILEVEL, LONGITUDINAL, AND STRUCTURAL EQUATION MODELS	A. Skrondal and S. Rabe-Hesketh
GRAPHICAL ANALYSIS OF MULTI-RESPONSE DATA	K. Basford and J. Tukey
INTRODUCTION TO COMPUTATIONAL BIOLOGY: MAPS, SEQUENCES, AND GENOMES	M. Waterman

Published titles

MARKOV CHAIN MONTE CARLO IN PRACTICE	W. Gilks, S. Richardson, and D. Spiegelhalter
MEASUREMENT ERROR AND MISCLASSIFICATION IN STATISTICS AND EPIDEMIOLOGY: IMPACTS AND BAYESIAN ADJUSTMENTS	P. Gustafson
MEASUREMENT ERROR: MODELS, METHODS, AND APPLICATIONS	J. P. Buonaccorsi
META-ANALYSIS OF BINARY DATA USING PROFILE LIKELIHOOD	D. Böhning, R. Kuhnert, and S. Rattanasiri
STATISTICAL ANALYSIS OF GENE EXPRESSION MICROARRAY DATA	T. Speed
STATISTICAL AND COMPUTATIONAL PHARMACOGENOMICS	R. Wu and M. Lin
STATISTICS IN MUSICOLOGY	J. Beran
STATISTICS OF MEDICAL IMAGING	T. Lei
STATISTICAL CONCEPTS AND APPLICATIONS IN CLINICAL MEDICINE	J. Aitchison, J.W. Kay, and I.J. Lauder
STATISTICAL AND PROBABILISTIC METHODS IN ACTUARIAL SCIENCE	P.J. Boland
STATISTICAL DETECTION AND SURVEILLANCE OF GEOGRAPHIC CLUSTERS	P. Rogerson and I. Yamada
STATISTICS FOR ENVIRONMENTAL BIOLOGY AND TOXICOLOGY	A. Bailer and W. Piegorsch
STATISTICS FOR FISSION TRACK ANALYSIS	R.F. Galbraith
VISUALIZING DATA PATTERNS WITH MICROMAPS	D.B. Carr and L.W. Pickle

Leabharlann James
Ollscoil na hÉireann, Gaillimh
D3510

Chapman & Hall/CRC
Interdisciplinary Statistics Series

STATISTICS OF MEDICAL IMAGING

Tianhu Lei

University of Pittsburgh
Pittsburgh, Pennsylvania, USA

CRC Press is an imprint of the
Taylor & Francis Group, an **informa** business
A CHAPMAN & HALL BOOK

CRC Press
Taylor & Francis Group
6000 Broken Sound Parkway NW, Suite 300
Boca Raton, FL 33487-2742

Printed in the United States of America on acid-free paper
Version Date: 20111028

International Standard Book Number: 978-1-4200-8842-7 (Hardback)

Library of Congress Cataloging-in-Publication Data

Lei, Tianhu.
Statistics of medical imaging / Tianhu Lei.
p. ; cm. -- (Chapman & Hall/CRC interdisciplinary statistics)
Includes bibliographical references and index.
ISBN 978-1-4200-8842-7 (hardccover : alk. paper)
I. Title. II. Series: Interdisciplinary statistics.
[DNLM: 1. Diagnostic Imaging. 2. Signal Processing, Computer-Assisted. 3. Statistics as Topic. WN 180]

616.07'54--dc23 2011043931

Visit the Taylor & Francis Web site at
http://www.taylorandfrancis.com

and the CRC Press Web site at
http://www.crcpress.com

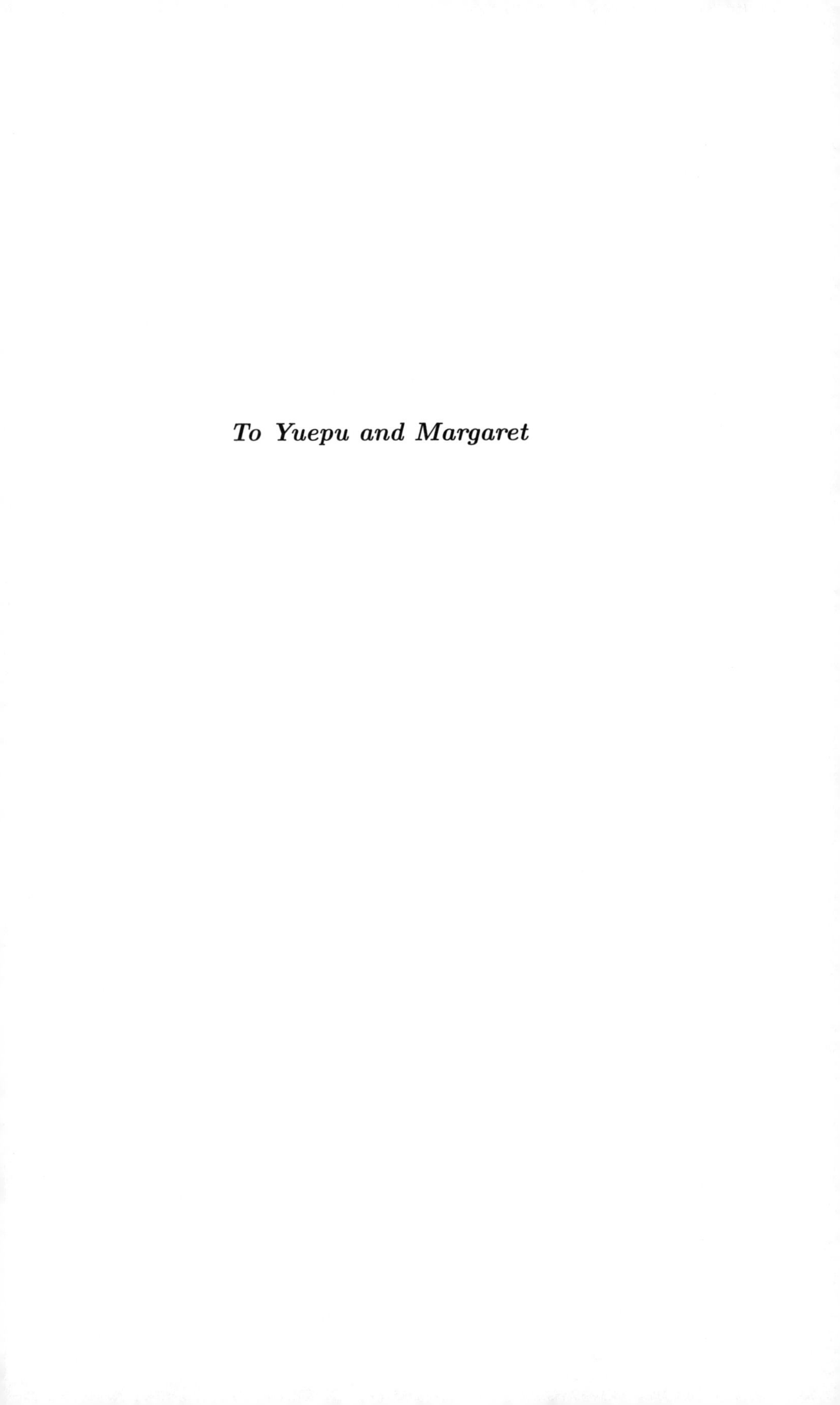

To Yuepu and Margaret

Contents

Preface

The past three decades have seen enormous developments in medical imaging technologies. X-ray computed tomography (CT), magnetic resonance imaging (MRI), positron computed tomography (PET), single photon emission computed tomography (SPECT), and ultrasonic imaging (US), etc. have revolutionized the diagnosis in clinical medicine and also become powerful tools for research in basic science. Physical principles and mathematical procedures of these technologies have been extensively studied during their development. However, less work has been done on their statistical aspect. In the history of technology development, statistical investigation into technology not only provides better understanding of the intrinsic features of the technology (**analysis**), but also leads to improved design of the technology (**synthesis**). This book attempts to fill this gap and to provide a theoretical framework for statistical investigation into medical technologies.

Medical imaging technologies encompass a broad array of disciplines: physics, chemistry, mathematics, engineering, computer science, etc. This book begins with descriptions of physical principles of medical imaging techniques, then derives statistical properties of the data (measurements) in the imaging domain based on these principles. In terms of these properties and mathematics of the image reconstruction, statistical properties of the data (pixel intensities) in the image domain are derived. By using these properties, stochastic image models are established and statistical image analysis methods are developed. The book ends with assessments of the performance evaluations of image analysis methods. The book essentially consists of three parts. The first part includes Chapters 1 through 4: imaging physics and mathematics. The second part consists of Chapters 5 through 8: imaging and image statistics. The third part has Chapters 9 through 12: statistical image analysis.

This book discusses two major medical imaging techniques: X-ray CT and MRI. For these two imaging techniques, the statistical investigation focuses on their basic imaging protocols. In X-ray CT, the study is limited to parallel and divergent projections and the convolution image reconstruction method. In MRI, the study is confined to the rectilinear k-space sampling and Fourier transform reconstruction method and the radial k-space sampling and projection reconstruction method. The purpose of this approach is to show the basic statistical properties of these two imaging techniques and to provide a method to conduct similar statistical investigations to more complicated imaging protocols. By revisiting X-ray CT and MRI and briefly reviewing PET and SPECT from the standpoint of tomography, the book shows that

X-ray CT, MRI, PET, and SPECT belong to a particular category of imaging non-diffraction computed tomography.

This book is partially based on the lecture notes developed for two graduate courses *Medical Imaging and Image Analysis* and *Biomedical Signal Processing* at the Department of Computer Science and Electrical Engineering and Department of Biological Resource Engineering at the University of Maryland, as well as for short courses at SPIE Medical Imaging conferences and the United States Patent and Trademark Office, and partially draws from research done at the University of Maryland and the University of Pennsylvania. In the book, statistical properties and their proofs are written in a theorem-proof format. To keep the compactness of proofs, only the major steps of proofs are given in sections, the details of these major steps are shown in the appendices. Sub-levels of some of these details are listed in Problems as readers' exercises. The list of references in each chapter by no means constitutes a complete bibliography on the topic. Those references are listed because I feel that they have been useful in teaching and research over the years.

The book is intended for graduates, seniors, and researchers from the fields of biomedical, electrical, and system engineering, with a strong interest in statistical image and signal analysis in medical imaging techniques. My students and colleagues who attended these courses deserve much credit for prompting the completion of this book. Specifically, I wish to thank Professors T.P.L. Roberts, F.W. Wehrli, H.K. Song, K.Ty Bae, D.P. Chakraborty, A.A. Amini, J.Z. Liang, J. Hsieh, G-S. Ying, M.H. Loew, J.M. Fitzpatrick, K. Hanson, R.F. Wagner, M. Sonka, D. Jaramillo, M.D. Schnall, R.N. Bryan, C.N. Dorny, I.G. Kazantsev, and L.M. Popescu for their support and useful discussions. I acknowledge that any errors that remain in this book are my own fault. I would also like to thank my colleagues: Professors W. Sewchand, C-I. Chang, J. Morris, T. Adali, D. Bruley, G.T. Herman, R.M. Lewitt, S. Matej, P.K. Saha, and J.K. Udupa. Finally, I am grateful to R. Calver, Senior Editor, at Chapman & Hall / CRC, for his patience, suggestions, and help from the very beginning of writing this book.

Tianhu Lei
Baltimore, Maryland

About the Author

Tianhu Lei is an associate professor at the University of Pittsburgh. He has previously worked at the University of Maryland, the University of Pennsylvania, and the Childrens Hospital of Philadelphia. He earned a Ph.D. in electric and systems engineering from the University of Pennsylvania

1

Introduction

The commonly used medical imaging techniques include but are not limited to X-ray computed tomography (X-ray CT), magnetic resonance imaging (MRI), positron emission tomography (PET), single photon emission computed tomography (SPECT), ultrasonic imaging (US), etc. These modern technologies not only revolutionize traditional diagnostic radiology in medicine, but also provide powerful tools for research in basic science.

1.1 Data Flow and Statistics

For each of these medical imaging techniques, we have observed its unique data flow from acquiring measurements to computing pixel intensities. Data may be real or complex valued, continuous or discrete. For instance, in the imaging domain of MRI, these data are thermal equilibrium macroscopic magnetization (TEMM), transverse precessing macroscopic magnetization (TPMM), free induction decay (FID) signal, phase sensitive detection (PSD) signal, analog-to-digital conversion (ADC) signal, k-space sample.

Most of medical imaging techniques were developed independently of each other. By comparing these techniques, we also find that their data flows may have some common features. For instance, in the imaging domain of X-ray CT, PET, and SPECT, we encounter the same procedures: photon emission, photon detection, and projection formation.

During the development and improvement of these technologies, it is necessary and reasonable to evaluate performances of these techniques and study some fundamental measures of image quality. For instance, we have to investigate the signal-to-noise ratio of the acquired data, the nature and variations of the signal components, the sources and limits of the noise components, etc.; in other words, the statistics of the data which normally consist of signal and noise components.

Medical imaging techniques encompass an array of disciplines: physics, chemistry, mathematics, computer science, and engineering. Statistical investigation into these techniques not only provides a better understanding of the intrinsic features of these techniques (**analysis**), but also leads to improved design of the techniques (**synthesis**).

For instance, in MRI, the spin noise may set a theoretical limit on the spatial resolution that the conventional MRI can achieve; statistics of Hamming and Hanning windows may provide insight into the filter design in Fourier transform image reconstruction in terms of the mainlobe width, the peak sidelobe level, and the sidelobe rolloff of their spectra, etc. Also, the similar statistical properties of X-ray CT and MR images suggest that these two imaging techniques may have some fundamental similarities. Indeed, they belong to a category of imaging, non-diffraction computed tomography, in which the interaction model and the external measurements are characterized by the straight line integral of some indexes of the medium and the image reconstruction is based on the Fourier slice theorem.

1.2 Imaging and Image Statistics

In this book, statistical properties of imaging data are described in a natural order as they are processed. In X-ray CT, they are in the order of photon emission $\Longrightarrow$ photon detection $\Longrightarrow$ projection formation. In MRI, they are in the order of TEMM $\Longrightarrow$ TPMM $\Longrightarrow$ MR signal (FID $\longrightarrow$ PSD $\longrightarrow$ ADC) $\Longrightarrow$ k-space sample. The means, variances, and correlations of imaging data are given in terms of conditions and settings of the imaging system. When the data travel in the space – time – (temporal or spatial) frequency domains, their statistics are evolving step by step.

Image reconstruction is an inevitable part of modern medical imaging technologies. From statistical standpoint, image reconstruction constitutes a transform from a set of random variables (e.g., projections or k-space samples) to an another set of random variables (e.g., pixel intensities of the image). This new set of random variables forms a spatial random process (often called a random field). The reconstructed image is a configuration of the random process and each pixel intensity is a value (of the corresponding random variable in the random process) in the state space. Statistical properties of data propagate from the imaging domain to the image domain through image reconstruction.

Statistical properties of the image are described at three levels: a single pixel, any two pixels, and a group of pixels (i.e., an image region) and are focused on the second-order statistics. They are (i) Gaussianity for any single pixel intensity, (ii) spatially asymptotic independence and exponential correlation coefficient for any two pixel intensities, and (iii) stationary and ergodic for pixel intensities in an image region. These basic statistics lead to that X-ray CT and MR image are embedded on a Markov random field (MRF) with the properly selected neighborhood system.

1.3 Statistical Image Analysis

Statistical properties of images provide sufficient information for establishing stochastic image models. Two Finite Normal Mixture (iFNM and cFNM) models are proposed. They are suitable for images with independent and correlated pixel intensities, respectively. Similarities and differences of these two models, their dependence on the signal-to-noise ratio of the image, and comparison with hidden Markov random field (hMRF) model have been discussed.

Imaging is commonly viewed as an operation or a process from data to the pictures, while image analysis refers to an operation or a process from pictures to the "*data.*" Here, "*data*" may include some image primitives such as edges or regions as well as some quantities and labels related to these primitives. Advanced image analysis techniques such as graph approach, classical snakes, and active contour approaches, level set methods, active shape model and active appearance model approaches, fuzzy connected object delineation method, and Markov random field (MRF) model based approach are widely used in medical image analysis.

This book proposed two stochastic model-based image analysis methods. They consist of three steps: detecting the number of image regions, estimating model parameters, and classifying pixels into image regions. These three steps form an unsupervised, data-driven approach. Two methods are implemented by expectation-maximization (EM) and the extended EM algorithms. In particular, a sensor array signal processing method for detecting the number of image regions, the determination of the order of neighborhood system of MRF, and a new procedure for clique potential design to perform maximum a posterior (MAP) operation are proposed.

With the rapid development of medical image analysis technologies, increasing interest has evolved toward the analyzing image analysis techniques, especially in the cases of the lack of the ground truth or a gold standard. A new protocol for evaluating the performance of FNM-EM-based image analysis method is proposed in this book. Performances in the detection, estimation, and classification are assessed. Probabilities of over- and under-detection of the number of image regions, Cramer-Rao bounds of variances of estimates of model parameters, and a misclassification probability for the Bayesian classifier are given. The strength of this protocol is that it not only provides the theoretically approachable accuracy limits of image analysis techniques, but also shows the practically achievable performance for the given images.

1.4 Motivation and Organization

1.4.1 Motivation

Physical principles and mathematical procedures of medical imaging techniques have been exhaustively studied during the past decades. The less work has been done in their statistical aspect. This book attempts to fill this gap and provides theoretical framework for the statistical investigation.

The philosophy of conducting statistical investigations in X-ray CT and MR imaging is as follows. The study starts from the very beginning of imaging process, for example, photon emission in X-ray CT and spin system in MRI, derives statistical properties of data in imaging domain step by step according to imaging procedures. Then through image reconstruction, the study derives statistical properties at various levels of image structure, for example, a single pixel, two pixels, an image region as well as neighborhood pixels. Finally, the study establishes stochastic image models based on image statistics and develops model-based image analysis techniques. Performance evaluation of image analysis techniques, in statistical content, is followed to assess the proposed methods.

For the various medical imaging modalities, this book is confined to two major imaging techniques: X-ray CT and MRI. For the variety of approaches in these two techniques, the book is restricted to the basic approaches. For instance, in X-ray CT, it is restricted to the parallel and divergent projections and the convolution image reconstruction method; in MRI, it is restricted in the rectilinear and radial k-space sampling and Fourier transform (FT) and projection reconstruction (PR) methods. The book focuses on imaging and image data processes.

This book is based mainly on the lecture notes developed for the graduate courses and partially on the results of research. It is intended for students in the engineering departments, especially in the field of signal and image processing. It is also intended to be useful for engineers, scientists, and researchers in the field of medical imaging as a reference book.

1.4.2 Organization

This book is essentially divided into three parts. Part I consists of Chapters 2, 3, and 4, and focuses on imaging physics and mathematics. Part II includes Chapters 5, 6, 7, and 8, and describes imaging and image statistics. Chapters 9, 10, 11, and 12 form Part III, and describe statistical image analysis. The book is organized as follows.

Chapter 2 describes physical principle and mathematical image reconstruction of X-ray CT.

Chapter 3 describes the physical principle and mathematical image reconstruction of MRI.

Chapter 4, by revisiting X-ray CT and MRI and briefly reviewing PET and SPECT, shows that they belong to non-diffraction CT.

Chapter 5 describes statistical properties in imaging domain of X-ray CT based on its imaging physics and mathematics.

Chapter 6 describes statistical properties in image domain of X-ray CT based on its statistics properties in imaging domain.

Chapter 7 describes statistical properties in imaging domain of MRI based on its imaging physics and mathematics.

Chapter 8 describes statistical properties in image domain of MRI based on its statistics properties in imaging domain.

Chapter 9 describes two stochastic models based on statistical properties in image domain of X-ray CT and MRI.

Chapter 10 describes an image analysis method for the images with independent pixel intensities, based on the model in Chapter 9.

Chapter 11 describes an image analysis method for the images with correlated pixel intensities, based on the model in Chapter 9.

Chapter 12 describes a protocol to evaluate the performance of image analysis techniques given in Chapters 10 and 11.

Each chapter has several appendices, that are used to give the detailed proofs of statistical properties or some related important issues in the chapter, for the purpose of keeping the text description in the chapter compact and in focus.

2

X-Ray CT Physics and Mathematics

2.1 Introduction

The word *tomography* consists of two Greek words *tomos* (slice) and *graphein* (draw). Historically, the term *tomography* has referred to a technique of X-ray photography in which only one plane of the internal structure inside the object is photographed in shape focus. Linear and transaxial tomography are two examples.

In the linear tomography as shown in Figure 2.1, an X-ray source and a photographic plate (which is parallel to the cross section of an object) are placed on two sides of the object. By moving the X-ray source at a fixed speed parallel to the cross section in one direction, and moving the plate at an appropriate speed in the opposite direction, a point in the cross section (denoted by •) is always projected onto the same point in the plate, but the point above or below the plate is projected onto different points in the plate. Thus, on the photographic plate the cross section stands out while the remainder of the object is blurred.

In transaxial tomography as shown in Figure 2.2, an object sits in a turntable in an upright position. The photographic plate is on a horizontal table next to the object. X-rays are directed obliquely through the object and projected onto the photographic plate. Both the object and the photographic plate are rotating in the same direction and at the same speed. Only those points actually on the focal section of the object remain in sharp focus throughout a rotation. Points that are above or below the focal section are blurred.

Nowadays, tomography refers to the cross-sectional imaging of an object from either transmission or reflection data collected by illuminating the object from many different directions.

Computed tomography (abbreviated as CT) is different from conventional tomography in that the images of the cross sections in CT are not influenced by the objects outside those sections.

In X-ray transmission CT (abbreviated as X-ray CT), a single X-ray source is contained in a tube and the detector device consists of an array of X-ray detectors. The X-ray tube and the collimator are on one side and the detector device and the data acquisition unit are on the other side of the object. Both

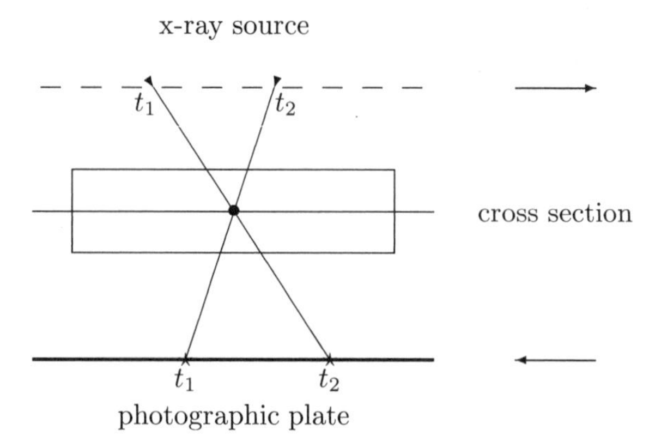

FIGURE 2.1
Linear tomography: t_1 and t_2 denote the time instants and $\longrightarrow$ represents the moving direction.

the object and this apparatus are stationary, and the apparatus is rotating around the object. An image of a slice of the object (which is determined by the X-ray beams from the source to the detectors) can be created [1–10].

X-ray CT technology has improved dramatically over the past 50 years and now CT scanners are an important and routinely used diagnostic imaging instrument. From the first X-ray CT scanner* to the modern 3D cone beam scanner, there may not be a clear definition of the generations of X-ray CT. However, based on X-ray generation, detection, data acquisition, and image reconstruction, the milestones in its research and development may include, but are not limited to, (1) the parallel beams (a translation-rotation mode), (2) the narrow divergent beams (a translation-rotation mode), (3) the wide divergent beams (a rotation mode), (4) the wide divergent beams with the closed detector ring (a rotation mode), (5) electron beams, (6) beams on a spiral or a helical path, (7) the cone beam geometry, as well as the micro-CT and PET-CT combined scanner, etc.

X-ray CT is closely related to a computational approach often known as *image reconstruction from projections*. As stated in [10], "*Image reconstruction from projections is the process of producing an image of a two-dimensional distribution (usually of some physical property) from estimation of its line integrals along a finite number of lines of known locations.*" In fact, X-ray CT is an example of the application of this approach. Over the past 50 years, the

*It utilized 180 views, 160 parallel projections in each view, algebraic reconstruction method, and took 2.5 hours to reconstruct a 64×64 slice image.

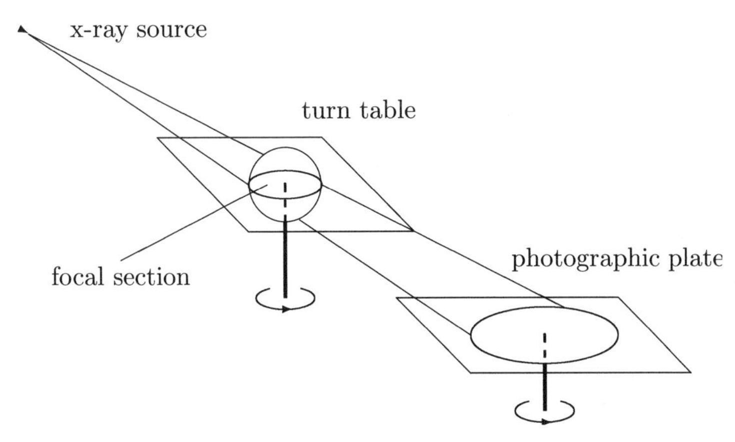

FIGURE 2.2
Transaxial tomography; → represents the rotating direction.

application of *image reconstruction from projections* have not only revolutionized the diagnostic medicine, but also covered divergent fields in science and technology, from electron microscopes (reconstructing molecular structures) to radio telescopes (reconstructing maps of radio emission from celestial objects).

Research and development of X-ray CT involve various fields of science and technology. They cover physics, chemistry, mathematics, computer science, and engineering. In 1901, the first Nobel prize was awarded to Wilhelm C. Röntgen, in physics, for his discovery of the X-ray. In 1979, the Nobel prize in physiology and medicine was awarded to Allan M. Cormack and Sir Godfrey N. Hounsfield for their development of computed tomography during the 1960s and 1970s. The 1982 Nobel prize in chemistry was given to Aaron Klug for the use of reconstruction from electron microscopic projections for the explanation of biologically important molecular complexes. The recipients of the 2003 Nobel prize in physiology and medicine, Paul C. Lauterbur and Peter Mansfield, used image reconstruction from projections in their early research.

More generally, according to the physical nature of source–medium interaction, transmission CT can be classified into two categories: (1) a nondiffraction CT, in which the interaction model and the external measurements are characterized by the line integrals of some index of the medium and the image reconstruction is based on the Fourier Slice Theorem [4, 6]; and (2) a diffraction CT imaging, in which the interaction and measurements are modeled with the wave equation and the tomographic reconstruction approach is based on the Fourier diffraction theorem [4, 11]. The former includes conventional X-ray CT, emission CT, ultrasound CT (refractive index CT and attenuation

CT), magnetic resonance imaging (MRI), etc. The latter includes acoustic, certain seismic, microwave, optical imaging, etc.

This chapter first describes the physical concepts of X-ray CT imaging. It covers photon emission, attenuation, and detection; relative linear attenuation coefficient; and projection formation (parallel and divergent). Then it discusses the mathematics of X-ray CT image reconstruction. Based on Fourier slice theorem, it shows two inverse radon formulas for parallel projections and one formula for divergent projections. Image reconstruction is limited in the convolution method, whose computational implementation consists of a double convolution and a projection. A signal processing paradigm of X-ray CT is given at the end of this chapter.

2.2 Photon Emission, Attenuation, and Detection

X-ray, made of photons that have no mass and no electrical charge, demonstrates wave–particle duality, that is, the properties of both waves and particles. X-rays have the wavelengths roughly 10^{-13}m to 10^{-8} m, or the frequencies 3×10^{16} Hz to 3×10^{21} Hz, with the energies 0.12 eV to 12 keV ("soft") or 12 eV to 120 keV ("hard").† X-ray is of electromagnetic in nature.

2.2.1 Emission

An X-ray tube is shown in Figure 2.3. The filament of the cathode is heated up to a certain temperature to overcome the binding energy of electrons to the metal of the filament. An electron cloud emerges around the hot filament. Those electrons are accelerated in the transit process from the cathode to the anode and then strike the focus volume of the anode. The shape and size of the electron beam are controlled by the focus cup. The anode is rotating. A special target material (tungsten) is embedded in the anode. The effective target area in the focus volume of the anode depends on the orientation of the anode surface with respect to the incoming high-speed electron beam.

When the accelerated electrons enter into the lattice atoms of the focus volume of the anode, the interactions between the high-speed electrons and the orbital electrons as well as the atomic nucleus cause the deceleration. As shown by classic electrodynamics, the deceleration of electrons radiates electromagnetic waves in the form of photons. Among several atomic processes caused by these interactions, two main physical events are X-ray fluorescence and Bremsstrahlung (in Germany, *brems* for braking, *strahlung* for radiation).

† "Hard" X-rays are mainly used for imaging, because they can penetrate solid objects.

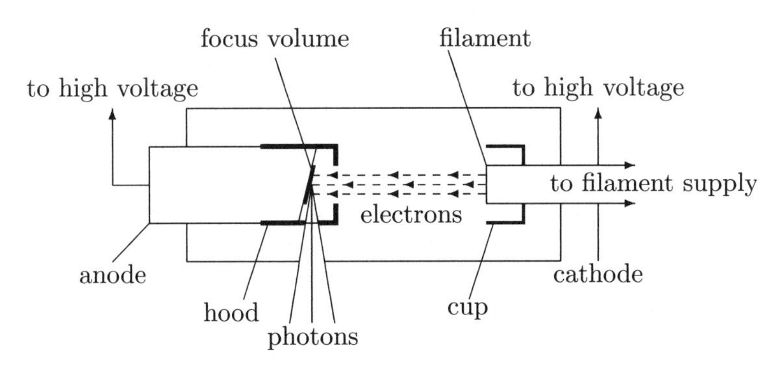

FIGURE 2.3
An illustration of X-ray tube and photon emission.

X-ray fluorescence represents a direct interaction of the high-speed electrons with electrons on the inner shell. When an electron on the inner shell is kicked out of the atom by this collision, an electron of one of the outer shells fills the vacant position on the inner shell. Because the potential energy of the inner shells is lower than that of the outer shells, photons are emitted in this process. X-ray fluorescence produces an emission spectrum of X-ray frequencies, called the characteristic line. In Bremsstrahlung, the high-speed electrons are scattered by the strong electric field near the nuclei with the high proton number and, hence, slowing down. This deceleration also emits photons. Because the scattering is a multiple process, the radiation spectrum of X-ray frequencies is continuous.

X-ray fluorescence and Bremsstrahlung generate X-rays and superimpose the characteristic lines on the continuous emission spectrum of X-ray frequencies. However, in total, X-ray fluorescence contributes far less than Bremsstrahlung to the radiation intensity, that is, the number of photons. X-rays produced by Bremsstrahlung are the most useful for medical and industrial applications.

2.2.2 Attenuation

The emitted photons travel through collimators, penetrate the medium, and arrive at the X-ray detectors. Between the X-ray tube and the detectors, several types of collimators are installed to control the size and shape of the X-ray beam, which may determine, for example, the slice thickness of an X-ray CT image, etc. X-ray beams are polychromatic in nature. By using several types of filtering, a more uniform beam (a monochromatic beam in the ideal case) can be obtained.

Photon–medium interaction can change the number of photons, the photon energy, and the travel direction. This phenomenon is also called the attenuation. Scattering and absorption are the two main mechanisms that cause X-ray attenuation. The measurement of the attenuation provides the basis for X-ray CT imaging.

Rayleigh scattering is an elastic scattering phenomena. This event occurs when the size of the scattering nucleus is less than the wavelength of the incident X-ray. In this case, the incident and the scattered X-rays have the equal wavelength but different travel directions. There is no energy transfer during the scattering. Other types of scattering may include Compton scattering. This event occurs when the incident photon collides with a weakly bound electron on the outer shell. The photon loses only a part of its energy in this case.

Photoelectric absorption occurs when the binding energy of atomic electrons is less than the energy of a photon. In this case, the electron on a lower shell is kicked off the atom and travels in the medium as a free photoelectron with a kinetic energy equal to the difference between the energy of the incident photon and the binding energy of the electron. This difference in energy is transferred to the lattice atoms locally in the form of heat.

2.2.3 Detection

In X-ray CT, the attenuated photons are not measured directly. They are detected via their interaction products. When photons enter the detectors, either crystal or ionizing gas produces light or electrical energy. The photon energy is then collected and converted to an electrical signal, which is further converted to a digital signal for image reconstruction.

Two types of detectors utilized in CT systems are the scintillation (a solid-state) detector and xenon gas (an ionizing) detector.

Scintillation detectors first utilize a crystal to convert the shortwave X-ray photons to longwave light. Then, a photodiode that is attached to the scintillation portion of the detector transforms the light energy into electrical energy. The strength of the detector signal is proportional to the number of attenuated photons.

The gas detector consists of a chamber made of a ceramic material with long thin ionization plates submersed in xenon gas. When photons interact with the charged plates, xenon gas ionization occurs. The ionization of ions produces an electrical current. Ionization of the plates and the electrical current rely on the number photons and ionizing the gas.

The Geiger-Muller counter used in the first tomographic experiment by Cormack and Hounsfield is a gas detector.

The performance of a detector is characterized by its efficiency and dynamic range. Detector efficiency describes the percent of incoming photons able to be converted to a useful electrical signal. Scintillation detectors can convert 99 to 100 percent of the attenuated photons, and gas detectors can convert 60 to 90 percent of the photons that enter the chamber. The dynamic range describes the ability of a detector to differentiate the range of X-ray intensities. Some detectors can detect 1,000,000 intensities at approximately 1,100 views per second.

2.3 Attenuation Coefficient

Photon–medium interactions are characterized by the attenuation. X-ray attenuation is often measured by attenuation coefficients. These coefficients can be mass attenuation coefficients, mass energy-absorption coefficient, linear attenuation coefficient, etc. Medical applications utilizing X-ray[‡] for forming images of tissues are based on linear attenuation coefficients.

2.3.1 Linear Attenuation Coefficient

A simple mechanism is shown in Figure 2.4a. Let ρ be the probability that a photon with the energy e, which enters a slab of the homogeneous tissue t with unit thickness in a direction perpendicular to the surface of the slab, is not absorbed or scattered in the slab. According to its physical meaning, the probability ρ is also called the *transmittance* of the tissue t at energy e. A single constant coefficient defined by

$$\mu(e,t) = -\ln \rho \tag{2.1}$$

is called the *linear attenuation coefficient* of tissue t at energy e and abbreviated as LAC. Linear attenuation coefficients of several materials are listed in Table 2.1. It is clear that the dimensionality of the linear attenuation coefficient is the inverse of the length.

TABLE 2.1
LACs of Several Materials (the unit is cm^{-1})

KeV	50	100	150
Air	0.00027	0.00020	0.00018
Water	0.22	0.17	0.15
Wood	0.120	0.085	0.080

[‡] In this chapter, the discussion is confined to the case of monochromatic X-ray.

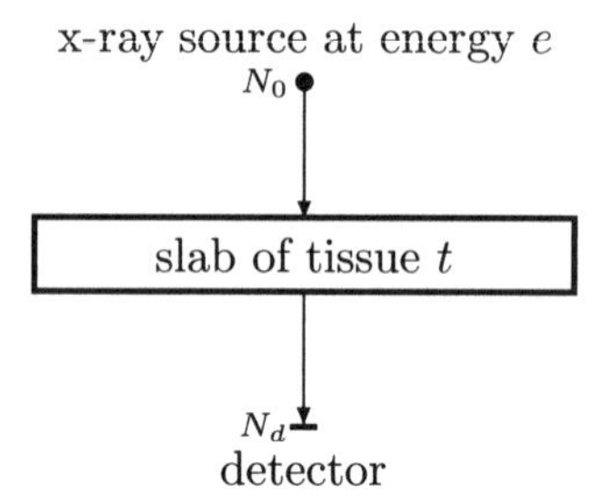

a) Illustration of probability ρ

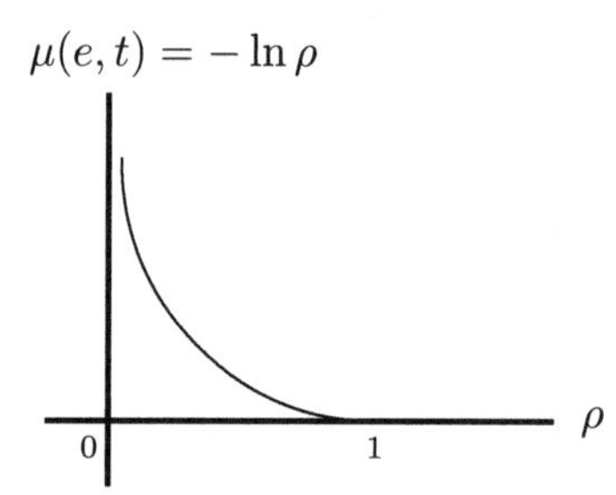

b) Relation between $\mu(e,t)$ and ρ

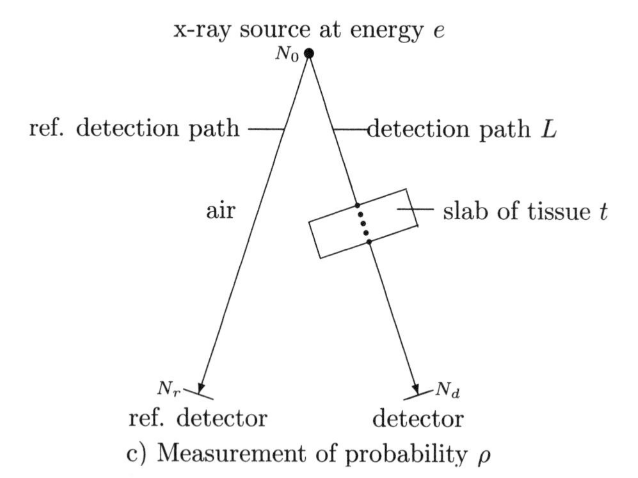

c) Measurement of probability ρ

FIGURE 2.4
An illustration of the linear attenuation coefficient of X-ray.

The relationship between the linear attenuation coefficient $\mu(e,t)$ and the probability ρ is shown in Figure 2.4b. ρ increasing leads to $\mu(e,t)$ decreasing. When $\rho = 1$, $\mu(e,t) = 0$, which implies that there is no attenuation; while $\rho = 0$, $\mu(e,t) \to \infty$, which implies that there is infinitely large attenuation.

Assume the attenuation of the air ignored or the entire mechanism of Figure 2.4a is placed in the free space. Let the numbers of photons transmitted by the X-ray source and received by the photon detector be N_0 and N_d, respectively. The probability ρ can be estimated by

$$\rho = \frac{N_d}{N_0}. \tag{2.2}$$

Hence, Eq. (2.1) can be rewritten as

$$\frac{N_d}{N_0} = e^{-\mu(e,t)}. \tag{2.3}$$

Appendix 1A shows that the definition in Eq. (2.1) and the expression in Eq. (2.3) can be derived from the reduction of photons when X-rays penetrate the slab.

A mechanism for the physical measurement of X-rays is shown in Figure 2.4c. It consists of two detection paths: detection and reference detection. That is, part of the X-ray beam travels through a homogeneous reference material such as air. The reference detection shows how many photons that leave the X-ray source are counted by the reference detector and compensates the fluctuations in the strength of the X-ray. Thus, let the numbers of photons counted by the detector and the reference detector be N_d and N_r, respectively. The probability ρ of Eq. (2.2) can be expressed by

$$\rho = \frac{N_d}{N_0} = \frac{N_d}{N_r}. \tag{2.4}$$

2.3.2 Relative Linear Attenuation Coefficient

In X-ray CT, X-ray measurement is performed in two processes: an actual measurement and a calibration measurement. Figure 2.5 shows a measurement mechanism that can be applied to both the actual and calibration processes. Mechanisms of the actual and calibration measurement processes are very similar. The only difference between these two processes is that an object (to be imaged) is placed in the detection path of the actual measurement process to partially replace the homogeneous reference material in the detection path of the calibration measurement process.

Let $\mu_a(e,t)$ and $\mu_c(e,t)$ be LACs (Eq. (2.1)) in the actual and calibration measurement processes, respectively. A quantity defined as the difference of these LACs in two processes,

$$\mu_a(e,t) - \mu_c(e,t) = -\ln \frac{\rho_a}{\rho_c}, \tag{2.5}$$

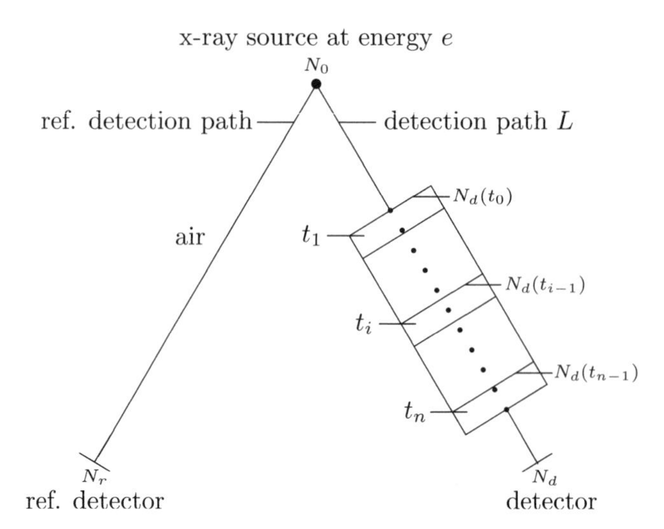

FIGURE 2.5
An illustration of the relative linear attenuation coefficient of X-ray.

is called the *relative linear attenuation coefficient* of tissue t at energy e, abbreviated as RLAC. In Eq. (2.5), ρ_a and ρ_c are the probabilities ρ (Eqs. (2.2) and (2.4)) in the actual and the calibration measurement processes, respectively. The dimensionality of the RLAC is the inverse of the length.

In the detection path of either the actual or calibration measurement in Figure 2.5, let the polar coordinate (r, ϕ) denote the location of a point in the object; $f(r, \phi)$ is used to represent RLAC at (r, ϕ) along the line L. That is,

$$f(r, \phi) = \mu_a(e, t) - \mu_c(e, t) = -\ln \frac{\rho_a}{\rho_c}. \tag{2.6}$$

The detection path is divided into n slabs of the equal (e.g., the unit) thickness. For simplicity, let the i-th slab represent the tissue type t_i $(i = 1, \cdots, n)$ and the top surfaces of i-th slab be located at (r_i, ϕ_i). Assume that the number of photons that enter the i-th slab in the detection path is $N_d(t_{i-1})$; then the probability $\rho(t_i)$ for the i-th slab is

$$\rho(t_i) = \frac{N_d(t_i)}{N_d(t_{i-1})}. \tag{2.7}$$

Because the linear attenuation coefficient represents the reduction of photons when they travel through the object, the overall LAC in the detection path is

$$\sum_{i=1}^{n} \mu(e, t_i) = -\sum_{i=1}^{n} \ln \rho(t_i) = -\ln \prod_{i=1}^{n} \rho(t_i)$$

$$= -\ln(\frac{N_d(t_1)}{N_d(t_0)} \cdot \frac{N_d(t_2)}{N_d(t_1)} \cdots \frac{N_d(t_n)}{N_d(t_{n-1})})$$

$$= -\ln \frac{N_d(t_n)}{N_d(t_0)} = -\ln \frac{N_d}{N_r} = -\ln \rho = \mu(e, t), \tag{2.8}$$

which implies that the overall LAC in the detection path can be expressed as the sum of the cascaded LACs at each point along the path.

Applying Eq. (2.8) to the actual and calibration measurement processes and computing the difference of two LACs, we have

$$\mu_a(e, t) - \mu_c(e, t) = \sum_{i=1}^{n} (\mu_a(e, t_i) - \mu_c(e, t_i)). \tag{2.9}$$

Based on the definition of RLAC (Eq. (2.5)), the left side of Eq. (2.9) is

$$\mu_a(e, t) - \mu_c(e, t) = -\ln \frac{\rho_a}{\rho_c}, \tag{2.10}$$

and the right side of Eq. (2.9) is

$$\sum_{i=1}^{n} (\mu_a(e, t_i) - \mu_c(e, t_i)) = -\sum_{i=1}^{n} \ln \frac{\rho_a(t_i)}{\rho_c(t_i)}. \tag{2.11}$$

By using Eq. (2.6), $-\ln \frac{\rho_a(t_i)}{\rho_c(t_i)}$ is replaced by $f(r_i, \phi_i)$ (over the slab of the unit thickness). Thus, when $n \longrightarrow \infty$, we have

$$-\sum_{i=1}^{n} \ln \frac{\rho_a(t_i)}{\rho_c(t_i)} = -\Delta l \sum_{i=1}^{n} f(r_i, \phi_i) \longrightarrow \int_L f(r, \phi) dz. \tag{2.12}$$

Therefore, by substituting Eqs. (2.10)-(2.12) into Eq. (2.9), we obtain

$$-\ln \frac{\rho_a}{\rho_c} = \int_L f(r, \phi) dz, \tag{2.13}$$

which implies that the overall RLAC in the detection path can be expressed as the line integral of RLAC at each point along the path.

2.4 Projections

2.4.1 Projection

In Figures 2.4 and 2.5, let N_{ad} and N_{ar} denote the numbers of photons N_d and N_r counted by the detector and the reference detector in the *actual measurement process*, respectively; and let N_{cd} and N_{cr} denote the numbers of photons N_d and N_r counted by the detector and the reference detector in the *calibration measurement process*, respectively. In X-ray CT, a quantity given by

$$-\ln \frac{N_{ad}/N_{ar}}{N_{cd}/N_{cr}} \tag{2.14}$$

is called the *projection* along the line L. In some literature, it is also called the ray or the beam. Although the projection is defined on a geometric line specified by the ray or the beam direction, it represents a physical concept and is determined by the physical measurements. Eq. (2.12) indicates that the projection is dimensionless, that is, it is just a number.

Figures 2.4 and 2.5 show the mechanisms of physical measurement of X-ray. Mathematically, as shown in Figure 2.6, let l be the distance from the origin of a rectangular coordinate system to a line L, and let θ be the angle between the normal direction of the line L and the positive direction of the X-axis of the coordinate system, the line L can be specified by

$$L: \quad r\cos(\theta - \phi) = l \ , \tag{2.15}$$

where (r, ϕ) is the polar coordinate of a point P on the line L.

From Eq. (2.4), $\frac{N_{ad}}{N_{ar}} = \rho_a$ and $\frac{N_{cd}}{N_{cr}} = \rho_c$. Thus, from Eqs. (2.14) and (2.13), we have

$$-\ln \frac{N_{ad}/N_{ar}}{N_{cd}/N_{cr}} = -\ln \frac{\rho_a}{\rho_c} = \int_L f(r, \phi) dz \ . \tag{2.16}$$

In order to directly link the physical measurement $-\ln \frac{N_{ad}/N_{ar}}{N_{cd}/N_{cr}}$ of Eq. (2.14) to its embedded geometry—the line L of Eq. (2.15), which is required by the

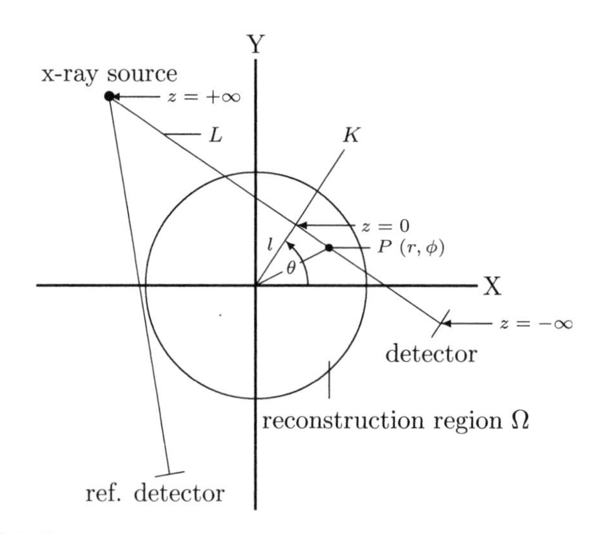

FIGURE 2.6
An illustration of the projection.

image reconstruction, and also for the purpose of simplicity, we use $p(l,\theta)$ to represent $-\ln \frac{N_{ad}/N_{ar}}{N_{cd}/N_{cr}}$. Thus, we obtain

$$p(l,\theta) = \int_L f(r,\phi)dz, \tag{2.17}$$

Eq. (2.17) indicates that the projection on the line L is the line integral of the relative linear attenuation coefficient of X-ray at the points along the L, that is, along the ray or the beam direction. This is the basis of straight ray computed tomography (non-diffraction computed tomography).

Eqs. (2.14) and (2.16), in fact, represent a double normalization procedure for N_{ad}. The first normalization (in both numerator and denominator) appears in both the actual and calibration measurement processes. It is a necessary and reasonable operation to estimate the transmittances ρ_a and ρ_c. The second normalization appears between the actual and calibration measurement processes. The double normalization not only reduces (or eliminates) the effect on $p(l,\theta)$ caused by the fluctuation in N_{ad} due to the photon emission in one projection and among projections, but also makes the definition of the projection $p(l,\theta)$ consistent with the definition of the relative linear attenuation coefficient $f(r,\phi)$ and, hence, establishes Eq. (2.17).

Eq. (2.6) shows that when the point (r,ϕ) is inside the reconstruction region (Figure 2.6), due to $\rho_a(r,\phi) \neq \rho_c(r,\phi)$, $f(r,\phi) \neq 0$; when (r,ϕ) is outside the reconstruction region, due to $\rho_a(r,\phi) = \rho_c(r,\phi)$, $f(r,\phi) = 0$. Thus, Eq. (1.17) can be further expressed as

$$p(l,\theta) = \int_{-\infty}^{+\infty} f(r,\phi)dz = \int_{-\infty}^{+\infty} f(\sqrt{l^2+z^2}, \theta + \tan^{-1}(\frac{z}{l}))dz, \tag{2.18}$$

which implies that the non-zero line integral of the relative linear attenuation coefficient is only performed inside the object (to be imaged).

Note, in Figure 2.6, $z = 0$ corresponds to the intersection of the line L and the line K, where $K \perp L$. In Eq. (2.18), $z = -\infty$ and $z = +\infty$ correspond to the locations of the detector and the source, respectively.

2.4.2 Parallel and Divergent Projections

A finite number of projections can be produced in the different ways, Parallel projection and Divergent projection are two basic modes for data acquisition in X-ray CT. They are shown in Figures 2.7 and 2.8, respectively.

2.4.2.1 Parallel Projections

As shown in Figure 2.7, the line L is specified by two parameters, θ and l, where θ is the angle between the normal direction of the line L and the positive direction of X-axis, l is the distance from the origin of the coordinate system to the line L. l and θ define a space, called (l,θ)-space.

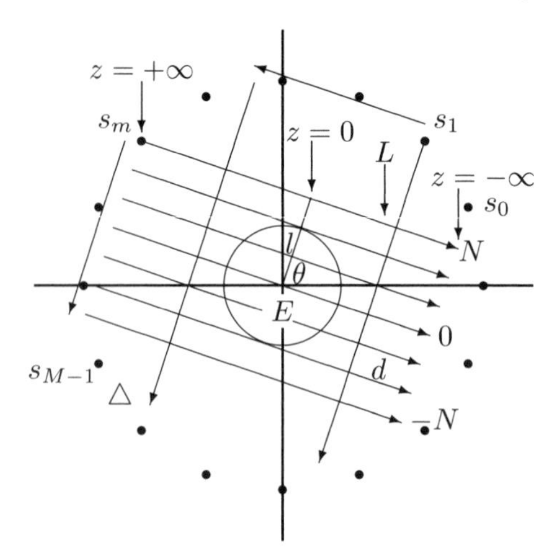

FIGURE 2.7
Parallel projections of the data collection in X-ray CT.

In Figure 2.7, a pair consisting of a single source and a single detector moves parallel (perpendicular to the line connecting the source and detector) to each other in $2N + 1$ steps with the spacing d, where $Nd > E$ and E is the radius of the circular reconstruction region Ω that contains the object to be imaged. After the data have been collected for these $2N + 1$ projections, which are considered in one view, the entire apparatus is rotated by an angle of Δ, and the data are again collected for $2N + 1$ projections of the next view. The procedure is repeated M times, where

$$M\Delta = \pi.$$

This is a *translation-rotation mode* that was used in the first generation of X-ray CT, as mentioned in Section 2.1.

Thus, parallel projections are defined on (l, θ)-space. When $p(l, \theta)$ represents the i-th projection in the m-th view of parallel mode, then it is denoted by $p(id, m\Delta)$ $(-N \leq i \leq N,\ 0 \leq m \leq M - 1)$.

2.4.2.2 Divergent Projections

As shown in Figure 2.8, the line L is specified by two parameters, β and σ, where β is the angle between the source direction (i.e., the line connecting the origin of the coordinate system and the source location) and the positive direction of y-axis, and σ is the angle between the source direction and the line L. Thus, projection can be expressed by $p(\sigma, \beta)$, where β determines the source location, that is the location of view, and σ determines the location of projection in that view. σ and β define a space called the (σ, β)-space.

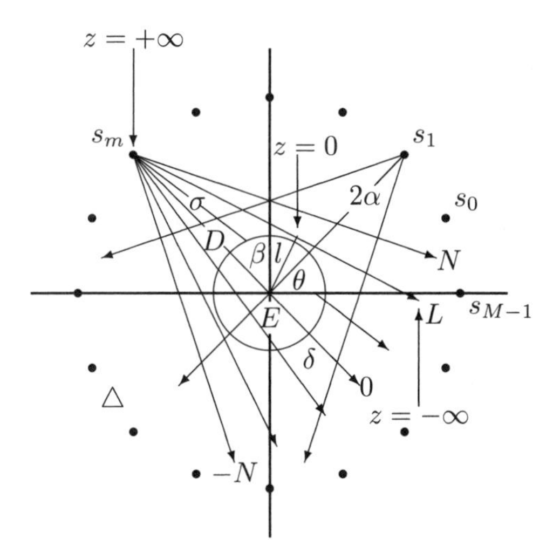

FIGURE 2.8
Divergent projections of the data collection in X-ray CT.

In Figure 2.8,[§] a single X-ray source faces a wide strip of detectors so that the angle subtended by the detector strip at the source location covers the whole reconstruction region Ω. The projections are taken for M equally spaced values of β with the angular spacing Δ, where

$$M\Delta = 2\pi,$$

and at each $m\Delta$ $(m = 0, 1, \cdots, M-1)$ the projections are sampled at $2N+1$ equally spaced angles with angle spacing δ, where $N\delta \geq \sin^{-1}(\frac{E}{D})$, and D is the distance from the origin of the coordinate system to the source location. This is a *rotation mode* that is used in the third generation of X-ray CT, mentioned in Section 2.1.

Thus, divergent projections are defined on (σ, β)-space. When $p(\sigma, \beta)$ represents the i-th projection in the m-th view of divergent mode, then it is denoted by $p(i\delta, m\Delta)$ $(-N \leq i \leq N,\ 0 \leq m \leq M-1)$.

[§]There are two types of divergent projections: equiangular and equispaced [4]. In this chapter, we only discuss the former.

2.5 Mathematical Foundation of Image Reconstruction

Two theorems described in this section provide a mathematical foundation of X-ray CT image reconstruction.¶ The first one is the Fourier slice theorem. It is also known as the projection slice theorem or central section theorem. The second one is the inverse Radon transform. It is also referred to as the Radon inversion formula.

This section first proves the Fourier slice theorem, and then derives the inverse Radon transform based on the first theorem. Two Radon inversion formulas (for parallel projections) are derived in this section, the third one (for divergent projections) is shown in Section 2.6.1.

In imaging theory, the spatial distribution of a physical property of interest of an object to be imaged is called the object function. In X-ray CT, it is the relative linear attenuation coefficient. The Fourier slice theorem and inverse Radon transform establish the relation between the object function and the projection.

2.5.1 Fourier Slice Theorem

Theorem 1. The 1-D Fourier transform of the projection at a given view equals the 2-D Fourier transform of the object function at the same view.

Proof

As shown in Figure 2.9, an x'-y' coordinate system is created by rotating an x-y coordinate system with an angle θ. The two coordinate systems are related by

$$\begin{pmatrix} x' \\ y' \end{pmatrix} = \begin{pmatrix} \cos\theta & \sin\theta \\ -\sin\theta & \cos\theta \end{pmatrix} \begin{pmatrix} x \\ y \end{pmatrix} \quad \text{or} \quad \begin{pmatrix} x \\ y \end{pmatrix} = \begin{pmatrix} \cos\theta & -\sin\theta \\ \sin\theta & \cos\theta \end{pmatrix} \begin{pmatrix} x' \\ y' \end{pmatrix}. \tag{2.19}$$

As a result, the line L: $r\cos(\theta - \phi) = l$ (Eq. (2.15)) in x-y coordinate system becomes the line $x' = l$ in the x'-y' system.

Let $f(x, y)$ be the object function. Let $p(x', \theta)$ be the projection on the line $x' = l$ in x'-y' system, that is, at the view θ in x-y system. From Eq. (2.17) or Eq. (2.18), we have

$$p(x', \theta) = \int_L f(r, \phi) dz = \int_{x'=l} f(x, y) dy' = \int_{-\infty}^{+\infty} f(x, y) dy'. \tag{2.20}$$

¶Rigorously speaking, this section is limited to the transform methods based image reconstruction.

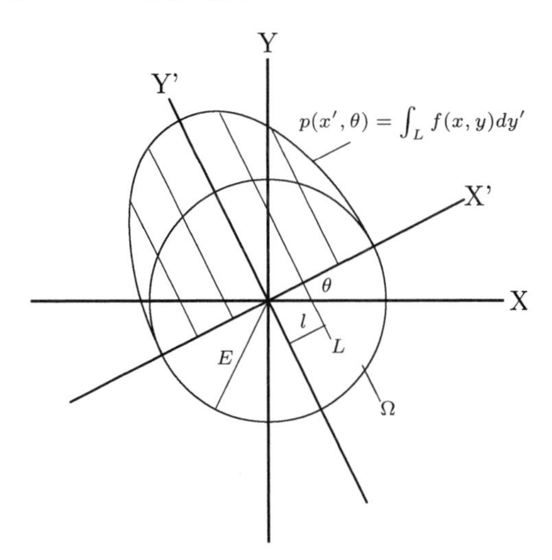

FIGURE 2.9
An illustration for Fourier slice theorem.

Let $P(f, \theta)$ denote the 1-D Fourier transform of $p(x', \theta)$; then

$$
\begin{aligned}
P(f, \theta) = \mathcal{F}_1\{p(x', \theta)\} &= \int_{-\infty}^{+\infty} p(x', \theta) e^{-i2\pi f x'} dx' \\
&= \int_{-\infty}^{+\infty} \int_{-\infty}^{+\infty} f(x, y) e^{-i2\pi f x'} dx' dy' \\
&= \int_{-\infty}^{+\infty} \int_{-\infty}^{+\infty} f(x, y) e^{-i2\pi f (x\cos\theta + y\sin\theta)} dx' dy' \\
&= \int_{-\infty}^{+\infty} \int_{-\infty}^{+\infty} f(x, y) e^{-i2\pi f (x\cos\theta + y\sin\theta)} dx dy, \qquad (2.21)
\end{aligned}
$$

where the Jacobian determinant is $J = |\frac{\partial(x', y')}{\partial(x, y)}| = 1$.

By defining a transform

$$f\cos\theta = f_x \quad \text{and} \quad f\sin\theta = f_y \qquad (2.22)$$

and letting $F(f_x, f_y)$ be the 2-D Fourier transform of $f(x, y)$, Eq. (2.21) becomes

$$
\begin{aligned}
P(f, \theta) &= \mathcal{F}_1\{p(x', \theta)\} \\
&= \int_{-\infty}^{+\infty} \int_{-\infty}^{+\infty} f(x, y) e^{-i2\pi (x f_x + y f_y)} dx dy \\
&= \mathcal{F}_2\{f(x, y)\} = F(f_x, f_y). \qquad (2.23)
\end{aligned}
$$

■

2.5.2 Inverse Radon Transform

Theorem 2. The Radon transform is given by Eq. (2.17) or Eq. (2.18)

$$p(l,\theta) = \int_L f(r,\phi)dz \tag{2.24}$$

and the Inverse Radon transform is given by

$$f(r,\phi) = \frac{1}{2\pi^2}\int_0^{\pi}\int_{-\infty}^{+\infty} \frac{1}{r\cos(\theta-\phi)-l}\cdot\frac{\partial p(l,\theta)}{\partial l}dld\theta. \tag{2.25}$$

or

$$f(r,\phi) = \frac{1}{4\pi^2}\int_0^{2\pi}\int_{-\infty}^{+\infty} \frac{1}{r\cos(\theta-\phi)-l}\cdot\frac{\partial p(l,\theta)}{\partial l}dld\theta. \tag{2.26}$$

Proof

From Eq. (2.23), we have

$$f(x,y) = \mathcal{F}_2^{-1}\{F(f_x,f_y)\} = \int_{-\infty}^{\infty}\int_{-\infty}^{\infty} F(f_x,f_y)e^{i2\pi(xf_x+yf_y)}df_xdf_y\ . \tag{2.27}$$

Using Eqs. (2.22) and (2.19), $xf_x + yf_y = f(x\cos\theta + y\sin\theta) = x'$, Eq. (2.27) becomes

$$f(x,y) = \int_0^{\pi}\int_{-\infty}^{\infty} |f|P(f,\theta)e^{i2\pi fx'}dfd\theta \tag{2.28}$$

$$= \int_0^{2\pi}\int_0^{\infty} fP(f,\theta)e^{i2\pi fx'}dfd\theta, \tag{2.29}$$

where the Jacobian determinant is $J = |\frac{\partial(f_x,f_y)}{\partial(f,\theta)}| = f$.

1) Let $q(x',\theta)$ denote the inner integral of Eq. (2.28)

$$q(x',\theta) = \int_{-\infty}^{\infty} (sgn(f)\cdot fP(f,\theta))e^{i2\pi fx'}df, \tag{2.30}$$

where $sgn(f)$ is the *sign* (*signum*) function defined by

$$sgn(f) = \begin{cases} 1 & (f>0) \\ -1 & (f<0). \end{cases} \tag{2.31}$$

By applying the Fourier transform of the Inverse function and Derivative theorem [12–14]

$$\mathcal{F}^{-1}\{sgn(f)\} = -\frac{1}{i\pi x'} \quad \text{and} \quad \mathcal{F}_1^{-1}\{fP(f,\theta)\} = \frac{1}{i2\pi}\frac{\partial p(x',\theta)}{\partial x'}, \tag{2.32}$$

Eq. (2.30) becomes

$$q(x',\theta) = \left(\frac{1}{i\pi x'}\right) \star \left(\frac{1}{i2\pi}\frac{\partial p(x',\theta)}{\partial x'}\right)$$
$$= \frac{1}{2\pi^2}\int_{-\infty}^{\infty}\frac{1}{x'-l}\frac{\partial p(l,\theta)}{\partial l}dl, \tag{2.33}$$

where $\star$ denotes the convolution and x' is given by Eq. (2.19). By substituting Eq. (2.33) into Eq. (2.28), noticing $f(x,y) = f(r,\phi)$ and $x' = r\cos(\theta-\phi)$, we obtain Eq. (2.25).

2) Let $q(x',\theta)$ denote the inner integral of Eq. (2.29)

$$q(x',\theta) = \int_{-\infty}^{\infty}(u(f)\cdot fP(f,\theta))e^{i2\pi fx'}df, \tag{2.34}$$

where $u(f)$ is the *unit step* function defined by

$$u(f) = \begin{cases} 1 & (f > 0) \\ 0 & (f < 0). \end{cases} \tag{2.35}$$

By applying the Fourier transform of the unit step function and the Derivative theorem [12–14]

$$\mathcal{F}^{-1}\{u(f)\} = \frac{1}{2}\delta(x') - \frac{1}{i2\pi x'} \quad \text{and} \quad \mathcal{F}_1^{-1}\{fP(f,\theta)\} = \frac{1}{i2\pi}\frac{\partial p(x',\theta)}{\partial x'}, \tag{2.36}$$

Eq. (2.34) becomes

$$q(x',\theta) = \left(\frac{1}{2}\delta(x') - \frac{1}{i2\pi x'}\right) \star \left(\frac{1}{i2\pi}\frac{\partial p(x',\theta)}{\partial x'}\right)$$
$$= \frac{1}{i4\pi}\delta(x') \star \frac{\partial p(x',\theta)}{\partial x'} + \frac{1}{4\pi^2}\frac{1}{x'} \star \frac{\partial p(x',\theta)}{\partial x'}$$
$$= \frac{1}{i4\pi}\frac{\partial p(x',\theta)}{\partial x'} + \frac{1}{4\pi^2}\int_{-\infty}^{\infty}\frac{1}{x'-l}\frac{\partial p(l,\theta)}{\partial l}dl, \tag{2.37}$$

where $\star$ denotes the convolution and x' is given by Eq. (2.19). By substituting Eq. (2.37) into Eq. (2.29), we have

$$f(x,y) = \frac{1}{i4\pi}\int_0^{2\pi}\frac{\partial p(x',\theta)}{\partial x'}d\theta + \frac{1}{4\pi^2}\int_0^{2\pi}\int_{-\infty}^{\infty}\frac{1}{x'-l}\frac{\partial p(l,\theta)}{\partial l}dld\theta. \tag{2.38}$$

Appendix 2B shows that

$$\int_0^{2\pi}\frac{\partial p(x',\theta)}{\partial x'}d\theta = 0; \tag{2.39}$$

thus, noticing $f(x,y) = f(r,\phi)$ and $x' = r\cos(\theta-\phi)$, Eq. (2.38) becomes Eq. (2.26). ■

2.6 Image Reconstruction

Methodology for X-ray CT image reconstruction can be classified into two categories: transform reconstruction methods and series-expansion reconstruction methods.

Transform reconstruction methods are based on analytic inversion formulas, which are manipulated into a variety of forms, depending on the underlying principles and data acquisition schemes. In transform reconstruction methods, the problem can be stated as "given the measured data $p(l, \theta)$ or $p(\sigma, \beta)$, estimate the object function $f(r, \phi)$." The estimate $\hat{f}(r, \phi)$ is determined in terms of (r, ϕ), $p(l, \theta)$, or $p(\sigma, \beta)$ through continuous functional operations. At the very end of image reconstruction, the inversion formula becomes discretized for the computational implementation. In the space domain, it uses Radon inversion formulas; in the frequency domain, it uses the Fourier slice theorem. The rho-filtered layergram method and the method involving expansion in angular harmonics are also examples of transform reconstruction methods.

Series-expansion reconstruction methods are fundamentally different from transform reconstruction methods. They are discretized at the very beginning of the image reconstruction by finding a finite set of numbers as estimates of the object function. It is embedded on the grids of square pixels, and the problem can be formulated as "given the measured vector $\mathbf{y}$, estimate the image vector $\mathbf{x}$, such that $\mathbf{y} = \mathbf{A}\mathbf{x}$," where $\mathbf{y}$ and $\mathbf{x}$ are the m- and n-dimensional vectors, respectively, and $\mathbf{A}$ is an $m \times n$-dimensional matrix. This estimation is done by requiring $\mathbf{x}$ to satisfy some specified optimization criteria. Series-expansion reconstruction methods are often known as algebraic methods, iterative algorithms, optimization theory techniques, etc., which are either non-iterative or iterative.

The most commonly used method in X-ray CT image reconstruction, particularly for parallel projections, is the convolution method derived from Radon inversion formulas. The reason for this is ease of implementation combined with good accuracy. This section discusses this transform reconstruction method only and shows that the implementation of this method finally leads to the operations consisting of a double convolution and a backprojection.

2.6.1 Convolution Method

2.6.1.1 Convolution Method for Parallel Projections

This section discusses the convolution method for parallel projections. It is based on the first Radon inversion formula Eq. (2.25).

Operator Expressions. For the convenience of analysis, we use operators to express Radon and Inverse Radon transforms. As shown in Figures 2.6 and

2.7, the Radon transform (Eq. (2.18)) can be written more accurately as

$$p(l,\theta) = \begin{cases} \int_{-\infty}^{+\infty} f(\sqrt{l^2+z^2}, \theta + \tan^{-1}(\frac{z}{l}))dz & (l \neq 0) \\ \\ \int_{-\infty}^{+\infty} f(z, \theta + \frac{\pi}{2})dz & (l = 0). \end{cases} \tag{2.40}$$

Let the operator $\mathcal{R}$ represent the Radon transform. Eq. (2.40) can be expressed as

$$p(l,\theta) = [\mathcal{R}f](l,\theta). \tag{2.41}$$

Similarly, let $\mathcal{D}_1$ represent the partial differentiation with respect to the first variable of a function of two real variables, that is,

$$[\mathcal{D}_1 p](l,\theta) = \frac{\partial p(l,\theta)}{\partial l} = p'(l,\theta), \tag{2.42}$$

let an operator $\mathcal{H}_1$ represent the Hilbert transform with respect to the first variable of a function of two real variables, that is,

$$[\mathcal{H}_1 p'](l',\theta) = -\frac{1}{\pi}\int_{-\infty}^{+\infty} \frac{p'(l,\theta)}{l'-l} dl = t(l',\theta), \tag{2.43}$$

let an operator $\mathcal{B}$ represent the backprojection, that is, given a function t of two variables, $\mathcal{B}t$ is another function of two polar variables whose value at any point (r,ϕ) is

$$[\mathcal{B}t](r,\phi) = \int_0^{\pi} t(r\cos(\theta-\phi),\theta)d\theta = s(r,\phi), \tag{2.44}$$

let an operator $\mathcal{N}$ represent the normalization, that is,

$$[\mathcal{N}s](r,\phi) = -\frac{1}{2\pi}s(r,\phi), \tag{2.45}$$

then, the Inverse Radon transform can be expressed as

$$\begin{aligned} f(r,\phi) &= \frac{1}{2\pi^2}\int_0^{\pi}\int_{-\infty}^{+\infty} \frac{1}{r\cos(\theta-\phi)-l}\frac{\partial p(l,\theta)}{\partial l} dl d\theta \\ &= \frac{1}{2\pi^2}\int_0^{\pi}\int_{-\infty}^{+\infty} \frac{1}{r\cos(\theta-\phi)-l}[\mathcal{D}_1 p](l,\theta) dl d\theta \\ &= -\frac{1}{2\pi}\int_0^{\pi}(-\frac{1}{\pi}\int_{-\infty}^{+\infty} \frac{[\mathcal{D}_1 p](l,\theta)}{r\cos(\theta-\phi)-l} dl) d\theta \\ &= -\frac{1}{2\pi}\int_0^{\pi}[\mathcal{H}_1\mathcal{D}_1 p](r\cos(\theta-\phi),\theta)d\theta \\ &= -\frac{1}{2\pi}[\mathcal{B}\mathcal{H}_1\mathcal{D}_1 p](r,\phi) \\ &= [\mathcal{N}\mathcal{B}\mathcal{H}_1\mathcal{D}_1 p](r,\phi). \end{aligned} \tag{2.46}$$

Thus, from Eqs. (2.41) and (2.46), the Inverse Radon transforms Eq. (2.25) can be expressed by the operators as

$$f(r,\phi) = [\mathcal{R}^{-1}p](r,\phi) \qquad \text{and} \qquad \mathcal{R}^{-1} = \mathcal{N}\mathcal{B}\mathcal{H}_1\mathcal{D}_1. \tag{2.47}$$

Eq. (2.47) implies that Inverse Radon transform in (l,θ)-space can be decomposed into four operations: (1) a partial differentiation with respect to l in each view θ, (2) a Hilbert transform with respect to l in each view θ, (3) a backprojection on $r\cos(\theta-\phi)$ over all views θ, and (4) a normalization on the backprojected data.

Hilbert Transform. The main problem in carrying out the Inverse Radon transform Eq. (2.46) is in implementing Hilbert transform, because Eq. (2.43) is an improper integral of both the first kind and second kind. The first kind of improper integral is that the integration is from $-\infty$ to $+\infty$. The second kind of improper integral is that the integration diverges at $l = l'$.

The Hilbert transform Eq. (2.43) is, in fact, a convolution of two functions, $p'(l,\theta)$ and $\rho(l) = -\frac{1}{\pi l}$:

$$[\mathcal{H}_1 p'](l',\theta) = p'(l,\theta) \star \rho(l), \tag{2.48}$$

where $\star$ denotes the convolution. By adopting the convolution approach, the following typical, mathematical handling can take place. Let a set of parameterized functions

$$\{\rho_A(l) | A > 0\} \tag{2.49}$$

be applied such that in the limiting case of $A \to \infty$,

$$p'(l,\theta) \star \rho(l) = \lim_{A\to\infty} (p'(l,\theta) \star \rho_A(l)), \tag{2.50}$$

then for a sufficiently large A, the Hilbert transform can be approximated by

$$[\mathcal{H}_1 p'](l',\theta) \simeq p'(l,\theta) \star \rho_A(l). \tag{2.51}$$

[1, 10] give a sufficient condition to Eq. (2.50). That is, if $p'(l,\theta)$ is reasonable at l and $\{\rho_A(l)|A > 0\}$ is a regularizing family, then

$$[\mathcal{H}_1 p'](l',\theta) = \lim_{A\to\infty} (p'(l,\theta) \star \rho_A(l)). \tag{2.52}$$

$p'(l,\theta)$ is said to be reasonable at l if

$$\begin{cases} p'(l,\theta) = 0 & (|l| > E) \\ \int_{-E}^{+E} p'(l,\theta)dl & \text{exists} \\ -\frac{1}{\pi}\lim_{\epsilon\to 0}\{\int_{-\infty}^{l'-\epsilon} \frac{p'(l,\theta)}{l'-l}dl + \int_{l'+\epsilon}^{+\infty} \frac{p'(l,\theta)}{l'-l}dl\} & \text{exists.} \end{cases} \tag{2.53}$$

A set of functions $\{\rho_A(l)|A > 0\}$ is said to be a regularizing family if for any reasonable function $p'(l, \theta)$ at l, Eq. (2.50) holds.

The first two formulas in Eq. (2.53) is to make the improper integral of the first kind in Eq. (2.43) exits. This is because

$$\int_{-\infty}^{+\infty} \frac{p'(l,\theta)}{l'-l} dl = \int_{|l|<E} \frac{p'(l,\theta)}{l'-l} dl + \int_{|l|>E} \frac{p'(l,\theta)}{l'-l} dl, \tag{2.54}$$

and due to $f(r, \phi) = 0$ $(|l| > E)$,

$$p'(l,\theta) = [\mathcal{D}_1 \mathcal{R} f](l,\theta) = [\mathcal{R} \mathcal{D}_1 f](l,\theta) = 0 \,. \tag{2.55}$$

The third formula in Eq. (2.53) is the Cauchy principal value (PV) of the improper integral of the second kind in Eq. (2.43). This condition assumes that this improper integral exists in the sense of PV.

By using the Liemann–Lebesgue lemma (for absolutely integrable functions) and the third formula of Eq. (2.53), [1, 10] show that

$$\rho_A(l) = -2 \int_0^{A/2} F_A(u) \sin(2\pi u l) du \tag{2.56}$$

can serve as a regularizing family of functions, where $F_A(u)$ is a real-valued integrable function that for $u \geq 0$

$$\begin{cases} 0 \leq F_A(u) \leq 1 \text{ and } F_A(u) = 0 & (u \geq \frac{A}{2}) \\ F_A(u_2) \geq F_A(u_1) & (u_2 < u_1) \\ \lim_{A\to\infty} F_A(u) = 1, & \end{cases} \tag{2.57}$$

$F_A(u)$ of Eq. (2.57) actually specifies a window with width A. For example, Hamming window given by

$$F_A(u) = \alpha + (1-\alpha)\cos(\frac{2\pi u}{A}) \tag{2.58}$$

is a such window.

Convolution for Parallel Projections. Based on Eqs. (2.53) and (2.56), it has been shown that Eq. (2.51) can be written as

$$\begin{aligned} [\mathcal{H}_1 \mathcal{D}_1 p](l',\theta) &= [\mathcal{H}_1 p'](l',\theta) \\ &\simeq p'(l',\theta) \star \rho_A(l') \\ &= \int_{-\infty}^{+\infty} p'(l,\theta)\rho_A(l'-l) dl \\ &= \int_{-\infty}^{+\infty} p(l,\theta)\rho'_A(l'-l) dl \\ &= p(l',\theta) \star \rho'_A(l'), \end{aligned} \tag{2.59}$$

where $\rho'_A(l'-l) = \frac{d\rho_A(l'-l)}{dl}$. Thus, Eq. (2.46) becomes

$$\begin{aligned} f(r,\phi) &= [\mathcal{N}\mathcal{B}\mathcal{H}_1\mathcal{D}_1 p](r,\phi) \\ &= [\mathcal{N}\mathcal{B}(p \star \rho'_A)](r,\phi) \\ &= [-\frac{1}{2\pi}\mathcal{B}(p \star \rho'_A)](r,\phi) \\ &= [\mathcal{B}(p \star (-\frac{1}{2\pi}\rho'_A))](r,\phi) \\ &= [\mathcal{B}(p \star q)](r,\phi) \\ &= \int_0^{\pi} p(l',\theta) \star q(l') d\theta \\ &= \int_0^{\pi}\int_{-\infty}^{+\infty} p(u,\theta) \star q(l'-u) du d\theta \ , \end{aligned} \tag{2.60}$$

where $q(l) = -\frac{1}{2\pi}\rho'_A(l)$ is called the convolution function. From Eq. (2.56), it is equal to

$$q(l) = 2\int_0^{A/2} uF_A(u)\cos(2\pi ul) du \tag{2.61}$$

with $l = r\cos(\theta - \phi)$.

In Eq. (2.60), the integration function $p(l',\theta) \star q(l')$ represents a convolution of the parallel projections $p(l',\theta)$ and the convolution function $q(l')$ inside one view θ, the integral represents a sum of these convolved projections over all views. Thus, the convolution method for the parallel projections consists of a convolution and a backprojection.

2.6.1.2 Convolution Method for Divergent Projections

This section discusses the convolution method for divergent projections. It is based on the second Radon inversion formula Eq. (2.26), ‖

$$f(r,\phi) = \frac{1}{4\pi^2}\int_0^{2\pi}\int_{-\infty}^{+\infty} \frac{1}{r\cos(\theta-\phi) - l} \cdot \frac{\partial p(l,\theta)}{\partial l} dl d\theta \ . \tag{2.62}$$

This formula is for parallel projections $p(l,\theta)$. In order to derive a formula to link the object function $f(r,\phi)$ and divergent projections $p(\sigma,\beta)$ which are defined in Section 2.4.2.2 and shown in Figures 2.6 and 2.8, parameters (l,θ) in Eq. (2.62) must be replaced by (σ,β).

Geometry. In Figure 2.10, two lines $L(\sigma,\beta)$ and $L'(\sigma',\beta)$ are in the same view β of the divergent projections; the line L' passes the point P at (r,ϕ). Let D denote the distance between the source location S and the origin O of

‖Based on the original Radon inversion formula in [15], [1, 10, 16] proved Eq. (2.26) in a different way.

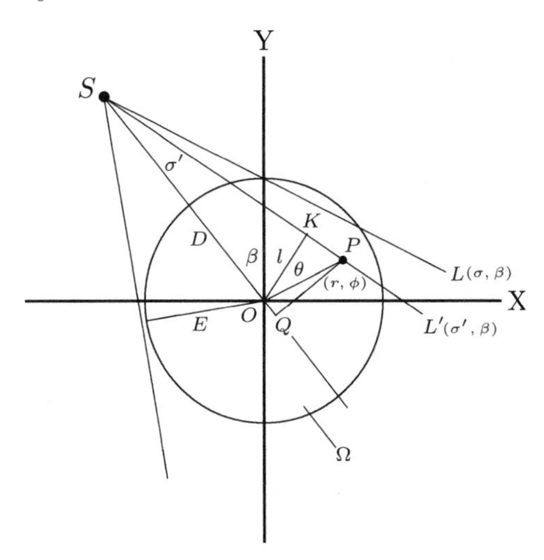

FIGURE 2.10
An illustration of the geometry for the convolution of divergent projections.

the coordinate system. It can be shown (Problem 2.4) that for any line $L(\sigma, \beta)$ in the view β,

$$D \sin \sigma = l \quad \text{and} \quad \sigma + \beta = \theta, \tag{2.63}$$

where $|\sigma'| \leq \sin^{-1}\left(\frac{E}{D}\right)$, and E is the radius of the circular reconstruction region Ω. Let D' denote the distance between the source location S and the point P. It can be shown (Problem 2.5) that for the line $L'(\sigma', \beta)$ in the view β,

$$\begin{aligned} D' &= \sqrt{(r\cos(\beta - \phi))^2 + (D + r\sin(\beta - \phi))^2} \\ \sigma' &= \tan^{-1}(r\cos(\beta - \phi)/(D + r\sin(\beta - \phi))) \\ r\cos(\theta - \phi) - l &= D' \sin(\sigma' - \sigma), \end{aligned} \tag{2.64}$$

where $|\sigma| \leq \sin^{-1}\left(\frac{E}{D}\right)$.

Projection. Let $p(l, \theta)$ and $p(\sigma, \beta)$ represent the same projection; then from Eq. (2.63) we have

$$p(l, \theta) = p(D \sin \sigma, \sigma + \beta) = p(\sigma, \beta) \ , \tag{2.65}$$

and in Eq. (2.62),

$$\frac{\partial p(l, \theta)}{\partial l} dl d\theta = \left(\frac{\partial p(l, \theta)}{\partial \sigma} \frac{\partial \sigma}{\partial l} + \frac{\partial p(l, \theta)}{\partial \beta} \frac{\partial \beta}{\partial l} \right) J d\sigma d\beta, \tag{2.66}$$

where $J = |\frac{\partial(l, \theta)}{\partial(\sigma, \beta)}| = \frac{1}{D \cos \sigma}$ is the Jacob determinant. It has been shown (Problem 2.6) that Eq. (2.66) can be written as

$$\frac{\partial p(l, \theta)}{\partial l} dl d\theta = (p_1'(\sigma, \beta) - p_2'(\sigma, \beta)) d\sigma d\beta, \tag{2.67}$$

where $p'_1(\sigma, \beta)$ and $p'_2(\sigma, \beta)$ are the partial derivatives of $p(\sigma, \beta)$ with respect to σ and β, respectively, that is,

$$p'_1(\sigma, \beta) = \frac{\partial p(\sigma, \beta)}{\partial \sigma} \quad \text{and} \quad p'_2(\sigma, \beta) = \frac{\partial p(\sigma, \beta)}{\partial \beta} . \tag{2.68}$$

Convolution for Divergent Projections. By substituting Eqs. (2.64) and (2.67) into Eq. (2.62), we obtain

$$\begin{aligned} f(r, \phi) &= \frac{1}{4\pi^2} \int_0^{2\pi} \int_{-\infty}^{+\infty} \frac{1}{D' \sin(\sigma' - \sigma)} (p'_1(\sigma, \beta) - p'_2(\sigma, \beta)) d\sigma d\beta \\ &= \frac{1}{4\pi^2} \int_0^{2\pi} \int_{-\infty}^{+\infty} \frac{1}{\sigma' - \sigma} \cdot \frac{\sigma' - \sigma}{D' \sin(\sigma' - \sigma)} (p'_1(\sigma, \beta) - p'_2(\sigma, \beta)) d\sigma d\beta \ . \end{aligned} \tag{2.69}$$

Let

$$P'_{1,2}(\sigma, \beta, \sigma') = \frac{\sigma' - \sigma}{D' \sin(\sigma' - \sigma)} (p'_1(\sigma, \beta) - p'_2(\sigma, \beta)). \tag{2.70}$$

Eq. (2.69) can be expressed as

$$f(r, \phi) = \frac{1}{4\pi^2} \int_0^{2\pi} \int_{-\infty}^{+\infty} \frac{P'_{1,2}(\sigma, \beta, \sigma')}{\sigma' - \sigma} d\sigma d\beta. \tag{2.71}$$

Eq. (2.71) is called an Inverse Radon transform for divergent projections. When $P'_{1,2}(\sigma, \beta, \sigma')$ is considered as a function of σ alone, the inner integral of Eq. (2.71) can be considered a Hilbert transform of $P'_{1,2}(\sigma, \beta, \sigma')$. Eq. (2.71) is similar to, but not exactly the same as, its parallel projection counterpart Eq. (2.43), because $P'_{1,2}(\sigma, \beta, \sigma')$ in Eq. (2.71) includes an additional parameters σ'.

Assume that $P'_{1,2}(\sigma, \beta, \sigma')$, as a function of σ, is reasonable at σ', and let $\{\rho_A | A > 0\}$ be a family of regularizing functions. Eq. (2.71) can be approximately written as

$$f(r, \phi) = \frac{1}{4\pi^2} \int_0^{2\pi} \int_{-\infty}^{+\infty} \rho_A(\sigma' - \sigma) P'_{1,2}(\sigma, \beta, \sigma') d\sigma d\beta, \tag{2.72}$$

where $\rho_A(\sigma) = -\frac{1}{\pi\sigma}$ and its parallel projection counterpart is given by Eq. (2.56). Based on Eq. (2.72), [16] proves that

$$f(r, \phi) = \frac{D}{4\pi^2} \int_0^{2\pi} \frac{1}{(D')^2} \int_{-\infty}^{+\infty} p(\sigma, \beta)[q_1(\sigma' - \sigma) \cos\sigma + q_2(\sigma' - \sigma) \cos\sigma'] d\sigma d\beta, \tag{2.73}$$

where

$$q_1(\sigma) = -\frac{\sigma \rho_A(\sigma)}{\sin^2 \sigma} \quad \text{and} \quad q_2(\sigma) = \frac{\rho_A(\sigma) + \sigma \rho'_A(\sigma)}{\sin \sigma}, \tag{2.74}$$

with $\rho'_A(\sigma) = \frac{d\rho_A(\sigma)}{d\sigma}$, and $p(\sigma, \beta)$ is the divergent projection given by Eq. (2.65).

Eq. (2.73) provides a convolution reconstruction formula for the divergent projections. Its inner integral is a sum of one weighted convolution of the projection $p(\sigma,\beta)$ with $q_1(\sigma)$ and the weight $\cos\sigma$ and one unweighted convolution of the projection $p(\sigma,\beta)$ with $q_2(\sigma)$. That is, the inner integral is essentially a convolution of the divergent projections and the convolution functions. The outer integral is a sum of these convolved projections over all views. Thus, similar to the convolution method for the parallel projections, the convolution method for the divergent projections consists of a convolution and a backprojection.

2.6.2 Computational Implementation

2.6.2.1 Parallel Projections

At the end of X-ray CT image reconstruction, the continuous operations of Eq. (2.60)—the convolution $p(l',\theta) \star q(l')$ and the integration $\int_0^{\pi} p(l',\theta) \star q(l')d\theta$—are discretized and are approximated by Riemann sums over projections at $(-Nd,\cdots,-d,0,d,\cdots,Nd)$ within one view and over all views on $(0,\Delta,\cdots,(M-1)\Delta)$, respectively. Convolution function $q(l^{'})$ is also discretized on the multiples of projection-spacing d. From Eq. (2.60), we have

$$p(l',\theta) \star q(l') = d \sum_{k=-N}^{N} p(kd,m\Delta)q(l'-kd) = t(l',m\Delta)\,, \tag{2.75}$$

and

$$f(r,\phi) = \int_0^{\pi} (p(l',\theta) \star q(l'))d\theta = \Delta \sum_{m=0}^{M-1} t(l',m\Delta)\,. \tag{2.76}$$

However, for the given grids $\{(r,\phi)\}$, that is, the centers of pixels, $l' = r\cos(\theta-\phi)$ may not be the multiples of the projection-spacing d,

$$t(l',m\Delta) \neq t(nd,m\Delta) \qquad (-N \le n \le N). \tag{2.77}$$

Thus, $t(l',m\Delta)$ must be approximated by some discretized $t(nd,m\Delta)$ $(-N \le n \le N)$, which are computed by

$$t(nd,m\Delta) = d \sum_{k=-N}^{N} p(kd,m\Delta)q((n-k)d); \tag{2.78}$$

$t(nd,m\Delta)$ of Eq. (2.77) are called the *convolved projections* in the m-th view. The procedure that approximates $t(l',m\Delta)$ using $t(nd,m\Delta)$ is known as *interpolation.* The resulting approximation is called the *interpolated data* in the m-th view, denoted by $s_m(r,\phi)$, that is,

$$s_m(r,\phi) = t(l',m\Delta)\,. \tag{2.79}$$

The commonly used interpolation methods are the linear interpolation, which uses the interpolation function given by

$$\psi(u) = \begin{cases} 1 - \frac{|u|}{d} \ (|u| < d) \\ \\ 0 \qquad\quad (|u| \geq d) \end{cases}, \tag{2.80}$$

and the nearest neighbor interpolation, which uses the interpolation function given by

$$\psi(u) = \begin{cases} 1 \quad (|u| < d/2) \\ \\ 0.5 \ (|u| = d/2) \ . \\ \\ 0 \quad (|u| > d/2) \end{cases} \tag{2.81}$$

When linear interpolation is applied, an integer n is selected such that $nd \leq l' < (n+1)d$ and $t(l', m\Delta)$ is approximated by a weighted summation of $t(nd, m\Delta)$ and $t((n+1)d, m\Delta)$, where the weights are proportional to the relative distances $\frac{(n+1)d-l'}{d}$ and $\frac{l'-nd}{d}$, respectively. When nearest neighbor interpolation is applied, an integer n is chosen such that $|nd - l'|$ is as small as possible and $t(l', m\Delta)$ is approximated by the value of $t(nd, m\Delta)$.

Appendix 2C shows that the interpolated data equal the convolution of the convolved projections $t(nd, m\Delta)$ in a view and proper interpolation function $\psi(u)$. That is,

$$s_m(r, \phi) = t(l', m\Delta) \simeq \sum_{n=-\infty}^{+\infty} t(nd, m\Delta)\psi(l' - nd)). \tag{2.82}$$

Thus, by substituting Eqs. (2.82) and (2.78) into Eq. (2.76), we obtain

$$\begin{aligned} f(r, \phi) &= \Delta[\sum_{m=0}^{M-1} s_m(r, \phi)] \\ &= \Delta[\sum_{m=0}^{M-1} (\sum_{n=-\infty}^{+\infty} t(nd, m\Delta)\psi(l' - nd))] \\ &= d\Delta\{\sum_{m=0}^{M-1} [\sum_{n=-\infty}^{+\infty} (\sum_{k=-N}^{N} p(kd, m\Delta)q((n-k)d))\psi(l' - nd)]\}, \end{aligned} \tag{2.83}$$

where $l' = r\cos(m\Delta - \phi)$.

Eq. (2.83) shows that computational implementation of the convolution method for parallel projections represented by Eq. (2.60) consists of three operations: (1) a convolution of the projections $p(nd, m\Delta)$ and the convolution function $q(nd)$ in one view $m\Delta$, (2) a convolution of the convolved projections $t(nd, m\Delta)$ and an interpolation functions $\psi(nd)$ in one view $m\Delta$, and (3) a backprojection of the interpolated data $s_m(r, \phi)$ over all views $m\Delta$ ($m = 0, \cdots, M-1$).

2.6.2.2 Divergent Projections

In the computational implementation of X-ray CT image reconstruction, the continuous operations of Eq. (2.73)—the inner and the outer integrations—are discretized and are approximated by Riemann sums over projections at $(-N\delta, \cdots, -\delta, 0, \delta, \cdots, N\delta)$ within one view and over all views on $(0, \Delta, \cdots, (M-1)\Delta)$, respectively.

Let $t(\sigma', m\Delta)$ denote the inner integration of Eq. (2.73); we have

$$t(\sigma', m\Delta) = \delta \sum_{k=-N}^{N} p(k\delta, m\Delta)[q_1(\sigma' - k\delta)\cos(k\delta) + q_2(\sigma' - k\delta)\cos\sigma'] \ , \tag{2.84}$$

and

$$f(r, \phi) = \frac{D}{4\pi^2}\Delta \sum_{m=0}^{M-1} \frac{1}{(D')^2} t(\sigma', m\Delta) \ , \tag{2.85}$$

where σ' and D' are given by

$$\sigma' = \tan^{-1}(r\cos(\beta - \phi)/D + r\sin(\beta - \phi)) \tag{2.86}$$

and

$$D' = \sqrt{(r\cos(\beta - \phi))^2 + (D + r\sin(\beta - \phi))^2} \ , \tag{2.87}$$

respectively.

Because for the given grids $\{(r, \phi)\}$, that is, the centers of pixels, σ' of Eq. (2.86) may not be the multiples of the projection-spacing δ, $t(\sigma', m\Delta)$ of Eq. (2.84) must be approximated by some discretized, convolved projections $t(n\delta, m\Delta)$ $(-N \leq n \leq N)$, which are computed by

$$t(n\delta, m\Delta) = \delta \sum_{k=-N}^{N} p(k\delta, m\Delta)[q_1((n-k)\delta)\cos(k\delta) + q_2((n-k)\delta)\cos(n\delta)]. \tag{2.88}$$

Similar to the discussion for parallel projections in Section 2.6.2.1, $t(\sigma', m\Delta)$ is approximated by an interpolation operation that is equivalent to a convolution of convolved projections with the proper interpolation function. That is,

$$s_m(r, \phi) = t(\sigma', m\Delta) \simeq \sum_{n=-\infty}^{+\infty} t(n\delta, m\Delta)\psi(\sigma' - n\delta), \tag{2.89}$$

where $\psi(\sigma')$ is an interpolation function and is defined by Eq. (2.80) or Eq. (2.81).

Thus, letting

$$W = \frac{D}{(2\pi D')^2} \tag{2.90}$$

and substituting Eqs. (2.88) and (2.89) into Eq. (2.85), we obtain

$$
\begin{aligned}
f(r,\phi) &= \Delta\{\sum_{m=0}^{M-1} W s_m(r,\phi)\} \\
&= \Delta\{\sum_{m=0}^{M-1} W[\sum_{n=-\infty}^{+\infty} t(n\delta, m\Delta)\psi(\sigma' - n\delta)]\} \\
&= \delta\Delta\{\sum_{m=0}^{M-1} W[\sum_{n=-\infty}^{+\infty} (\sum_{k=-N}^{N} p(k\delta, m\Delta) \\
&(q_1((n-k)\delta)\cos(k\delta) + q_2((n-k)\delta)\cos(n\delta)))\psi(\sigma' - n\delta)]\}, \quad (2.91)
\end{aligned}
$$

where σ' is given by Eq. (2.86).

Eq. (2.91) indicates that computational implementation of the convolution method for divergent projections represented by Eq. (2.73) consists of three operations: (1) a convolution of the projections $p(nd, m\Delta)$ and the convolution function $q_1(nd)$ and $q_2(nd)$ in one view $m\Delta$, (2) a convolution of the convolved projections $t(nd, m\Delta)$ and an interpolation functions $\psi(nd)$ in one view $m\Delta$, and (3) a backprojection of the interpolated data $s_m(r,\phi)$ over all views $m\Delta$ $(m = 0, \cdots, M-1)$.

In summary, Eqs. (2.83) and (2.91) show the computational implementations of the Inverse Radon transform-based X-ray CT image reconstruction methods for the parallel and the divergent projections, respectively. Although two equations have some differences, they are essentially the same. Their common features are that these computational implementations consist of a double convolution and a backprojection. Thus, the image reconstruction approach implemented by Eqs. (2.83) and (2.91) has been given the name of the convolution method.

2.7 Appendices

2.7.1 Appendix 2A

This appendix proves Eq. (2.1).

Proof.

Figure 2.11 shows a physical mechanism that illustrates the reduction of photons when they penetrate the object. In Figure 2.11, a slat of the homogeneous medium with thickness Δl is placed between the X-ray source and a detector. The X-ray is monochromatic and its beam direction is perpendicular to the surface of the slat. Let the number of photons arriving at the surface located at $l - \Delta l$ be $n(l - \Delta l)$ and the number of photons departing from

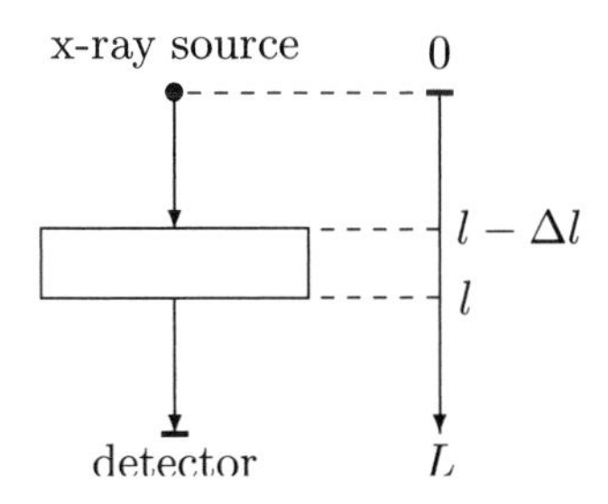

FIGURE 2.11
Physical mechanisms for illustrating the attenuation coefficient.

the surface located at l be $n(l)$. A single constant *coefficient* μ that is able to describe the reduction of photons due to the scattering and the absorption when X-rays penetrate the object can be defined by

$$n(l) = n(l - \Delta l) - \mu n(l - \Delta l)\Delta l. \tag{2.92}$$

This equation can be rewritten as

$$\frac{n(l) - n(l - \Delta l)}{n(l - \Delta l)} = -\mu \Delta l. \tag{2.93}$$

In the limiting case of $\Delta l \rightarrow 0$, we have $l - \Delta l \rightarrow l$, $\Delta l \rightarrow dl$ and $n(l) - n(l - \Delta l) \rightarrow dn(l)$. Thus, the above equation becomes

$$\frac{dn(l)}{n(l)} = -\mu dl. \tag{2.94}$$

The solution of this differential equation is

$$n(l) = e^{-\mu l + C}, \tag{2.95}$$

where the constant C is determined by the initial condition $n(0)$, which leads to $n(0) = e^C$. Thus, Eq. (2.96) becomes

$$\frac{n(l)}{n(0)} = e^{-\mu l}. \tag{2.96}$$

Assume that the attenuation of the air is ignored or the entire physical mechanism of Figure 2.11 is placed in a free space. Then, $n(0) = n(l - \Delta l)$. Thus, the left side of the above equation becomes $\frac{n(l)}{n(0)} = \frac{n(l)}{n(l - \Delta l)}$, which represents a ratio between the number of photons leaving the slab and the number of photons coming to the slab. That is, this ratio can be considered the probability of the photons that enter the slab without being scattered and absorbed. Thus, let $\Delta l = 1$; $\frac{n(l)}{n(l-\Delta l)}$ is the probability ρ of Eq. (2.1), and hence, Eq(2.97) becomes

$$\rho = e^{-\mu}, \tag{2.97}$$

that is,

$$\mu = -\ln \rho. \tag{2.98}$$

where μ is the Linear attenuation coefficient. In order to show its dependence on the energy of the X-ray and the property of the object (e.g., the tissues), μ is denoted by $\mu(e, t)$, where e and t denote the energy and the tissue, respectively. Eq. (2.97) is known as Lambert-Beer's law. ■

2.7.2 Appendix 2B

This appendix proves Eq. (2.39).

Proof.
Eq. (2.39) can be rewritten as

$$\int_0^{2\pi} \frac{\partial p(x',\theta)}{\partial x'} d\theta = \int_0^{\pi} \frac{\partial p(x',\theta)}{\partial x'} d\theta + \int_0^{\pi} \frac{\partial p(x',\theta+\pi)}{\partial x'} d\theta. \tag{2.99}$$

It can be shown that the Radon transform given by Eq. (2.17) or Eq. (2.18) have the following properties

$$p(x',\theta) = p(-x',\theta \pm \pi) = p(x',\theta + 2n\pi) , \tag{2.100}$$

where n is an integer. These properties are called symmetry and periodicity.
Thus, the first integral on the right side of Eq. (2.100) becomes

$$\int_0^{\pi} \frac{\partial p(x',\theta)}{\partial x'} d\theta = \int_0^{\pi} \frac{\partial p(-x',\theta+\pi)}{\partial x'} d\theta = -\int_0^{\pi} \frac{\partial p(x',\theta+\pi)}{\partial x'} d\theta. \tag{2.101}$$

By substituting Eq. (2.102) into Eq. (2.100), we prove Eq. (2.39). ■

2.7.3 Appendix 2C

This appendix proves that the interpolated data equals the convolution of the convolved projections $t(nd, m\Delta)$ in a view and proper interpolation function $\psi(u)$.

Proof.
At the m-th view, for a pixel centered at (r, ϕ), $l' = r\cos(m\Delta - \phi)$. Suppose that $nd \leq l' \leq (n+1)d$. The distances between the interpolation

data $s_m(r,\phi) = t(l^{'}, m\Delta)$ and the n-th and $(n+1)$-th convolved projections $t(nd, m\Delta)$ and $t((n+1)d, m\Delta)$ are $(l' - nd)$ and $((n+1)d - l^{'})$, respectively.

1) *For the linear interpolation.* Based on Eq. (2.80), the non-zero interval of $\psi(u)$ is $|u| \leq d$, which implies that only two projections are involved in the interpolation. Thus, we have

$$\begin{aligned} t(l', m\Delta) &= \frac{(n+1)d - l'}{d} t(nd, m\Delta) + \frac{l' - nd}{d} t((n+1)d, m\Delta) \\ &= t(nd, m\Delta)(1 - \frac{|l' - nd|}{d}) + t((n+1)d, m\Delta)(1 - \frac{|l' - (n+1)d|}{d}) \\ &= \sum_{n=-\infty}^{+\infty} t(nd, m\Delta)\psi(l^{'} - nd). \end{aligned} \tag{2.102}$$

2) *For the nearest neighbor interpolation.* Based on Eq. (2.81), the non-zero interval of $\psi(u)$ is $|u| \leq \frac{d}{2}$, which implies that only one, at most two, projections are involved in the interpolation. Thus, we have

$$\begin{array}{ll} if\ |l' - (n+1)d| < d/2, & then\ \ t(l', m\Delta) = t((n+1)d, m\Delta)\ , \\ if\ |l' - nd| = d/2, & then\ \ t(l', m\Delta) = \frac{1}{2}(t(nd, m\Delta) + t((n+1)d, m\Delta))\ , \\ if\ |l' - nd| < d/2, & then\ \ t(l', m\Delta) = t(nd, m\Delta)\ . \end{array} \tag{2.103}$$

Clearly, in the above all cases, $s_m(r,\phi) = t(l^{'}, m\Delta) = \sum_{n=-\infty}^{+\infty} t(nd, m\Delta)\psi(l^{'} - nd)$. ■

Problems

2.1. In X-ray CT, physical measurements of X-ray include actual measurement and calibration measurement processes. What is the role of the calibration measurement?

2.2. Both the actual measurement and the calibration measurement processes have a detection path and a reference detection path. What is the role of the reference detection?

2.3. Prove Eq. (2.59), that is, prove $[\mathcal{H}_1\mathcal{D}_1 p](l', \theta) = p(l', \theta) \star \rho'_A(l')$.

2.4. Prove Eq. (2.63).

2.5. Prove three formulas in Eq. (2.64). (Hint, See Figure 2.10. Prove $\angle OPQ = \beta - \phi$ first.)

2.6. Derive Eq. (2.67).

2.7. Verify the symmetry and periodicity of Radon transform, Eq. (2.101).

References

[1] Herman, G.: *Image Reconstruction from Projections.* Academic Press, New York (1980).

[2] Macovski, A.: Physical problems of computerized tomography. *IEEE Proc.* **71**(3) (1983).

[3] Lee, H., (ed), G.W.: *Imaging Technology.* IEEE Press, New York (1986).

[4] Kak, A., Slaney, M.: *Principles of Computerized Tomographic Imaging.* IEEE Press, New York (1988).

[5] Jain, A.K.: *Fundamentals of Digital Image Processing.* Prentice Hall, Englewood Cliffs, NJ (1989).

[6] Cho, Z.H., Jones, J.P., Singh, M.: *Foundations of Medical Imaging.* John Wiley & Sons, Inc., New York (1993).

[7] Hsieh, J.: *Computed Tomography, Principles, Design, Artifacts and Recent Advances.* SPIE, Bellingham (2004).

[8] Kalender, W.: *Computed Tomography: Fundamentals, System Technology, Image Quality, Applications.* Publics MCD Verlag, Germany (2005).

[9] Buzug, T.: *Computed Tomography.* Springer, Berlin (2008).

[10] Herman, G.: *Fundamentals of Computerized Tomography.* Springer, London (2009).

[11] Stark, H.: *Image Recovery: Theory and Application.* Academic Press, New York (1987).

[12] Barrett, H., Swindell, W.: *Radiological Imaging.* Academic Press, Hoboken, NJ. (1981).

[13] Papoulis, A.: *The Fourier Integral and Its Applications.* McGraw-Hill Book Company Inc., New York (1962).

[14] Papoulis, A.: *Signal Analysis.* McGraw-Hill Book Company Inc., New York (1977).

[15] Radon, J.: Über die bestimmung von funktionen durch ihre integralwerte längs gewisser mannigfaltigkeiten. *Berichte Saechsische Akademie der Wissenschaften* **69** (1917) 262–277.

[16] Herman, G., Naparstek, A.: Fast image reconstruction based on a Radon inversion formula appropiate for rapidly collected data. *SIAM J. Appl. Math.* **33**(3) (1977) 511–533.

3

MRI Physics and Mathematics

3.1 Introduction

3.1.1 History

Magnetic resonance imaging (MRI) is based on the nuclear magnetic resonance (NMR) phenomenon, which was first observed by Edward M. Purcell (United States) and Felix Bloch (United States) independently in 1946. They found that nuclei absorb radio waves at specified frequencies. This finding provided chemists and physicists with a way to probe molecular structures and diffusion. They received the Nobel Prize in Physics in 1952 for this discovery [1, 2]. In 1972, the first magnetic resonance (MR) image (a cross-sectional image of two water tubes) using the spatial information encoding principle was reported. The first MR images of the human head were published in 1978, with body scans following soon afterward. During the 1970s, most research in MRI took place in academia, primarily in the United Kingdom. In the 1980s, industry joined forces with universities, investing substantial resources to develop MRI systems. MRI scanners were first indicated for clinical use in 1983.

Since then, the image quality of MRI has improved dramatically and MRI scanners have proliferated throughout the world. With the ever-improving technology to produce images at higher resolution (micro imaging), higher speed (fast imaging), and higher information content (combined anatomical, metabolic, and functional imaging), the impact of MRI has been revolutionary not only in diagnostic radiology, but also in biology and neuroscience. Consequently, the Nobel Prize in Chemistry was awarded to Richard R. Ernst (Switzerland) in 1991 for Fourier transform nuclear magnetic resonance spectroscopy, and Kurt Wuthrich (Switzerland) in 2002 for the development of nuclear magnetic resonance spectroscopy in determining the three-dimensional structure of biological macromolecules in solution [3, 4]. The Nobel Prize in Medicine was given to Paul C. Lauterbur (United States) and Sir Peter Mansfield (United Kingdom) in 2003 for the discoveries leading to MRI [5, 6].

3.1.2 Overview

MRI encompasses an array of disciplines: physics, chemistry, mathematics, and engineering. For the purpose of description, a "simple big picture" of the principles of MRI is outlined here. When an object is placed in a static, external magnetic field which is in the longitudinal direction, all nuclear spins are oriented with the field direction at either a parallel or an anti-parallel, and at the same time precess around this external magnetic field at the Larmor frequency. The net magnetization (defined as a vector sum of spin moments in an unit volume) of these uneven populations generates a thermal equilibrium macroscopic magnetization. When a pulse radiofrequency field is applied in a transverse direction, the thermal equilibrium macroscopic magnetization is excited and perturbed from its equilibrium state and flipped towards to the transverse plane. After the radiofrequency pulse is removed and sufficient time is given, the precessing macroscopic magnetization undergoes a longitudinal relaxation and a transverse relaxation to return to its equilibrium state. As a consequence, a real-valued physical signal, often known as Free Induction Decay, is induced in the receiver coil.

In addition to the static field and the radiofrequency field, a specially designed, time-varying pulse gradient field is applied immediately after the radiofrequency field is removed. Spatial information is then encoded into the Free Induction Decay signal, which is followed by a complex-valued baseband signal produced by a quadratic phase sensitive detector and a discrete, complex-valued baseband signal generated by an analog-digital converter. The time integral of field gradient at a direction multiplied by the gyromagnetic ratio of nuclear spins gives a spatial frequency at that direction, which defines k-space. The sampling schedule of analog-digital converter is controlled by the timing diagram of the gradient pulse sequences. Thus, the discrete, complex-valued baseband signal at a given time is mapped to a sample at the corresponding spatial frequency in k-space.

The dynamics of macroscopic magnetization are described by Bloch equation. By ignoring the transverse relaxation term in the solution of Bloch equation or by using a simple physical model of Bloch equation, MR signal equation is formed, which shows that k-space sample is a two-dimensional, inverse Fourier transform of the thermal equilibrium macroscopic magnetization. When k-space is sampled rectilinearly, from MR signal equation, an MR image is reconstructed via Fourier transform and implemented by DFT (discrete Fourier transform); when k-space is sampled radially, by using Fourier Slice Theorem, an MR image is reconstructed via Inverse Radon transform and implemented by FBP (filtered backprojection) algorithm.

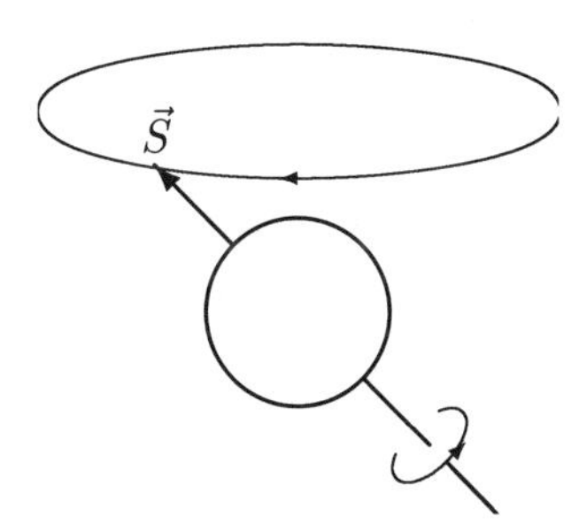

FIGURE 3.1
A spin and the spin angular momentum $\vec{S}$.

3.2 Nuclear Spin and Magnetic Moment

Nuclei with an odd number of protons and/or an odd number of neutrons possess a *spin angular momentum*, and therefore demonstrate the nuclear magnetic resonance (NMR) phenomenon. These nuclei can be imaged as spinning charged particles and are often referred to as *spins*, which are shown in Figure 3.1. An ensemble of such nuclei of the same type in the sample is referred to as a nuclear *spin system*. As a moving charged particle, the spin creates a tiny magnetic field around it, which is represented by a *magnetic dipole moment*.

Spin angular momentum and magnetic dipole moment are vector quantities, denoted by $\vec{S}$ and $\vec{\mu}$, and are related by

$$\vec{\mu} = \gamma \vec{S}, \tag{3.1}$$

where γ is the *gyromagnetic ratio*. It is a physical property of the nucleus of the atom. Different chemical elements, even the different isotopes of the same element, demonstrate different gyromagnetic ratios. Gyromagnetic ratios of some nuclear species are listed in Table 3.1.

TABLE 3.1
Gyromagnetic Ratios

Nucleus		$\frac{\gamma}{2\pi}$ (MHz/T)
Hydrogen	^{1}H	42.576
Carbon	^{13}C	10.705
Fluorine	^{19}F	40.054
Phosphorus	^{31}P	17.235

Let μ be the magnitude of $\vec{\mu}$; Quantum theory shows,

$$\mu = \gamma\hbar\sqrt{\mathrm{I}(\mathrm{I}+1)}, \tag{3.2}$$

where $\hbar = \frac{h}{2\pi}$ and h is Planck's constant (6.626×10^{-34} J $\cdot$ s), and I is the *spin quantum number*, which takes the value of the zero and the non-negative values of integers and half-integers ($0, \frac{1}{2}, 1, \frac{3}{2}, 2, \cdots$) [7–9].

A spin system is called a spin-I system when I is assigned to one of the following values: (1) zero if the atomic mass number is even and the charge number is also even, (2) an integer value if the atomic mass number is even and the charge number is odd, and (3) a half-integer value if the atomic mass number is odd. Thus, when NMR phenomena occur, I must be non-zero. ^{1}H, ^{13}C, ^{19}F, and ^{31}P are all spin-$\frac{1}{2}$ systems.

Because the magnetic dipole moment $\vec{\mu}$, the spin angular momentum $\vec{S}$, and other quantities in MRI are vectors, two coordinate systems, often known as the reference frames in MRI, are introduced. Let $\{\vec{i}, \vec{j}, \vec{k}\}$ be a set of unit directional vectors at the X-, Y-, and Z-axes of a Cartesian coordinate system {X,Y,Z}. $\{\vec{i}, \vec{j}, \vec{k}\}$ or {X,Y,Z} defines a frame known as the *fixed reference frame*. With respect to $\{\vec{i}, \vec{j}, \vec{k}\}$, a set of vectors $\{\vec{i}(t), \vec{j}(t), \vec{k}(t)\}$ is defined by

$$\begin{pmatrix} \vec{i}(t) \\ \vec{j}(t) \\ \vec{k}(t) \end{pmatrix} = \begin{pmatrix} \cos\omega t & -\sin\omega t & 0 \\ \sin\omega t & \cos\omega t & 0 \\ 0 & 0 & 1 \end{pmatrix} \begin{pmatrix} \vec{i} \\ \vec{j} \\ \vec{k} \end{pmatrix} \triangleq R(\omega t) \begin{pmatrix} \vec{i} \\ \vec{j} \\ \vec{k} \end{pmatrix}. \tag{3.3}$$

Let the unit directional vector $\{\vec{i}(t), \vec{j}(t), \vec{k}(t)\}$ be at the X$'$-, Y$'$-, and Z$'$-axes of a Cartesian coordinate system $\{\mathrm{X}', \mathrm{Y}', \mathrm{Z}'\}$. $\{\vec{i}(t), \vec{j}(t), \vec{k}(t)\}$ or $\{\mathrm{X}', \mathrm{Y}', \mathrm{Z}'\}$ defines a frame known as the *rotating reference frame*. $R(\omega t)$ is referred to as the rotation matrix. Clearly, $\{\mathrm{X}', \mathrm{Y}', \mathrm{Z}'\}$ is rotating about the $\vec{k}$-direction clockwise at the angular frequency ω. $\vec{k}$ and $\vec{k}(t)$ specify the *longitudinal direction*; $\vec{i}$ and $\vec{j}$ or $\vec{i}(t)$ and $\vec{j}(t)$ define the *transverse plane*. These two reference frames are shown in Figure 3.2.

3.3 Alignment and Precession

When a spin system is placed in a static, external magnetic field $\vec{B}_o$, two physical phenomena occur: *alignment* and *precession*.

3.3.1 Alignment

In the absence of an external magnetic field, nuclear spins in a sample are oriented randomly. However, when the sample is placed in a static, external

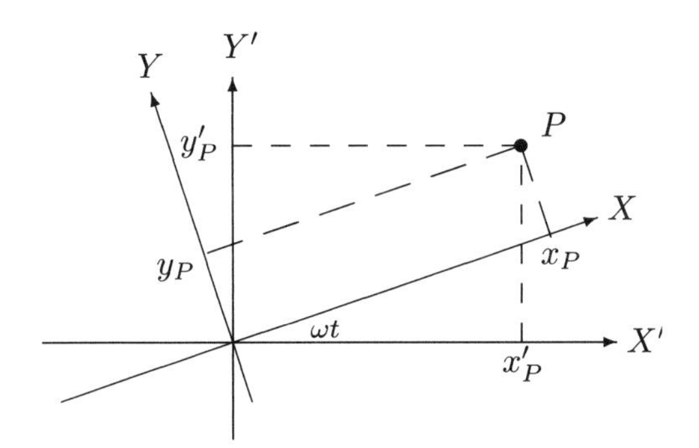

FIGURE 3.2
The fixed reference frame {X,Y,Z} and the rotating reference frame {X', Y', Z'} have the same vertical axis. The horizontal plane {X', Y'} is rotating about the vertical axis in the clockwise direction.

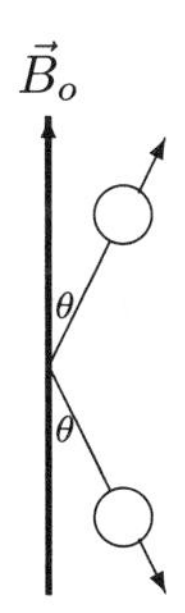

FIGURE 3.3
The alignment of nuclear spins in a static, external magnetic field.

magnetic field

$$\vec{B}_o = B_o\vec{k} \; , \tag{3.4}$$

nuclear spins in the sample will align themselves in one of two directions: with $\vec{B}_o$ (parallel) or against $\vec{B}_o$ (anti-parallel). In fact, spin *alignment* is not exactly parallel or anti-parallel to $\vec{B}_o$. Instead, the spin is oriented at an angle θ to $\vec{B}_o$ as shown in Figure 3.3 and therefore precesses around $\vec{B}_o$.

By the quantum model, the magnetic dipole moment $\vec{\mu} = \mu_x\vec{i} + \mu_y\vec{j} + \mu_z\vec{k}$ may take one of several possible orientations. In other words, the magnitude of its z-component can be

$$\mu_z = I_z(\gamma\hbar) \; , \tag{3.5}$$

where I_z takes one of (2I+1) values: $-$I, $(-\text{I}+1)$, $\cdots$, 0, $\cdots$, $(\text{I}-1)$, I, and is called the *magnetic quantum number*. Thus, the angle θ between $\vec{\mu}$ and $\vec{B}_o$

can be calculated by

$$\theta = \cos^{-1}(\frac{I_z}{\sqrt{\mathrm{I}(\mathrm{I}+1)}}). \tag{3.6}$$

Let $\vec{\mu}_{xy} = \mu_x\vec{i} + \mu_y\vec{j}$ be the transverse component of $\vec{\mu}$. The counterpart of its complex representation is

$$\vec{\mu}_{xy} \longrightarrow \mu_x + i\mu_y = \mu_{xy}e^{i\phi_{\mu_{xy}}}, \tag{3.7}$$

where $i = \sqrt{-1}$, $\mu_{xy} = \sqrt{\mu_x^2 + \mu_y^2}$, and $\phi_{\mu_{xy}} = \tan^{-1}(\frac{\mu_y}{\mu_x})$. Thus, we have

$$\mu_{xy} = \gamma\hbar\sqrt{\mathrm{I}(\mathrm{I}+1) - I_z^2}. \tag{3.8}$$

As an example, for ^{1}H, a spin-$\frac{1}{2}$ system, $I_z = -\frac{1}{2}$ or $+\frac{1}{2}$; therefore, $\theta \simeq 54.73°$ and $\mu_{xy} = \frac{\gamma\hbar}{\sqrt{2}} \simeq 1.245 \times 10^{-7}$ eV/T.

3.3.2 Precession

Spin *precession* is a type of rotation and can be viewed as a gyroscope. According to classical mechanics, the rate of change of spin angular momentum $\frac{d\vec{S}}{dt}$ equals the torque $\vec{\mu} \times \vec{B}_o$ that is imposed on the magnetic dipole moment $\vec{\mu}$ that is placed in the magnetic field $\vec{B}_o$:

$$\frac{d\vec{S}}{dt} = \vec{\mu} \times \vec{B}_o. \tag{3.9}$$

where "×" denotes the vector cross-product. Eq. (3.9) leads to

$$\frac{d\vec{\mu}}{dt} = \vec{\mu} \times \gamma\vec{B}_o. \tag{3.10}$$

The solution to the differential equation (3.10) is

$$\mu_{xy}(t) = \mu_{xy}(0)e^{-i\gamma B_o t} \quad \text{and} \quad \mu_z(t) = \mu_z(0), \tag{3.11}$$

where $\mu_{xy}(0)$ and $\mu_z(0)$ are the initial values of $\mu_{xy}(t)$ and $\mu_z(t)$. The physical interpretation of this mathematical solution is that $\vec{\mu}$ precesses about $\vec{B}_o$ at the frequency γB_o. This frequency is known as the Larmor frequency ω_o and defined by

$$\omega_o = \gamma B_o. \tag{3.12}$$

An example: Let $B_o = 1$ T, then $f = \frac{\omega_o}{2\pi} = \frac{\gamma}{2\pi}$. For different nuclear species, $\frac{\gamma}{2\pi}$ is shown in Table 3.1.

3.4 Macroscopic Magnetization

3.4.1 Macroscopic Magnetization

The vector sum of individual magnetic dipole moments $\vec{\mu}_i$ in a unit volume of the sample forms a *macroscopic magnetization* expressed by $\vec{M}$

$$\vec{M} = \sum_{i=1}^{n} \vec{\mu}_i \ , \tag{3.13}$$

where n is the number of nuclear spins in the unit volume. In the absence of a static, external magnetic field, due to the random orientations of nuclear spins, the net macroscopic magnetization is zero, while in the presence of a such field, due to the alignment and precession of nuclear spins, the net macroscopic magnetization becomes non-zero. Macroscopic magnetization is a bulk property of the spin system and can be utilized to describe the molecular structure and environment surrounding nuclei.

Due to the linear relationship between $\vec{M}$ and $\vec{\mu}$ as shown in Eq. (3.13), the counterpart of Eq. (3.10) for the macroscopic magnetization is

$$\frac{d\vec{M}}{dt} = \vec{M} \times \gamma \vec{B}_o \ , \tag{3.14}$$

which characterizes the precession for the macroscopic magnetization when the static field $\vec{B}_o$ is applied only. Similar to $\vec{\mu}$, $\vec{M} = M_x\vec{i}+M_y\vec{j}+M_z\vec{k}$ can be expressed as $\vec{M} = \vec{M}_{xy}+M_z\vec{k}$, and its transverse component $\vec{M}_{xy} = M_x\vec{i}+M_y\vec{j}$ has a corresponding complex notation:

$$\vec{M}_{xy} \longrightarrow M_x + iM_y = M_{xy}e^{i\phi_{M_{xy}}} , \tag{3.15}$$

where $M_{xy} = \sqrt{M_x^2 + M_y^2}$ and $\phi_{M_{xy}} = \tan^{-1}(\frac{M_y}{M_x})$.

When the macroscopic magnetization $\vec{M}$ is at an angle θ to the static, external magnetic field $\vec{B}_o$, the energy of this spin system is

$$E = -\vec{M} \cdot \vec{B}_o = -MB_o \cos\theta = -M_z B_o \ , \tag{3.16}$$

where "$\cdot$" denotes the vector dot product, and M and B_o denote the magnitudes of $\vec{M}$ and $\vec{B}_o$. When $\theta = 0$, the system is at its lowest energy state, the thermal equilibrium state, and the corresponding $\vec{M}$ is called *thermal equilibrium macroscopic magnetization*, abbreviated as TEMM and denoted by $\vec{M}_z^o$.

3.4.2 Thermal Equilibrium Macroscopic Magnetization

Let M_z^o be the magnitude of $\vec{M}_z^o$; Appendix 3A shows that for a spin-I system,

$$M_z^o = \frac{\gamma^2\hbar^2 B_o n I(I+1)}{3\kappa T}, \tag{3.17}$$

where κ is Boltzmann's constant and T is absolute temperature. Eq. (3.17) indicates that for a given static, external magnetic field, TEMM is determined by both the spin system itself (via n) and the environment thermal agitation (via T).

For hydrogen ^{1}H, its magnetic dipole moment μ_z can be calculated using Eq. (3.5). The nuclear spin of ^{1}H has a bi-level of energy. From Eqs. (3.5) and (3.16), its low and high energy are

$$e_l = -\frac{1}{2}\gamma\hbar B_o \quad \text{and} \quad e_h = +\frac{1}{2}\gamma\hbar B_o, \tag{3.18}$$

respectively and the energy difference between these two levels is

$$\delta e = e_h - e_l = \gamma\hbar B_o. \tag{3.19}$$

Let n_l and n_h be the number of nuclear spins in the directions of parallel and anti-parallel to $\vec{B}_o$ in a unit volume of the sample, $n_l + n_h = n$. The difference between these two populations of nuclear spins, $n_l - n_h$, generates a net macroscopic magnetization—TEMM:

$$\vec{M}_z^o = \mu_z(n_l - n_h)\vec{k}. \tag{3.20}$$

Its magnitude is

$$M_z^o = \frac{1}{2}(n_l - n_h)\gamma\hbar. \tag{3.21}$$

By applying Eq. (3.17) to hydrogen ^{1}H (i.e., I $= \frac{1}{2}$) and letting it equal Eq. (3.21), we obtain a ratio of the population difference and the total population of nuclear spins in a unit volume of the sample,

$$\frac{n_l - n_h}{n} = \frac{\gamma\hbar B_o}{2\kappa T} \triangleq \epsilon, \tag{3.22}$$

where ϵ is a function of the strength of the static, external magnetic field B_o and the environment thermal agitation T. Eq. (3.22) leads to

$$n_l = \frac{1+\epsilon}{2}n \quad \text{and} \quad n_h = \frac{1-\epsilon}{2}n. \tag{3.23}$$

Under normal conditions and the ordinary settings of MRI as shown in Table 3.2, Eq. (3.22) gives

$$\epsilon \simeq 3.4 \times 10^{-6}.$$

Most commonly used methods take the first-order approximation of the Boltzmann distribution [7–10] to derive the value of ϵ and/or other measures related to n_l, n_h, and n. For example,

(a) [11] shows that for the typical settings, $\frac{n_-}{n_+} \simeq 0.999993$ ($n_+ = n_l$ and $n_- = n_h$);

TABLE 3.2
The Physical Constants and Some Typical Settings

Gyromagnetic ratio	$\gamma/2\pi$	42.576×10^6 Hz/T (^{1}H)
Planck's constant	h	6.626×10^{-34} J·s
Boltzmann's constant	κ	1.381×10^{-23} J/K
Number of spins/(mm)^3	n	6.69×10^{19} (H_2O)
Absolute temperature	T	300K
Static magnetic field	B_0	1 Tesla

(b) [10] shows an excess (i.e., $n_l - n_h$) of 5 out of every 10^6 spins at 1.5 T;

(c) [12] shows that at 1 T and 300 K, $\frac{N_\uparrow - N_\downarrow}{N_s} \simeq 0.000003$ ($N_\uparrow = n_l$, $N_\downarrow = n_h$, and $N_s = n$);

(d) [13] shows that at 1 T and for κT in the body approximately equal to $\frac{1}{40}$ eV, the ratio of the mean populations of two spin states is given by $\frac{\tilde{n}_1}{\tilde{n}_2} \simeq 1.00000693$ ($\tilde{n}_1 \to n_l$ and $\tilde{n}_2 \to n_h$).

It has been verified that the numbers in the above (a)–(d) are identical to or consistent with the value of ϵ derived by the method that makes Eq. (3.17) equal to Eq. (3.21).

Eq. (3.17) shows that TEMM is a function of the spin density n, the strength of the static, external magnetic field B_o, and the temperature T. Eq. (3.21) shows that TEMM is determined by the uneven populations of two spin states. Eq. (3.22) shows that these two interpretations are equivalent. Although the population difference between two spin states is very small (about three in a million spins in the sample at 1 T and at room temperature), it is this uneven spin population that generates the TEMM, which is one that can be measured and actually observed in MRI.

TEMM represents the electrical current density, that is, its physical unit is A/m (A - *ampere* and m - *meter*).* To verify this, a simple way is to use Eq. (3.21), where the units of γ and $\hbar$ are $s^{-1} \cdot T^{-1}$ and J·s (s - *second*, T - *Tesla*, J - *Joule*). Thus, the physical unit of TEMM M_z^o is

$$\begin{aligned} &\mathrm{m}^{-3} \cdot (\mathrm{s}^{-1} \cdot \mathrm{T}^{-1}) \cdot (\mathrm{J} \cdot \mathrm{s}) = \mathrm{m}^{-3} \cdot \mathrm{T}^{-1} \cdot \mathrm{J} \\ &= \mathrm{m}^{-3} \cdot (\mathrm{kg} \cdot \mathrm{s}^{-2} \cdot \mathrm{A}^{-1})^{-1} \cdot (\mathrm{kg} \cdot \mathrm{m}^2 \cdot \mathrm{s}^{-2}) = \mathrm{A/m}\,. \end{aligned} \tag{3.24}$$

*The meter-kilogram-second (MKS) system; an International System of Units is used in this book.

3.5 Resonance and Relaxation

In addition to the main static magnetic field $\vec{B}_o$, a radiofrequency (RF) field $\vec{B}_1$ is also applied in MRI. When the RF field turns on and off, the spin system undergoes a forced precession known as *resonance* and a free precession known as *relaxation*, respectively.

3.5.1 Resonance

Resonance is a physical state that can be defined in different ways. For example, it can be viewed as a state that occurs when the system vibrates at a certain frequency, or it may involve the transfer of energy between two systems with the same certain frequency. By applying a pulsed RF field $\vec{B}_1$ in one transverse direction (say $\vec{i}$)

$$\vec{B}_1 = B_1\vec{i} \tag{3.25}$$

at the Larmor frequency to a spin system that is in a static magnetic field, some nuclear spins absorb energy from the RF pulse, become excited, and undergo transitions from the low energy state to the high energy state. This phenomenon is known as *nonselective excitation* and this state is called magnetic *resonance*.

The excitation of nuclear spins changes the TEMM. However, as a macroscopic magnetization, it behaves differently from individual spins. TEMM does not necessarily adopt an anti-parallel orientation, but instead it is perturbed from its equilibrium state and spirally flipped toward to the transverse plane. The macroscopic magnetization in this transition process is called the *precession macroscopic magnetization*, abbreviated as PMM. Section 3.7.1 will mathematically prove that in this nonselective excitation process, PMM undergoes a forced precession about $\vec{B}_1$ in the rotating reference frame. Analogous to the relationship between the static field $\vec{B}_o$ and the Larmor frequency ω_o: $\omega_o = \gamma B_o$, the relationship between the RF field $\vec{B}_1$ and the radiofrequency ω_1 is

$$\omega_1 = \gamma B_1. \tag{3.26}$$

Thus, the angle that PMM rotates about $\vec{B}_1$ (i.e., the angle between the PMM direction and the $\vec{k}(t)$ direction in the Y′-Z′ plane of the rotating reference frame) is given by

$$\alpha = \omega_1 \tau_p = \gamma B_1 \tau_p, \tag{3.27}$$

where τ_p is the duration of RF pulse. α is known as the *flip angle* and is determined by the strength of RF field B_1 and the duration of RF pulse τ_p.

In addition to inducing nuclear spins to absorb energy and change their energy status, an RF pulse also forces nuclear spins to precess in phase. This is so called *phase coherence* that is maintained within RF pulse duration τ_p.

With the RF field, a z-gradient field[†] caused by z-gradient $G_z(t)\vec{k}$

$$\vec{B}_{G_z} = B_{G_z}\vec{k} = G_z(t)z\vec{k} \tag{3.28}$$

is applied at the longitudinal direction. As a result, nuclear spins in some restricted regions in the sample are excited. This physical phenomenon is known as *selective excitation.* Section 3.7.2 will show that selective excitation will make the slice selection.

Resonance occurs in the forced precession of nuclear spins. The forced precession is characterized by

$$\frac{d\vec{M}}{dt} = \vec{M} \times \gamma\vec{B}, \tag{3.29}$$

where $\vec{B} = \vec{B}_o + \vec{B}_1$ or $\vec{B} = \vec{B}_o + \vec{B}_1 + \vec{B}_{G_z}$ for the nonselective or selective excitation, respectively.

3.5.2 Relaxation

After the excitation magnetic fields—RF field $\vec{B}_1$ and the slice-selection excitation field $\vec{B}_{G_z}$—are removed and sufficient time is given, according to the laws of thermodynamics, PMM begins to return to its thermal equilibrium state. This return process is known as *relaxation.* In this relaxation process, PMM undergoes a free precession about $\vec{B}_o$ in the fixed reference frame.

Relaxation consists of longitudinal and transverse relaxations. The former is a recovery of the longitudinal precession macroscopic magnetization abbreviated as LPMM and is also known as *spin-lattice* relaxation, because it is a process for nuclear spins to release their absorbed energy (the dissipation of energy on a subatomic level) during RF excitation to their environment—the lattice. The magnitude of LPMM increases during this relaxation and achieves a maximum at the end. The longitudinal relaxation can be characterized by

$$\frac{d\vec{M}_z}{dt} = -\frac{\vec{M}_z - \vec{M}_z^o}{T_1}, \tag{3.30}$$

where M_z^o is the TEMM in the presence of $\vec{B}_o$ only and can be calculated by Eq. (3.17) or Eq. (3.21), and T_1 is the spin-lattice relaxation time constant that governs the evolution of M_z toward its equilibrium value M_z^o.

[†]In MRI, the gradient field $\vec{B}_G$ is a special magnetic field whose longitudinal component B_{G_z} varies linearly at the gradient directions and whose transverse components B_{G_x} and B_{G_y} are often ignored—because of the very strong main static magnetic field $\vec{B}_o$ in the longitudinal direction. The gradient directions are often defined as x-, y-, and z-directions.

Transverse relaxation is a destruction of the transverse precession macroscopic magnetization abbreviated as TPMM and is also known as *spin-spin* relaxation, as it is a process involving the interactions among individual nuclear spins. Due to the spins interaction and the local magnetic field inhomogeneities, nuclear spins lose phase coherence and begin dephasing as the transverse relaxation takes place. The magnitude of TPMM decreases progressively and achieves a minimum, zero, at the end. The transverse relaxation can be characterized by

$$\frac{d\vec{M}_{xy}}{dt} = -\frac{\vec{M}_{xy}}{T_2}, \tag{3.31}$$

where T_2 is the spin-spin relaxation time constant that governs the evolution of M_{xy} toward its equilibrium value of zero.

T_2 must be shorter than T_1: $T_2 < T_1$, because when all nuclear spins are aligned in the longitudinal direction, there is no transverse component.

Transverse relaxation induces an *electromagnetic force* (emf) in the surrounding coil that, with the thermal noise voltage, produces a *free induction decay* (FID) signal.

During the free precession period, x- and y-gradient fields caused by x- and y-gradients $G_x(t)\vec{i}$ and $G_y(t)\vec{j}$, $\vec{B}_{G_z} = G_x(t)x\vec{k}$ and $\vec{B}_{G_z} = G_y(t)y\vec{k}$, are applied at the longitudinal direction to encode spatial localization information.

3.6 Bloch Eq. and Its Solution

The dynamic behavior of PMM $\vec{M}(t)$ can be phenomenologically described by the Bloch equation

$$\frac{d\vec{M}}{dt} = \vec{M} \times \gamma\vec{B} - \frac{M_x\vec{i} + M_y\vec{j}}{T_2} - \frac{M_z - M_z^o}{T_1}\vec{k}, \tag{3.32}$$

where $\vec{M} = M_x\vec{i} + M_y\vec{j} + M_z\vec{k}$ is PMM, $M_x\vec{i} + M_y\vec{j}$ is TPMM, $M_z\vec{k}$ is LPMM, M_z^o is TEMM, T_1 and T_2 are the spin-lattice and the spin-spin relaxation time constants, γ is the gyromagnetic ratio, and $\vec{B}(t) = B_x\vec{i} + B_y\vec{j} + B_z\vec{k}$ is the magnetic field applied to the spin system that consists of three types of fields: (1) static field $\vec{B}_o$, (2) RF field $\vec{B}_1$, and (3) gradient field $\vec{B}_G$. The mathematical solutions of the Bloch equation are given under different conditions. As shown by Eqs. (3.29), (3.30), and (3.31), the first term on the right side of Eq. (3.32) characterizes the forced precession process—excitation; the second and third terms characterize the free precession process—relaxation.

3.6.1 Homogeneous Sample and Uniform Magnetic Field

In this case, the sample is composed of one type of nuclear spin, which is often known as an *isochromat*. RF field and gradient field are not applied, that is, the magnetic field $\vec{B}$ consists of the static field $\vec{B} = B_o\vec{k}$ only. Without relaxation, the Bloch equation (3.32) becomes

$$\frac{d\vec{M}(t)}{dt} = \vec{M}(t) \times \gamma B_o \vec{k}. \tag{3.33}$$

By converting the vector cross product to the pure matrix multiplication (see Appendix 3B), the cross product of Eq. (3.33) can be rewritten as the multiplication of a skew-symmetric matrix of γB_o and the vector $\vec{M}(t)$:

$$\begin{pmatrix} dM_x(t)/dt \\ dM_y(t)/dt \\ dM_z(t)/dt \end{pmatrix} = \begin{pmatrix} 0 & \gamma B_o & 0 \\ -\gamma B_o & 0 & 0 \\ 0 & 0 & 0 \end{pmatrix} \begin{pmatrix} M_x(t) \\ M_y(t) \\ M_z(t) \end{pmatrix}. \tag{3.34}$$

The solution to Eq. (3.34) is

$$\begin{array}{rl} M_x(t) = & M_x(0)\cos\omega_o t + M_y(0)\sin\omega_o t \\ M_y(t) = & -M_x(0)\sin\omega_o t + M_y(0)\cos\omega_o t \\ M_z(t) = & M_z(0)\ , \end{array} \tag{3.35}$$

where $M_x(0)$, $M_y(0)$, and $M_z(0)$ are the initial values of $M_x(t)$, $M_y(t)$, and $M_z(t)$. Eq. (3.35) represents a rotation of TPMM around $\vec{k}$-direction at the angular frequency ω_o in the fixed reference frame.

With the relaxation, the Bloch equation (3.32) becomes

$$\frac{d\vec{M}(t)}{dt} = \vec{M}(t) \times \gamma \vec{B}_o - \frac{M_x(t)\vec{i} + M_y(t)\vec{j}}{T_2} - \frac{M_z(t) - M_z^o}{T_1}\vec{k}. \tag{3.36}$$

Similarly, by converting the vector cross product to the pure matrix multiplication, Eq. (3.36) becomes

$$\begin{pmatrix} dM_x(t)/dt \\ dM_y(t)/dt \\ dM_z(t)/dt \end{pmatrix} = \begin{pmatrix} -1/T_2 & \gamma B_o & 0 \\ -\gamma B_o & -1/T_2 & 0 \\ 0 & 0 & -1/T_1 \end{pmatrix} \begin{pmatrix} M_x(t) \\ M_y(t) \\ M_z(t) \end{pmatrix} + \begin{pmatrix} 0 \\ 0 \\ M_z^o/T_1 \end{pmatrix}. \tag{3.37}$$

The solution to Eq. (3.37) is

$$\begin{array}{rl} M_x(t) = & (\ \ M_x(0)\cos\omega_o t + M_y(0)\sin\omega_o t)e^{-t/T_2} \\ M_y(t) = & (-M_x(0)\sin\omega_o t + M_y(0)\cos\omega_o t)e^{-t/T_2} \\ M_z(t) = & M_z^o + (M_z(0) - M_z^o)e^{-t/T_1}, \end{array} \tag{3.38}$$

where M_z^o is TEMM. Eq. (3.38) represents an inverse spiral motion of PMM around the $\vec{k}$-direction at an angular frequency ω_o in the fixed reference frame.

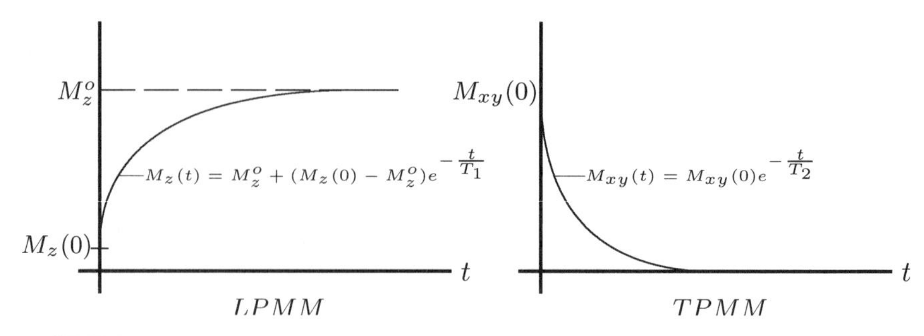

FIGURE 3.4
The dynamic behavior of the longitudinal and the transverse macroscopic magnetization.

3.6.2 Complex Representation

Similar to Eq. (3.15), TPMM $\vec{M}_{xy}(t) = M_x(t)\vec{i} + M_y(t)\vec{i}$ has a complex representation $M_{xy}(t) = M_x(t) + iM_y(t)$. Thus, from Eq. (3.37), the Bloch equation (3.36) can be decomposed into two equations, one for TPMM and another for LPMM:

$$\begin{aligned}\frac{dM_{xy}(t)}{dt} &= -(\frac{1}{T_2} + i\omega_o)M_{xy}(t) \\ \frac{dM_z(t)}{dt} &= -\frac{M_z(t) - M_z^o}{T_1}.\end{aligned} \tag{3.39}$$

The solutions to Eq. (3.39) are

$$\begin{aligned}M_{xy}(t) &= M_{xy}(0)e^{-\frac{t}{T_2}}e^{-i\omega_o t} \\ M_z(t) &= M_z^o + (M_z(0) - M_z^o)e^{-\frac{t}{T_1}},\end{aligned} \tag{3.40}$$

where $M_{xy}(0)$ and $M_z(0)$ the initial values of $M_{xy}(t)$ and $M_z(t)$. Eq. (3.40) is shown in Figure 3.4.

3.6.3 Heterogeneous Sample and Nonuniform Magnetic Field

In this case, in order to demonstrate the inhomogeneity of the spin system, nonuniformity of the magnetic field, and the spatial dependence, both the spatial argument $\mathbf{r} = (x, y, z)$ and the temporal argument t are inserted into Eq. (3.32). The Bloch equation becomes

$$\frac{d\vec{M}(\mathbf{r},t)}{dt} = \vec{M}(\mathbf{r},t) \times \gamma\vec{B}(\mathbf{r},t) - \frac{M_x(\mathbf{r},t)\vec{i} + M_y(\mathbf{r},t)\vec{j}}{T_2(\mathbf{r})} - \frac{M_z(\mathbf{r},t) - M_z^o}{T_1(\mathbf{r})}\vec{k}. \tag{3.41}$$

Although the magnetic field $\vec{B}(\mathbf{r},t)$ varies spatially, due to the very strong main static magnetic field $\vec{B}_o$, its orientation is still in the $\vec{k}$-direction, that is,

$$\vec{B}(\mathbf{r},t) = (B_o + \Delta B(\mathbf{r},t))\vec{k}, \tag{3.42}$$

where $\Delta B(\mathbf{r},t)$ characterizes the nonuniformity of the magnetic field. In practice, it is normally caused by the inhomogeneities δB_o of B_o, the gradient field B_G, and the chemical shift ω_{cs}.[‡] In the ideal case, $\delta B_o = 0$. Thus, by ignoring the chemical shift, $\Delta B(\mathbf{r},t)$ mainly represents a spatially dependent, time-varying gradient field, that is,

$$\Delta B(\mathbf{r},t) = B_G = \vec{G}(\mathbf{r},t) \cdot \mathbf{r}, \tag{3.43}$$

where $\vec{G}(\mathbf{r},t)$ is the field gradient vector.

By applying the complex representation $M_{xy}(\mathbf{r},t)$ and using Eq. (3.37), the Bloch equation (3.37) can be decomposed into two equations:

$$\begin{aligned} \frac{dM_{xy}(\mathbf{r},t)}{dt} &= -(\frac{1}{T_2(\mathbf{r})} + i(\omega_o + \Delta\omega(\mathbf{r},t)))M_{xy}(\mathbf{r},t) \\ \frac{dM_z(\mathbf{r},t)}{dt} &= -\frac{M_z(\mathbf{r},t) - M_z^o}{T_1}, \end{aligned} \tag{3.44}$$

where

$$\Delta\omega(\mathbf{r},t) = \gamma \Delta B(\mathbf{r},t) \; . \tag{3.45}$$

The first equation of Eq. (3.44) is for TPMM and its solution is

$$M_{xy}(\mathbf{r},t) = M_{xy}(\mathbf{r},0)e^{-\frac{t}{T_2(\mathbf{r})}}e^{-i\omega_o t}e^{-i\int_0^t \Delta\omega(\mathbf{r},\tau))d\tau}, \tag{3.46}$$

where $M_{xy}(\mathbf{r},0)$ is the initial value of the complex representation $M_{xy}(\mathbf{r},t)$.

For the static gradient $\vec{G}(\mathbf{r},t) = \vec{G} = G_x\vec{i} + G_y\vec{j} + G_z\vec{k}$, $\Delta\omega(\mathbf{r},t) = \gamma\vec{G} \cdot \mathbf{r}$, Eq. (3.46) becomes

$$M_{xy}(\mathbf{r},t) = M_{xy}(\mathbf{r},0)e^{-\frac{t}{T_2(\mathbf{r})}}e^{-i\omega_o t}e^{-i\gamma\vec{G}\cdot\mathbf{r}t}. \tag{3.47}$$

For the time-varying gradient, $\vec{G}(\mathbf{r},t) = \vec{G}(t) = G_x(t)\vec{i} + G_y(t)\vec{j} + G_z(t)\vec{k}$, $\Delta\omega(\mathbf{r},t)) = \gamma\vec{G}(t) \cdot \mathbf{r}$, Eq. (3.46) becomes

$$M_{xy}(\mathbf{r},t) = M_{xy}(\mathbf{r},0)e^{-\frac{t}{T_2(\mathbf{r})}}e^{-i\omega_o t}e^{-i\gamma\int_0^t \vec{G}(\tau)\cdot\mathbf{r}d\tau}. \tag{3.48}$$

The second equation of Eq. (3.44) is for LPMM and is not affected by the complex representation of TPMM; therefore, its solution remains the same as shown in Eq. (3.40).

[‡]Chemical shift is a small displacement ω_{cs} of the resonance frequency ω_o due to shield caused by the orbital motion of the surrounding electrons in response to the main B_o field.

3.7 Excitation

When an RF field is turned on in the presence of a static field, all nuclear spins in the sample are excited and undergo a forced precession. This kind of excitation is known as nonselective. Generally, when an RF field is turned on with an additional gradient field, the nuclear spins in a restricted region (typically a plane) in the sample are excited. This type of excitation is known as selective.

3.7.1 Nonselective Excitation

An RF field is turned on by applying an amplitude-modulated RF pulse in one transverse direction at a carrier frequency ω,

$$\vec{B}_1(t) = 2B_1(t)\cos\omega t\vec{i}, \tag{3.49}$$

where $B_1(t)$ is the amplitude modulation function. This linearly polarized field can be decomposed into two circularly polarized fields as

$$\vec{B}_1(t) = B_1(t)(\cos\omega t\vec{i} - \sin\omega t\vec{j}) + B_1(t)(\cos\omega t\vec{i} + \sin\omega t\vec{j}). \tag{3.50}$$

The first and second terms on the right side of Eq. (3.50) represent a left-handed and a right-handed rotating circularly polarized field, respectively. Because the PMM of the sample and the left-handed field are rotating in the same direction, spins are more responded (resonant) to the left-handed rotating field and are less affected by the other. Thus, the effective RF field is

$$\vec{B}_1(t) = B_1(t)(\cos\omega t\vec{i} - \sin\omega t\vec{j}), \tag{3.51}$$

and the total magnetic field experienced by nuclear spins in the sample is

$$\vec{B}(t) = \vec{B}_1(t) + \vec{B}_o = B_1(t)\cos\omega t\vec{i} - B_1(t)\sin\omega t\vec{j} + B_o\vec{k}, \tag{3.52}$$

which is shown in Figure 3.5.

Because the modulation function $B_1(t)$ turns the RF field on in a time interval shorter than the relaxation time constant T_2, hence T_1, the relaxation terms in the Bloch equation can be ignored. Thus, in nonselective excitation, the Bloch equation becomes

$$\frac{\vec{M}(\mathbf{r},t)}{dt} = \vec{M}(\mathbf{r},t) \times \gamma\vec{B}(t). \tag{3.53}$$

Therefore, in the fixed reference frame, Eq. (3.53) can be explicitly expressed as

$$\begin{pmatrix} dM_x(\mathbf{r},t)/dt \\ dM_y(\mathbf{r},t)/dt \\ dM_z(\mathbf{r},t)/dt \end{pmatrix} = \begin{pmatrix} 0 & \omega_o & \omega_1(t)\sin\omega t \\ -\omega_o & 0 & \omega_1(t)\cos\omega t \\ -\omega_1(t)\sin\omega t & -\omega_1(t)\cos\omega t & 0 \end{pmatrix} \begin{pmatrix} M_x(\mathbf{r},t) \\ M_y(\mathbf{r},t) \\ M_z(\mathbf{r},t) \end{pmatrix}, \tag{3.54}$$

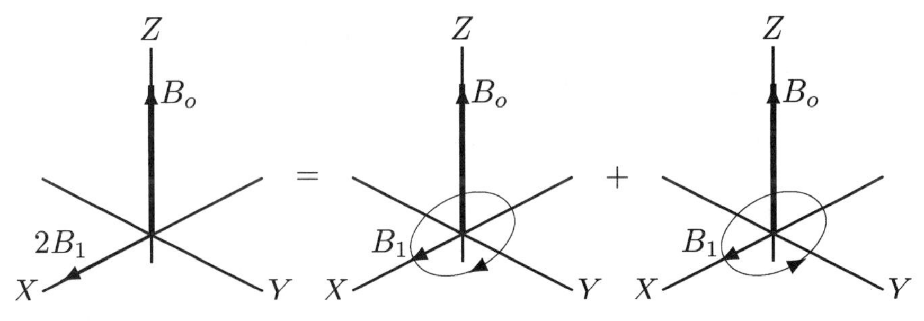

FIGURE 3.5
A linear polarized field is decomposed into two counter-rotating circularly polarized fields.

where $\omega_o = \gamma B_o$ and $\omega_1(t) = \gamma B_1(t)$.

Let
$$\vec{M}'(\mathbf{r},t) = M_{x'}(\mathbf{r},t)\vec{i}(t) + M_{y'}(\mathbf{r},t)\vec{j}(t) + M_{z'}(\mathbf{r},t)\vec{k}(t) \tag{3.55}$$
represent PMM in the rotating reference frame. Using the rotation matrix Eq. (3.3), we have

$$\begin{aligned} M_{x'}(\mathbf{r},t) &= M_x(\mathbf{r},t)\cos\omega t - M_y(\mathbf{r},t)\sin\omega t \\ M_{y'}(\mathbf{r},t) &= M_x(\mathbf{r},t)\sin\omega t + M_y(\mathbf{r},t)\cos\omega t \\ M_{z'}(\mathbf{r},t) &= M_z(\mathbf{r},t). \end{aligned} \tag{3.56}$$

Thus, by taking the derivatives of $M_{x'}(\mathbf{r},t)$, $M_{y'}(\mathbf{r},t)$, and $M_{z'}(\mathbf{r},t)$ of Eq. (3.56) with respect to t and then using Eq. (3.54), we obtain

$$\begin{aligned} \frac{dM_{x'}(\mathbf{r},t)}{dt} &= \frac{dM_x(\mathbf{r},t)}{dt}\cos\omega t - \frac{dM_y(\mathbf{r},t)}{dt}\sin\omega t \\ &\quad -\omega M_x(\mathbf{r},t)\sin\omega t - \omega M_y(\mathbf{r},t)\cos\omega t \\ &= \omega_o M_y(\mathbf{r},t)\cos\omega t + \omega_o M_x(\mathbf{r},t)\sin\omega t \\ &\quad -\omega M_x(\mathbf{r},t)\sin\omega t - \omega M_y(\mathbf{r},t)\cos\omega t \\ &= (\omega_o - \omega)M_{y'}(\mathbf{r},t), \end{aligned} \tag{3.57}$$

$$\begin{aligned} \frac{dM_{y'}(\mathbf{r},t)}{dt} &= \frac{dM_x(\mathbf{r},t)}{dt}\sin\omega t + \frac{dM_y(\mathbf{r},t)}{dt}\cos\omega t \\ &\quad +\omega M_x(\mathbf{r},t)\cos\omega t - \omega M_y(\mathbf{r},t)\sin\omega t \\ &= \omega_o M_y(\mathbf{r},t)\sin\omega t + \omega_1(t)M_z(\mathbf{r},t)\sin^2\omega t \\ &\quad -\omega_o M_x(\mathbf{r},t)\cos\omega t + \omega_1(t)M_z(\mathbf{r},t)\cos^2\omega t \\ &\quad +\omega M_x(\mathbf{r},t)\cos\omega t - \omega M_y(\mathbf{r},t)\sin\omega t \\ &= -(\omega_o - \omega)M_{x'}(\mathbf{r},t) + \omega_1(t)M_z(\mathbf{r},t), \end{aligned} \tag{3.58}$$

$$\frac{dM_{z'}(\mathbf{r},t)}{dt} = \frac{dM_z(\mathbf{r},t)}{dt}$$

$$= -\omega_1(t)M_x(\mathbf{r},t)\sin\omega t - \omega_1(t)M_y(\mathbf{r},t)\cos\omega t$$
$$= -\omega_1(t)M_{y'}(\mathbf{r},t)\ . \tag{3.59}$$

Eqs. (3.57) through (3.59) can be expressed as

$$\begin{pmatrix} dM_{x'}(\mathbf{r},t)/dt \\ dM_{y'}(\mathbf{r},t)/dt \\ dM_{z'}(\mathbf{r},t)/dt \end{pmatrix} = \begin{pmatrix} 0 & (\omega_o - \omega) & 0 \\ -(\omega_o - \omega) & 0 & \omega_1(t) \\ 0 & -\omega_1(t) & 0 \end{pmatrix} \begin{pmatrix} M_{x'}(\mathbf{r},t) \\ M_{y'}(\mathbf{r},t) \\ M_{z'}(\mathbf{r},t) \end{pmatrix}. \tag{3.60}$$

Let

$$\vec{B}'(t) = B_{x'}(t)\vec{i}(t) + B_{y'}(t)\vec{j}(t) + B_{z'}(t)\vec{k}(t) \tag{3.61}$$

be the counterpart of $\vec{B}(t)$ of Eq. (3.52) in the rotating reference frame. By applying the rotation matrix Eq. (3.3) to Eq. (3.52), we have

$$\begin{aligned} B_{x'}(t) &= B_1(t) \\ B_{y'}(t) &= 0 \\ B_{z'}(t) &= B_o, \end{aligned} \tag{3.62}$$

that is,

$$\vec{B}'(t) = B_1(t)\vec{i}(t) + B_o\vec{k}(t). \tag{3.63}$$

By introducing an effective magnetic field $\vec{B}_{eff}(t)$, which is defined in the rotating reference frame by

$$\vec{B}_{eff}(t) = \vec{B}'(t) - \frac{\omega}{\gamma}\vec{k}(t) = B_1(t)\vec{i}(t) + \frac{\omega_o - \omega}{\gamma}\vec{k}(t), \tag{3.64}$$

then, in terms of $\vec{M}'(\mathbf{r},t)$ of Eq. (3.55) and $\vec{B}_{eff}(t)$ of (3.64), the Eq. (3.60) can be written as

$$\frac{d\vec{M}'(\mathbf{r},t)}{dt} = \vec{M}'(\mathbf{r},t) \times \gamma\vec{B}_{eff}(t), \tag{3.65}$$

which is the Bloch equation in the rotating reference frame for nonselective excitation

Consider a simple case. When RF frequency equals the Larmor frequency and RF field is constant: $\omega = \omega_o$ and $\omega_1(t) = \omega_1$, the solution to Eq. (3.65) or Eq. (3.60) is

$$\begin{aligned} M_{x'}(\mathbf{r},t) &= \quad M_{x'}(\mathbf{r},0) \\ M_{y'}(\mathbf{r},t) &= \quad M_{y'}(\mathbf{r},0)\cos\omega_1 t + M_{z'}(\mathbf{r},0)\sin\omega_1 t \\ M_{z'}(\mathbf{r},t) &= -M_{y'}(\mathbf{r},0)\sin\omega_1 t + M_{z'}(\mathbf{r},0)\cos\omega_1 t, \end{aligned} \tag{3.66}$$

where $M_{x'}(\mathbf{r},0)$, $M_{y'}(\mathbf{r},0)$, and $M_{z'}(\mathbf{r},0)$ are the initial values of $M_{x'}(\mathbf{r},t)$, $M_{y'}(\mathbf{r},t)$, and $M_{z'}(\mathbf{r},t)$.

Eq. (3.35) shows that when a static field $\vec{B}_o$ is applied in the $\vec{k}$-direction, PMM rotates about the $\vec{k}$-direction in the fixed reference frame at the frequency $\omega_o = \gamma B_o$. Comparing with Eq. (3.35), Eq. (3.66) shows that when

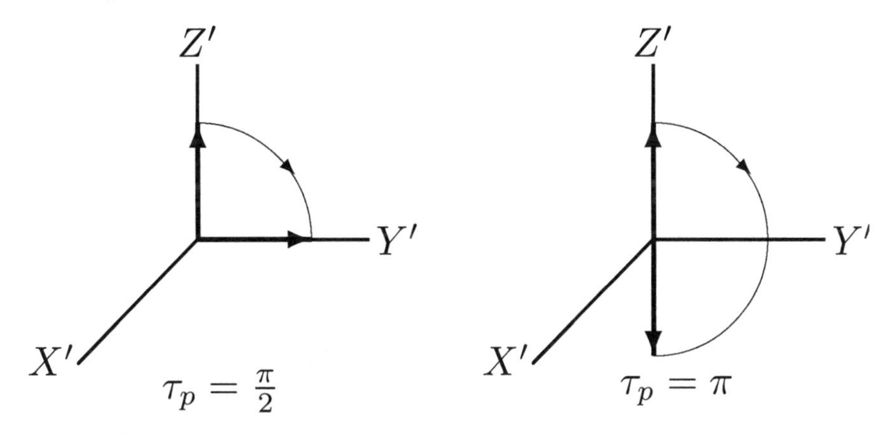

FIGURE 3.6
$(\frac{\pi}{2})_{x'}$ and $(\pi)_{x'}$ excitation pulses flip TEMM.

an RF field $\vec{B}_1(t) = 2B_1 \cos\omega t\vec{i}$ is applied in the $\vec{i}$-direction in the fixed reference frame at the Larmor frequency ω_o, the resultant PMM $\vec{M}'(\mathbf{r},t)$ will rotate about the $\vec{i}(t)$-direction in the rotating reference frame at the frequency $\omega_1 = \gamma B_1$.

Let the width of the amplitude modulation function $B_1(t)$ be τ_p. When τ_p is chosen such that the flip angle $\alpha = \omega_1\tau_p = \gamma B_1\tau_p = \frac{\pi}{2}$ or π, the RF pulse is called the $(\frac{\pi}{2})_{x'}$ or $(\pi)_{x'}$ pulse; here the subscript x' indicates that the RF field is placed in the $\vec{i}(t)$-direction. From Eq. (3.66), after a $(\frac{\pi}{2})_{x'}$ pulse,

$$M_{x'}(\mathbf{r},\tau_p) = M_{x'}(\mathbf{r},0)\ , \quad M_{y'}(\mathbf{r},\tau_p) = M_{z'}(\mathbf{r},0)\ , \quad M_{z'}(\mathbf{r},\tau_p) = -M_{y'}(\mathbf{r},0),$$

and after a $(\pi)_{x'}$ pulse

$$M_{x'}(\mathbf{r},\tau_p) = M_{x'}(\mathbf{r},0)\ , \quad M_{y'}(\mathbf{r},\tau_p) = -M_{y'}(\mathbf{r},0)\ , \quad M_{z'}(\mathbf{r},\tau_p) = -M_{z'}(\mathbf{r},0),$$

which are shown in Figure 3.6.

3.7.2 Selective Excitation

The previous section showed that when an RF field $\vec{B}_1(t) = 2B_1(t)\cos\omega t\vec{i}$ is on in the presence of a static field $\vec{B}_o$, all nuclear spins in the sample are excited, and this nonselective excitation rotates PMMs about the $\vec{i}(t)$-direction in the rotating reference frame. This section shows that when an RF field is on in the presence of an additional static gradient field, for example, $\vec{B}_G = B_{G_z}\vec{k}(t) = G_z z\vec{k}(t)$, only those spins at a certain plane in the sample whose resonance frequency equals the RF frequency ω are excited, and this selective excitation localizes the excited plane, in other words, makes a slice selection.

Because the RF pulse is short, the relaxation terms in the Bloch equation are ignored. Thus, the Bloch equation in the rotating reference frame for the selective excitation is the same as Eq. (3.65), which is for the nonselective excitation

$$\frac{\vec{M}'(\mathbf{r},t)}{dt} = \vec{M}'(\mathbf{r},t) \times \gamma \vec{B}_{eff}(t), \tag{3.67}$$

but $\vec{B}_{eff}(t)$ is replaced by

$$\vec{B}_{eff}(t) = B_1(t)\vec{i}(t) + (B_o + B_{G_z} - \frac{\omega}{\gamma})\vec{k}(t), \tag{3.68}$$

which includes an additional static gradient field $B_{G_z}\vec{k}(t)$ (see Eq. (3.28)). Eq. (3.67), via Eq. (3.68), can be explicitly expressed as

$$\begin{pmatrix} dM_{x'}(\mathbf{r},t)/dt \\ dM_{y'}(\mathbf{r},t)/dt \\ dM_{z'}(\mathbf{r},t)/dt \end{pmatrix} = \begin{pmatrix} 0 & (\omega_o + \omega_{G_z} - \omega) & 0 \\ -(\omega_o + \omega_{G_z} - \omega) & 0 & \omega_1(t) \\ 0 & -\omega_1(t) & 0 \end{pmatrix} \begin{pmatrix} M_{x'}(\mathbf{r},t) \\ M_{y'}(\mathbf{r},t) \\ M_{z'}(\mathbf{r},t) \end{pmatrix}, \tag{3.69}$$

where $\omega_{G_z} = \gamma B_{G_z} = \gamma G_z z$ is called the z-gradient frequency.

In MRI physics, $\vec{G} = G_x\vec{i} + G_y\vec{j} + G_z\vec{k}$ represents a gradient vector and $\vec{G}\cdot\mathbf{r} = G_x x + G_y y + G_z z$ is the gradient field at the location $\mathbf{r}$. In mathematics,

$$\vec{G}\cdot\mathbf{r} = G_x x + G_y y + G_z z = p \tag{3.70}$$

represents a plane, which is shown in Figure 3.7. This plane is perpendicular to the gradient vector $\vec{G}$ and is located at a distance $\frac{p}{G}$ from the origin $\mathbf{r} = (0,0,0)$, here $G = \sqrt{G_x^2 + G_y^2 + G_z^2}$. Thus, two planes $\vec{G}\cdot\mathbf{r} = p_1$ and $\vec{G}\cdot\mathbf{r} = p_2$ define a slice that is perpendicular to the vector $\vec{G}$ and has the thickness $\frac{|p_1 - p_2|}{G}$. In the case of selective excitation, because $\vec{G} = G_z\vec{k}$, the plane $\vec{G}\cdot\mathbf{r} = p$ becomes $z = \frac{p}{G_z}$, which is parallel to the transverse plane and located at $z = \frac{p}{G_z}$.

When the RF frequency ω is tuned to the Larmor frequency $\omega_o + \gamma G_z z_o$ of the central plane of a slice (here $z_o = \frac{p_1 + p_2}{2}$ is the location of the central plane in the slice), Eq. (3.69) becomes

$$\begin{pmatrix} dM_{x'}(\mathbf{r},t)/dt \\ dM_{y'}(\mathbf{r},t)/dt \\ dM_{z'}(\mathbf{r},t)/dt \end{pmatrix} = \begin{pmatrix} 0 & \gamma G_z(z - z_o) & 0 \\ -\gamma G_z(z - z_o) & 0 & \omega_1(t) \\ 0 & -\omega_1(t) & 0 \end{pmatrix} \begin{pmatrix} M_{x'}(\mathbf{r},t) \\ M_{y'}(\mathbf{r},t) \\ M_{z'}(\mathbf{r},t) \end{pmatrix}, \tag{3.71}$$

By assuming that the RF pulse is "weak" such that the flip angle $\alpha < \frac{\pi}{6}$, we can approximate $M_{z'}(\mathbf{r},t) \simeq M_z^o(\mathbf{r})$ and $\frac{dM_{z'}(\mathbf{r},t)}{dt} \simeq 0$. Thus, Eq. (3.71) can be approximated by

$$\begin{pmatrix} dM_{x'}(\mathbf{r},t)/dt \\ dM_{y'}(\mathbf{r},t)/dt \\ dM_{z'}(\mathbf{r},t)/dt \end{pmatrix} \simeq \begin{pmatrix} 0 & \gamma G_z(z - z_o) & 0 \\ -\gamma G_z(z - z_o) & 0 & \omega_1(t) \\ 0 & 0 & 0 \end{pmatrix} \begin{pmatrix} M_{x'}(\mathbf{r},t) \\ M_{y'}(\mathbf{r},t) \\ M_z^o(\mathbf{r}) \end{pmatrix}, \tag{3.72}$$

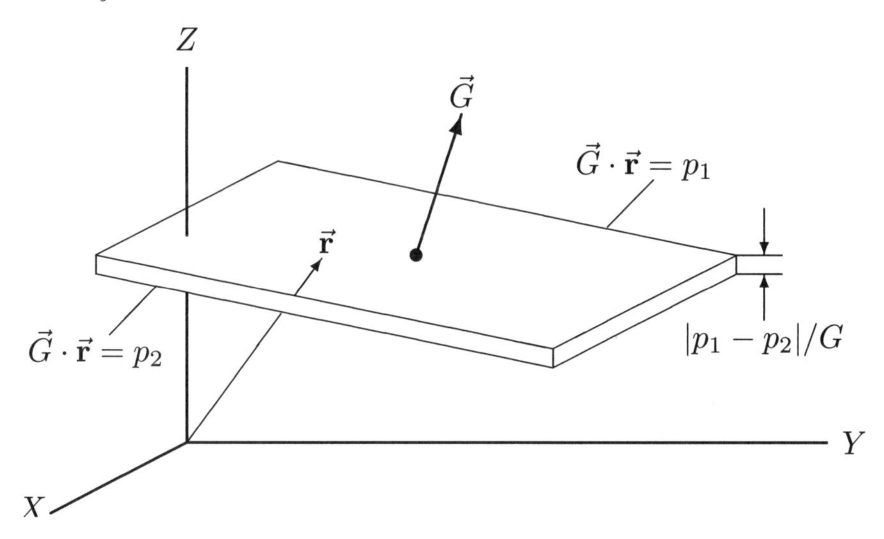

FIGURE 3.7
Mathematical expressions of the plane and the slice.

This approximation removes the interaction between TPMM and LPMM.

Similar to Section 3.6.2, TPMM in the rotating reference frame $\vec{M}'_{xy}(\mathbf{r},t) = M_{x'}(\mathbf{r},t)\vec{i}(t) + M_{y'}(\mathbf{r},t)\vec{j}(t)$ has a complex representation $M'_{xy}(\mathbf{r},t) = M_{x'}(\mathbf{r},t) + iM_{y'}(\mathbf{r},t)$. Thus, the first two equations in Eq. (3.72) are equivalent to

$$\frac{dM'_{xy}(\mathbf{r},t)}{dt} = -i\gamma G_z(z - z_o)M'_{xy}(\mathbf{r},t) + i\omega_1(t)M_z^o(\mathbf{r}). \tag{3.73}$$

By assuming that the initial condition $M'_{xy}(\mathbf{r},0) = 0 + i0$, we have shown that the solution to Eq. (3.73) is

$$M'_{xy}(\mathbf{r},t) = i\gamma M_z^o(\mathbf{r})e^{-i\gamma G_z(z-z_o)t}\int_0^t B_1(\tau)e^{i\gamma G_z(z-z_o)\tau}d\tau. \tag{3.74}$$

Because the amplitude modulation function $B_1(\tau)$ is pulsed and symmetric to its center point:

$$B_1(\tau) = 0 \ \ (\tau < 0 \ \text{ or } \ \tau > \tau_p) \quad \text{and} \quad B_1(-\tau + \frac{\tau_p}{2}) = B_1(\tau + \frac{\tau_p}{2}), \tag{3.75}$$

Eq. (3.74) at the end of the selective excitation can be written as

$$M'_{xy}(\mathbf{r},\tau_p) = i\gamma M_z^o(\mathbf{r})e^{-i\gamma G_z(z-z_o)\tau_p/2}\int_{-\tau_p/2}^{\tau_p/2} B_1(\tau + \frac{\tau_p}{2})e^{-i\gamma G_z(z-z_o)\tau}d\tau\,. \tag{3.76}$$

Because $\omega_{G_z} = \gamma G_z z$ (Eq. (3.69)), $\gamma G_z(z - z_o) = \omega_{G_z} - \omega_{G_{z_o}} = \gamma(B_{G_z} - B_{G_{z_o}})$ $= \Delta\omega_{G_z} = 2\pi f$, and f is the frequency. Thus Eq. (3.76) becomes

$$M'_{xy}(\mathbf{r},\tau_p) = i\gamma M_z^o(\mathbf{r})e^{-i2\pi f\tau_p/2}\mathcal{F}_1^{-1}\{B_1(\tau + \frac{\tau_p}{2})\}|_{f=\frac{1}{2\pi}\gamma G_z(z-z_o)}, \tag{3.77}$$

where $\mathcal{F}_1^{-1}$ denotes the one-dimensional inverse Fourier transform.

The left side of Eq. (3.77) is the profile of TPMM in the excited slice and the right side is the inverse Fourier transform of the modulation function of the RF excitation pulse multiplied by TEMM and a phase factor (shift). Thus, Eq. (3.77) establishes a relation between the excited slice TPMM profile and the envelope of the RF excitation pulse. This relation provides useful insight into the design of $B_1(\tau)$ for the given or expected slice TPMM profile. More conveniently, we can use the magnitude of the slice TPMM profile:

$$|M'_{xy}(\mathbf{r}, \tau_p)| = \gamma M_z^o(\mathbf{r}) |\mathcal{F}_1^{-1}\{B_1(\tau + \frac{\tau_p}{2})\}|. \tag{3.78}$$

Let Δz be the slice thickness. The quantity defined by

$$r_{|M'_{xy}|} = \frac{\int_{-\Delta z/2}^{\Delta z/2} |M'_{xy}(\mathbf{r}, \tau_p)| dz}{\int_{-\infty}^{\infty} |M'_{xy}(\mathbf{r}, \tau_p)| dz} \tag{3.79}$$

represents a ratio between the TPMM within the slice and the total TPMM excited by RF (in magnitude). It provides measures on (1) the relation between $r_{|M'_{xy}|}$ and various Δz for a given RF excitation pulse and (2) the relation between $r_{|M'_{xy}|}$ and various RF excitation pulses for a fixed slice thickness Δz. In the following, two examples are used for illustrating these relations.

Example 1. Let the RF excitation pulse $B_1(\tau)$ take a Gaussian shape such as

$$B_1(\tau + \frac{\tau_p}{2}) = e^{-\pi \tau^2}, \tag{3.80}$$

then

$$\mathcal{F}_1^{-1}\{B_1(\tau + \frac{\tau_p}{2})\} = e^{-\pi f^2}. \tag{3.81}$$

By substituting Eq. (3.81) into Eq. (3.78), the slice TPMM profile has a Gaussian shape

$$|M'_{xy}(\mathbf{r}, t)| = \gamma M_z^o(\mathbf{r})\, e^{-\pi f^2}. \tag{3.82}$$

Appendix 3C shows that both $e^{-\pi \tau^2}$ and $e^{-\pi f^2}$ are the probability density function (pdf) of Gaussian distribution with the zero mean and the variance $\sigma^2 = \frac{1}{2\pi}$. Appendix 3C also shows that when the slice is defined by $-2\sigma < f < 2\sigma$, then the slice thickness Δz and the ratio $r_{|M'_{xy}|}$ are

$$\Delta z = \frac{4\sqrt{2\pi}}{\gamma G_z} \quad \text{and} \quad r_{|M'_{xy}|} > 0.95, \tag{3.83}$$

that is, more than 95% of TPMM is located within the slice.

Example 2. Let the RF excitation pulse $B_1(\tau)$ take a sinc shape such as

$$B_1(\tau + \frac{\tau_p}{2}) = sinc(\tau \Delta f) = \frac{\sin(\pi \tau \Delta f)}{\pi \tau \Delta f}, \tag{3.84}$$

where we set $\Delta f = \frac{1}{\tau_p}$; then

$$\mathcal{F}_1^{-1}\{B_1(\tau + \frac{\tau_p}{2})\} = rect(\frac{f}{\Delta f}) = \begin{cases} \frac{1}{\Delta f} \ (|f| \leq \frac{\Delta f}{2}) \\ 0 \ \ (|f| > \frac{\Delta f}{2}). \end{cases} \tag{3.85}$$

By substituting Eq. (3.85) into Eq. (3.78), the slice TPMM profile has a rectangular shape

$$|M'_{xy}(\mathbf{r}, \tau_p)| = \gamma M_z^o(\mathbf{r}) \ rect(\frac{f}{\Delta f}). \tag{3.86}$$

Appendix 3C shows that when the slice is defined by $-\frac{\Delta f}{2} < f < +\frac{\Delta f}{2}$, then the slice thickness Δz and the ratio $r_{|M'_{xy}|}$ are

$$\Delta z = \frac{1}{\frac{1}{2\pi}\gamma G_z \tau_p} \quad \text{and} \quad r_{|M'_{xy}|} = 1.00, \tag{3.87}$$

that is, 100% of TPMM is located within this slice.

3.7.2.1 Discussion

In practice, $B_1(\tau)$ cannot be the perfect Gaussian or sinc shape because the duration of RF pulse must be finite. This fact implies that Eq. (3.80) and Eq. (3.84) must be multiplied by a rect function. As a result, Eq. (3.81) and Eq. (3.85) will be replaced by a convolution with the sinc function. Thus, the selective excitation cannot be uniform across the slice thickness and can also spread to the neighboring slices.

Various functions can be used for $B_1(\tau)$ to achieve the well-defined slice TPMM profile. Some of these functions are discussed in Appendix 3C.

The phase factor $e^{-i\gamma G_z(z-z_o)\tau_p/2}$ in Eq. (3.76) (or $e^{-i2\pi f \tau_p/2}$ in Eq. (3.77)) exists across the slice thickness. The phase dispersion is a linear function of z and will cause signal loss. This dephasing effect can be eliminated by applying a linear z-gradient immediately after selective excitation, which has the same magnitude as the slice selection z-gradient G_z but with the opposite polarity and lasts for only half the duration of the selective excitation $\frac{\tau_p}{2}$. Thus, this post-excitation refocusing produces a rephasing factor $e^{i\gamma G_z(z-z_o)\tau_p/2}$ that exactly cancels the dephasing factor $e^{-i\gamma G_z(z-z_o)\tau_p/2}$.

Eq. (3.74) also provides insight into the relation between nonselective and selective excitations. In Eq. (3.74), let $G_z = 0$ (i.e., without applying z-gradient) or $z = z_o$ (i.e., at the central plane of the slice), we have

$$M'_{xy}(\mathbf{r}, t) = iM_z^o(\mathbf{r}) \int_0^t \gamma B_1(\tau) d\tau = iM_z^o(\mathbf{r}) \int_0^t \omega_1(\tau) d\tau = iM_z^o(\mathbf{r})\alpha. \tag{3.88}$$

Because α is small, $\alpha \simeq \sin\alpha$. Thus, $M'_{xy}(\mathbf{r}, t) \simeq iM_z^o(\mathbf{r})\sin\alpha$, that is the outcome of the nonselective excitation, and i indicates that this $M'_{xy}(\mathbf{r}, t)$ is at the $\vec{j}(t)$-direction because it rotates about the $\vec{i}(t)$-direction during excitation in the rotating reference frame.

3.8 Induction

3.8.1 Signal Detection

According to Faraday's law of electromagnetic induction, PMM $\vec{M}(\mathbf{r},t)$ produces an electromagnetic force (emf), denoted by $s_r(t)$, in the surrounding receiver coil, and $s_r(t)$ is determined by the rate of change of the magnetic flux $\Phi(t)$ through the receiver coil:

$$s_r(t) = -\frac{\partial \Phi(t)}{\partial t}. \tag{3.89}$$

Let $\vec{B}_r(\mathbf{r}) = B_{r_x}(\mathbf{r})\vec{i} + B_{r_y}(\mathbf{r})\vec{j} + B_{r_z}(\mathbf{r})\vec{k}$ be indicative of the sensitivity of the receiver coil. In other words, $\vec{B}_r(\mathbf{r})$ represents a magnetic field at the location $\mathbf{r}$ produced by a hypothetical unit direct current flowing in the receiver coil. According to the principle of reciprocity, the magnetic flux $\Phi(t)$ is given by

$$\Phi(t) = \int_V \vec{B}_r(\mathbf{r}) \cdot \vec{M}(\mathbf{r},t) d\mathbf{r}, \tag{3.90}$$

where V denotes the volume of the sample.

Thus, the emf $s_r(t)$ induced in the receiver coil is

$$s_r(t) = -\frac{\partial}{\partial t}\int_V (B_{r_x}(\mathbf{r})M_x(\mathbf{r},t) + B_{r_y}(\mathbf{r})M_y(\mathbf{r},t) + B_{r_z}(\mathbf{r})M_z(\mathbf{r},t))d\mathbf{r} \; . \tag{3.91}$$

Because LPMM $M_z(\mathbf{r},t)$ varies slowly compared to TPMM $M_{xy}(\mathbf{r},t)$, $\frac{\partial M_z(\mathbf{r},t)}{\partial t}$ is ignored in computing $s_r(t)$. This leads to

$$s_r(t) = -\int_V (B_{r_x}(\mathbf{r})\frac{\partial M_x(\mathbf{r},t)}{\partial t} + B_{r_y}(\mathbf{r})\frac{\partial M_y(\mathbf{r},t)}{\partial t})d\mathbf{r}. \tag{3.92}$$

Similar to the complex expression $M_{xy}(\mathbf{r},t)$ of Section 3.6.2, the transverse component of $\vec{B}_r(\mathbf{r})$, $\vec{B}_{r_{xy}}(\mathbf{r}) = B_{r_x}(\mathbf{r})\vec{i} + B_{r_y}(\mathbf{r})\vec{j}$, can be expressed as $B_{r_{xy}}(\mathbf{r}) = B_{r_x}(\mathbf{r}) + iB_{r_y}(\mathbf{r}) = |B_{r_{xy}}(\mathbf{r})|e^{i\phi_{r_{xy}}(\mathbf{r})}$; here $|B_{r_{xy}}(\mathbf{r})|$ and $\phi_{B_{r_{xy}}}(\mathbf{r})$ are the magnitude and phase of $B_{r_{xy}}(\mathbf{r})$. By converting the vector dot product to the complex multiplication,

$$\vec{B}_{r_{xy}}(\mathbf{r}) \cdot \frac{\partial}{\partial t}\vec{M}_{xy}(\mathbf{r},t) = \Re\{B_{r_{xy}}(\mathbf{r})(\frac{\partial}{\partial t}M_{xy}(\mathbf{r},t))^*\}, \tag{3.93}$$

where $*$ and $\Re$ denote the complex conjugate and the real part of the complex quantity, Eq. (3.93) can be rewritten as

$$s_r(t) = -\Re\{\int_V B_{r_{xy}}(\mathbf{r})\frac{\partial}{\partial t}M^*_{xy}(\mathbf{r},t)d\mathbf{r}\}. \tag{3.94}$$

By substituting Eq. (3.46) into Eq. (3.94), we have

$$
\begin{aligned}
s_r(t) &= -\Re\{\int_V B_{r_{xy}}(\mathbf{r})M_{xy}^*(\mathbf{r},0)(-\frac{1}{T_2(\mathbf{r})} + i(\omega_o + \Delta\omega(\mathbf{r},t))) \\
&\quad e^{-\frac{t}{T_2(\mathbf{r})}} e^{i(\omega_o t + \int_0^t \Delta\omega(\mathbf{r},\tau)d\tau)} d\mathbf{r}\} \\
&= -\Re\{\int_V |B_{r_{xy}}(\mathbf{r})|\ |M_{xy}(\mathbf{r},0)|(-\frac{1}{T_2(\mathbf{r})} + i(\omega_o + \Delta\omega(\mathbf{r},t))) \\
&\quad e^{-\frac{t}{T_2(\mathbf{r})}} e^{i(\omega_o t + \int_0^t \Delta\omega(\mathbf{r},\tau)d\tau + \phi_{B_{r_{xy}}}(\mathbf{r}) - \phi_{M_{xy}}(\mathbf{r},0))} d\mathbf{r}\}, \qquad (3.95)
\end{aligned}
$$

where $\Delta\omega(\mathbf{r},t)$ is given by Eq. (3.45), and $|M_{xy}(\mathbf{r},0)|$ and $\phi_{M_{xy}}(\mathbf{r},0)$ are the magnitude and the phase of $M_{xy}(\mathbf{r},0)$, respectively.

For many applications in practice, $\omega_o \gg \Delta\omega(\mathbf{r},t)$ and $\omega_o \gg \frac{1}{T_2(\mathbf{r})}$; thus $-\frac{1}{T_2(\mathbf{r})} + i(\omega_o + \Delta\omega(\mathbf{r},t)) \simeq \omega_o e^{i\frac{\pi}{2}}$. As a result, Eq. (3.95) becomes

$$
\begin{aligned}
s_r(t) &= \omega_o \int_V |B_{r_{xy}}(\mathbf{r})|\ |M_{xy}(\mathbf{r},0)| e^{-\frac{t}{T_2(\mathbf{r})}} \\
&\cos(\omega_o t + \int_0^t \Delta\omega(\mathbf{r},\tau)d\tau + \phi_{B_{r_{xy}}}(\mathbf{r}) - \phi_{M_{xy}}(\mathbf{r},0) - \frac{\pi}{2}) d\mathbf{r}. \qquad (3.96)
\end{aligned}
$$

$s_r(t)$ is a real-valued physical signal. Because it is generated through the induction of TPMM in the free precession and exponentially decayed, it is commonly called *free induction decay* (FID). Rigorously speaking, as described in Section 3.5.2, $s_r(t)$ of Eq. (3.96) represents the signal component of the FID signal; it and a noise counterpart (which is discussed in Chapter 7) together form the FID signal. Eq. (3.96) is often known as MR signal equation (in terms of $s_r(t)$).

3.8.2 Signal Demodulation

In practice, as shown in Figure 3.8, the FID signal $s_r(t)$ is split into two channels: the in-phase denoted by I and the quadrature denoted by Q. In these two channels, $s_r(t)$ is multiplied by $\cos\omega_o t$ and $\sin\omega_o t$ separately, then passes through the low-pass filters with the impulse response function $h(t)$, and finally mixed. This type of signal processing procedure is called quadrature *phase sensitive detection* (PSD). The quadrature PSD results in a frequency shift of $s_r(t)$ by ω_o and produces a baseband signal $s_c(t)$ (i.e., without the phase factor $\omega_o t$ - demodulation).

Let $s_I(t)$ and $s_Q(t)$ be the outputs of the in-phase and quadrature channels, respectively. They are

$$
\begin{aligned}
s_I(t) &= (s_r(t)\cos\omega_o t) * h(t) \\
&= (\omega_o \int_V |B_{r_{xy}}(\mathbf{r})|\ |M_{xy}(\mathbf{r},0)| e^{-\frac{t}{T_2(\mathbf{r})}} \cos\omega_o t \\
&\cos(\omega_o t + \int_0^t \Delta\omega(\mathbf{r},\tau)d\tau + \phi_{B_{r_{xy}}}(\mathbf{r}) - \phi_{M_{xy}}(\mathbf{r},0) - \frac{\pi}{2}) d\mathbf{r}) * h(t)
\end{aligned}
$$

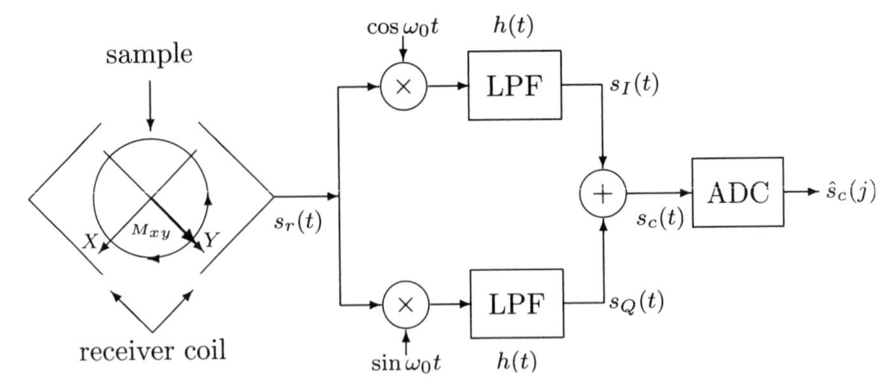

FIGURE 3.8
MR signal detection and demodulation. In this simplified signal flow diagram, $s_r(t)$, $s_c(t)$, and $\hat{s}_c(j)$ represent the pure signal components; the corresponding noise components are not included.

$$
\begin{aligned}
&= (\frac{\omega_o}{2}\int_V |B_{r_{xy}}(\mathbf{r})|\,|M_{xy}(\mathbf{r},0)|e^{-\frac{t}{T_2(\mathbf{r})}} \\
&\quad [\cos(2\omega_o t + \int_0^t \Delta\omega(\mathbf{r},\tau)d\tau + \phi_{B_{r_{xy}}}(\mathbf{r}) - \phi_{M_{xy}}(\mathbf{r},0) - \frac{\pi}{2}) \\
&\quad + \cos(\int_0^t \Delta\omega(\mathbf{r},\tau)d\tau + \phi_{B_{r_{xy}}}(\mathbf{r}) - \phi_{M_{xy}}(\mathbf{r},0) - \frac{\pi}{2})]d\mathbf{r}) * h(t) \\
&= \frac{\omega_o}{2}\int_V |B_{r_{xy}}(\mathbf{r})|\,|M_{xy}(\mathbf{r},0)|e^{-\frac{t}{T_2(\mathbf{r})}} \\
&\quad \cos(\int_0^t \Delta\omega(\mathbf{r},\tau)d\tau + \phi_{B_{r_{xy}}}(\mathbf{r}) - \phi_{M_{xy}}(\mathbf{r},0) - \frac{\pi}{2})d\mathbf{r}, \qquad (3.97)
\end{aligned}
$$

and

$$
\begin{aligned}
s_Q(t) &= (s_r(t)\sin\omega_o t) * h(t) \\
&= (\omega_o \int_V |B_{r_{xy}}(\mathbf{r})|\,|M_{xy}(\mathbf{r},0)|e^{-\frac{t}{T_2(\mathbf{r})}} \sin\omega_o t \\
&\quad \cos(\omega_o t + \int_0^t \Delta\omega(\mathbf{r},\tau)d\tau + \phi_{B_{r_{xy}}}(\mathbf{r}) - \phi_{M_{xy}}(\mathbf{r},0) - \frac{\pi}{2})d\mathbf{r}) * h(t) \\
&= (\frac{\omega_o}{2}\int_V |B_{r_{xy}}(\mathbf{r})|\,|M_{xy}(\mathbf{r},0)|e^{-\frac{t}{T_2(\mathbf{r})}} \\
&\quad [\sin(2\omega_o t + \int_0^t \Delta\omega(\mathbf{r},\tau)d\tau + \phi_{B_{r_{xy}}}(\mathbf{r}) - \phi_{M_{xy}}(\mathbf{r},0) - \frac{\pi}{2}) \\
&\quad - \sin(\int_0^t \Delta\omega(\mathbf{r},\tau)d\tau + \phi_{B_{r_{xy}}}(\mathbf{r}) - \phi_{M_{xy}}(\mathbf{r},0) - \frac{\pi}{2})]d\mathbf{r}) * h(t) \\
&= -\frac{\omega_o}{2}\int_V |B_{r_{xy}}(\mathbf{r})|\,|M_{xy}(\mathbf{r},0)|e^{-\frac{t}{T_2(\mathbf{r})}}
\end{aligned}
$$

$$\sin(\int_0^t \Delta\omega(\mathbf{r},\tau)d\tau + \phi_{B_{r_{xy}}}(\mathbf{r}) - \phi_{M_{xy}}(\mathbf{r},0) - \frac{\pi}{2})d\mathbf{r} \; . \tag{3.98}$$

Thus, the output of the quadrature PSD is

$$\begin{aligned} s_c(t) &= s_I(t) + is_Q(t) \\ &= \frac{\omega_o}{2}\int_V |B_{r_{xy}}(\mathbf{r})|\,|M_{xy}(\mathbf{r},0)|e^{-\frac{t}{T_2(\mathbf{r})}} \\ &\quad [\cos(\int_0^t \Delta\omega(\mathbf{r},\tau)d\tau + \phi_{B_{r_{xy}}}(\mathbf{r}) - \phi_{M_{xy}}(\mathbf{r},0) - \frac{\pi}{2}) \\ &\quad -i\sin(\int_0^t \Delta\omega(\mathbf{r},\tau)d\tau + \phi_{B_{r_{xy}}}(\mathbf{r}) - \phi_{M_{xy}}(\mathbf{r},0) - \frac{\pi}{2})]d\mathbf{r} \\ &= \frac{\omega_o}{2}\int_V |B_{r_{xy}}(\mathbf{r})|\,|M_{xy}(\mathbf{r},0)|e^{-\frac{t}{T_2(\mathbf{r})}} \\ &\quad e^{-i(\int_0^t \Delta\omega(\mathbf{r},\tau)d\tau + \phi_{B_{r_{xy}}}(\mathbf{r}) - \phi_{M_{xy}}(\mathbf{r},0) - \frac{\pi}{2})}d\mathbf{r}. \end{aligned} \tag{3.99}$$

Using $|B_{r_{xy}}(\mathbf{r})|e^{-i\phi_{B_{r_{xy}}}(\mathbf{r})} = B^*_{r_{xy}}(\mathbf{r})$ and $|M_{xy}(\mathbf{r},0)|e^{i\phi_{M_{xy}}(\mathbf{r},0)} = M_{xy}(\mathbf{r},0)$, Eq. (3.99) becomes

$$s_c(t) = \frac{\omega_o}{2}e^{i\frac{\pi}{2}}\int_V B^*_{r_{xy}}(\mathbf{r})\; M_{xy}(\mathbf{r},0)e^{-\frac{t}{T_2(\mathbf{r})}}\,e^{-i\int_0^t \Delta\omega(\mathbf{r},\tau)d\tau}d\mathbf{r}. \tag{3.100}$$

For the homogeneous sample and the uniform reception field, $T_2(\mathbf{r})$ and $B_{r_{xy}}(\mathbf{r})$ are independent of the location $\mathbf{r}$, i.e., they are constant over the sample volume V: $T_2(\mathbf{r}) = T_2$ and $B_{r_{xy}}(\mathbf{r}) = B_{r_{xy}}$. For $t \ll T_2$, $e^{-\frac{t}{T_2}} \simeq 1$. Thus, letting $c = \frac{\omega_o}{2}e^{i\frac{\pi}{2}}B^*_{r_{xy}}$ and substituting $\Delta\omega(\mathbf{r},\tau)$ by $\gamma\vec{G}(\tau)\cdot\mathbf{r}$ (see Eqs. (3.47) and (3.48)), Eq. (3.100) can be simplified as

$$s_c(t) \simeq c\int_V M_{xy}(\mathbf{r},0)e^{-i\gamma\int_0^t \vec{G}(\tau)\cdot\mathbf{r}d\tau}d\mathbf{r}. \tag{3.101}$$

The continuous, complex-valued baseband signal $s_c(t)$ of Eq. (3.101) is generated by the quadrature PSD and called PSD signal. Similar to the notation of Eq. (3.96), $s_c(t)$ of Eq. (3.101) actually represents the signal component of PSD signal.

Eq. (3.101) reveals Fourier formulation relation between $s_c(t)$ and $M_{xy}(\mathbf{r},0)$, and is also known as the MR signal equation (in terms of $s_c(t)$). TPMM $M_{xy}(\mathbf{r},0)$ and TEMM $M_z^o(\mathbf{r})$ are linked by Eq. (3.88). When the flip angle $\alpha = \frac{\pi}{2}$, $M_{xy}(\mathbf{r},0) = iM_z^o(\mathbf{r})$. Thus, Eq. (3.101), in fact, reveals the Fourier formulation relation between $s_c(t)$ and TEMM.

$s_c(t)$ is further sampled at the Nyquist frequency for the given bandwidth of the anti-aliasing filter to yield a discrete, complex-valued baseband signal $\hat{s}_c(j)$,

$$\hat{s}_c(j) \simeq c\int_V M_{xy}(\mathbf{r},0)e^{-i\gamma\int_0^j \vec{G}(\tau)\cdot\mathbf{r}d\tau}d\mathbf{r}, \tag{3.102}$$

where j denotes the time instant. The sampling is carried out by Analog-Digital converter (ADC), therefore, $\hat{s}_c(j)$ is called ADC signal. Following to the notation of Eq. (3.101), $\hat{s}_c(j)$ of Eq. (3.102) only represents the signal component of ADC signal. Eq. (3.102) is the discrete version of MR signal equation (in terms of $\hat{s}_c(j)$).

$s_r(t)$, $s_c(t)$, and $\hat{s}_c(j)$ are MR signals at the different stages of signal detection and formulation module. They represent the electrical voltage, i.e., their physical units are V (V - Volt). To verify this, a simple way is to use Eqs. (3.86) and (3.87), where the units of $\vec{B}(\mathbf{r})$ and $\vec{M}(\mathbf{r},t)$ are $\mathrm{Wb}/(\mathrm{A}\cdot\mathrm{m}^2)$ and A/m (Wb - weber). Thus, the physical unit of $s_r(t)$ is

$$\mathrm{s}^{-1}\cdot\mathrm{m}^3\cdot\mathrm{Wb}/(\mathrm{A}\cdot\mathrm{m}^2)\cdot\mathrm{A/m} = \mathrm{Wb/s} = \mathrm{V}. \tag{3.103}$$

Clearly, $s_c(t)$ and $\hat{s}_c(j)$ have the same physical unit as $s_r(t)$.

3.8.3 Spatial Localization

Section 3.7.2 shows that the z-gradient and RF excitation pulse select a slice. After these excitations are turned off, the nuclear spin system transfers itself from the forced precession period to the free precession period, and the spins undergo through processes of the longitudinal and transverse relaxations. As mentioned in Section 3.5.2, immediately after the excitation, the remaining x- and y-gradients, G_x and G_y, are applied at the $\vec{i}$- and $\vec{j}$-directions in the transverse plane to encode spatial localization information.

When y-gradient G_y is imposed, both the magnetic field strength and Larmor frequency vary linearly with y-location:

$$\omega_y = \gamma(B_o + G_y y). \tag{3.104}$$

By turning this gradient on, nuclear spins at the different y locations precess with different frequencies. When y-gradient is turned off after a period t_y, these nuclear spins return to precessing at the same frequency, but have the different phases ϕ_y:

$$\phi_y = \omega_y t_y = \gamma(B_o + G_y y)t_y. \tag{3.105}$$

Thus, the nuclear spin at a different y-location has its own unique phase; therefore, y-location is encoded into the phase. This process is often called *phase encoding*.

Similarly, when x-gradient G_x is imposed, both magnetic field strength and Larmor frequency vary linearly with x-location:

$$\omega_x = \gamma(B_o + G_x x). \tag{3.106}$$

By turning this gradient on, nuclear spins at the different x locations precess with different frequencies. The resultant signal during this period is the sum of all signal components with different frequencies that come from different range bins in the x-direction. Thus, this gradient links spatial location x with the frequency, and this process is often called *frequency encoding.*

Intuitively, the functions of three gradients can be thought of as the selective excitation z-gradient selects a slice, the phase encoding y-gradient cuts that slice into strips, and the frequency encoding x-gradient cuts those strips into cubes (pixels or voxels).

3.9 k-Space and k-Space Sample

k-Space is very useful in MRI. Not only does it provide a basis for establishing a new version of MR signal equation that explicitly demonstrates the Fourier transform relation between TPMM and k-space sample, but it also gives a clear interpretation of the connection between the spatial localization (phase encoding and frequency encoding) and Fourier transform.

A k-space sample is an alternative representation of an ADC signal in terms of the spatial frequency. In this section, the underlying mechanism that transforms the ADC signal to a k-space sample is revealed, followed by a review of the commonly used k-space sampling schemes: rectilinear and radial. Then, the sampling requirements for these schemes are discussed.

3.9.1 Concepts

For simplicity in introducing the concept of k-space, let the gradient vector be constant and the slice be selected at $z = z_o$ with the unit thickness. Thus $\vec{G}(\tau) \cdot \mathbf{r} = G_x x + G_y y + G_z z_o$, and Eq. (3.102) becomes

$$\hat{s}_c(j) \simeq c' \int_V M_{xy}(\mathbf{r}, 0) e^{-i\gamma(G_x x + G_y y)j} d\mathbf{r}, \tag{3.107}$$

where $c' = c e^{-i\gamma G_z z_o j}$ is a constant and c is given by Eq. (3.101). By defining the quantities

$$k_x(j) = \frac{1}{2\pi}\gamma G_x j \quad \text{and} \quad k_y(j) = \frac{1}{2\pi}\gamma G_y j, \tag{3.108}$$

a relation between the one-dimensional time index j and the two-dimensional coordinates, $k_x(j)$ and $k_y(j)$, is established.

The physical unit of $k_x(j)$ and $k_y(j)$ is m^{-1} (m - meter), that is, the inverse of the length. To verify this, a simple way is to use Eq. (3.108). The unit of $\gamma G_x j$ (or $\gamma G_y j$) is $(\mathrm{s}^{-1}\mathrm{T}^{-1})(\mathrm{Tm}^{-1})\mathrm{s} = \mathrm{m}^{-1}$. Thus, $k_x(j)$ and $k_y(j)$ represent the time-varying spatial frequency and define a space called *k-space.*

In the X-Y plane, when the gradient $\vec{G}_\theta$ is defined at the direction specified by an angle θ from the x-direction, then, similar to $k_x(j)$ and $k_y(j)$ of

Eq. (3.108), the spatial frequency $k(j)$ caused by the gradient $\vec{G}_\theta$ is $\frac{1}{2\pi}\gamma G_\theta j$. Because $G_x = G_\theta \cos\theta$ and $G_y = G_\theta \sin\theta$, θ is given by $\tan^{-1}(\frac{G_y}{G_x})$. When the time instant j is within the bounded pairwise time intervals, the ratio $\frac{G_y}{G_x}$ may stay the same, while between these time intervals, the ratio $\frac{G_y}{G_x}$ can vary. Therefore, the relation between the angle θ and the time instant j can be characterized by a staircase function. In order to show $\theta \sim j$ dependence, the notation $\theta(j)$ is used. Thus, a pair of two polar coordinates

$$k(j) = \frac{1}{2\pi}\gamma G_\theta j \quad \text{and} \quad \theta(j) = \tan^{-1}(\frac{G_y}{G_x}) \tag{3.109}$$

also specifies a location in k-space.

For the slice at $z = z_o$, Eq. (3.107) implicitly adopts $\mathbf{r} = (x, y)$, $d\mathbf{r} = dxdy$, and $V \to S$; here, S denotes the area of the slice under the integration. By using $k_x(j)$ and $k_y(j)$ of Eq. (3.108), Eq. (3.107) can be rewritten as

$$\begin{aligned} \hat{s}_c(j) &\simeq c' \int_S M_{xy}(x, y, 0) e^{-i2\pi(k_x(j)x + k_y(j)y)} dxdy \\ &= c' \mathcal{F}_2\{M_{xy}(x, y, 0)\} \triangleq \mathcal{M}(k_x(j), k_y(j)), \end{aligned} \tag{3.110}$$

where $\mathcal{F}_2$ denotes the two-dimensional Fourier transform, and $\mathcal{M}(k_x(j), k_y(j))$ is called the *k-space sample*. Rigorously speaking, this $\mathcal{M}(k_x(j), k_y(j))$ is the signal component of the k-space sample, because, as indicated in Section 3.8.2, $\hat{s}_c(j)$ only represents the signal component of the ADC signal. In the polar coordinate system, the k-space sample is expressed by $\mathcal{M}(k(j), \theta(j))$.

Eq. (3.108) and Eq. (3.109) map a time index j to a k-space location $(k_x(j), k_y(j))$ or $(k(j), \theta(j))$. Eq. (3.110) maps the corresponding ADC signal $\hat{s}_c(j)$ in the time domain to the k-space sample $\mathcal{M}(k_x(j), k_y(j))$, or $\mathcal{M}(k(j), \theta(j))$ in the spatial frequency domain. Eq. (3.110) is a k-space version of the MR signal equation. It also shows that the physical unit of the k-space sample $\mathcal{M}(k_x(j), k_y(j))$ or $\mathcal{M}(k(j), \theta(j))$ is the same as the ADC signal $\hat{s}_c(j)$: V (*Volt*). Generally, when the gradients G_x and G_y are time varying, the spatial frequencies $k_x(t)$ and $k_x(t)$ of Eq. (3.108) are defined by

$$k_x(t) = \frac{\gamma}{2\pi}\int_0^t G_x(\tau)d\tau \quad \text{and} \quad k_y(t) = \frac{\gamma}{2\pi}\int_0^t G_y(\tau)d\tau. \tag{3.111}$$

3.9.2 Sampling Protocols

The commonly used k-space sampling protocols include rectilinear sampling and radial sampling, which lead to two different MR image reconstruction methods: Fourier transform and projection reconstruction.

3.9.2.1 Rectilinear Sampling

Figures 3.9 and 3.10 demonstrate the rectilinear k-space sampling scheme. Figure 3.9 shows a timing sequence. A $\frac{\pi}{2}$ RF excitation pulse and a z-gradient

G_z select a slice. The negative lobe of the z-gradient is used for refocusing (see Section 3.7.2.1). Both y-gradient G_y and x-gradient G_x are turned on for a period t_y. G_y, which will be changed in the next period t_y, is for the phase encoding. The negative G_x, which will not be changed in the next period t_y, is used for covering the desired area of the k-space. At the end of period t_y, G_y is turned off that completes the phase encoding; G_x is changed to the positive that carries out the frequency encoding and reads out the FID signal. From Eq. (3.107), this readout signal is a sum of all signal components

$$M_{xy}(x, y, 0)e^{-i\gamma(G_x xj + G_y y t_y)}$$

across the sample.

Corresponding to the timing sequence in Figure 3.9, Figure 3.10.a shows the k-space sample trajectory for the rectilinear sampling. For the positive G_y and the negative G_x, the trajectory begins at the origin and moves along a line in the second quadrant (with an angle θ in the x-direction). The slope of this line is determined by the ratio $\frac{G_y}{G_x}$. This process corresponds to phase encoding. At the end of period t_y, it moves right along the k_x direction. This process corresponds to the frequency encoding and readout of the FID signal.

On the next period t_y, the amplitude G_y changes (as shown by $\updownarrow$ in Figure 3.9) but the amplitude G_x remains the same. Thus, the angle θ changes. A change in the amplitude G_y leads to a different line in the k-space—the horizontal dashed line in Figure 3.10a. By choosing an appropriate set of G_y (i.e., the period t_y will be repeated for certain times), a desired rectangular region in k-space can be covered.

The k-space data and the sampled k-space data in rectilinear sampling are expressed by $\mathcal{M}(k_x(t), k_y(t))$ and $\mathcal{M}(k_x(j), k_y(j))$, respectively. When the sampling indexed by j occurs at a Cartesian grid as shown in Figure 3.10b, k-space samples are often expressed by $\mathcal{M}(m\Delta k_x, n\Delta k_y)$; here, m and n are integers, Δk_x and Δk_y are horizontal and vertical sampling intervals. Clearly, $\mathcal{M}(m\Delta k_x, n\Delta k_y)$ is a k-space sample at the m-th frequency encoding after the n-th phase encoding.

3.9.2.2 Radial Sampling

Figures 3.11 and 1.12 demonstrate the radial k-space sampling scheme. Figure 3.11 shows a timing sequence. An RF excitation pulse and a z-gradient G_z select a slice. The negative z-gradient is used for refocusing (see Section 3.7.2.1). Both y-gradient G_y and x-gradient G_x are turned on for a period t_y. At the end of the period t_y, both G_y and G_x change their polarities, perform the frequency encoding, and read out the FID signal. From Eq. (3.107), this readout signal is the sum of all signal components

$$M_{xy}(x, y, 0)e^{-i\gamma(G_x xj + G_y yj)}$$

across the sample, which is different from that in rectilinear sampling.

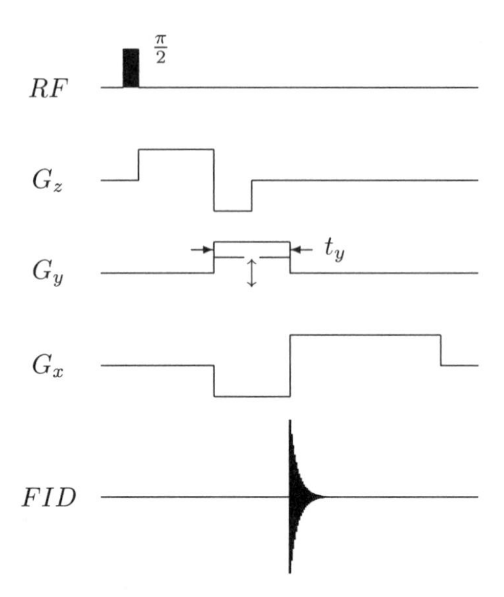

FIGURE 3.9
A timing diagram in the rectilinear k-space sampling.

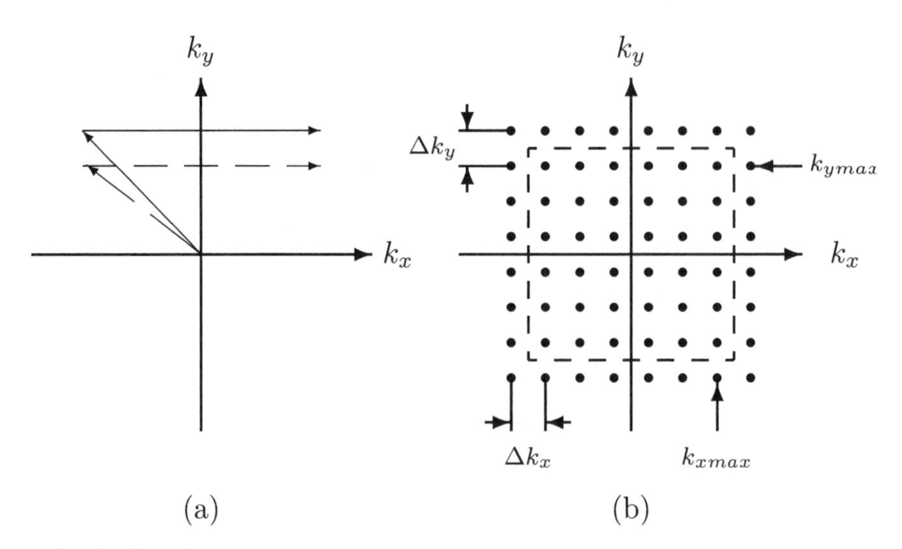

FIGURE 3.10
The trajectory (a) and the sampling (b) in the rectilinear k-space sampling.

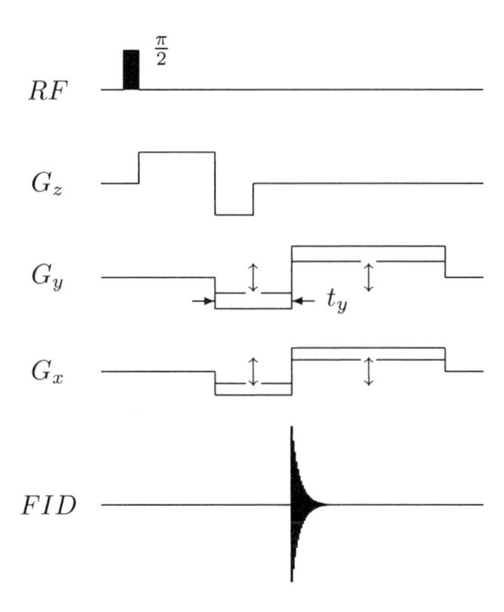

FIGURE 3.11
MR signal detection and demodulation in the radial k-space sampling.

Corresponding to the timing sequence in Figure 3.11, Figure 3.12a shows the k-space sample trajectory for the radial sampling. For the negative G_y and G_x, the trajectory begins at the origin and moves along a line in the third quadrant(with an angle θ in the x-direction). The slope of this line is determined by the ratio $\frac{G_y}{G_x}$. At the end of period t_y, the trajectory turns back, passes through the origin, and moves along the same direction but in the first quadrant as shown in Figure 3.12a. This process corresponds to G_y and G_x frequency encoding. The FID signal is read out during the positive lobe of G_x in Figure 3.11.

On the next period t_y, both G_y and G_x change their amplitudes as shown in Figure 3.11 by $\updownarrow$. A change in the ratio $\frac{G_y}{G_x}$ results in a different line in the k-space as shown in Figure 3.12a by the tilted dashed line. By choosing an appropriate set of G_y and G_x such that θ changes from 0 to π (i.e., t_y is repeated for certain times), a desired circular region in k-space will be covered.

The k-space data and the sampled k-space data in polar sampling are expressed by $\mathcal{M}(k(t), \theta(t))$ and $\mathcal{M}(k(j), \theta(j))$, respectively. When the sampling indexed by j occurs at a polar grid as shown in Figure 3.12.b, k-space samples are often expressed by $\mathcal{M}(m\Delta k, n\Delta\theta)$, where m and n are integers, and Δk and $\Delta\theta$ are the radial and angular sampling intervals. Clearly, $\mathcal{M}(m\Delta k, n\Delta\theta)$ is a k-space sample at the m-th frequency encoding at the n-th readout.

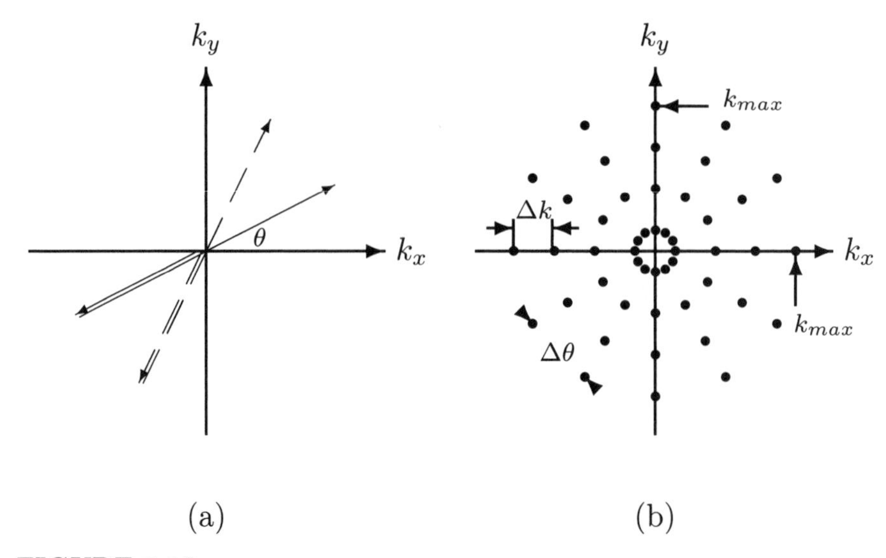

FIGURE 3.12
The trajectory (a) and the sampling (b) in the radial k-space sampling.

3.9.3 Sampling Requirements

3.9.3.1 Rectilinear Sampling

Figure 3.13 shows the details of the timing sequence in Figure 3.9 for phase encoding, frequency encoding, and the readout sampling. In Figure 3.13, G_y and t_y, G_x and t_x denote the amplitude and the time period for the phase and frequency encoding, respectively. ΔG_y and Δt represent the incremental magnitude of the y-gradient and the sampling interval of the readout signal. Let n_{pe} and n_{fe} be the numbers of the phase encoding and frequency encoding (in each phase encoding). Then

$$(n_{pe}-1)\Delta G_y = 2G_y \quad \text{and} \quad (n_{fe}-1)\Delta t = t_x, \tag{3.112}$$

which is equivalent to

$$n_{pe} = \frac{2G_y}{\Delta G_y} + 1 \quad \text{and} \quad n_{fe} = \frac{t_x}{\Delta t} + 1. \tag{3.113}$$

At the time instant j that corresponds to the m-th frequency encoding after the nth phase encoding, we have

$$k_x(j) = \frac{\gamma}{2\pi} G_x(m\Delta t) \quad \text{and} \quad k_y(j) = \frac{\gamma}{2\pi}(n\Delta G_y)t_y, \tag{3.114}$$

where $-\frac{n_{fe}}{2} \leq m \leq \frac{n_{fe}-1}{2}$ and $-\frac{n_{pe}}{2} \leq n \leq \frac{n_{pe}-1}{2}$. Eq. (3.114) leads to

$$\Delta k_x = \frac{\gamma}{2\pi} G_x \Delta t \quad \text{and} \quad \Delta k_y = \frac{\gamma}{2\pi} \Delta G_y t_y \,. \tag{3.115}$$

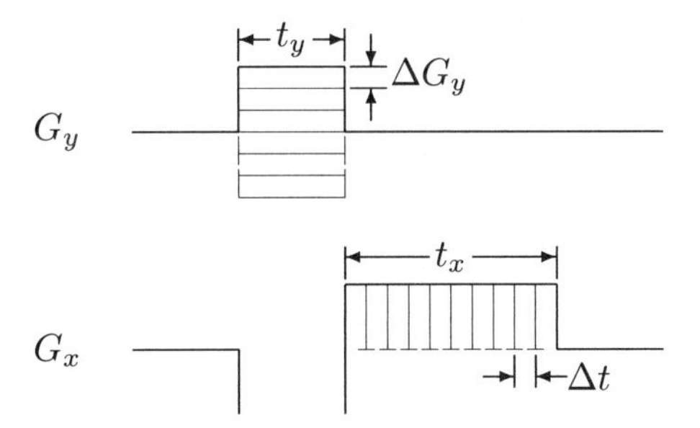

FIGURE 3.13
Phase encoding, frequency encoding, and readout sampling.

Figure 3.13 shows that the sampling intervals Δk_x and Δk_y represent a scaled area between two adjacent vertical lines in the frequency encoding period and a scaled area between two adjacent horizontal lines in the phase encoding period, respectively.

Eq. (3.110) indicates that TPMM $M_{xy}(x, y, 0)$ and k-space samples $\mathcal{M}(k_x, k_y)$ constitute a pair of Fourier transform (ignoring the constant c') and TPMM $M_{xy}(x, y, 0)$ is the space-limited. These two conditions enable us to use the Nyquist Shannon sampling theorem, which states that a temporal frequency band-limited signal can be reconstructed from uniform samples in the frequency domain at a sampling rate equal to or greater than twice the frequency band (or with the sampling interval equal to or less than the reciprocal of the sampling rate).

In the object domain, let TPMM $M_{xy}(x, y, 0)$ be confined in a rectangular area $FOV_x \times FOV_y$; here, FOV_x and FOV_y denote the dimensions of the field of view in the x- and y-directions, respectively. Thus, in order to reconstruct TPMM $M_{xy}(x, y, 0)$, the sampling intervals Δk_x and Δk_y in the k-space domain (which is equivalent to the sampling interval Δt for the ADC signal $\hat{s}_c(j)$ in the time domain) must satisfy

$$\Delta k_x \leq \frac{1}{FOV_x} \quad \text{and} \quad \Delta k_y \leq \frac{1}{FOV_y}. \tag{3.116}$$

Using Eq. (3.115), the sampling requirements for the rectilinear sampling are

$$\Delta t \leq \frac{2\pi}{\gamma G_x FOV_x} \quad \text{and} \quad \Delta G_y \leq \frac{2\pi}{\gamma t_y FOV_y}. \tag{3.117}$$

Let $\delta x \times \delta y$ denote the spatial resolution (i.e., the pixel size) of the reconstructed MR image. Due to $FOV_x = I \cdot \delta x$ and $FOV_y = J \cdot \delta y$ (I and J are the numbers of pixels in each row and each column of the image), Eq. (3.116) leads to

$$\delta x \leq \frac{1}{k_{xmax}} \quad \text{and} \quad \delta y \leq \frac{1}{k_{ymax}}. \tag{3.118}$$

3.9.3.2 Radial Sampling

Based on Eqs. (3.108) and (3.109), $\mathcal{F}_2\{M_{xy}(x,y,0)\}$ in Eq. (3.110) can be rewritten as

$$\mathcal{F}_2\{M_{xy}(x,y,0)\} = \int M_{xy}(x,y,0)e^{-i2\pi k(j)(x\cos\theta(j)+y\sin\theta(j))}dxdy, \quad (3.119)$$

where $k(j) = \sqrt{k_x^2(j) + k_y^2(j)}$ and $\theta(j) = \tan^{-1}(\frac{k_y(j)}{k_x(j)})$.

Let the coordinate system $X' - Y'$ be generated by rotating the coordinate system X-Y with an angle $\theta(j)$ (which is given by Eq. (3.109)) such that the x'-direction coincides with the gradient $\vec{G}_\theta(j)$ direction. The coordinates (x',y') and (x,y) are related by

$$\begin{pmatrix} x' \\ y' \end{pmatrix} = \begin{pmatrix} \cos\theta(j) & \sin\theta(j) \\ -\sin\theta(j) & \cos\theta(j) \end{pmatrix} \begin{pmatrix} x \\ y \end{pmatrix}.$$

By using new coordinates (x',y'), Eq. (3.119) becomes

$$\mathcal{F}_2\{M_{xy}(x',y',0)\} = \int M_{xy}(x',y',0)e^{-i2\pi k(j)x'}dx'dy'. \quad (3.120)$$

By defining the *measured projection* $p(x',\theta(j))$ as

$$p(x',\theta(j)) = \int M_{xy}(x',y',0)dy', \quad (3.121)$$

Eq. (3.120) becomes

$$\mathcal{F}_2\{M_{xy}(x',y',0)\} = \int p(x',\theta(j))e^{-i2\pi k(j)x'}dx' = \mathcal{F}_1\{p(x',\theta(j))\}, \quad (3.122)$$

where $\mathcal{F}_1$ denotes the one-dimensional Fourier transform with respect to x'.

Eq. (3.122) indicates that the one-dimensional Fourier transform of the projection $p(x',\theta(j))$ of TPMM $M_{xy}(x',y',0)$ at a given direction $\theta(j)$ equals the two-dimensional Fourier transform of the object function (i.e., TPMM $M_{xy}(x',y',0)$) in that direction).[§] This is the well-known Fourier slice theorem, which is illustrated in Figure 3.14.

In the object domain, let TPMM $M_{xy}(x',y',0)$ be confined in a circular area with diameter FOV_r. Then, in each direction $\theta(j)$, in order to avoid the aliasing artifact in the radial direction (i.e., x'-direction) of the reconstructed TPMM $M_{xy}(x',y',0)$, according to the Nyquist–Shannon sampling theorem, the sampling interval Δk (which is equivalent to the sampling interval Δt for ADC signal $\hat{s}(j)$ in the time domain) must satisfy

$$\Delta k \leq \frac{1}{FOV_r}. \quad (3.123)$$

[§]The coordinates (x',y') implicitly specify this direction.

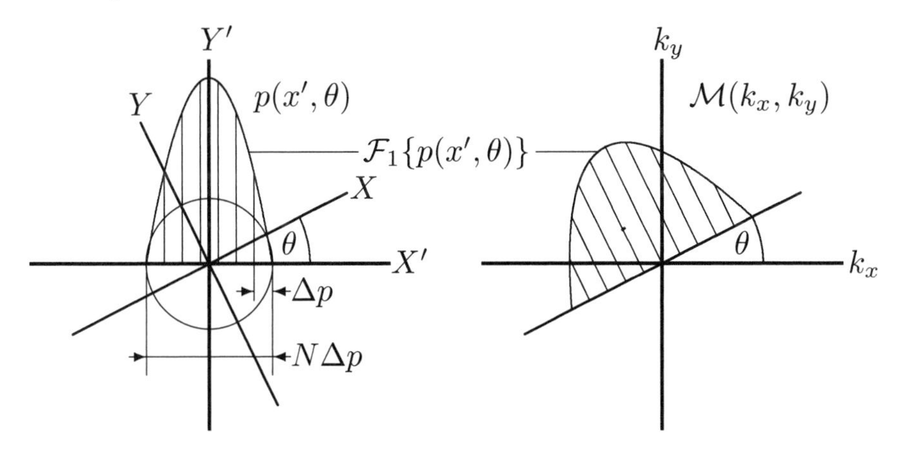

FIGURE 3.14
An illustration of the Fourier slice theorem.

$\mathcal{F}_2\{M_{xy}(x', y', 0)\}$ in Eq. (3.122) can be expressed either rectilinearly by $\mathcal{M}(k_{x'}(j), k_{y'}(j))$ or radially by $\mathcal{M}(k(j), \theta(j))$. Radial samples are often used for the projection reconstruction method, which is based on the Inverse Radon transform and implemented by filtered backprojection (FBP) algorithm. In the numerical implementation of FBP, radial samples $\mathcal{M}(k(j), \theta(j))$ and measured projections $p(x', \theta(j))$ are discretized both radially and angularly as $\mathcal{M}(n\Delta k, m\Delta\theta)$ and $p(n\Delta p, m\Delta\theta)$ $(-\frac{N}{2} \leq n \leq \frac{N-1}{2}, 0 \leq m \leq M-1)$. Here Δk and $\Delta\theta$ are the radial and angular spacings between two adjusting k-space samples, and Δp is the spacing between two adjacent measured projections.

Thus, from Eq. (3.122), the measured projections $p(k\Delta p, m\Delta\theta)$ at the direction specified by the angle $m\Delta\theta$ and the radial k-space samples $\mathcal{M}(n\Delta k, m\Delta\theta)$ at the time instant j, which corresponds to the n-th frequency encoding at the m-th readout are linked by

$$p(k\Delta p, m\Delta\theta) = \Delta k \sum_{n=-N/2}^{(N-1)/2} \mathcal{M}(n\Delta k, m\Delta\theta) e^{i2\pi k\Delta p \, n\Delta k}, \tag{3.124}$$

where Δk is restricted by Eq. (3.123). To ensure the accuracy of the interpolation required by FBP, N and Δp must be chosen so that $\frac{1}{N\Delta p}$ is significantly smaller than Δk [14], that is,

$$\frac{1}{N\Delta p} < \frac{1}{FOV_r}. \tag{3.125}$$

In Figure 3.15, $\widehat{\Delta s}$ and Δs denote the arc and the chord spanned by an angle interval $\Delta\theta$ on the circumference of a circle. Δs and Δk represent the azimuthal and radial sampling interval and are orthogonal. Δk is fixed and Δs varies with circles of different diameters. Here, only the most outward

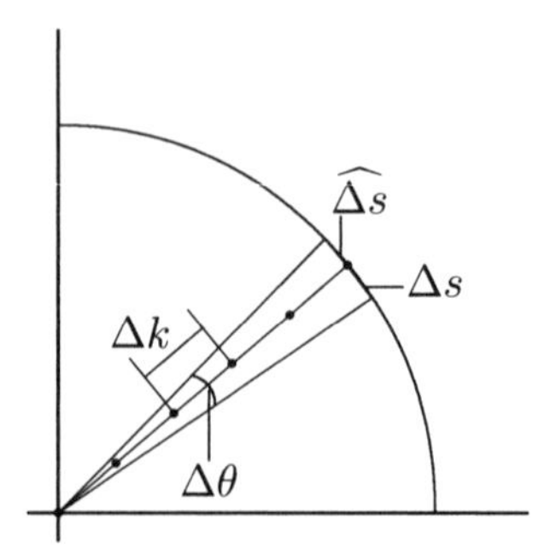

FIGURE 3.15
An illustration of the relationship between M and N in FBP. Because M is large, the arc $\widehat{\Delta s}$ and the chord Δs are very close. They are nearly overlapped on each other and are not virtually distinguished in this figure.

Δs (which corresponds to the circle with the diameter $N\Delta k$) is chosen for consideration, because it is the largest.

Similar to the Δk_x and Δk_y in the rectilinear sampling, two orthogonal sampling intervals Δk and Δs are considered the same, that is, $\Delta k = \Delta s$. Also, because M is usually very large, $\Delta s \simeq \widehat{\Delta s}$. Thus, $\frac{2M\Delta s}{N\Delta k} \simeq \frac{2M\widehat{\Delta s}}{N\Delta k} = \pi$ which leads to M and N in FBP being related by

$$\frac{M}{N} \simeq \frac{\pi}{2}. \tag{3.126}$$

This relation is confirmed in references [12, 14, 15]. Thus, from Eqs. (3.125) and (3.126), the angular sampling interval $\Delta\theta$ in the polar sampling scheme will be

$$\Delta\theta = \frac{\pi}{M} \simeq \frac{2}{N} < \frac{2\Delta p}{FOV_r} . \tag{3.127}$$

Therefore, the sampling requirements for the polar sampling scheme are

$$\Delta k \leq \frac{1}{FOV_r} \quad \text{and} \quad \Delta\theta < \frac{2\Delta p}{FOV_r} . \tag{3.128}$$

3.10 Image Reconstruction

Corresponding to two k-space sampling schemes (rectilinear and radial), two methods—Fourier transform (FT) and projection reconstruction (PR)—are commonly used for MR image reconstruction. In the two-dimensional case, they are abbreviated as 2DFT and 2DPR, respectively.

3.10.1 Fourier Transform

3.10.1.1 Mathematical Description

As shown in Section 3.9.2.1, $\mathcal{M}(k_x(t), k_y(t))$ and $\mathcal{M}(k_x(j), k_y(j))$ represent the continuous and discrete rectilinear k-space data, and $\mathcal{M}(m\Delta k_x, n\Delta k_y)$ represents the rectilinear k-space samples. In the k-space version of the MR signal equation (3.110), ignoring the constant c', replacing $M_{xy}(x, y, 0)$ by $M_{xy}(x, y)$ (0 implies $t = 0$) and $\mathcal{M}(k_x(j), k_y(j))$ by $\mathcal{M}(m\Delta k_x, n\Delta k_y)$, we have

$$M_{xy}(x, y) = \mathcal{F}_2^{-1}\{\mathcal{M}(m\Delta k_x, n\Delta k_y)\}, \tag{3.129}$$

where $\mathcal{F}_2^{-1}$ denotes a two-dimensional inverse Fourier transform.

Eq. (3.129) shows that an appropriate set of k-samples $\mathcal{M}(m\Delta k_x, n\Delta k_y)$ (i.e., a set of the ADC signal $\hat{s}_c(j)$) can be acquired to produce an image $X(x, y)$ which is an estimate of the TPMM $M_{xy}(x, y, 0)$. Let $(x, y) = \mathbf{r}$; Eq. (3.88) indicates that TPMM $M_{xy}(\mathbf{r}, 0)$ and TEMM $M_z^o(\mathbf{r})$ are related by $M_{xy}(\mathbf{r}, 0) = iM_z^o(\mathbf{r})\sin\alpha$ (α - the flip angle). Therefore, the MR image generated by 2DFT essentially represents the spatial distribution of TEMM.

Based on the Fourier transform relation between TPMM and k-space samples given by Eq. (3.129) and considering the requirements of data acquisition and image quality, a framework for 2DFT can be mathematically described by

$$M_{xy}(x, y) = \mathcal{F}_2^{-1}\{\mathcal{M}(k_x(t), k_y(t)) \cdot \frac{1}{\Delta k_x \Delta k_y}$$
$$comb\left(\frac{k_x(t)}{\Delta k_x}, \frac{k_y(t)}{\Delta k_y}\right) rect\left(\frac{k_x(t)}{W_{k_x}}, \frac{k_y(t)}{W_{k_y}}\right) filt\left(\frac{k_x(t)}{W_{k_x}}, \frac{k_y(t)}{W_{k_y}}\right)\}. \tag{3.130}$$

In Eq. (3.130), *comb* function (also known as "the bed of nails") is given by

$$comb\left(\frac{k_x(t)}{\Delta k_x}, \frac{k_y(t)}{\Delta k_y}\right) = \Delta k_x \Delta k_y \sum_{m,n=-\infty}^{\infty} \delta(k_x(t) - m\Delta k_x, k_y(t) - n\Delta k_y), \tag{3.131}$$

where $\delta(k_x(t) - m\Delta k_x, k_y(t) - n\Delta k_y)$ is the 2-D Dirac delta function defined by

$$\delta(k_x(t) - m\Delta k_x, k_y(t) - n\Delta k_y) = \begin{cases} \text{non } 0 & (k_x(t) = m\Delta k_x \text{ and } k_y(t) = n\Delta k_y) \\ 0 & (\text{otherwise}), \end{cases} \tag{3.132}$$

and *rect* is a rectangular window function (also known as the scope function) given by

$$rect\left(\frac{k_x(t)}{W_{k_x}}, \frac{k_y(t)}{W_{k_y}}\right) = \begin{cases} 1 & (|k_x(t)| \leq \frac{W_{k_x}}{2} \text{ and } |k_y(t)| \leq \frac{W_{k_y}}{2}) \\ 0 & (\text{otherwise}), \end{cases} \tag{3.133}$$

where W_{k_x} and W_{k_y} are the widths of the window shown in Figure 3.10;

$filt$ is the filter function (also known as the apodization function) given by

$$filt\left(\frac{k_x(t)}{W_{k_x}}, \frac{k_y(t)}{W_{k_y}}\right) = \Phi\left(\frac{k_x(t)}{W_{k_x}}, \frac{k_y(t)}{W_{k_y}}\right) \cdot rect\left(\frac{k_x(t)}{W_{k_x}}, \frac{k_y(t)}{W_{k_y}}\right), \quad (3.134)$$

where the function Φ can take various forms, which are real, even, and normalized, such as Hamming and Hanning functions:

$$\Phi\left(\frac{k_x(t)}{W_{k_x}}, \frac{k_y(t)}{W_{k_y}}\right) = (\alpha + (1-\alpha)\cos(\frac{2\pi k_x(t)}{W_{k_x}}))(\alpha + (1-\alpha)\cos(\frac{2\pi k_y(t)}{W_{k_y}})); \quad (3.135)$$

$\alpha = 0.54$ for the Hamming function and $\alpha = 0.50$ for the Hanning function.

Thus, Eq. (3.130) can be decomposed into three steps, which are

• Step 1: *Sampling.* It is given by

$$\mathcal{M}_1(m\Delta k_x, n\Delta k_y) = \mathcal{M}(k_x(t), k_y(t)) \cdot \frac{1}{\Delta k_x \Delta k_y} comb\left(\frac{k_x(t)}{\Delta k_x}, \frac{k_y(t)}{\Delta k_y}\right). \quad (3.136)$$

Using Eq. (3.131), we have

$$\mathcal{M}_1(m\Delta k_x, n\Delta k_y) = \mathcal{M}(k_x(t), k_y(t)) \sum_{m,n=-\infty}^{\infty} \delta(k_x(t) - m\Delta k_x, k_y(t) - n\Delta k_y).$$

This sampling operation (from the continuous k-space data to the discrete samples) is carried out by the rectilinear sampling scheme described in Section 3.9.2.1. The sampling requirements are specified by Eqs. (3.116) or (3.117), which are mainly for eliminating the *aliasing artifact.*

• Step 2: *Truncating.* It is given by

$$\mathcal{M}_2(m\Delta k_x, n\Delta k_y) = \mathcal{M}_1(m\Delta k_x, n\Delta k_y) \cdot rect\left(\frac{k_x(t)}{W_{k_x}}, \frac{k_y(t)}{W_{k_y}}\right). \quad (3.137)$$

This truncation operation is necessary because only the finite k-space samples can be utilized in the practical computation of image reconstruction. With this step, the finite k-space samples $\mathcal{M}_2(m\Delta k_x, n\Delta k_y)$ ($-\frac{I}{2} \leq m \leq \frac{I-1}{2}$, $-\frac{J}{2} \leq n \leq \frac{J-1}{2}$, I, J, m, and n - integers) are extracted from the entire discrete k-space domain $\mathcal{M}_1(m\Delta k_x, n\Delta k_y)$.

• Step 3: *Filtering.* It is given by

$$\mathcal{M}(m\Delta k_x, n\Delta k_y) = \mathcal{M}_2(m\Delta k_x, n\Delta k_y) \cdot filt\left(\frac{k_x(t)}{W_{k_x}}, \frac{k_y(t)}{W_{k_y}}\right), \quad (3.138)$$

It is well known that the Fourier transform of the truncated data always produces *Gibbs ringing artifact.* In order to remove or reduce this artifact, a spatial filter must be applied to $\mathcal{M}_2(m\Delta k_x, n\Delta k_y)$. The $filt$ function decays smoothly to the zero at $m = -\frac{I}{2}, \frac{I-1}{2}$ and $n = -\frac{J}{2}, \frac{J-1}{2}$. $\mathcal{M}(m\Delta k_x, n\Delta k_y)$ is finally used in Eq. (3.129).

3.10.1.2 Computational Implementation

FOV_x and FOV_y—the dimensions of the field of view (Section 3.9.3.1)—specify the size of the reconstructed MR image. Let δx and δy be the Fourier pixel size [12]; the numbers of pixels in each row and each column of the reconstructed image are $I = \frac{FOV_x}{\delta x}$ and $J = \frac{FOV_y}{\delta y}$, respectively. Eq. (3.129) or Eq. (3.130) can be numerically implemented by 2-D DFT. Thus, the pixel values $X(x, y)$, that is, the estimates of $M_{xy}(x, y)$, are computed by the inverse 2-D DFT of k-space samples $\mathcal{M}(m\Delta k_x, n\Delta k_y)$.

When $I \times J$ rectilinear k-space samples $\mathcal{M}(m\Delta k_x, n\Delta k_y)$ are acquired, for the pixels whose centers are at the locations $(x = k\delta x, y = l\delta y)$, based on Eq. (3.116) with the equal-sign, we have

$$\begin{aligned} Y(k\delta x, l\delta y) &= \Delta k_x \Delta k_y \sum_{m=-I/2}^{(I-1)/2} \sum_{n=-J/2}^{(J-1)/2} \\ &\quad \mathcal{M}(m\Delta k_x, n\Delta k_y)\Phi(m\Delta k_x, n\Delta k_y) e^{i2\pi m\Delta k_x k\delta x} e^{i2\pi n\Delta k_y l\delta y} \\ &= \frac{1}{I \cdot J \cdot \delta x \cdot \delta y} \sum_{m=-I/2}^{(I-1)/2} \sum_{n=-J/2}^{(J-1)/2} \\ &\quad \mathcal{M}(m\Delta k_x, n\Delta k_y)\Phi(m\Delta k_x, n\Delta k_y) e^{i2\pi km/I} e^{i2\pi ln/J}, \end{aligned} \quad (3.139)$$

where $\Phi(m\Delta k_x, n\Delta k_y)$ is the filter function given by Eq. (3.134), $-\frac{I}{2} \leq k \leq \frac{I-1}{2}$, $-\frac{J}{2} \leq l \leq \frac{J-1}{2}$. For pixels with unity size (i.e., $\delta x = \delta y = 1$) and using the abbreviation $\mathcal{M}(m, n)$ and $\Phi(m, n)$ for $\mathcal{M}(m\Delta k_x, n\Delta k_y)$ and $\Phi(m\Delta k_x, n\Delta k_y)$, respectively, Eq. (139) becomes

$$X(k, l) = \frac{1}{I \cdot J} \sum_{m=-I/2}^{(I-1)/2} \sum_{n=-J/2}^{(J-1)/2} \mathcal{M}(m, n)\Phi(m, n) e^{i2\pi km/I} e^{i2\pi ln/J}.$$

3.10.2 Projection Reconstruction

The FT method dominates in the field of MR image reconstruction. The main reason is its convenience and efficiency in image reconstruction afforded by the rectilinear sampling on the Cartesian grids in the k-space. However, the PR method is also very useful in some applications, particularly due to its analogy with X-ray computed tomography (CT).

3.10.2.1 Mathematical Description

As shown in Figure 3.16, L is a straight line, l is the distance from the origin $(0, 0)$ of the X-Y coordinate system to the line L, and θ is the angle between the positive direction of the X-axis and the line which is perpendicular to the line L and passes through the origin $(0, 0)$. Let P be a point on the line L. The coordinate of P can be either in the polar (r, θ) or in the rectilinear (x, y). It can be shown that

$$x \cos\theta + y \sin\theta - l = 0.$$

Thus, the line L can be uniquely specified by (l, θ) and explicitly expressed by the above equation.

Let $X(x, y)$ be a 2-D spatial function and let $p(l, \theta)$ be the integration of $X(x, y)$ along the line $L(l, \theta)$ defined by

$$p(l, \theta) = \int_{L(l,\theta)} X(x, y) ds.$$

It is known that $X(x, y)$ and $p(l, \theta)$ are linked by the Inverse Radon transform (IRT) given by

$$X(x, y) = \frac{1}{2\pi^2} \int_0^{\pi} \int_{-\infty}^{\infty} \frac{\partial p(l, \theta)}{\partial l} \frac{1}{x \cos\theta + y \sin\theta - l} dl d\theta. \tag{3.140}$$

In the imaging, $X(x, y)$ is often referred to as the *object function* and $p(l, \theta)$ is commonly called the *projection*. For example, in X-ray computed tomography (CT), $X(x, y)$ represents the relative linear attenuation coefficient (RLAC) and $p(l, \theta)$ represents the projection: either parallel, or divergent, or the cone beam rays. When $p(l, \theta)$ is formulated, $X(x, y)$ will be determined. This approach leads to a typical image reconstruction of X-ray CT.

In MRI, the object function is the transverse precession macroscopic magnetization (TPMM) $M_{xy}(x, y)$. In a rotated coordinate system as specified in Section 3.9.3.2 (which is slightly different from the rotating reference frame defined by Eq. (3.3)), it is expressed as $M_{xy}(x', y')$, where $x' = x \cos\theta + y \sin\theta$ and $y' = -x \sin\theta + y \cos\theta$. The projection—often called as the *measured projection*—is defined by Eq. (3.121). It is the integration of $M_{xy}(x', y')$ over y' along the line $x' = l$:

$$p(x', \theta) = \int_{x'=l} M_{xy}(x', y') dy'.$$

Thus, $M_{xy}(x', y')$ and $p(x', \theta)$ are linked by the Inverse Radon transform. That is, $M_{xy}(x', y')$ can be reconstructed by using Eq. (3.140), where it is denoted by $X(x, y)$.

Eq. (3.121) gives a theoretical definition of the measured projection $p(x', \theta)$. When the $\mathcal{F}_2\{M_{xy}(x', y')\}$ in Eq. (3.122) is expressed by the radial k-space sample data $\mathcal{M}(k, \theta)$, an operational definition of the measured projection is

$$p(x', \theta) = \mathcal{F}_1^{-1}\{\mathcal{M}(k, \theta)\}, \tag{3.141}$$

which directly links the measured projection $p(x', \theta)$ with k-space sample $\mathcal{M}(k, \theta)$. In Eq. (3.141), the subscript 1 in $\mathcal{F}_1^{-1}$ denotes a 1-D Fourier transform with respect to the first variable k in $\mathcal{M}(k, \theta)$. In this book, $\mathcal{F}_k^{-1}$ is also used as the same notation.

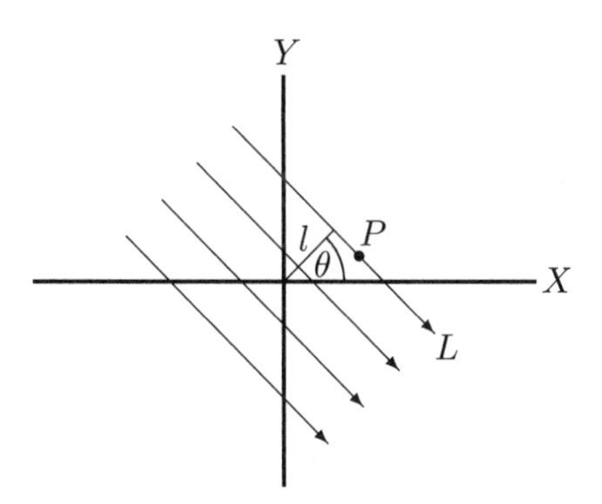

FIGURE 3.16
The geometry of Radon space.

By using $x' = x\cos\theta + y\sin\theta$ and $\frac{\partial p(l,\theta)}{\partial l} = p'(l,\theta)$, the inner integral in Eq. (3.140), with the coefficient $\frac{1}{2\pi^2}$, can be written as

$$\frac{1}{2\pi^2}\int_{-\infty}^{\infty}\frac{\partial p(l,\theta)}{\partial l}\frac{1}{x'-l}dl = \frac{1}{2\pi^2}\int_{-\infty}^{\infty}\frac{p'(l,\theta)}{x'-l}dl \triangleq t(x',\theta). \tag{3.142}$$

Based on the methods used to compute the projection $t(x',\theta)$, IRT of Eq. (3.140) can be implemented in two ways [10–12, 14]: (1) filtering by Fourier transform, where $t(x',\theta)$ is called the filtered projection, and (2) filtering by the convolution, where $t(x',\theta)$ is called the convolved projection.

(1) *Filtering by Fourier transform.* Eq. (3.142) can be rewritten as

$$t(x',\theta) = \frac{p'(x',\theta)}{i2\pi} \star \frac{-1}{i\pi x'},$$

where $\star$ denotes the convolution. Thus, the IRT of Eq. (3.140) consists of two steps.

• Step 1: *Filtering at each view.* Based on Eq. (3.141) and using the Fourier transform of the derivative theorem [16, 17], we have

$$\mathcal{F}\left\{\frac{p'(x',\theta)}{i2\pi}\right\} = k\mathcal{M}(k,\theta). \tag{3.143}$$

By using the Fourier transform of the Inverse function [16, 17], we have

$$\mathcal{F}\left\{\frac{-1}{i\pi x'}\right\} = sign(k) = \begin{cases} -1 \ (k<0) \\ \\ 1 \ (k>0) \end{cases}. \tag{3.144}$$

Thus, the filtered projection $t(x',\theta)$ can be computed by

$$t(x',\theta) = \mathcal{F}_1^{-1}\{k\mathcal{M}(k,\theta)\cdot sign(k)\} = \mathcal{F}_1^{-1}\{\mathcal{M}(k,\theta)|k|\}. \tag{3.145}$$

In order to limit the unboundness of $|k|$ filter at high frequency, $|k|$ is multiplied by the bandlimited functions such as $rect(\frac{k}{W_k})$ or $rect$-modulated cos, $sinc$, and Hamming window, etc.; here W_k denotes the bandwidth. Therefore, the filtered projection $t(x',\theta)$ is computed by

$$t(x',\theta) = \mathcal{F}_1^{-1}\{\mathcal{M}(k,\theta)|k|\ rect(\frac{k}{W_k})\}. \tag{3.146}$$

• Step 2: *Backprojection over all views.* By substituting Eq. (3.146) into Eq. (3.140), we obtain

$$X(x,y) = \int_0^{\pi} t(x',\theta)d\theta = \int_0^{\pi} \mathcal{F}_1^{-1}\{\mathcal{M}(k,\theta)|k|\ rect(\frac{k}{W_k})\}d\theta. \tag{3.147}$$

With filtering by the transform method, FBP consists of a 1-D Fourier transform of the weighted k-space samples at each view and a backprojection over all views.

(2) *Filtering by the convolution.* Eq. (3.142) can be rewritten as

$$-2\pi t(x',\theta) = -\frac{1}{\pi}\int_{-\infty}^{\infty} \frac{p'(x',\theta)}{x'-l}dl = p'(x',\theta) \star \frac{-1}{\pi x'} = \mathcal{H}_1\{p'(x',\theta)\},$$

where $\star$ denotes the convolution, $\mathcal{H}_1$ represents the Hilbert transform with respect to the first variable x' of $p'(x',\theta)$. Let $\rho(x') = \frac{-1}{\pi x'}$ and $\{\rho_A(x')\}$ be a set of functions of $\rho(x')$ parametrized at $A > 0$. As shown in X-ray CT image reconstruction [14], when $p(x',\theta)$ is reasonable and $\{\rho_A(x')\}$ is from a regulating family, that is, $\lim_{A\to\infty} p'(x',\theta) \star \rho_A(x') = \mathcal{H}_1\{p'(x',\theta)\}$, then we have $\mathcal{H}_1\{p'(x',\theta)\} = \lim_{A\to\infty} p(x',\theta) \star \rho'_A(x')$; here, $\rho'_A(x') = \frac{\partial \rho_A(x')}{\partial x'}$. By defining a convolution function $q(x') = -\frac{1}{2\pi}\rho'_A(x')$, we obtain

$$t(x',\theta) = \lim_{A\to\infty} p(x',\theta) \star q(x').$$

The more detailed discussion on the filter function $q(x')$ can be found in [14, 18, 19]. Thus, the IRT of Eq. (3.140) consists of three steps.

• Step 1: *Filtering at each view.* For a sufficiently large A, the convolved projection $t(x',\theta)$ can be computed by

$$t(x',\theta) = p(x',\theta) \star q(x'). \tag{3.148}$$

• Step 2: *Interpolation at each view.* For any specified n ($-N/2 \le n \le (N-1)/2$), since $x' \neq n\Delta p$ with the probability one, 1-D interpolation must be applied to approximate x' with some $n\Delta p$. Among a variety of signal interpolation schemes, the most effective one may be the nearest neighbor interpolation given by

$$t(x',\theta) = t(k\Delta p,\theta) \quad \text{if} \quad k = Arg\ \min_n |x' - n\Delta p|. \tag{3.149}$$

It has been shown [14, 18, 19] that the nearest neighbor interpolation Eq. (3.149) is equivalent to a convolution of the convolved projections $t(x',\theta)$ in a view with an interpolation function $\psi(x')$ defined by

$$\psi(x') = \begin{cases} 1 & (|x'| < \frac{\Delta p}{2}) \\ 0.5 & (|x'| = \frac{\Delta p}{2}) \\ 0 & (|x'| > \frac{\Delta p}{2}). \end{cases} \tag{3.150}$$

Thus, the *interpolated data* $s_\theta(x,y)$ at the view θ is

$$s_\theta(x,y) = t(x',\theta) \star \psi(x'), \tag{3.151}$$

• Step 3: *Backprojection over all views.* By substituting Eqs. (3.148) and (3.151) into Eq. (3.140), we obtain

$$X(x,y) = \int_0^\pi s_\theta(x,y)d\theta = \int_0^\pi [(p(x',\theta) \star q(x')) \star \psi(x')]d\theta. \tag{3.152}$$

With filtering by the convolution method, FBP consists of a double convolution at each view and a backprojection over all views, in terms of the measured projections.

Two steps of filtering by the Fourier transform and three steps of filtering by convolution form a framework of the FBP algorithm for the 2DPR of the MR image. Eqs. (3.147) and (3.152) show that an appropriate set of k-space data can be acquired to produce an image $X(x,y)$, which (via Eq. (3.121)) is an estimate of TPMM $M_{xy}(x',y',0)$. Due to the intrinsic link between TPMM and TEMM described by Eq. (3.88), the MR image generated by FBP (a 2DPR method) essentially represents the spatial distribution of TEMM.

3.10.2.2 Computational Implementation

(1) 2DPR via filtering by Fourier transform. Eq. (3.147) can be numerically implemented by an inverse 1-D DFT over k in each view and a backprojection over all views. When $M \times N$ radial k-space samples $\mathcal{M}(n\Delta k, m\Delta\theta)$ are acquired, for the pixels whose centers are at the locations $(x = k\delta x, y = l\delta y)$, we have

$$X(k\delta x, l\delta y) = \Delta\theta\Delta k \sum_{m=0}^{M-1} \sum_{n=-N/2}^{(N-1)/2} \mathcal{M}(n\Delta k, m\Delta\theta)|n\Delta k| e^{i2\pi x' n\Delta k}, \tag{3.153}$$

where $x' = k\delta x \cos m\Delta\theta + l\delta y \sin m\Delta\theta$ is defined in Eq. (3.120).

Based on Eqs. (3.125) and (3.128) with the equal sign, we have $\Delta k = \frac{1}{N\Delta p}$. From Eq. (3.127), we have $\Delta\theta = \frac{\pi}{M}$. For pixels with the unity size (i.e., $\delta x = \delta y = 1$), using the abbreviations $\mathcal{M}(n,m)$ for $\mathcal{M}(n\Delta k, m\Delta\theta)$, and substituting the above Δk and $\Delta\theta$ into Eq. (3.153), we obtain

$$X(k,l) = \frac{c_1}{M \cdot N} \sum_{m=0}^{M-1} \sum_{n=-N/2}^{(N-1)/2} \mathcal{M}(n,m)|n| e^{i2\pi n x'(k,l,m)},$$

where N is the number of the radial samples at each view, M is the number of views (i.e., the number of the angle samples), $c_1 = \frac{\pi}{\Delta p}$, Δp is the spacing between two adjacent measured projections (Section 3.9.3.2), and $x'(k,l,m) = k\cos\frac{m\pi}{M} + l\sin\frac{m\pi}{M}$.

Discussion. Let FOV_r and FOV_x, FOV_y be the field of view of a circular image and its insider square $I \times I$ image. Under the conditions $\delta x = \delta y = 1$, $FOV_r = \sqrt{FOV_x^2 + FOV_y^2} = \sqrt{(I\delta x)^2 + (I\delta y)^2} = \sqrt{2}I$. Because $FOV_r = N\Delta p$ (Eq. (3.125) with the equal sign), we have $N\Delta p = \sqrt{2}I$, which implies that the spacing Δp is less than or, at least, equal to the diagonal size $\sqrt{2}$ of the unity pixel ($\delta x = \delta y = 1$) because $N \geq I$.

(2) 2DPR via filtering by the convolution. Eq. (3.152) can be numerically implemented by a convolution with the convolution function (in each view), a convolution with the interpolation function (in each view), and a backprojection over all views, in terms of the measured projections, as follows.

$$X(x,y) = \Delta p \Delta\theta \{ \sum_{m=0}^{M-1} [\sum_{n=-\infty}^{\infty} (\sum_{l=-N/2}^{(N-1)/2} p(l\Delta p, m\Delta\theta) q((n-l)\Delta p)) \psi(x' - n\Delta p)] \} , \tag{3.154}$$

where $x' = x\cos m\Delta\theta + y\sin m\Delta\theta$ and the measured projection $p(l\Delta p, m\Delta\theta)$ is given in Eq. (3.124).

3.11 Echo Signal

In the previous sections, analyses on magnetizations, MR signals, k-space samples, and image reconstructions were based on ideal conditions. In practice, however, non-ideal conditions such as the nonuniform magnetic field and the heterogeneous sample always exist. This section discusses these nonidealities. As we will see, these discussions will lead to some important topics in MRI, for example, T_2^* decay, various echoes, the true T_2 decay, etc.

3.11.1 T_2^* Decay

FID signal $s_r(t)$ is given by Eq. (3.96), where $\Delta\omega(\mathbf{r},\tau)$ represents the off-resonance frequency. Eq. (3.45) shows $\Delta\omega(\mathbf{r},t) = \gamma\Delta B(\mathbf{r},t)$. Three factors: (1) the inhomogeneities δB_o of the main magnetic field B_o, (2) the gradient field B_G, and (3) the chemical shift ω_{cs}/γ, can affect $\Delta B(\mathbf{r},t)$. Thus, $\Delta\omega(\mathbf{r},t)$ can be expressed as

$$\Delta\omega(\mathbf{r},t) = \gamma\delta B_o(\mathbf{r}) + \gamma B_G(\mathbf{r},t) + \omega_{cs}(\mathbf{r}). \tag{3.155}$$

Eq. (3.155) implicitly indicates that $\delta B_o(\mathbf{r})$ and $\omega_{cs}(\mathbf{r})$ are spatially variant and $B_G(\mathbf{r},t)$ is spatio-temporally variant.

When $\Delta\omega(\mathbf{r},t) = 0$, then for an (one) isochromat, Eq. (3.96) becomes

$$s_r(t) = \omega_o \int_V |B_{r_{xy}}(\mathbf{r})|\,|M_{xy}(\mathbf{r},0)|e^{-\frac{t}{T_2}}$$
$$\cos(\omega_o t + \phi_{B_{r_{xy}}}(\mathbf{r}) - \phi_{M_{xy}}(\mathbf{r},0) - \frac{\pi}{2})d\mathbf{r}. \tag{3.156}$$

It implies that under ideal conditions, the composite signal components of $s_r(t)$ (i.e., the items inside the integral) sinusoidally oscillate with resonance frequency ω_o and decay with intrinsic transverse relaxation time constant T_2.

When $\Delta\omega(\mathbf{r},t) \neq 0$, then for an (one) isochromat, Eq. (3.96) remains

$$s_r(t) = \omega_o \int_V |B_{r_{xy}}(\mathbf{r})|\,|M_{xy}(\mathbf{r},0)|e^{-\frac{t}{T_2}}$$
$$\cos(\omega_o t + \int_0^t \Delta\omega(\mathbf{r},\tau)d\tau + \phi_{B_{r_{xy}}}(\mathbf{r}) - \phi_{M_{xy}}(\mathbf{r},0) - \frac{\pi}{2})d\mathbf{r}\ . \tag{3.157}$$

It implies that under the practical conditions, due to the non-zero off-resonance frequency $\Delta\omega(\mathbf{r},t)$, the composite signal components of $s_r(t)$ sinusoidally oscillate with various frequencies. Therefore, they lose the phase coherence and undergo a destructive dephasing process. As a result of the integration of these dephasing components over the volume V, the FID signal $s_r(t)$ of Eq. (3.157) decays with effective reduced time constants, which is much faster than its counterpart of Eq. (3.156).

The first effective reduced time constant results from the inhomogeneities δB_o of the main magnetic field B_o, denoted by T_2^*, and is given by

$$\frac{1}{T_2^*} = \frac{1}{T_2} + \gamma\delta B_o. \tag{3.158}$$

T_2^* is then further rapidly reduced by the gradient field B_G and becomes T_2^{**}, which is given by

$$\frac{1}{T_2^{**}} = \frac{1}{T_2^*} + \gamma G R, \tag{3.159}$$

where G is the gradient and R is the radius of the sample. It is clear that

$$T_2^{**} < T_2^* < T_2. \tag{3.160}$$

The impacts of T_2^* and T_2^{**} on the FID signal decays are shown in Figure 3.17.

Although each nuclear spin in the sample experiences a decay with its intrinsic transverse relaxation time constant T_2, the FID signal $s_r(t)$, as a function of TPMM $M_{xy}(\mathbf{r},0)$ which is a collective representation of spins in the unit volume $d\mathbf{r}$ over the sample volume, decays with the effective reduced time constants T_2^* or T_2^{**}.

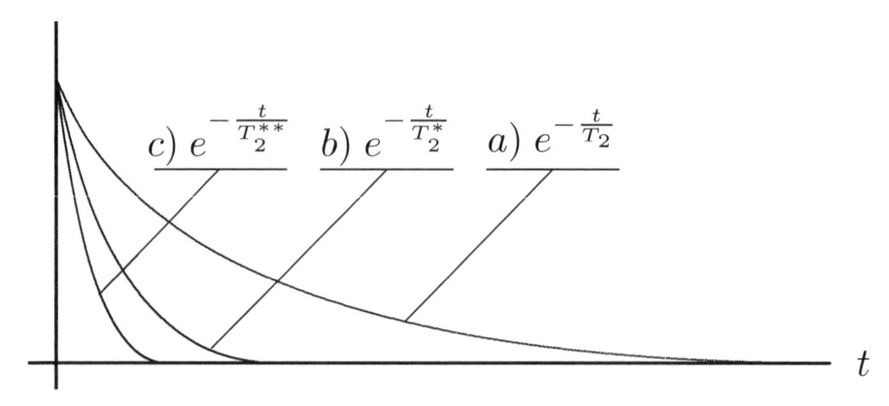

FIGURE 3.17
Three types of decays of FID signal: (a) with the intrinsic T_2, (b) with T_2^* caused by the inhomogeneities δB_o of the main magnetic field B_o, and (c) with T_2^{**} caused by the gradient field B_G.

3.11.2 Echoes

A single population of nuclear spins (i.e., an isochromat) has an intrinsic transverse relaxation time constant T_2. Different isochromats have different values of T_2. By measuring values of T_2 from the ideal FID signals, the isochromats that normally correspond to different tissues can be distinguished. In practice, however, as shown in the previous section, FID signals decay with effective reduced time constants T_2^* or T_2^{**} that are the field and/or the sample dependent. Thus, in order to distinguish different isochromats, the true T_2 decay is required.

From Eq. (3.155), the phase shift from $\omega_o t$ (i.e., the phase accumulation over the time t) of FID signal caused by three types of off-resonance frequencies is

$$\int_0^t \Delta\omega(\mathbf{r},\tau)d\tau = \gamma\delta B_o(\mathbf{r})t + \gamma\int_0^t B_G(\mathbf{r},\tau)d\tau + \omega_{cs}(\mathbf{r})t \triangleq \Delta\phi(\mathbf{r},t). \quad (3.161)$$

In Eq. (3.161), $\delta B_o(\mathbf{r})$ and $\omega_{cs}(\mathbf{r})$ are spatially variant, and $B_G(\mathbf{r},\tau)$ is spatiotemporally variant but controllable. The phase shift caused by spatially variant factors can be removed instantly in different ways, thereby recovering the phase coherence. The resultant signals are different from the FID signal and are called *echo signals.* Echo signals can be generated either by gradient field reversal or by multiple RF pulses. The former are called gradient echoes and the latter are known as RF echoes.

3.11.2.1 Gradient Echoes

By ignoring the inhomogeneities of the main magnetic field and the chemical shift, that is, in Eq. (3.161), letting $\delta B_o(\mathbf{r}) = 0$ and $\omega_{cs}(\mathbf{r}) = 0$, then $\Delta\phi(\mathbf{r},t) =$

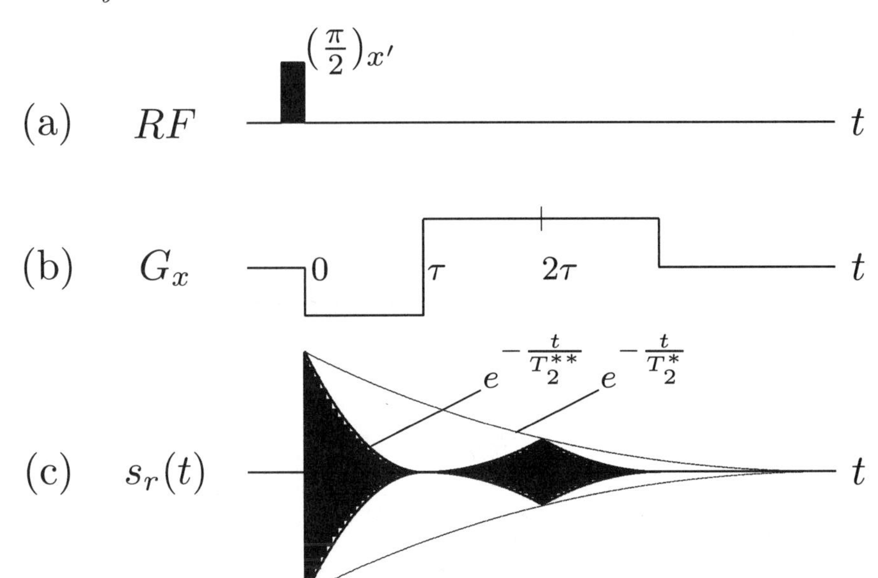

FIGURE 3.18
Gradient echo generation: (a) an excitation RF pulse $(\frac{\pi}{2})_{x'}$, (b) the gradient G_x, and (c) the decays of FID signal with T_2^* and T_2^{**}.

$\gamma \int_0^t B_G(\mathbf{r}, \tau)d\tau$. As shown in Figure 3.18, the readout gradient is $\vec{G}(\tau) = G_x\vec{i}$.

Because of the spatio-temporally variant gradient field, all spins in the sample lose phase coherence after the excitation RF pulse $(\frac{\pi}{2})_{x'}$ is turned off. As shown in Figure 3.18, from the time $t = 0$, spins begin to dephase. During the time period $0 < t < \tau$, the FID signal $s_r(t)$ decays with an effective reduced time constant T_2^{**} and the phase shift is given by

$$\Delta\phi(\mathbf{r}, t) = -\gamma G_x x t. \tag{3.162}$$

Also as shown in Figure 3.18, at time $t = \tau$, the readout gradient $\vec{G}_x$ reverses its polarity and spins begin to rephase. During the time period $\tau < t < 2\tau$, spins gradually regain phase coherence and the phase shift is given by

$$\Delta\phi(\mathbf{r}, t) = -\gamma G_x x\tau + \gamma G_x x(t - \tau). \tag{3.163}$$

Spin rephasing results in a regrowth of the signal. At time $t = 2\tau$, Eq. (3.163) becomes

$$\Delta\phi(\mathbf{r}, t) = 0, \tag{3.164}$$

which implies that the spins are rephased completely. Thus, a signal occurs at time $t = 2\tau$, and is called the *gradient echo*.

The phase shift $\Delta\phi(\mathbf{r}, t)$ as a function of the time t for three locations ($t = -x,\ 0,\ x$) is plotted in Figure 3.19, which gives the phase progressions of spins.

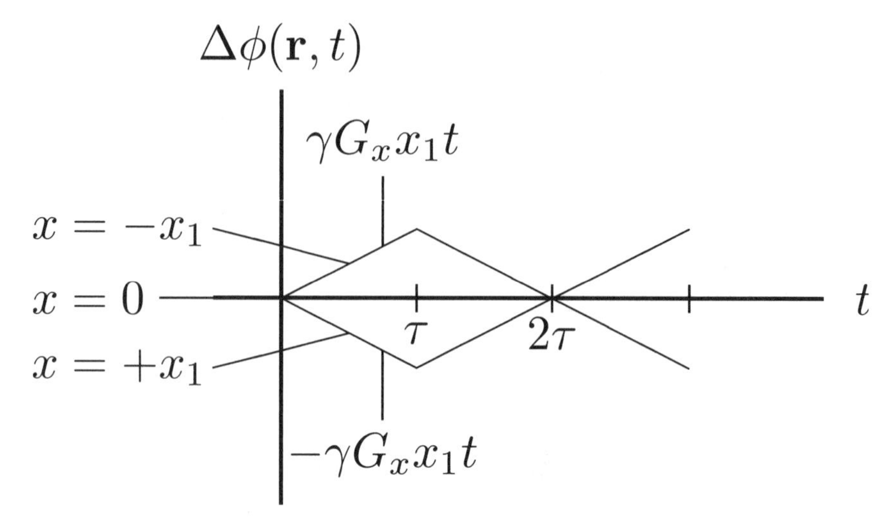

FIGURE 3.19
Phase progressions of spins at three locations, $x = x_1,\ 0,\ -x_1$, in the gradient echo.

For time $t > 2\tau$, as the gradient remains on, spins begin to dephase again. However, by repetitively switching gradient polarity as shown in Figure 3.20, multiple gradient echoes are generated at time $t = n \cdot 2\tau$ $(n > 1)$ within the limit of T_2^* decay.

Changing the polarity of the readout gradient inverts spins from the dephasing process to the rephasing process, thereby, producing a gradient echo signal at certain time instants. This signal, in fact, is a two-side, nearly symmetric FID signal with T_2^{**} decay and its peak amplitudes have T_2^* decay.

3.11.2.2 Spin Echoes

By ignoring the gradient field and the chemical shift, that is, in Eq. (3.161), letting $B_G(\mathbf{r}, \tau) = 0$ and $\omega_{cs}(\mathbf{r}) = 0$, then $\Delta\phi(\mathbf{r}, t) = \gamma \delta B_o(\mathbf{r}) t$.

Figure 3.21 shows a two-pulse sequence—an excitation pulse $(\frac{\pi}{2})_{x'}$ followed by a time delay τ and then a refocusing RF pulse $(\pi)_{x'}$—denoted by

$$(\frac{\pi}{2})_{x'} - \tau - (\pi)_{x'} - 2\tau.$$

The RF pulse $(\frac{\pi}{2})_{x'}$ flips TEMM onto the y'-axis, which becomes TPMM as shown in Figure 3.22a.

Because of the spatially variant inhomogeneities δB_o of the main magnetic field B_o, all spins in the sample lose phase coherence after the excitation RF pulse $(\frac{\pi}{2})_{x'}$ is turned off. As shown in Figure 3.22b, from the time $t = 0$ (here, the pulse width $\tau_{\pi/2}$ of $(\frac{\pi}{2})_{x'}$ is ignored because $\tau_{\pi/2} \ll \tau$. A more detailed discussion is given in Appendix 3D.), TPMM begins to fan out and dephase.

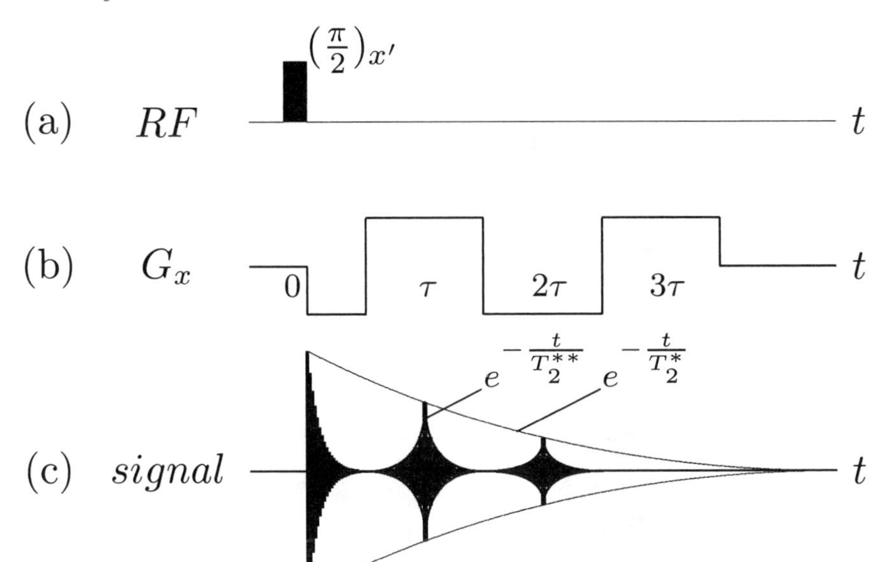

FIGURE 3.20
The gradient echo train generated by switching the polarity of the gradient $\vec{G}_x$.

During the time period $0 < t < \tau$, the FID signal $s_r(t)$ decays with an effective reduced time constant T_2^* and the phase shift is given by

$$\Delta\phi(\mathbf{r},t) = \gamma\delta B_o(\mathbf{r})t. \tag{3.165}$$

Immediately before time τ, the pre-pulse phase shift is given by $\Delta\phi(\mathbf{r},\tau^-) = \gamma\delta B_o(\mathbf{r})\tau$.¶

At the time $t = \tau$, the refocusing RF pulse $(\pi)_{x'}$ reverses all spins by the phase π, i.e., rotates TPMM to the other side of the transverse plane, as shown in Figure 3.22.c. Immediately after $(\pi)_{x'}$ pulse (Here, the pulse width τ_π of $(\pi)_{x'}$ is ignored because $\tau_\pi \ll \tau$. A more detailed discussion is given in Appendix 3D.), the post-pulse phase shift is given by $\Delta\phi(\mathbf{r},\tau^+) = \pi - \Delta\phi(\mathbf{r},\tau^-) = \pi - \gamma\delta B_o(\mathbf{r})\tau$.

From time $t = \tau$, TPMM continues the precession as before the refocusing RF pulse $(\pi)_{x'}$; therefore, TPMM begins to rephase as shown in Figure 3.22.c. During the time period $\tau < t < 2\tau$, spins gradually regain phase coherence and the phase shift is given by

$$\Delta\phi(\mathbf{r},t) = \Delta\phi(\mathbf{r},\tau^+) + \gamma\delta B_o(\mathbf{r})(t-\tau). \tag{3.166}$$

¶When τ is properly selected, TPMM completely vanishes and $s_r(t)$ disappears before time $t = \tau$.

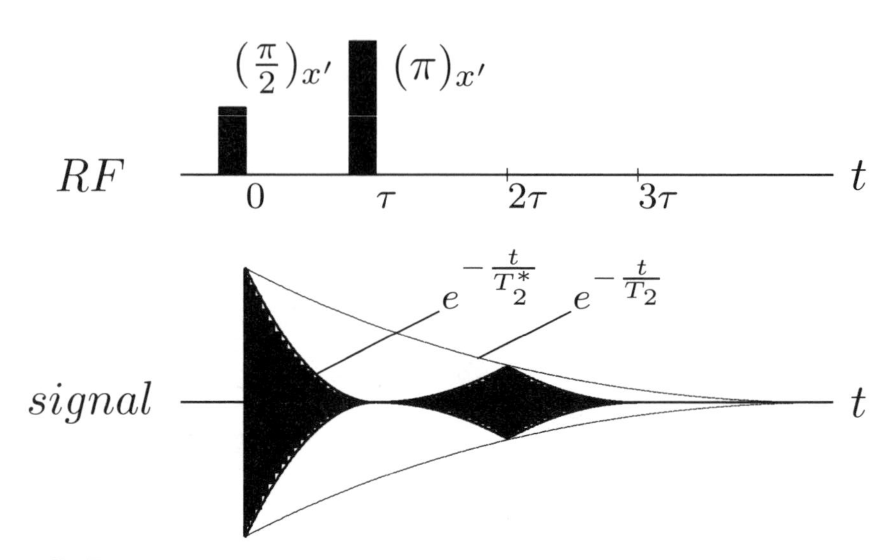

FIGURE 3.21
Spin echo generation by RF pulses $(\frac{\pi}{2})_{x'} - \tau - (\pi)_{x'} - 2\tau$.

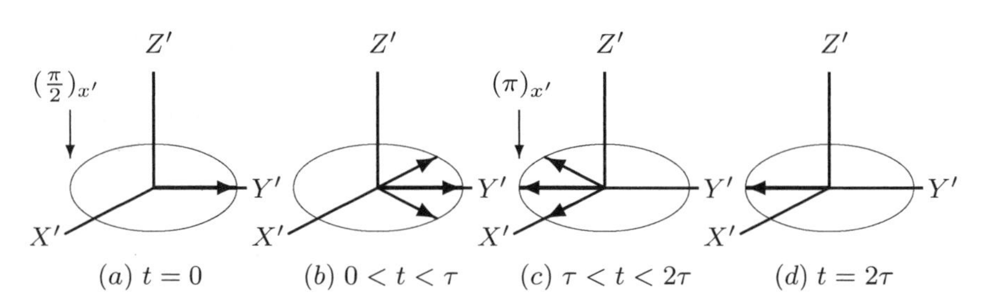

FIGURE 3.22
Phase diagram of TPMM: (a) at $t = 0$, TEMM is flipped by an excitation RF pulse $(\frac{\pi}{2})_{x'}$ onto the y'-direction and becomes a TPMM; (b) for $0 < t < \tau$, TPMM begins to fan out and dephase; (c) for $\tau < t < 2\tau$, TPMM is reversed by a refocusing RF pulse $(\pi)_{x'}$ and continues precession; and (d) TPMM is rephased in the $-y'$-direction and an echo is formed.

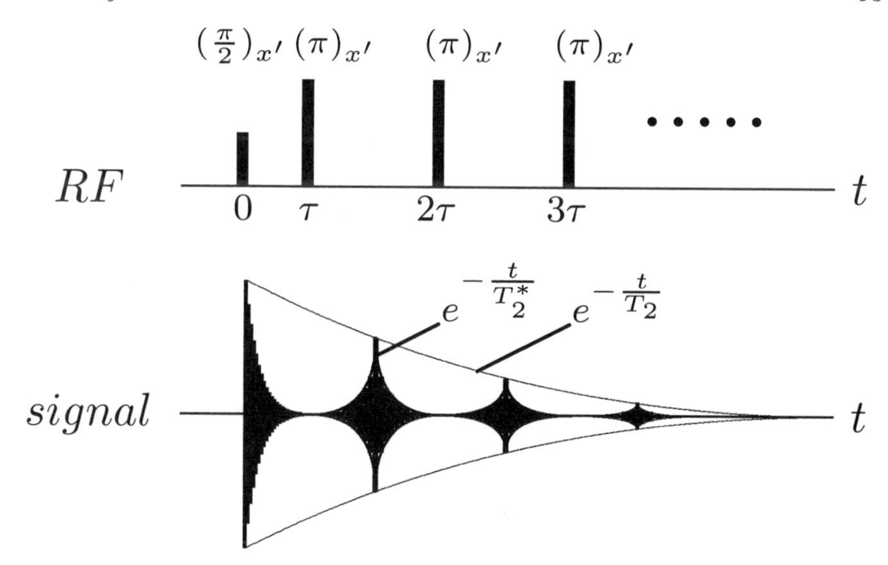

FIGURE 3.23
The spin echo train generated by the RF pulse sequence $(\frac{\pi}{2})_{x'} - \tau - (\pi)_{x'}^{(1)} - 2\tau - \cdots - (\pi)_{x'}^{(N)} - 2\tau$.

Spin rephasing results in a regrowth of the signal. At time $t = 2\tau$, the phase shift becomes

$$\Delta\phi(\mathbf{r}, 2\tau^-) = \Delta\phi(\mathbf{r}, \tau^+) + \gamma\delta B_o(\mathbf{r})(2\tau - \tau) = \pi, \tag{3.167}$$

as shown in Figure 3.22d, which implies that all spins are rephased completely. Thus a signal occurs at time $t = 2\tau$ and is known as *spin echo*. 2τ is commonly called the *echo time* and denoted by T_E.

For the time $t > 2\tau$, as the relaxation continues, spins begin to dephase again. However, by repetitively applying the refocusing RF pulse $(\pi)_{x'}$ at time $t = (2n-1)\tau$ $(n > 1)$ as shown in Figure 3.23, multiple spin echoes are generated at time $t = n \cdot 2\tau$ $(n > 1)$ within the limit of T_2 decay.

Applying the refocusing RF pulse $(\pi)_{x'}$ inverts spins from the dephasing process to the rephasing process, thereby, producing a spin echo signal at certain time instants. This signal, in fact, is a two-side, nearly symmetric FID signal with T_2^* decay and its peak amplitudes have T_2 decay.

3.11.3 T$_2$ Decay

Section 3.11.2.2 provides a graphic analysis of the generation of spin echo signals. This section presents a rigorous analysis of the evolution of the spin echoes generated by the RF pulse sequence

$$(\frac{\pi}{2})_{x'} - \tau - (\pi)_{x'}^{(1)} - 2\tau \ \cdots \ (\pi)_{x'}^{(N)} - 2\tau.$$

It shows that at time $t = n \cdot T_E$ ($n = 1, \cdots, N$ and $T_E = 2\tau$—the echo time), TPMMs $M_{x'y'}(\mathbf{r}, t)$ at the different locations $\mathbf{r}$ in the sample regain phase coherence; therefore, at these time instants, FID signals are not degraded by the inhomogeneities of the main magnetic field. Because the gradient field and the chemical shift are ignored in spin echoes, the peak amplitudes of spin echoes that occur at time $t = n \cdot T_E$ have the true T_2 decay.

The analysis adopts the following approach:

(a) The entire process of the spin echo generation is decomposed into several individual processes.

(b) Each of these processes represents either a forced precession or a free precession.

(c) The solutions for PMM in the forced and free precession are given by Eqs. (3.66) and (3.38), respectively.‖

(d) The values of PMM at the end of the preceding process serve as the initial values of PMM in the immediately following process.

By using this approach, the analytical expressions of PMM in the spin echo generation can be obtained.

The beginning portion of this RF pulse sequence is $(\frac{\pi}{2})_{x'} - \tau - (\pi)^{(1)}_{x'} - 2\tau$ and the remaining portion is the repetitive $(\pi)^{(n)}_{x'} - 2\tau$ ($n = 2, \cdots, N$). Appendix 3D shows that $(\frac{\pi}{2})_{x'} - \tau - (\pi)^{(1)}_{x'} - 2\tau$ pulse sequence generates two forced precessions and two free precessions and $(\pi)^{(2)}_{x'} - 2\tau$ pulse generates one forced precession and one free precession. These six precession processes are in the order of (1) an excitation, (2) a relaxation (rephasing), (3) a refocusing, (4) a relaxation (dephasing and rephasing), (5) a refocusing, and (6) a relaxation (dephasing and rephasing).

By analyzing each of these processes, Appendix 3D proves that when $\tau_{\pi/2} << \tau$ and $\tau_\pi << \tau$, that is, the widths of RF pulses are ignored, then

(1) TPMMs $M_{x'y'}(\mathbf{r}, t)$ at the different locations $\mathbf{r}$ in the sample undergo the dephasing and rephasing in the time periods $((n-1)T_E, (n-\frac{1}{2})T_E)$ and $((n-\frac{1}{2})T_E, (n)T_E)$, respectively.

(2) At the time $t = n \cdot T_E$ ($n = 1, \cdots, N$), as shown by Eq. (3.161), $\Delta\phi(\mathbf{r}, t) = 0$, the rephasing is completed, TPMMs $M_{x'y'}(\mathbf{r}, t)$ regain the phase coherence, and $|M_{x'y'}(\mathbf{r}, t)| \propto e^{-t/T_2}$.

Thus, Eq. (3.96) (i.e., Eq. (3.157) becomes Eq. (3.156), and the peak amplitudes $s_r(nT_E)$ of FID signal have the true T_2 decay.

When the sample consists of only one isochromat, $T_2(\mathbf{r}) = T_2$. At time $t = n \cdot T_E$, Eq. (3.156) can be rewritten as

$$s_r(nT_E) = \omega_o(\int_V |B_{r_{xy}}(\mathbf{r})|\, |M_{xy}(\mathbf{r}, 0)|$$

$$\cos(\omega_o nT_E + \phi_{B_{r_{xy}}}(\mathbf{r}) - \phi_{M_{xy}}(\mathbf{r}, 0) - \frac{\pi}{2})d\mathbf{r})e^{-\frac{nT_E}{T_2}}, \quad (3.168)$$

‖See Section 3.5.2; PMM consists of TPMM and LPMM.

which implies that the compound signal components of the FID signal sinusoidally oscillate with the resonance frequency ω_o and the FID signal decays with an intrinsic transverse relaxation time constant T_2.

Although the above analysis is developed for the $(\frac{\pi}{2})_{x'} - \tau - (\pi)_{x'}^{(1)} - 2\tau \cdots (\pi)_{x'}^{(N)} - 2\tau$ RF pulse sequence, the idea and method used in this approach can be applied to arbitrary $\alpha_1 - \tau - \alpha_2$ ($\alpha_1 \neq \frac{\pi}{2}$, $\alpha_2 \neq \pi$, and not restricted at x'-axis) pulse sequences.

3.12 Appendices

3.12.1 Appendix 3A

This appendix proves that for a spin-I system, its TEMM is given by Eq. (3.17). Following Abragam's guidance [9], $M_z^o(\mathbf{r})$ can be derived from the populations of energy levels of nuclear spins.

The Boltzmann distribution for energies is given by

$$\frac{n_i}{n} = \frac{\exp(-e_i/\kappa T)}{\sum_i \exp(-e_i/\kappa T)}, \tag{3.169}$$

where n_i is the number of spins at equilibrium temperature T in a state i that has energy e_i, n is the total number of spins in the system, and κ is the Boltzmann constant (1.380×10^{-23} J $\cdot$ K^{-1}).

Let a spin system of population n be placed in a static magnetic field $\vec{B}_o = B_o\vec{k}$. Because different orientations of the nuclear spins with respect to the field, described by different quantum number I_z of the spin quantized along the field, correspond to different magnetic energies, a net macroscopic magnetization appears [7–10].

According to the Boltzmann distribution, for this spin system, n_i is given by

$$n_{I_z} = n\frac{\exp(-e_{I_z}/\kappa T)}{\sum_{I_z=-I}^{I} \exp(-e_{I_z}/\kappa T)}, \tag{3.170}$$

where $e_{I_z} = \gamma\hbar I_z B_o$ and the net macroscopic magnetization $M_z^o(\mathbf{r})$ is

$$M_z^o(\mathbf{r}) = \sum_{I_z=-I}^{I} n_{I_z}(\gamma\hbar I_z) = \gamma\hbar n\frac{\sum_{I_z=-I}^{I} I_z \exp(\gamma\hbar B_o I_z/\kappa T)}{\sum_{I_z=-I}^{I} \exp(\gamma\hbar B_o I_z/\kappa T)}. \tag{3.171}$$

Because $\gamma\hbar B_o/\kappa T$ is a very small number, the exponential term in the Boltzmann distribution can be approximated by its linear expansion. Thus, the denominator in Eq. (3.171) is

$$\sum_{I_z=-I}^{I} \exp(\gamma\hbar B_o I_z/\kappa T) \simeq \sum_{I_z=-I}^{I} (1 + \gamma\hbar B_o I_z/\kappa T) = 2I + 1, \tag{3.172}$$

and the numerator in Eq. (3.171) is

$$\sum_{I_z=-I}^{I} I_z \exp(\gamma\hbar B_o I_z/\kappa T) \simeq \sum_{I_z=-I}^{I} I_z(1+\gamma\hbar B_o I_z/\kappa T) = \frac{1}{3}I(I+1)(2I+1)\frac{\gamma\hbar B_o}{\kappa T}. \tag{3.173}$$

By substituting Eqs. (3.172) and (3.173) into (3.171), Eq. (3.17) is proved.

3.12.2 Appendix 3B

This appendix shows that a vector cross product can be converted to a pure matrix multiplication. Let

$$\vec{a} = \begin{pmatrix} a_1 \\ a_2 \\ a_3 \end{pmatrix} \quad \text{and} \quad \vec{b} = \begin{pmatrix} b_1 \\ b_2 \\ b_3 \end{pmatrix} \tag{3.174}$$

be two column vectors, the cross product $\vec{a} \times \vec{b}$ and $\vec{b} \times \vec{a}$ can be expressed as the multiplication of a skew-symmetric matrix $\mathbf{A}$ and the vector $\vec{b}$ as follows.

$$\vec{a} \times \vec{b} = \mathbf{A}\vec{b} = \begin{pmatrix} 0 & -a_3 & a_2 \\ a_3 & 0 & -a_1 \\ -a_2 & a_1 & 0 \end{pmatrix} \begin{pmatrix} b_1 \\ b_2 \\ b_3 \end{pmatrix} \tag{3.175}$$

or

$$\vec{b} \times \vec{a} = \mathbf{A}^T\vec{b} = \begin{pmatrix} 0 & a_3 & -a_2 \\ -a_3 & 0 & a_1 \\ a_2 & -a_1 & 0 \end{pmatrix} \begin{pmatrix} b_1 \\ b_2 \\ b_3 \end{pmatrix} \tag{3.176}$$

where $\mathbf{A}^T$ is the transpose matrix of $\mathbf{A}$.

Verification. The definition of the vector cross product is given by

$$\vec{a} \times \vec{b} = \begin{pmatrix} \vec{i} & \vec{j} & \vec{k} \\ a_1 & a_2 & a_3 \\ b_1 & b_2 & b_3 \end{pmatrix} = (a_2b_3 - a_3b_2)\vec{i} + (a_3b_1 - a_1b_3)\vec{j} + (a_1b_2 - a_2b_1)\vec{k}. \tag{3.177}$$

Eq. (3.175) gives

$$\vec{a} \times \vec{b} = \mathbf{A}\vec{b} = \begin{pmatrix} a_2b_3 - a_3b_2 \\ a_3b_1 - a_1b_3 \\ a_1b_2 - a_2b_1 \end{pmatrix}. \tag{3.178}$$

Eq. (3.177) and Eq. (3.178) are identical. Similarly, we can also verify Eq. (3.176) which is used in this book for the Bloch equation, where $\vec{b} = \vec{M}$ and $\vec{a} = \gamma\vec{B}$.

3.12.3 Appendix 3C

First, we prove that a Gaussian RF pulse produces a Gaussian slice TPMM profile; the slice thickness is $\Delta z = 4\sqrt{2\pi}/\gamma G_z$ and the ratio $r_{|M'_{xy}|} > 0.95$.

• Let $p(x)$ be the pdf of $N(0, \frac{1}{2\pi})$, that is, a Gaussian distribution with zero mean and variance $\sigma^2 = \frac{1}{2\pi}$; then

$$p(x) = \frac{1}{\sqrt{2\pi}\sigma} e^{-\frac{x^2}{2\sigma^2}} = \frac{1}{\sqrt{2\pi}(1/\sqrt{2\pi})} e^{-\frac{x^2}{2(1/2\pi)}} = e^{-\pi x^2}. \tag{3.179}$$

Thus, both $e^{-\pi\tau^2}$ (Eq. (3.80)) and $e^{-\pi f^2}$ (Eq. (3.81)) are pdf of a Gaussian distribution with $N(0, \frac{1}{2\pi})$. This implies that a Gaussian RF excitation pulse (Eq. (3.80)) produces a Gaussian slice TPMM profile (Eq. (3.82)).

• When $-2\sigma < f < 2\sigma$ is chosen to define a slice, then based on Eqs. (1.76) and (3.77), the slice thickness can be determined by z, which satisfies $\frac{1}{2\pi}\gamma G_z(z - z_o) = \pm 2\sigma$. Thus $\Delta z = 4\sigma/\frac{1}{2\pi}\gamma G_z$. Due to $\sigma = \frac{1}{\sqrt{2\pi}}$, $\Delta z = 4\sqrt{2\pi}/\gamma G_z$.

• It is known that for Gaussian pdf $p(x)$ of Eq. (3.179)

$$\int_{-2\sigma}^{2\sigma} p(x)dx = \int_{-2\sigma}^{2\sigma} e^{-\pi x^2} dx > 0.95. \tag{3.180}$$

Thus, using Eq. (3.82), the ratio $r_{|M'_{xy}|}$ defined by Eq. (3.79) for this case is

$$r_{|M'_{xy}|} = \frac{\int_{z_o-2\sigma/\frac{1}{2\pi}\gamma G_z}^{z_o+2\sigma/\frac{1}{2\pi}\gamma G_z} |M'_{xy}(\mathbf{r}, \tau_p)|dz}{\int_{-\infty}^{\infty} |M'_{xy}(\mathbf{r}, \tau_p)|dz} = \frac{\int_{-2\sigma}^{2\sigma} e^{-\pi f^2} df}{\int_{-\infty}^{\infty} e^{-\pi f^2} df} > 0.95. \tag{3.181}$$

It implies that more than 95% of TPMM is located within the slice.

Second, we prove that for the slice TPMM profile with the rectangular shape as shown in Eg.(3.86), which is produced by an RF excitation pulse with sinc shape as shown in Eq. (3.84), the slice thickness is $\Delta z = 1/\frac{1}{2\pi}\gamma G_z \tau_p$ and the ratio $r_{|M'_{xy}|} = 1.00$.

• When $-\frac{\Delta f}{2} < f < \frac{\Delta f}{2}$ ($\Delta f = \frac{1}{\tau_p}$) is chosen to define a slice, then based on Eqs. (1.76) and (3.77), the slice thickness can be determined by z, which satisfies $\frac{1}{2\pi}\gamma G_z(z - z_o) = \pm\frac{\Delta f}{2}$. Thus $\Delta z = \Delta f/\frac{1}{2\pi}\gamma G_z = 1/\frac{1}{2\pi}\gamma G_z \tau_p$.

• Using Eq. (3.86), the ratio $r_{|M'_{xy}|}$ defined by Eq. (3.79) for this case is

$$r_{|M'_{xy}|} = \frac{\int_{z_o-0.5/\frac{1}{2\pi}\gamma G_z\tau_p}^{z_o+0.5/\frac{1}{2\pi}\gamma G_z\tau_p} |M'_{xy}(\mathbf{r}, \tau_p)|dz}{\int_{-\infty}^{\infty} |M'_{xy}(\mathbf{r}, \tau_p)|dz} = \frac{\int_{-\Delta f/2}^{\Delta f/2} rect(\frac{f}{\Delta f})df}{\int_{-\infty}^{\infty} rect(\frac{f}{\Delta f})df} = 1.00. \tag{3.182}$$

It implies that 100% of TPMM is located within the slice.

Third, we prove that for the slice TPMM profile with sinc shape which is produced by an RF excitation pulse with rectangular shape, the slice thickness is $\Delta z = 2/\frac{1}{2\pi}\gamma G_z \tau_p$ and the ratio $r_{|M'_{xy}|} > 0.90$.

Let the RF excitation pulse $B_1(\tau)$ take a rectangular shape

$$B_1(\tau + \frac{\tau_p}{2}) = rect(\frac{\tau}{\tau_p}) = \begin{cases} \frac{1}{\tau_p} & (|\tau| \leq \frac{\tau_p}{2}) \\ 0 & (|\tau| > \frac{\tau_p}{2}); \end{cases} \tag{3.183}$$

then

$$\mathcal{F}_1^{-1}\{B_1(\tau + \frac{\tau_p}{2})\} = \frac{\sin(\pi f \tau_p)}{\pi f \tau_p} = sinc(f\tau_p). \tag{3.184}$$

By substituting Eq. (3.184) into Eq. (3.78), the slice TPMM profile has a sinc shape

$$|M'_{xy}(\mathbf{r}, \tau_p)| = \gamma M_z^o(\mathbf{r})|\ sinc(f\tau_p)|\ . \tag{3.185}$$

• When a pair of points at the f-axis ($f > 0$ and $f < 0$) that are the first zero cross of $sinc(f\tau_p)$ is chosen to define a slice, then based on Eqs. (1.76) and (3.77), the slice thickness can be determined by z which satisfies $f\tau_p = \pm 1$, that is, $\frac{1}{2\pi}\gamma G_z(z - z_o) = \pm\frac{1}{\tau_p} = f$. Thus $z = z_o \pm 1/\frac{1}{2\pi}\gamma G_z \tau_p$; therefore, the slice thickness is $\Delta z = 2/\frac{1}{2\pi}\gamma G_z \tau_p$.

• Using the formula $\int_0^\infty sinc(x)dx = \frac{1}{2}$, we have

$$\int_{-\infty}^{\infty} sinc(f\tau_p)df = 2\int_0^{\infty} sinc(f\tau_p)df = \frac{1}{\tau_p}. \tag{3.186}$$

Using the formula $Si(y) = \int_0^y \frac{\sin x}{x}dx = y - \frac{1}{3}\cdot\frac{y^3}{3!} + \frac{1}{5}\cdot\frac{y^5}{5!} - \cdots$, we have $\int_0^1 sinc(x)dx = \frac{1}{\pi}Si(\pi) \simeq 1 - \frac{\pi^2}{18}$.** Thus

$$\int_{-1/\tau_p}^{1/\tau_p} sinc(f\tau_p)df = 2\int_0^{1/\tau_p} sinc(f\tau_p)df = \frac{2}{\pi\tau_p}Si(\pi) \simeq \frac{2}{\tau_p}(1-\frac{\pi^2}{18}). \tag{3.187}$$

Thus, by using Eq. (3.185) as well as Eqs. (3.186) and (3.187), the ratio $r_{|M'_{xy}|}$ defined by Eq. (3.79) for this case is

$$r_{|M'_{xy}|} = \frac{\int_{z_o-1/\frac{1}{2\pi}\gamma G_z\tau_p}^{z_o+1/\frac{1}{2\pi}\gamma G_z\tau_p} |M'_{xy}(\mathbf{r},\tau_p)|dz}{\int_{-\infty}^{\infty}|M'_{xy}(\mathbf{r},\tau_p)|dz} = \frac{\int_{-1/\tau_p}^{1/\tau_p}|sinc(f\tau_p)|df}{\int_{-\infty}^{\infty}|sinc(f\tau_p)|df} > 2(1-\frac{\pi^2}{18}) \simeq 0.9 \tag{3.188}$$

It implies that more than 90% of TPMM is located within the slice.††

**Which is based on the first two terms of Taylor series expansion of $\frac{\sin(x)}{x}$, therefore "$\simeq$" actually implies ">."

††In Eq. (3.188), we use $\int_{-\infty}^{\infty}|sinc(f\tau_p)|df \simeq |\int_{-\infty}^{\infty} sinc(f\tau_p)df| = \frac{1}{\tau_p}$. The more accurate results can be obtained by numerical computation.

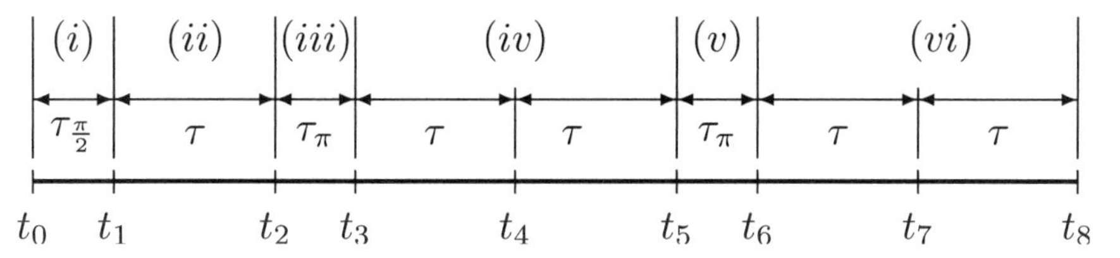

FIGURE 3.24
$(\frac{\pi}{2})_{x'} - \tau - (\pi)^{(1)}_{x'} - 2\tau - (\pi)^{(2)}_{x'} - 2\tau$ pulse sequence generates a process that consists of six sub-processes: (i) an excitation $(t_0 - t_1)$, (ii) a relaxation $(t_1 - t_2)$, (iii) a refocusing $(t_2 - t_3)$, (iv) a relaxation (t_3, t_5), (v) a refocusing (t_5, t_6), and (vi) a relaxation (t_6, t_8).

3.12.4 Appendix 3D

This appendix proves that for the spin echoes generated by the RF pulse sequence $(\frac{\pi}{2})_{x'} - \tau - (\pi)^{(1)}_{x'} - 2\tau \ \cdots \ (\pi)^{(N)}_{x'} - 2\tau$, TPMMs $M_{x'y'}(\mathbf{r}, t)$ at the different locations $\mathbf{r}$ in the sample undergo dephasing and rephasing in the time periods $((n-1)T_E, (n-\frac{1}{2})T_E)$ and $((n-\frac{1}{2})T_E, (n)T_E)$, respectively; and at the time $t = n \cdot T_E$ ($n = 1, \cdots, N$ and $T_E = 2\tau$), the rephasing is completed, TPMMs regain phase coherence and $|M_{x'y'}(\mathbf{r}, t)| \propto e^{-t/T_2}$.

It is sufficient to prove the above properties for the RF pulse sequence

$$(\frac{\pi}{2})_{x'} - \tau - (\pi)^{(1)}_{x'} - 2\tau - (\pi)^{(2)}_{x'} - 2\tau, \tag{3.189}$$

because $(\pi)^{(n)}_{x'} - 2\tau$ ($n > 2$) is just repetitions of $(\pi)^{(2)}_{x'} - 2\tau$.

As indicated in Section 3.11.2.2, the RF pulse sequence $(\frac{\pi}{2})_{x'} - \tau - (\pi)^{(1)}_{x'} - 2\tau - (\pi)^{(2)}_{x'} - 2\tau$ generates three forced precession and three free precession processes. As shown in Figure 3.24, these six processes are in the order of (i) an excitation (t_0, t_1), (ii) a relaxation: dephasing (t_1, t_2), (iii) a refocusing (t_2, t_3), (iv) a relaxation: rephasing (t_3, t_4) and dephasing (t_4, t_5), (v) a refocusing (t_5, t_6), and (vi) a relaxation: rephasing (t_6, t_7) and dephasing (t_7, t_8).

At each time point t_i ($i = 1, \cdots, 8$), t_i^- and t_i^+ denote the end time instant of the preceding process and the start time instant of the immediately following process. $\tau_{\pi/2}$ and τ_π denote the width of RF pulses $(\frac{\pi}{2})_{x'}$ and $(\pi)^{(n)}_{x'}$ ($n = 1, 2$), $\tau_{\pi/2} \ll \tau$ and $\tau_\pi \ll \tau$.

(i) In the period $t_0 - t_1$. When an excitation RF pulse $(\frac{\pi}{2})_{x'}$ is applied at the time $t = t_0 = 0$, due to $\omega_1 \tau_{\pi/2} = \frac{\pi}{2}$, Eq. (3.66) shows

$$\begin{array}{rl} M_{x'}(\mathbf{r}, t_1^-) = & M_{x'}(\mathbf{r}, 0) \\ M_{y'}(\mathbf{r}, t_1^-) = & M_{z'}(\mathbf{r}, 0) \\ M_{z'}(\mathbf{r}, t_1^-) = & -M_{y'}(\mathbf{r}, 0), \end{array} \tag{3.190}$$

where $M_{x'}(\mathbf{r}, 0)$, $M_{y'}(\mathbf{r}, 0)$, and $M_{z'}(\mathbf{r}, 0)$ are the initial values of the components of PMM $\vec{M}'(\mathbf{r}, t) = M_{x'}(\mathbf{r}, t)\vec{i}(t) + M_{y'}(\mathbf{r}, t)\vec{j}(t) + M_{z'}(\mathbf{r}, t)\vec{k}(t)$

(Eq. (1.5.5)). If the spin system is at the thermal equilibrium state (it is always the case in practice) immediately before the excitation, then, Eq. (3.190) becomes

$$\begin{array}{l} M_{x'}(\mathbf{r},t_1^-) = 0 \\ M_{y'}(\mathbf{r},t_1^-) = M_z^o(\mathbf{r}) \\ M_{z'}(\mathbf{r},t_1^-) = 0, \end{array} \tag{3.191}$$

where $M_z^o(\mathbf{r})$ denotes the TEMM at $\mathbf{r}$. The PMM of Eq. (3.191) is shown in Figure 3.22a.

(ii) In the period $t_1 - t_2$. After the excitation RF pulse $(\frac{\pi}{2})_{x'}$ is turned off, spins undergo relaxation which is characterized by Eq. (3.38). Because $\vec{M}'(\mathbf{r},t_1^+) = \vec{M}'(\mathbf{r},t_1^-)$, the initial values in Eq. (3.38) are given by Eq. (3.191). Thus, for $t_1 < t < t_2$, PMM $\vec{M}'(\mathbf{r},t)$ is described by

$$\begin{array}{l} M_{x'}(\mathbf{r},t) = M_z^o(\mathbf{r}) \sin\omega_o(t-t_1)\ e^{-(t-t_1)/T_2} \\ M_{y'}(\mathbf{r},t) = M_z^o(\mathbf{r}) \cos\omega_o(t-t_1)\ e^{-(t-t_1)/T_2} \\ M_{z'}(\mathbf{r},t) = M_z^o(\mathbf{r})(1 - e^{-(t-t_1)/T_1})\ , \end{array} \tag{3.192}$$

and TPMM $M_{x'y'}(\mathbf{r},t) = M_{x'}(\mathbf{r},t) + iM_{y'}(\mathbf{r},t)$ is given by

$$M_{x'y'}(\mathbf{r},t) = i\ M_z^o(\mathbf{r})e^{-(t-t_1)/T_2}e^{-i\omega_o(t-t_1)}. \tag{3.193}$$

In Eq. (3.193), as the time t ($t_1 < t < t_2$) approaches t_2, the phase shift defined by Eq. (3.161) $|\Delta\phi(\mathbf{r},t)| = \omega_o(t-t_1)$ becomes larger and approaches its maximum $\omega_o\tau$. Thus, Eq. (3.193) represents a dephasing process.

After τ delay, that is, at the time $t = t_2^-$, Eq. (3.192) shows PMM

$$\begin{array}{l} M_{x'}(\mathbf{r},t_2^-) = M_z^o(\mathbf{r}) \sin\omega_o\tau\ e^{-\tau/T_2} \\ M_{y'}(\mathbf{r},t_2^-) = M_z^o(\mathbf{r}) \cos\omega_o\tau\ e^{-\tau/T_2} \\ M_{z'}(\mathbf{r},t_2^-) = M_z^o(\mathbf{r})(1 - e^{-\tau/T_1}), \end{array} \tag{3.194}$$

and Eq. (3.193) shows TPMM

$$M_{x'y'}(\mathbf{r},t_2^-) = iM_z^o(\mathbf{r})e^{-\tau/T_2}e^{-i\omega_o\tau}. \tag{3.195}$$

(iii) In the period $t_2 - t_3$. When the refocusing RF pulse $(\pi)_{x'}^{(1)}$ is applied at time $t = t_2$, due to $\omega_1\tau_\pi = \pi$, Eq. (3.66) shows that

$$\begin{array}{l} M_{x'}(\mathbf{r},t_3^-) = \ \ M_{x'}(\mathbf{r},t_2^+) \\ M_{y'}(\mathbf{r},t_3^-) = -M_{y'}(\mathbf{r},t_2^+) \\ M_{z'}(\mathbf{r},t_3^-) = -M_{z'}(\mathbf{r},t_2^+). \end{array} \tag{3.196}$$

Because $\vec{M}'(\mathbf{r},t_2^+) = \vec{M}'(\mathbf{r},t_2^-)$, Eq. (3.196) becomes

$$\begin{array}{l} M_{x'}(\mathbf{r},t_3^-) = \ \ M_z^o(\mathbf{r}) \sin\omega_o\tau\ e^{-\tau/T_2} \\ M_{y'}(\mathbf{r},t_3^-) = -M_z^o(\mathbf{r}) \cos\omega_o\tau\ e^{-\tau/T_2} \\ M_{z'}(\mathbf{r},t_3^-) = -M_z^o(\mathbf{r})(1 - e^{-\tau/T_1}). \end{array} \tag{3.197}$$

(iv) In the period $t_3 - t_5$. After the $(\pi)_{x'}^{(1)}$ RF pulse is turned off, spins undergo the relaxation that is characterized by Eq. (3.38). Because of $\vec{M}'(\mathbf{r}, t_3^+) = \vec{M}'(\mathbf{r}, t_3^-)$, the initial values in Eq. (3.38) are given by Eq. (3.197). Thus, for $t_3 < t < t_4$, PMM $\vec{M}'(\mathbf{r}, t)$ is described by

$$\begin{aligned} M_{x'}(\mathbf{r}, t) &= \quad M_z^o(\mathbf{r}) \sin \omega_o(\tau - (t - t_3))\, e^{-(\tau+(t-t_3))/T_2} \\ M_{y'}(\mathbf{r}, t) &= -M_z^o(\mathbf{r}) \cos \omega_o(\tau - (t - t_3))\, e^{-(\tau+(t-t_3))/T_2} \\ M_{z'}(\mathbf{r}, t) &= \quad M_z^o(\mathbf{r})(1 - (2 - e^{-\tau/T_1})e^{-(t-t_3)/T_1}), \end{aligned} \tag{3.198}$$

and TPMM $M_{x'y'}(\mathbf{r}, t)$ is given by

$$M_{x'y'}(\mathbf{r}, t) = -i\, M_z^o(\mathbf{r}) e^{-(\tau+(t-t_3))/T_2} e^{i\omega_o(\tau-(t-t_3))}. \tag{3.199}$$

In Eq. (3.199), as the time t ($t_3 < t < t_4$) approaches t_4, the phase shift defined by Eq. (3.161) $\Delta\phi(\mathbf{r}, t) = \omega_o(\tau - (t - t_3))$ becomes smaller and approaches the zero. Thus, Eq. (3.199) represents a rephasing process.

After τ delay, that is, at the time $t = t_4^-$, Eq. (3.198) shows PMM

$$\begin{aligned} M_{x'}(\mathbf{r}, t_4^-) &= \quad 0 \\ M_{y'}(\mathbf{r}, t_4^-) &= -M_z^o(\mathbf{r})\, e^{-2\tau/T_2} \\ M_{z'}(\mathbf{r}, t_4^-) &= \quad M_z^o(\mathbf{r})(1 - e^{-\tau/T_1})^2, \end{aligned} \tag{3.200}$$

and Eq. (3.199) shows TPMM

$$M_{x'y'}(\mathbf{r}, t_4^-) = -iM_z^o(\mathbf{r})e^{-2\tau/T_2} \simeq -iM_z^o(\mathbf{r})e^{-t_4^-/T_2}, \tag{3.201}$$

which indicates that the spin rephasing is completed at $t = t_4^-$ because of the phase shift $\Delta\phi(\mathbf{r}, t_4^-) = 0$. $|M_{x'y'}(\mathbf{r}, t_4^-)| \simeq M_z^o(\mathbf{r})e^{-t_4^-/T_2}$ implies that TPMM has a true T_2 decay. In Eq. (3.201), $-i$ indicates that TPMM $M_{x'y'}(\mathbf{r}, t_4^-)$ is at the $-\vec{j}(t)$-direction as shown in Figure 3.22d.

After $t > t_4$, spins continue relaxation as they do before the refocusing RF pulse $(\pi)_{x'}^{(1)}$. Thus, for $t_4 < t < t_5$, due to $t - t_3 = t - (t_4 - \tau) = t - t_4 + \tau$, Eq. (3.198) becomes

$$\begin{aligned} M_{x'}(\mathbf{r}, t) &= -M_z^o(\mathbf{r}) \sin \omega_o(t - t_4)\, e^{-(2\tau+(t-t_4))/T_2} \\ M_{y'}(\mathbf{r}, t) &= -M_z^o(\mathbf{r}) \cos \omega_o(t - t_4)\, e^{-(2\tau+(t-t_4))/T_2} \\ M_{z'}(\mathbf{r}, t) &= \quad M_z^o(\mathbf{r})(1 - (2 - e^{-\tau/T_1})e^{-(\tau+(t-t_4))/T_1}), \end{aligned} \tag{3.202}$$

and TPMM $M_{x'y'}(\mathbf{r}, t)$ is given by

$$M_{x'y'}(\mathbf{r}, t) = -i\, M_z^o(\mathbf{r}) e^{-(2\tau+(t-t_4))/T_2} e^{-i\omega_o(t-t_4)}. \tag{3.203}$$

In Eq. (3.203), as the time t ($t_4 < t < t_5$) approaches t_5, the phase shift $|\Delta\phi(\mathbf{r}, t)| = \omega_o(t - t_4)$ becomes larger and approaches its maximum $\omega_o\tau$. Thus, Eq. (3.203) represents a dephasing process.

After τ delay, that is, at the time $t = t_5^-$, Eq. (3.202) shows PMM

$$\begin{aligned} M_{x'}(\mathbf{r}, t_5^-) &= -M_z^o(\mathbf{r}) \sin \omega_o \tau \ e^{-3\tau/T_2} \\ M_{y'}(\mathbf{r}, t_5^-) &= -M_z^o(\mathbf{r}) \cos \omega_o \tau \ e^{-3\tau/T_2} \\ M_{z'}(\mathbf{r}, t_5^-) &= \ \ M_z^o(\mathbf{r})(1 - (2 - e^{-\tau/T_1})e^{-2\tau/T_1}), \end{aligned} \tag{3.204}$$

and Eq. (3.203) shows TPMM

$$M_{x'y'}(\mathbf{r}, t_5^-) = -iM_z^o(\mathbf{r})e^{-3\tau/T_2}e^{-i\omega_o\tau}. \tag{3.205}$$

(v) In the period $t_5 - t_6$. When the refocusing RF pulse $(\pi)_{x'}^{(2)}$ is applied at the time $t = t_5$, due to $\omega_1\tau_\pi = \pi$, Eq. (3.66) shows that

$$\begin{aligned} M_{x'}(\mathbf{r}, t_6^-) &= \ \ M_{x'}(\mathbf{r}, t_5^+) \\ M_{y'}(\mathbf{r}, t_6^-) &= -M_{y'}(\mathbf{r}, t_5^+) \\ M_{z'}(\mathbf{r}, t_6^-) &= -M_{z'}(\mathbf{r}, t_5^+). \end{aligned} \tag{3.206}$$

Because of $\vec{M}'(\mathbf{r}, t_5^+) = \vec{M}'(\mathbf{r}, t_5^-)$, Eq. (3.206) becomes

$$\begin{aligned} M_{x'}(\mathbf{r}, t_6^-) &= -M_z^o(\mathbf{r}) \sin \omega_o \tau \ e^{-3\tau/T_2} \\ M_{y'}(\mathbf{r}, t_6^-) &= \ \ M_z^o(\mathbf{r}) \cos \omega_o \tau \ e^{-3\tau/T_2} \\ M_{z'}(\mathbf{r}, t_6^-) &= -M_z^o(\mathbf{r})(1 - (2 - e^{-\tau/T_1})e^{-2\tau/T_1}). \end{aligned} \tag{3.207}$$

(vi) In the period $t_6 - t_8$. After the refocusing RF pulse $(\pi)_{x'}^{(2)}$ is turned off, spins undergo the relaxation that is characterized by Eq. (3.38). Because $\vec{M}'(\mathbf{r}, t_6^+) = \vec{M}'(\mathbf{r}, t_6^-)$, the initial values in Eq. (3.38) are given by Eq. (3.207). Thus, for $t_6 < t < t_7$, PMM $\vec{M}'(\mathbf{r}, t)$ is described by

$$\begin{aligned} M_{x'}(\mathbf{r}, t) &= -M_z^o(\mathbf{r}) \sin \omega_o(\tau - (t - t_6)) \ e^{-(3\tau + (t - t_6))/T_2} \\ M_{y'}(\mathbf{r}, t) &= \ \ M_z^o(\mathbf{r}) \cos \omega_o(\tau - (t - t_6)) \ e^{-(3\tau + (t - t_6))/T_2} \\ M_{z'}(\mathbf{r}, t) &= \ \ M_z^o(\mathbf{r})(1 - (2 - (2 - e^{-\tau/T_1})e^{-2\tau/T_1})e^{-(t - t_6)/T_1}), \end{aligned} \tag{3.208}$$

and TPMM $M_{x'y'}(\mathbf{r}, t)$ is given by

$$M_{x'y'}(\mathbf{r}, t) = i \ M_z^o(\mathbf{r})e^{-(3\tau + (t - t_6))/T_2}e^{i\omega_o(\tau - (t - t_6))}. \tag{3.209}$$

In Eq. (3.209), as the time t ($t_6 < t < t_7$) approaches t_7, the phase shift defined by Eq. (3.161) $\Delta\phi(\mathbf{r}, t) = \omega_o(\tau - (t - t_6))$ becomes smaller and approaches the zero. Thus, Eq. (3.209) represents a rephasing process.

After τ delay,that is, at the time $t = t_7^-$, Eq. (3.208) shows PMM

$$\begin{aligned} M_{x'}(\mathbf{r}, t_7^-) &= 0 \\ M_{y'}(\mathbf{r}, t_7^-) &= M_z^o(\mathbf{r}) \ e^{-4\tau/T_2} \\ M_{z'}(\mathbf{r}, t_7^-) &= M_z^o(\mathbf{r})(1 - e^{-2\tau/T_1})(1 - e^{-\tau/T_1})^2, \end{aligned} \tag{3.210}$$

and Eq. (3.209) shows TPMM

$$M_{x'y'}(\mathbf{r}, t_7^-) = iM_z^o(\mathbf{r})e^{-4\tau/T_2} \simeq iM_z^o(\mathbf{r})e^{-t_7^-/T_2}, \tag{3.211}$$

which indicates that the spin rephasing is completed at $t = t_7^-$ because of the phase shift $\Delta\phi(\mathbf{r}, t_7^-) = 0$. $|M_{x'y'}(\mathbf{r}, t_7^-)| \simeq M_z^o(\mathbf{r})e^{-t_7^-/T_2}$ implies that TPMM has a true T_2 decay. In Eq. (3.211), i indicates that TPMM $M_{x'y'}(\mathbf{r}, t_7^-)$ is at the $\vec{j}(t)$-direction (not shown in Figure 3.22).

After $t > t_7$, spins continue relaxation as they do before the RF pulse $(\pi)_{x'}^{(2)}$. Thus, for $t_7 < t < t_8$, due to $t - t_6 = t - (t_7 - \tau) = t - t_7 + \tau$, Eq. (3.208) becomes

$$\begin{aligned} M_{x'}(\mathbf{r}, t) &= M_z^o(\mathbf{r}) \sin \omega_o(t - t_7)\ e^{-(4\tau+(t-t_7))/T_2} \\ M_{y'}(\mathbf{r}, t) &= M_z^o(\mathbf{r}) \cos \omega_o(t - t_7)\ e^{-(4\tau+(t-t_7))/T_2} \\ M_{z'}(\mathbf{r}, t) &= M_z^o(\mathbf{r})(1 - (2 - (2 - e^{-\tau/T_1})e^{-2\tau/T_1})e^{-(\tau+(t-t_7))/T_1}), \end{aligned} \tag{3.212}$$

and TPMM $M_{x'y'}(\mathbf{r}, t)$ is given by

$$M_{x'y'}(\mathbf{r}, t) = i\ M_z^o(\mathbf{r})e^{-(4\tau+(t-t_7))/T_2}e^{-i\omega_o(t-t_7)}. \tag{3.213}$$

In Eq. (3.213), as the time t ($t_7 < t < t_8$) approaches t_8, the phase shift $|\Delta\phi(\mathbf{r}, t)| = \omega_o(t - t_7)$ becomes larger and approaches its maximum $\omega_o\tau$. Thus, Eq. (3.213) represents a dephasing process.

From Figure 3.24, we find that the peak amplitudes of the n-th spin echo occurs at time instants

$$t_{3n+1} = \tau_{\pi/2} + n(\tau_\pi + T_E). \tag{3.214}$$

It is easy to verify that the peak amplitudes of the first, second, and third spin echoes are at $t_4 = \tau_{\pi/2} + (\tau_\pi + T_E)$, $t_7 = \tau_{\pi/2} + 2(\tau_\pi + T_E)$, and $t_{10} = \tau_{\pi/2} + 3(\tau_\pi + T_E)$. Under the conditions $\tau_{\pi/2} << \tau$ and $\tau_\pi << \tau$ (i.e., ignoring the widths of RF pulses), Eq. (3.214) can be approximated by $t_{3n+1} \simeq nT_E$. Thus, based on Eqs. (3.201) and (3.211), we have

$$|M_{x'y'}(\mathbf{r}, t_{3n+1})| = M_z^o(\mathbf{r})e^{-nT_E/T_2} \simeq M_z^o(\mathbf{r})e^{-t_{3n+1}/T_2} \propto e^{-t_{3n+1}/T_2}. \tag{3.215}$$

Eq. (3.215) indicates that in order to have the true T_2 decay, the FID signal $s_r(t)$ should be sampled at time $\tau_{\pi/2} + n(\tau_\pi + T_E)$, not at nT_E.

Problems

3.1. Section 3.2 gives two statements on nuclear magnetic resonance (NMR) phenomena. The first one says, "Nuclei with an odd number of protons and/or an odd number of neutrons demonstrate NMR phenomena." The second one says that "When NMR phenomena occurs, I (the spin quantum number) must be non-zero." Elaborate that these two statements are identical.

3.2. The spin precession is characterized by Eq. (3.10) in terms of the magnetic dipole moment $\vec{\mu}$. (a) Prove that the solution to Eq. (3.10) is Eq. (3.11). (b) Discuss the physical meaning of this solution. Repeat (a) and (b) for Eq. (3.14); that is, find the solution of Eq. (3.14) and interpret it.

3.3. Section 3.4.2 shows that under the normal conditions and the ordinary settings of MRI as shown in Table 3.2, $\epsilon \simeq 3.4 \times 10^{-6}$. References [7–10] used the first-order approximation of Boltzmann distribution and derive the values of ϵ and/or other measures related to n_l, n_h, and n, which are given in (a)–(d) in that section. Prove that values listed in (a)–(d) are identical to or consistent with $\epsilon \simeq 3.4 \times 10^{-6}$.

3.4. Solve Eq. (3.34) to show that its solution is Eq. (3.35).

3.5. Solve Eq. (3.37) to show that its solution is Eq. (3.38).

3.6. Show that by using the complex representation, Eq. (3.37) can be decomposed into two equations in Eq. (3.39).

3.7. Solve Eq. (3.39) to show that its solution is Eq. (3.40).

3.8. Solve the first equation of Eq. (3.44) to show that its solution is Eq. (3.46).

3.9. By following the instructions in Section 3.7.1, verify the Eqs. (3.57), (3.58), and (3.59).

3.10. Verify Eqs. (3.65) and (3.66).

3.11. Prove Eq. (3.73) and show that its solution is Eq. (3.74).

3.12. Prove Eqs. (1.76) and (3.77).

3.13. Verify Eq. (3.93).

3.14. Mathematically explain k-space trajectory shown in Figures 3.10a and 1.12a. (Hint: using Eqs. (3.108) and (3.109) for the rectilinear and radial sampling, respectively, and considering k_x and k_y as well as k and θ as functions of both the gradient and the time.)

3.15. Derive Eq. (3.198) and Eq. (3.208) in Appendix 3D.

3.16. In Appendix 3D, TPMMs reach their peak amplitudes at t_4 and t_7. Derive the corresponding LPMMs $M_{z'}(\mathbf{r}, t_4^-)$ in Eq. (3.200) and $M_{z'}(\mathbf{r}, t_7^-)$ in Eq. (3.210).

3.17. Appendix 3D shows the evolution of spin echoes generated by the RF pulse sequence

$$(\frac{\pi}{2})_{x'} - \tau - (\pi)^{(1)}_{x'} - 2\tau - (\pi)^{(2)}_{x'} - 2\tau.$$

Show the corresponding results for the RF pulse sequence

$$(\frac{\pi}{2})_{y'} - \tau - (\pi)^{(1)}_{y'} - 2\tau - (\pi)^{(2)}_{y'} - 2\tau.$$

References

[1] Purcell, E., Torrey, H., Pound, R.: Resonance absorption by nuclear magnetic moments in a solid. *Phys. Rev.* **69** (1946) 37–38.

[2] Bloch, F.: Nuclear induction. *Phys. Rev.* **70**(7 and 8) (1946) 460–474.

[3] Kumar, A., Welti, D., Ernst, R.: NMR Fourier zeugmatography. *J. Magn. Res.* **18** (1975) 69–75.

[4] Wuthrich, K., Wider, G., Wagner, G., Braun, W.: Sequential resonance assignments as a basis for determination of spatial protein structures by high resolution proton nuclear magnetic resonance. *J. Mol. Biol.* **155** (1982) 311–319.

[5] Lauterbur, P.: Image formation by induced local interactions: Examples employing nuclear magnetic resonance. *Nature* **242** (1973) 190–191.

[6] Mansfield, P., Grannell, P.: NMR 'diffraction' in solids? *J. Phys C: Solid State Phys.* **6** (1973) 422–427.

[7] Slichter, C.: *Principle of Magnetic Resonance.* Springer-Verlag, Berlin (1980).

[8] Evans, R.: *The Atomic Nucleus.* Robert E. Krieger Publishing Company, Florida (1982).

[9] Abragam, A.: *Principles of Nuclear Magnetism.* Oxford University Press, New York (1983).

[10] Haacke, E., Brown, R., Thompson, M., Venkatesan, R.: *Magnetic Resonance Imaging: Physical Principles and Sequence Design.* John Wiley & Sons Inc., New York (1999).

[11] Nishimura, D.: *Principles of Magnetic Resonance Imaging.* Stanford University, Palo Alto, CA. (1996).

[12] Liang, Z.P., Lauterbur, P.: *Principles of Magnetic Resonance Imaging, A Signal Processing Perspective.* IEEE Press, New York (2000).

[13] Wagner, R., Brown, D.: Unified SNR analysis of medical imaging systems. *Phys. Med. Biol.* **30**(6) (1985) 489–518.

[14] Herman, G.: *Image Reconstruction from Projections.* Academic Press, New York (1980).

[15] Kak, A., Slaney, M.: *Principles of Computerized Tomographic Imaging.* IEEE Press, New York (1988).

[16] Bracewell, R.N.: *The Fourier Transform and Its Applications.* McGraw-Hill Book Company, New York (1978).

[17] Barrett, H., Swindell, W.: *Radiological imaging.* Academic Press, Hoboken, NJ. (1981).

[18] Lei, T., Sewchand, W.: Statistical approach to x-ray CT imaging and its applications in image analysis. Part 1: Statistical analysis of x-ray CT imaging. *IEEE Trans. on Medical Imaging.* **11**(1) (1992) 53–61.

[19] Cho, Z.H., Jones, J.P., Singh, M.: *Foundations of Medical Imaging.* John Wiley & Sons, Inc., New York (1993).

4

Non-Diffraction Computed Tomography

4.1 Introduction

In X-ray CT imaging, the measurements are photons, which demonstrate wave-particle duality, that is, the properties of both waves and particles. X-rays have wavelengths roughly 10^{-13} m to 10^{-8} m, or frequencies of 10^{16} Hz to 10^{21} Hz. In MR imaging, the measurements are free induction decay (FID) signals, which are radio frequency (RF) signals with the wavelengths roughly 10^{0} m to 10^{2} m, or the frequencies 10^{6} Hz to 10^{8} Hz. X-ray and FID signal are electromagnetic in nature. In this chapter, imaging sources in several medical imaging techniques are characterized as electromagnetic (EM) waves.

When EM waves impinge on an object, or an object is immersed in the EM field, several physical phenomena occur on the object: its surface, inside, and surrounding. These phenomena include, but are not limited to, absorption, diffraction, non-diffraction, reflection, refraction, scattering, etc. Many of these phenomena can be utilized for imaging the object: its shape or surface or the internal structure.

In medical applications, these imaging techniques are X-ray CT [1–10], MRI [11–20] positron emission tomography (PET) [4, 6, 21–24], single photon emission computed tomography (SPECT) [4, 6, 22, 24, 25], ultrasonic (US) [4, 6, 26–29], etc. Although these techniques were developed based on different physical phenomena and principles, according to the nature of source-medium interaction, they can be classified into a category of imaging, transmission computed tomography.* Transmission CT can be further divided into two groups: (1) a non-diffraction CT imaging, in which the interaction model and the external measurements are characterized by the straight line integrals of some indexes of the medium and the image reconstruction is based on Fourier Slice theorem [4, 6], and (2) a diffraction CT imaging, in which the interaction and measurements are modeled with the wave equation and the tomographic reconstruction approach is based on the Fourier diffraction theorem [4, 30]. The former includes X-ray CT, MRI, emission CT, ultrasonic CT (e.g., refrac-

*Its counterpart is called the reflection computed tomography, which is outside the scope of this book.

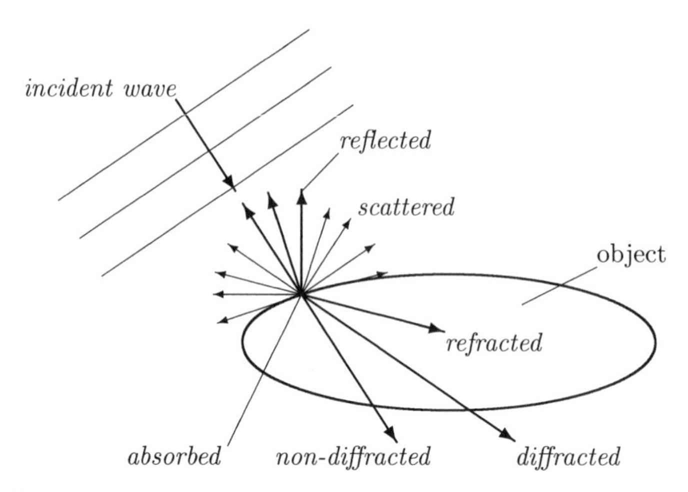

FIGURE 4.1
Physical phenomena caused by interaction between EM wave and an object.

tive index CT and attenuation CT), etc. The latter includes acoustic, certain seismic, microwave, and optical imaging, etc.

In this chapter, we first use the inverse scattering problem as an example to demonstrate how the interactions between the incident EM wave and the object can be used to generate the image of the object. Then we revisit X-ray CT and MRI and briefly review emission CT from a specific standpoint, and show that they belong to a category of imaging—the non-diffraction computed tomography. This insight may explain why X-ray CT and MRI have very similar statistical properties that are described in the remaining chapters of this book.

4.2 Interaction between EM Wave and Object

When EM waves impinge on an object, due to the interaction between the wave and the object, the following physical phenomena occur at the object, which are shown in Figure 4.1. All these physical phenomena can be used for imaging the shape of the object, its external surface, and internal structure, and have practical applications:

1. Absorption
2. Diffraction

3. Non-diffraction
4. Reflection
5. Refraction
6. Scattering

For the given EM wave and the fixed points on the surface of the object, the diffracted, reflected, and refracted waves are in specified directions, while the scattered waves are in all directions. Thus, when the detectors (or receivers) are randomly placed around the object, the diffracted, reflected and refracted waves are received with probability zero, and the scattered waves are received with probability one. The inverse scattering problem in the imaging arises from scattered EM waves.

4.3 Inverse Scattering Problem

4.3.1 Relationship between Incident and Scattered Waves

The inverse scattering problem may be the simplest imaging principle. We use it as an example. In Figure 4.2, S and V denote the surface and volume of the object, $\vec{I}$ is the unit vector in the direction of the incident wave, $\vec{r}$ is the vector from the origin of the coordinate system to a point on the surface

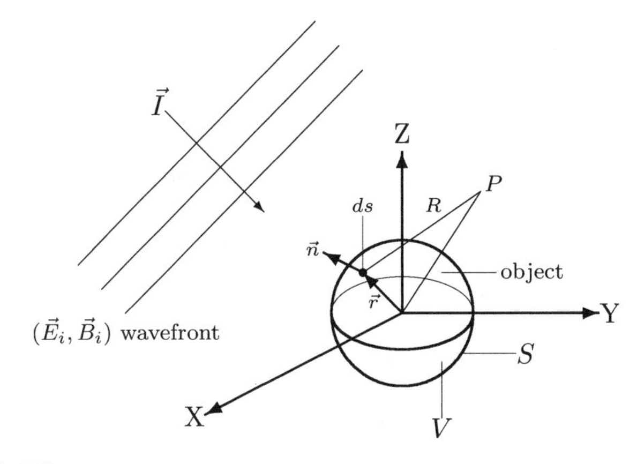

FIGURE 4.2
Incident and scattered waves.

S, P is a point in space, $\vec{n}$ is the unit vector outward normal to the surface S at the end point of the vector $\vec{r}$, $ds(\vec{r})$ is the area element that contains the end point of the vector $\vec{r}$, and R denotes the distance between the end point of the vector $\vec{r}$ and the point P.

Under the following conditions:

1. The incident EM wave is transverse, plane, and time-harmonic,

$$\vec{E}(\vec{r},t) = \vec{E}_0 \, e^{i(K\vec{I}\cdot\vec{r}-\omega t)} \quad \text{and} \quad \vec{B}(\vec{r},t) = \vec{B}_0 \, e^{i(K\vec{I}\cdot\vec{r}-\omega t)}, \tag{4.1}$$

   where $\vec{E}_0$ and $\vec{B}_0$ are constant in both space and time. Let λ and f be the wavelength and frequency of the incident wave, $K = \frac{2\pi}{\lambda}$ the wave number[†], $\omega = 2\pi f$, and $\cdot$ in $\vec{I}\cdot\vec{r}$ denotes the inner product,

2. The wavefront of the incident wave is infinitely wide and the object is immersed in the far-field of this wave,

3. Only the E-field component of the incident wave is considered, and

4. The object is convex,

it has been shown that the E-field components of the backscattered wave at point P and the incident wave at point $\vec{r}$ have the relationship

$$\vec{E}_s \simeq \frac{\rho}{2\sqrt{\pi}R}\vec{E}_0 \, e^{i(2KR-\omega t)}, \tag{4.2}$$

where ρ is defined by

$$\rho = \frac{-iK}{\sqrt{\pi}} \int_{S:\, \vec{I}\cdot\vec{n}<0} \vec{I}\cdot\vec{n} \, e^{i2K\vec{I}\cdot\vec{r}} ds(\vec{r}), \tag{4.3}$$

where S is the surface illuminated by the incident wave, $\vec{I}\cdot\vec{n} < 0$ and $\vec{I}\cdot\vec{n} > 0$ denote the portions of S that produce the back- and the forward- scattering, respectively.

From Eq. (4.2), we have

$$4\pi R^2 |\vec{E}_s|^2 = |\rho|^2 |\vec{E}_0|^2. \tag{4.4}$$

$4\pi R^2$ is the total area of the spherical surface with radius R. If $|\vec{E}_s|^2$ is thought of as the power density on this spherical surface, then Eq. (4.4) shows that the total power that penetrates this surface is equal to a portion of the power that is generated by a source $\vec{E}_0$ inside this sphere. This portion is determined by ρ, and $|\rho|^2$ is called the power cross section. From Eq. (4.3), ρ is a function of K. Define the wavenumber vector $\vec{P}$ by[‡]

$$\vec{P} = -2K\vec{I}. \tag{4.5}$$

ρ will be a function of $\vec{P}$ and is denoted by $\rho(\vec{P})$.

[†] K is sometimes called the spatial frequency.

[‡] $\vec{P}$ is sometimes called the spatial frequency vector.

4.3.2 Inverse Scattering Problem Solutions

• *Theoretical solution* of the inverse scattering problem. Using $\vec{P}$ to replace $\vec{I}$, Eq. (4.3) becomes

$$\rho(\vec{P}) = \frac{i}{2\sqrt{\pi}} \int_{S:\ \vec{P}\cdot\vec{n}>0} \vec{P}\cdot\vec{n}\, e^{-i\vec{P}\cdot\vec{r}} ds(\vec{r}). \tag{4.6}$$

Using Eq. (4.6), Appendix 4A shows that

$$2\sqrt{\pi}\frac{\rho(\vec{P}) + \rho^*(-\vec{P})}{|\vec{P}|^2} = \int\int\int_V e^{-i\vec{P}\cdot\vec{r}} dv(\vec{r}), \tag{4.7}$$

where V is the volume enclosed by the surface S, $dv(\vec{r})$ is the volume element that contains the end point of the vector $\vec{r}$, and $*$ denotes the complex conjugate.

Define

$$\Gamma(\vec{P}) = 2\sqrt{\pi}\frac{\rho(\vec{P}) + \rho^*(-\vec{P})}{|\vec{P}|^2}, \tag{4.8}$$

and

$$\gamma(\vec{r}) = \begin{cases} 1 & \vec{r} \in V \\ 0 & \vec{r} \overline{\in} V \end{cases}; \tag{4.9}$$

then

$$\Gamma(\vec{P}) = \int\int\int_V \gamma(\vec{r})\, e^{-i\vec{P}\cdot\vec{r}} dv(\vec{r}), \tag{4.10}$$

that is,

$$\gamma(\vec{r}) = (2\pi)^{-3} \int\int\int_{\mathcal{P}} \Gamma(\vec{P})\, e^{i\vec{P}\cdot\vec{r}} d\vec{P}, \tag{4.11}$$

where $\mathcal{P}$ is the wavenumber space: its directions occupy 360° solid angles and its magnitudes range from 0 to ∞.

Eqs. (4.10) and (4.11) show that $\gamma(\vec{r})$ and $\Gamma(\vec{P})$ form a pair of three-dimensional Fourier transform

$$\gamma(\vec{r}) \stackrel{\mathcal{F}_3}{\longleftrightarrow} \Gamma(\vec{P}). \tag{4.12}$$

$\gamma(\vec{r})$ defined by Eq. (4.9) is an indicator function of the geometry of the object. $\Gamma(\vec{P})$ is, from Eqs. (4.8) and (4.3), measurable. Thus, Eqs. (4.11) shows that if all $\Gamma(\vec{P})$ are available, then $\gamma(\vec{r})$—the geometry of the object - will be uniquely determined.

• *Practical solution* of inverse scattering problem. In practice, the backscattering data can only be acquired at limited directions and magnitudes; that is,

$\Gamma(\vec{P})$ can only be obtained in a limited portion of the space $\mathcal{P}$, say $\mathcal{D}$, $\mathcal{D} \subset \mathcal{P}$. Define an indicator function $K(\vec{P})$ by

$$K(\vec{P}) = \begin{cases} 1 & \vec{P} \in \mathcal{D} \\ 0 & \vec{P} \overline{\in} \mathcal{D} \end{cases} ; \tag{4.13}$$

then the actually available $\Gamma(\vec{P})$ is

$$\hat{\Gamma}(\vec{P}) = K(\vec{P})\Gamma(\vec{P}) \, . \tag{4.14}$$

Thus, the object geometry that is determined by $\hat{\Gamma}(\vec{P})$ will be

$$\hat{\gamma}(\vec{r}) = k(\vec{r}) \star \gamma(\vec{r}), \tag{4.15}$$

where $\star$ denotes the convolution and $k(\vec{r})$ is given by

$$k(\vec{r}) \stackrel{\mathcal{F}_3}{\longleftrightarrow} K(\vec{P}). \tag{4.16}$$

Eq. (4.15) indicates that $\hat{\gamma}(\vec{r})$ is an estimate of $\gamma(\vec{r})$ and can be obtained by

$$\hat{\gamma}(\vec{r}) = \mathcal{F}_3^{-1}\{\hat{\Gamma}(\vec{P})\}. \tag{4.17}$$

The above discussion shows that the geometry of an object can be determined using backscattered measurements. This procedure is known as the inverse backscattering problem. One of its well-known applications is the Turntable Data Imaging experiment. In this experiment, an object model and a transmitter/receiver (T/R) are located in the same X-Y plane of the coordinate system. The object model is rotating at constant speed around the Z-axis. T/R transmits signals and receives the echos returned by the object model. Using the inverse backscattering solution, an image of the shape of the object model can be created. In this experiment, $\mathcal{P}$ is limited in the X-Y plane and the wavenumber is limited by the frequency band of the T/R.

4.4 Non-Diffraction Computed Tomography

In this section we revisit X-ray CT and MRI and briefly review emission CT from a specific standpoint. We will show that in these imaging techniques, the interaction model and the external measurements are characterized by straight line integrals of some indexes of the object and the image reconstruction is based on Fourier slice theorem. That is, they belong to the non-diffraction computed tomography.

4.4.1 X-Ray Computed Tomography

Physical measurements in X-ray CT are in photons. The "external measurements" directly used in the image reconstruction are the projections. The definition of the projection is given by Eq. (2.14) and denoted by $p(l,\theta)$

$$p(l,\theta) = -\ln \frac{N_{ad}/N_{ar}}{N_{cd}/N_{cr}}, \tag{4.18}$$

where N_{ad} and N_{ar} are the numbers of photons counted by the detector and the reference detector in the actual measurement process, respectively, and N_{cd} and N_{cr} are the counterparts of N_{ad} and N_{ar} in the calibration measurement process, respectively.

Chapter 2 showed that $\frac{N_{ad}}{N_{ar}} = \rho_a$ and $\frac{N_{cd}}{N_{cr}} = \rho_c$. ρ_a and ρ_c are the transmittances along the straight line in the actual and calibration measurement processes, respectively. $-\ln \rho_a$ and $-\ln \rho_c$ are the linear attenuation coefficient (LAC) in these two processes. $-\ln \frac{N_{ad}/N_{ar}}{N_{cd}/N_{cr}} = (-\ln \rho_a) - (-\ln \rho_c)$ is the relative linear attenuation coefficient (RLAC) along a straight line. It has been shown that RLAC along a straight line specified by a pair of parameters (l,θ), denoted by $p(l,\theta)$, is the line integral of RLAC at each point (r,ϕ) at this straight line, denoted by $f(r,\phi)$. Eqs. (2.16) and (2.17) show that

$$p(l,\theta) = \int_L f(r,\phi) dz, \tag{4.19}$$

where the straight line L is specified by

$$L: \quad r\cos(\theta - \phi) = l. \tag{4.20}$$

$p(l,\theta)$ is the "external measurement." $f(r,\phi)$ is an "index of the object," and is called the object function of X-ray CT.

Eq. (4.18) actually represents a double normalization procedure for N_{ad}. The first normalization appears in its numerator (for the actual measurement process) and its denominator (for the calibration measurement process) separately. It is a necessary and reasonable operation to estimate the transmittances ρ_a and ρ_c. The second normalization appears between the numerator and denominator of Eq(4.18). The double normalization not only reduces (or eliminates) the effect on $p(l,\theta)$ caused by the fluctuation in N_{ad} due to the photon emission in one projection and among projections, but also makes the definition of the projection $p(l,\theta)$ consistent with the definition of the relative linear attenuation coefficient $f(r,\phi)$ and hence establishes Eq. (4.19).

Let X-Y be a rectangular coordinate system that corresponds to a radial coordinate system where the line $L: \ (l,\theta)$ and RLAC $f(r,\phi)$ are defined. By rotating the X-Y coordinate system with an angle θ, a new X′-Y′ rectangular coordinate system is created. It is easy to verify that (1) the line $L: \ r\cos(\theta - \phi) = l$ becomes the line $x' = l$ and the projection $p(l,\theta)$ becomes the projection

$p(x',\theta)$, and (2) the projection $p(x',\theta)$ still equals the line integral of the object function $f(x,y)$, that is,

$$p(x',\theta) = \int_L f(r,\phi)dz = \int_{x'=l} f(x,y)dy' = \int_{-\infty}^{+\infty} f(x,y)dy'. \tag{4.21}$$

Eqs. (2.21) through (2.23) show that

$$\mathcal{F}_1\{p(x',\theta)\} = \mathcal{F}_2\{f(x,y)\}, \tag{4.22}$$

which is Fourier slice theorem for X-ray CT image reconstruction.

Eq. (4.19) is also known as the Radon transform. Based on Eq. (4.22), three formulas of the Inverse Radon transform have been derived. They are given in Eqs. (2.25) and (2.26) for the parallel projection and Eq. (2.71) for the divergent projection. These formulas lead to filtered backprojection (FBP) image reconstruction algorithm.

Photons travel in the straight lines. Projections are formulated along the straight lines. Projection ("the external measurements") is a line integral of the object function RLAC ("the index of the object"). Image reconstruction is based on the Fourier slice theorem. Thus, X-ray CT is a typical non-diffraction CT imaging.

4.4.2 Magnetic Resonance Imaging

As shown in Chapter 3, the thermal equilibrium macroscopic magnetization (TEMM, denoted by $M_z^o(x',y')$) of a spin system is a physical quantity that can be actually measured and observed in MR imaging. It is an index of the object, we call it the object function of MRI. The measurement in MRI that is directly used in the image reconstruction is the k-space sample, that is, an alternative representation of the discrete, complex-valued, baseband signal. All these relations are depicted by one formula—Eq. (3.110).

Chapter 3 described two basic k-space sample acquisition schemes and two typical image reconstruction protocols. The following combinations—(1) the rectilinear k-space sampling and Fourier transform (FT) reconstruction, and (2) the radial k-space sampling and the projection reconstruction (PR)—are two approaches widely used in MRI. In order to show the tomography nature of MRI, we elaborate the second approach.

By rotating a rectangular coordinate system X-Y by an angle θ at which the radial k-space sampling takes place and using X$'$-Y$'$ to denote this rotated rectangular coordinate system, Eq. (3.121) gives a theoretical definition of the measured projection $p(x',\theta)$

$$p(x',\theta) = \int_{x'=l} M_{xy}(x',y')dy', \tag{4.23}$$

where $M_{xy}(x',y')$ represents the transverse precession macroscopic magnetization (TPMM). The relation between TPMM $M_{xy}(x',y')$ and TEMM

$M_z^o(x', y')$ is given by Eq. (3.88), which can be rewritten as

$$M_{xy}(x', y') = i\ M_z^o(x', y') \sin\alpha, \tag{4.24}$$

where α is the flip angle during excitation, i ($\sqrt{-1}$) denotes that $M_{xy}(x', y')$ is in the $\vec{j}(t)$ direction of the rotating coordinate system X′-Y′. Substituting Eq. (4.24) into Eq. (4.23), we have

$$p(x', \theta) = i\ \sin\alpha \int_{x'=l} M_z^o(x', y')dy', \tag{4.25}$$

which shows that the "external measurement," the projection $p(x', \theta)$, is a straight line integral of the object function TEMM $M_z^o(x', y')$, the "index of the object."

By using the coordinate transform

$$\begin{pmatrix} x' \\ y' \end{pmatrix} = \begin{pmatrix} \cos\theta(j) & \sin\theta(j) \\ -\sin\theta(j) & \cos\theta(j) \end{pmatrix} \begin{pmatrix} x \\ y \end{pmatrix} \tag{4.26}$$

and the definition of k-space sample (Eq. (3.110)), we obtain Eq. (3.122), that is,

$$\mathcal{F}_1\{p(x', \theta)\} = \mathcal{F}_2\{M_{xy}(x', y')\}, \tag{4.27}$$

which is the Fourier slice theorem for MR image reconstruction and is depicted in Figure 3.14.

Eq. (4.23) is the Radon transform for MRI. Based on Eq. (4.27), the formula for the inverse Radon transform for MRI is

$$M_{xy}(x, y) = \frac{1}{2\pi^2} \int_0^{\pi} \int_{-\infty}^{\infty} \frac{\partial p(l, \theta)}{\partial l} \frac{1}{x' - l} dl d\theta, \tag{4.28}$$

where $x' = x\cos\theta + y\sin\theta$. The inner integral (including the coefficient $\frac{1}{2\pi^2}$) is defined as

$$t(x', \theta) = \frac{1}{2\pi^2} \int_{-\infty}^{\infty} \frac{\partial p(l, \theta)}{\partial l} \frac{1}{x' - l} dl d\theta. \tag{4.29}$$

The implementation of Eq. (4.28) can be carried out by two approaches: (1) filtering by Fourier transform where $t(x', \theta)$ is called the filtered projection, and (2) filtering by convolution where $t(x', \theta)$ is known as the convolved projection.

• In the filtering by Fourier transform approach, the first step is to compute the filtered projection $t(x', \theta)$ by

$$t(x', \theta) = \frac{p'(x', \theta)}{i2\pi} \star \frac{-1}{i\pi x'} = \mathcal{F}_1^{-1}\{\mathcal{M}(k, \theta)|k|\}, \tag{4.30}$$

where $\star$ denotes the convolution, and $\mathcal{M}(k, \theta)$ is the radial k-space sample. To derive Eq. (4.30), an operational definition of the measured projection $p(x', \theta)$ is used:

$$p(x', \theta) = \mathcal{F}_1^{-1}\{\mathcal{M}(k, \theta)\}. \tag{4.31}$$

The second step is to compute the backprojection by

$$M_{xy}(x,y) = \int_0^{\pi} t(x',\theta)d\theta. \tag{4.32}$$

• In the filtering by the convolution approach, the first step is to compute the convolved projection $t(x',\theta)$ by

$$t(x',\theta) = p'(x',\theta) \star q(x'), \tag{4.33}$$

where $q(x')$ is the convolution function. The second step is to compute the interpolated data $s_\theta(x,y)$ by

$$s_\theta(x,y) = t(x',\theta) \star \psi(x'), \tag{4.34}$$

where $\psi(x')$ is an interpolation function. The third step is to compute the backprojection by

$$M_{xy}(x,y) = \int_0^{\pi} s_\theta(x,y)d\theta. \tag{4.35}$$

Thus, filtering by the convolution approach leads to a typical FBP algorithm.

Although there is no physical measurement along the straight line in MRI, a straight line integral of k-space samples (Eq. (4.31)) can be properly formulated such that it is a line integral of the object function (Eq. (4.23)). As a result, MR image reconstruction (PR method) is embedded on the Fourier slice theorem and is implemented by a typical tomographic reconstruction algorithm. In this sense, MRI belongs to non-diffraction imaging.

4.4.3 Emission Computed Tomography

A brief review of emission CT is given in this section. This is not intended to provide a systematic description of imaging principles of these techniques and the details of image reconstruction methods used in these techniques. The purpose of this review is to show their non-diffraction imaging nature. It focuses on the fact that in these imaging techniques, the interaction model and the external measurements are characterized by the straight line integrals of some indexes of the objects to be imaged. Emission CT includes positron emission tomography (PET), single-photon emission computed tomography (SPECT), and conventional or planar imaging.

4.4.3.1 Positron Emission Tomography

Although the possibility of positron imaging was discovered in the 1950s, real tomographic imaging occurred after 1972 when X-ray CT was developed. This history may hint that PET and X-ray CT have some intrinsic similarities.

PET utilizes the short-lived radionuclides (also known as positron emitting compounds) generated by a cyclotron. These radionuclides are suitable for (1) radiopharmaceuticals that can be administered the human body, and (2) devices that can detect the annihilation photons and process signals to generate images of the distribution of positron concentrations or their activities.

When a positron is emitted and combined with a nearby electron, two photons of 511 keV are produced. Because these two annihilation photons are generated simultaneously and travel in opposite directions, they are often called simultaneity and collinearity, respectively. When two detectors are placed exactly on the opposite sides along the direction of photon travel, the collinearity of the two annihilation photons provides possibilities to identify the activity (annihilation event) or existence of the positron emitter (concentration). When two annihilation photons arrive at these two detectors within a "coincidence resolving time" (i.e., 10^{-8} seconds or less), the coincidence detection circuits can sense annihilation photons simultaneously and record an event.

It is clear that the total number of coincident events counted by a given pair of detectors is proportional to the integral of the concentration of the positron emitting radionuclides over the column or strip joining these two detectors, and constitutes a measure of the integrated radioactivity, the line-integral projection data.

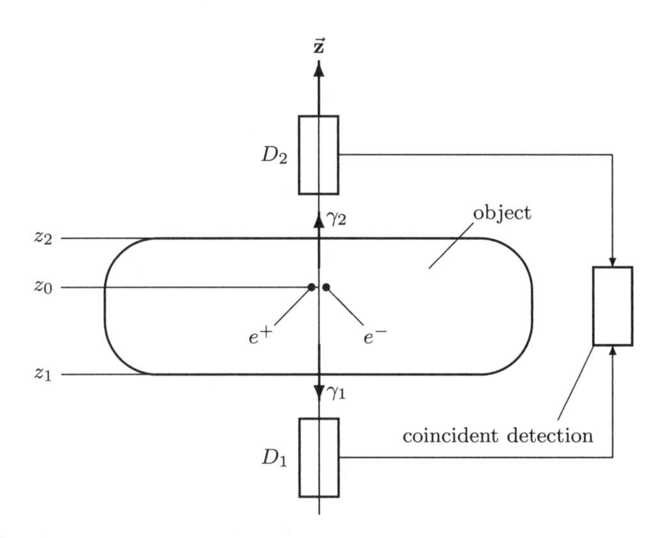

FIGURE 4.3
Illustration of the coincident detection of PET.

In Figure 4.3, e^+ and e^- denote a positron and an electron, γ_1 and γ_2 denote two annihilation photons, D_1 and D_2 represent two detectors, z_i $(i = 0, 1, 2)$

represent three locations on the axis $\vec{\mathbf{z}}$, which is along the direction of photon travel. From Eq. (2.13), the probabilities of photons γ_1 and γ_2 arriving at the detectors D_1 and D_2 are

$$\rho_1 = \exp(-\int_{z_1}^{z_0} f(r,\phi)dz) \quad \text{and} \quad \rho_2 = \exp(-\int_{z_0}^{z_2} f(r,\phi)dz), \tag{4.36}$$

respectively; here, $f(r,\phi)$ represents the linear attenuation coefficient [§] at a point (r,ϕ) in the axis $\vec{\mathbf{z}}$ ((r,ϕ) represents the polar coordinate). Thus, the probability of the annihilation photons being simultaneously detected by the coincident detection is the product of ρ_1 and ρ_2, that is,

$$\rho = \rho_1 \cdot \rho_2 = \exp(-\int_{z_1}^{z_2} f(r,\phi)dz). \tag{4.37}$$

Eq. (4.36) gives a very important outcome. First, it shows that the probability of the coincident detection of the annihilation photons only depends on the attenuation of photons (511 keV) propagating from z_1 to z_2. Second, this attenuation is the same, that is, it does not depend on where the positron annihilation occurs on the line from D_1 to D_2. Rewriting Eq. (4.37), we have

$$-\ln\rho = \int_{z_1}^{z_2} f(r,\phi)dz. \tag{4.38}$$

Similar to X-ray CT (see Eq. (2.16)), ρ is measurable. Thus, we obtain that the "external measurement" characterized by $-\ln\rho$ is a straight line integral of the object function $f(r,\phi)$, the "index of the object."

4.4.3.2 Single-Photon Emission Computed Tomography

SPECT utilizes the decay of radioactive isotopes to generate images of the distribution of the isotope concentration. In the form of radiopharmaceuticals, these isotopes are administered to the human body. Thus, SPECT images can show both anatomic structures and functional states of tissue and organ systems.

SPECT is different from PET, which utilizes the positron and annihilation photons, SPECT is based on detecting individual photons emitted by the radionuclide. Localizing gamma photon emission activity is performed by a narrow mechanical collimator. The first SPECT image was demonstrated in the early 1960s using backprojection reconstruction.

Figure 4.4 is used to illustrate the relations among the concentration of the radioactive isotope, the attenuation of the object, and the measurement, that is, the projection. In Figure 4.4, the line L is specified by a pair of parameters (l,θ). The detector is located at the end of the line L. Let $p(l,\theta)$ denote the

[§]It is not the relative linear attenuation coefficient in X-ray CT.

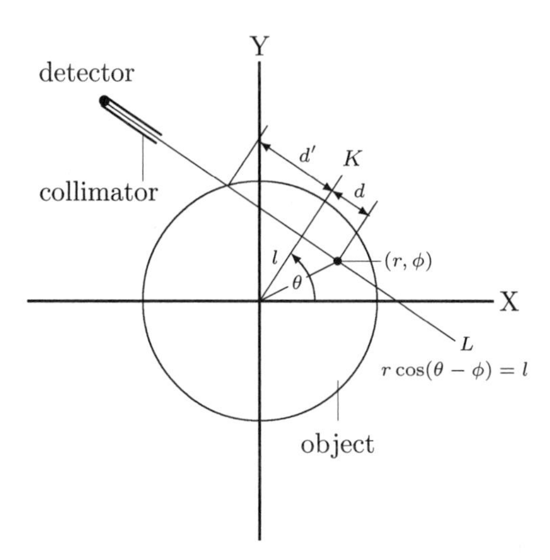

FIGURE 4.4
Illustration of the detection of SPECT.

projection along the line L, that is, at the location l in the view θ. The line K passes through the origin of the X-Y rectangular coordinate system and is perpendicular to the line L: $K \perp L$. d' denotes the distance from the line K to the edge of the object along the line L. For simplicity in discussion, the object is assumed to be convex.

Let (r, ϕ) represent the polar coordinate, $f(r, \phi)$ denote the concentration of the radioactive isotope, and μ be the linear attenuation coefficient. From Eq. (2.13), the original measurement at the line L

$$p'(l, \theta) = \int_L f(r, \phi) e^{-\mu(d'+d)} dz. \tag{4.39}$$

For a given (l, θ), d' is determined by the external shape of the object, and d is given by

$$d = r \sin(\theta - \phi), \tag{4.40}$$

and the projection at the line L is defined as

$$p(l, \theta) = p'(l, \theta) e^{\mu d'} = \int_L f(r, \phi) e^{-\mu d} dz. \tag{4.41}$$

Eq. (4.41) is called the exponential Radon transform. [31] shows that $f(r, \phi)$

can be estimated by

$$f(r,\phi) = \int_0^{2\pi} (\int_{-\infty}^{\infty} p(r\cos(\theta-\phi)-l,\theta)q(l)dl)e^{-\mu r\sin(\theta-\phi)}d\theta, \qquad (4.42)$$

where $q(l)$ is a convolution function that is chosen under certain conditions.

Eq. (4.41) shows that the "external measurement," the projection $p(l,\theta)$, is a straight line integral of the "index of the object," the object function $f(r,\phi)$. Eq. (4.42) essentially represents a filtered backprojection operation. The inner integral—the convolution of the projection and a convolution function—generates a filtered projection in each view. Then, the weighted (by $e^{-\mu r\sin(\theta-\phi)}$) filtered projection forms the backprojection contribution in each view. These are very similar to the FPB operation used in X-ray CT.

4.5 Appendix

4.5.1 Appendix 4A

This appendix proves Eq. (4.7).

Proof

Because

$$\rho(\vec{P}) = \frac{i}{2\sqrt{\pi}} \int_{\vec{P}\cdot\vec{n}\geq 0} \vec{P}\cdot\vec{n}\, e^{-i\vec{P}\cdot\vec{r}} ds(\vec{r}), \qquad (4.43)$$

then

$$\rho^*(-\vec{P}) = \frac{i}{2\sqrt{\pi}} \int_{-\vec{P}\cdot\vec{n}\geq 0} \vec{P}\cdot\vec{n}\, e^{-i\vec{P}\cdot\vec{r}} ds(\vec{r}), \qquad (4.44)$$

where $*$ denotes the complex conjugate. Thus

$$\rho(\vec{P}) + \rho^*(-\vec{P}) = \frac{i}{2\sqrt{\pi}} \int\int_S \vec{P}\cdot\vec{n}\, e^{-i\vec{P}\cdot\vec{r}} ds(\vec{r}). \qquad (4.45)$$

Let

$$\vec{P}e^{-i\vec{P}\cdot\vec{r}} = \vec{u} \quad \text{and} \quad \vec{n}ds(\vec{r}) = \vec{ds}, \qquad (4.46)$$

and using divergence theorem

$$\int\int_S \vec{u}\cdot\vec{ds} = \int\int\int_V \nabla\cdot\vec{u}dv, \qquad (4.47)$$

Eq. (4.45) can be written as

$$\rho(\vec{P}) + \rho^*(-\vec{P}) = \frac{i}{2\sqrt{\pi}} \int\int\int_V$$
$$(\tfrac{\partial}{\partial x}(p_x e^{-i\vec{P}\cdot\vec{r}}) + \tfrac{\partial}{\partial y}(p_y e^{-i\vec{P}\cdot\vec{r}}) + \tfrac{\partial}{\partial z}(p_z e^{-i\vec{P}\cdot\vec{r}}))dv(\vec{r}), \qquad (4.48)$$

where $\vec{P} = p_x\vec{i} + p_y\vec{j} + p_z\vec{k}$.

Because $\vec{P}$ is fixed in the space and

$$\vec{P} \cdot \vec{r} = xp_x + yp_y + zp_z, \tag{4.49}$$

therefore

$$\begin{aligned} \tfrac{\partial}{\partial x}(p_x e^{-i\vec{P}\cdot\vec{r}}) &= -ip_x^2 e^{-i\vec{P}\cdot\vec{r}} \\ \tfrac{\partial}{\partial y}(p_y e^{-i\vec{P}\cdot\vec{r}}) &= -ip_y^2 e^{-i\vec{P}\cdot\vec{r}} \\ \tfrac{\partial}{\partial z}(p_z e^{-i\vec{P}\cdot\vec{r}}) &= -ip_z^2 e^{-i\vec{P}\cdot\vec{r}}. \end{aligned} \tag{4.50}$$

Substituting Eq. (4.50) into Eq. (4.48), we obtain

$$\rho(\vec{P}) + \rho^*(-\vec{P}) = \frac{|\vec{P}|^2}{2\sqrt{\pi}} \int\int\int_V e^{-i\vec{P}\cdot\vec{r}} dv(\vec{r}). \tag{4.51}$$

■

Problems

4.1. Prove and interpret Eq. (4.21).

4.2. Prove Eqs. (4.40) and (4.41) regardless that the point (r, ϕ) is below or above the intersection point of lines L and K (along the line L).

4.3. Prove and interpret Eq. (4.45).

References

[1] Herman, G.: *Image Reconstruction from Projections.* Academic Press, New York (1980).

[2] Macovski, A.: Physical problems of computerized tomography. *IEEE Proc.* **71**(3) (1983).

[3] Lee, H., (Ed), G.W.: *Imaging Technology.* IEEE Press, New York (1986).

[4] Kak, A., Slaney, M.: *Principles of Computerized Tomographic Imaging.* IEEE Press, New York (1988).

[5] Jain, A.K.: *Fundamentals of Digital Image Processing.* Prentice Hall, Englewood Cliffs, NJ (1989).

[6] Cho, Z.H., Jones, J.P., Singh, M.: *Foundations of Medical Imaging.* John Wiley & Sons, Inc., New York (1993).

[7] Hsieh, J.: *Computed Tomography, Principles, Design, Artifacts and Recent Advances.* SPIE, Bellingham (2004).

[8] Kalender, W.: *Computed Tomography: Fundamentals, System Technology, Image Quality, Applications.* Publics MCD Verlag, Germany (2005).

[9] Buzug, T.: *Computed Tomography.* Springer, Berlin (2008).

[10] Herman, G.: *Fundamentals of Computerized Tomography.* Springer, London (2009).

[11] Bloch, F.: Nuclear induction. *Phys. Rev.* **70**(7 and 8) (1946) 460–474.

[12] Purcell, E., Torrey, H., Pound, R.: Resonance absorption by nuclear magnetic moments in a solid. *Phys. Rev.* **69** (1946) 37–38.

[13] Kumar, A., Welti, D., Ernst, R.: NMR Fourier zeugmatography. *J. Magn. Res.* **18** (1975) 69–75.

[14] Wuthrich, K., Wider, G., Wagner, G., Braun, W.: Sequential resonance assignments as a basis for determination of spatial protein structures by high resolution proton nuclear magnetic resonance. *J. Mol. Biol.* **155** (1982) 311–319.

[15] Lauterbur, P.: Image formation by induced local interactions: Examples employing nuclear magnetic resonance. *Nature* **242** (1973) 190–191.

[16] Mansfield, P., Grannell, P.: NMR 'diffraction' in solids? *J. Phys C: Solid State Phys.* **6** (1973) 422–427.

[17] Slichter, C.: *Principles of Magnetic Resonance.* Springer-Verlag, Berlin (1980).

[18] Haacke, E., Brown, R., Thompson, M., Venkatesan, R.: *Magnetic Resonance Imaging: Physical Principles and Sequence Design.* John Wiley & Sons Inc., New York (1999).

[19] Nishimura, D.: *Principles of Magnetic Resonance Imaging.* Stanford University, Palo Alto, California (1996).

[20] Liang, Z.P., Lauterbur, P.: *Principles of Magnetic Resonance Imaging, A Signal Processing Perspective.* IEEE Press, New York (2000).

[21] Bendriem, B., (Eds), D.T.: *The Theory and Practice of 3D PET.* Kluwer Academic Publishers, Dordrecht, The Netherlands (2010).

[22] Wernick, M., (Ed), J.A.: *Emission Tomography - The fundamentals of PET and SPECT.* Elsevier Academic Press, Amsterdam, The Netherlands (2004).

[23] Bailey, D., Townsend, D., Valk, P., (Eds), M.M.: *Positron Emission Tomography: Basic Sciences.* Springer-Verlag, London, United Kingdom (2005).

[24] Sorenson, J., Phelps, M.: *Physics in Nuclear Medicine.* Grune & Stratton, Inc., Orlando, Florida (1987).

[25] Barrett, H., Myers, K.: *Foundations of Image Science.* John Wiley & Sons, Inc., New York (2004).

[26] Glover, G., Sharp, J.: Reconstruction of ultrasound propagation speed distribution in soft tissue: Time-of-flight tomography. *IEEE Trans. Sonic Ultrason.* **24** (1977) 229–234.

[27] Miller, J., O'Donnell, M., Mimbs, J., Sobel, B.: Ultrasonic attenuation in normal and ischemic myocardium. *Proc. Second Int. Symp. on Ultrasonic Tissue Characterization,* National Bureau of Standards (1977).

[28] Shutilov, V.: *Fundamental Physics of Ultrasonics.* Gordon and Breach, New York, New Jersey (1988).

[29] Hill, C.: *Physical Principles of Medical Ultrasonics.* Ellis Horwood, Chichester, United Kingdom (1986).

[30] Stark, H.: *Image Recovery: Theory and Application.* Academic Press, New York (1987).

[31] Tretiak, O.J., Metz, C.: The exponential Radon transform. *SIAM J. Appl. Math.* **39** (1980). 341–354

5

Statistics of X-Ray CT Imaging

5.1 Introduction

As shown in Chapter 2, X-ray CT images are reconstructed from projections, while the projections are formulated from photon measurements. In X-ray CT imaging, photon measurements and projections are called CT imaging data, abbreviated as CT data.*

Similar to any type of realistic data, each type of CT data consists of its signal and noise components. In photon measurements, the instrumental and environmental noise form the noise component of photon measurements, which is random. The emitted and detected photons (numbers) form the signal component of the photon measurements, which is also random, due to the intrinsic variations in the emission and detection processes.

The parallel projection is a translation-rotation mode (Sections 2.3 and 2.4). Within one view, the photon measurements and the projections are acquired and formulated sequentially in time from one projection location to another. The divergent projection is a rotation mode (Sections 2.3 and 2.4). Within one view, the photon measurements and the projections are acquired and formulated simultaneously for all projections. Over all views, both parallel and divergent projections are collected sequentially in time from one view location to another. Thus, projections are spatio-temporal in nature. Because the time interval of CT data collection, particularly in the dievergent projection, is very short, the time argument in CT data is excluded in the process of image reconstruction.

This chapter describes the statistics of both signal and noise components of each type of CT data, and is focused on their second-order statistics. Based on the physical principles of X-ray CT described in Chapter 2 and according to CT data acquisition procedures, the statistical description of X-ray CT imaging is progresses in the following order: photon measurement (emission $\rightarrow$ detection $\rightarrow$ emission and detection) $\Longrightarrow$ projection.

This chapter also provides signal processing paradigms for the convolution image reconstruction method for the parallel and divergent projections. Then,

*Imaging is often referred to as a process or an operation from the imaging data (e.g., the acquired measurements) to the reconstructed pictures.

it gives a statistical interpretation to CT image reconstruction. CT image reconstruction can be viewed as a transform from a set of random variables (projections) to another set of random variables (pixel intensities). These new random variables form a spatial random process, also known as a random field. Statistics of CT data in the imaging domain propagate to the statistics in the image domain through image reconstruction.

Discussions in this chapter are confined to the monochromatic X-ray, the basic parallel and divergent projections, and the convolution image reconstruction method [1–10].

5.2 Statistics of Photon Measurements

This section describes statistics of the photon measurements in terms of their signal and noise components.

5.2.1 Statistics of Signal Component

Physical principles of photon emission, attenuation, and detection are described in Section 2.2; some parameters in these processes are given in Section 2.3. In this section, statistics of the signal component of photon measurements are analyzed in the following order: photon emission $\rightarrow$ photon detection $\rightarrow$ photon emission and detection, based on their physical principles.

5.2.1.1 Photon Emission

In the process of photon emission, let t_i and $(t_i, t_{i+1}]$ $(i = 0, 1, 2, \cdots)$ denote the time instant and the time interval, respectively; $n(t_i, t_{i+1}]$ represents the number of photons emitted in the interval $(t_i, t_{i+1}]$. $n(t_i, t_{i+1}]$ is a random variable. $P(n(t_i, t_{i+1}] = m)$ denotes the probability that m photons are emitted in the time interval $(t_i, t_{i+1}]$.

Generally, similar to other particle emissions in physics, photon emission in X-ray CT is considered to follow the Poisson law. Specifically, the photon emission in X-ray CT satisfies the following four conditions:

1) $P(n(t_i, t_{i+1}] = m)$ depends on m and the interval $\tau_i = t_{i+1} - t_i$ only; it does not depend on the time instant t_i.

2) For the nonoverlapping intervals $(t_i, t_{i+1}]$, random variables $n(t_i, t_{i+1}]$ $(i = 0, 1, 2, \cdots)$ are independent

3) There are only finite numbers of photons emitted in a finite interval, $P(n(t_i, t_{i+1}] = \infty) = 0$. Also, $P(n(t_i, t_{i+1}] = 0) \neq 1$.

4) The probability that more than one photon is emitted in an interval

$\tau_i = t_{i+1} - t_i$ approaches zero as $\tau_i \to 0$. That is,

$$\lim_{\tau \to 0} \frac{1 - P(n(t_i, t_{i+1}] = 0) - P(n(t_i, t_{i+1}] = 1)}{\tau_i} = 0.$$

Therefore, photon emission in X-ray CT forms a Poisson process [11, 12].

Let λ be the average number of photons emitted per unit time by a stable X-ray source and X represents the random variable $n(t_i, t_{i+1}]$. The probability mass function (pmf) of the number of emitted photons, $P_X(m) = P(n(t_i, t_{i+1}] = m)$, is given by

$$P_X(m) = e^{-\lambda} \frac{\lambda^m}{m!} \qquad (0 < m < \infty), \tag{5.1}$$

where m is an integer. X has a Poisson distribution, denoted by $X \sim P(\lambda)$. Its mean and variance are

$$\mu_X = \lambda \qquad \text{and} \qquad \sigma_X^2 = \lambda. \tag{5.2}$$

5.2.1.2 Photon Detection

Suppose the probability that one of the photons emitted by a stable X-ray source is counted by the detector (without having been absorbed or scattered when it passes through the object) is ρ. Then, the probability that n out of m emitted photons are counted by this detector (without having been absorbed or scattered when it passes through the object) follows an mth-order binomial law. This probability, denoted by $P(n|m)$, is given by

$$P(n|m) = \frac{m!}{n!(m-n)!} \rho^n (1-\rho)^{m-n} \qquad (0 < n < m), \tag{5.3}$$

where n and m are integers.

Let Y represent the number of emitted photons that are counted by the detector without having been absorbed or scattered when passing through the object. Y has a binomial distribution, denoted by $Y \sim B(m, \rho)$. Its mean and variance are

$$\mu_Y = m\rho \qquad \text{and} \qquad \sigma_Y^2 = m\rho(1-\rho). \tag{5.4}$$

It is known that when m is sufficiently large, ρ is sufficiently small, and $m\rho = \lambda$, the Poisson distribution $P(\lambda)$ is the limiting case of binomial distribution $B(m, \rho)$ as $m \to \infty$ [11, 12].

5.2.1.3 Photon Emission and Detection

Combining the photon emission (Eq. (5.1)) and the photon detection (Eq. (5.3)), the probability that the number of photons emitted from a steady

X-ray source and counted by a detector is

$$
\begin{aligned}
P_Y(n) &= \sum_{m=0}^{\infty} P(n|m)P_X(m) \\
&= \sum_{m=0}^{\infty} \frac{m!}{n!(m-n)!}\rho^n(1-\rho)^{m-n} \cdot e^{-\lambda}\frac{\lambda^m}{m!} \\
&= \frac{\rho^n\lambda^n}{n!}e^{-\lambda}\sum_{m=0}^{\infty}\frac{\lambda^{(m-n)}}{(m-n)!}(1-\rho)^{m-n} \\
&= \frac{(\lambda\rho)^n}{n!}e^{-\lambda}\sum_{k=0}^{\infty}\frac{\lambda^k}{k!}(1-\rho)^k \\
&= \frac{(\lambda\rho)^n}{n!}e^{-\lambda}e^{\lambda(1-\rho)} \\
&= e^{-\lambda\rho}\frac{(\lambda\rho)^n}{n!} \qquad (0 < n < \infty),
\end{aligned} \tag{5.5}
$$

where n is an integer [1, 13].

Eq. (5.5) represents the pmf of a Poisson distribution $Y \sim P(\lambda\rho)$. Its mean and variance are

$$\mu_Y = \lambda\rho \qquad \text{and} \qquad \sigma_Y^2 = \lambda\rho\,. \tag{5.6}$$

It is known [11, 12] that when $\lambda\rho$ is sufficiently large, the Gaussian distribution $N(\lambda\rho, \lambda\rho)$ is an excellant approximation of the Poisson distribution $P(\lambda\rho)$.† That is, $P(\lambda\rho) \longrightarrow N(\lambda\rho, \lambda\rho)$ as $\lambda\rho \to \infty$. In X-ray CT, $\lambda\rho$ is always very large; thus, $Y \sim N(\mu_Y, \sigma_Y^2)$; the mean μ_Y and the variance σ_Y^2 are given by Eq. (5.6); and the probability density function (pdf) is given by

$$p_Y(y) = \frac{1}{\sqrt{2\pi\sigma_Y^2}}\exp(-\frac{(y-\mu_Y)^2}{2\sigma_Y^2}). \tag{5.7}$$

The number Y of detected X-ray photons forms the signal component of the photon measurement in X-ray CT. It is characterized by a Gaussian random variable. Because Y cannot be negative and infinite, Y must be limited in a finite range $[0, M]$. M depends on the physical settings and conditions of the X-ray production system. Because $\mu_Y = \lambda\rho$ is usually very large, it is possible to make $M = 2\mu_Y - 1$. Thus, $Y \sim N(\mu_Y, \sigma_Y^2)$ over $[0, M]$ is symmetric with respect to the mean μ_Y.

†Bernoulli-Poisson theorem [14] or DeMoivre-Laplace theorem [12] shows that the binomial distribution $B(n, p)$ can also be accurately approximated by a Gaussian distribution $N(np, np(1-p))$ as n approaches infinity.

5.2.2 Statistics of Noise Component

The noise component of the photon measurement in X-ray CT, denoted by N, mainly consists of the noise caused by detection system electronics. At the lower frequencies, it is mainly characterized by flicker noise with the $\frac{1}{f}$ spectrum (f denotes the frequency); at higher frequencies, it is dominated by the white noise with the flat spectrum. Overall, the noise in X-ray CT is characterized by a Gaussian random variable with zero mean and variance σ_N^2, its pdf is given by

$$p_N(n) = \frac{1}{\sqrt{2\pi\sigma_N^2}} \exp(-\frac{n^2}{2\sigma_N^2}). \tag{5.8}$$

That is, $N \sim N(0, \sigma_N^2)$. It is additive and independent of the signal component of photon measurement Y.

5.2.3 Statistics of Photon Measurements

The photon measurement in X-ray CT, denoted by Z, consists of two components: signal Y and noise N.

$$Z = Y + N. \tag{5.9}$$

Because Y and N are characterized by two independent Gaussian random variables, Z is a Gaussian randow variable with the mean μ_Z and variance σ_Z^2 given by

$$\mu_Z = \mu_Y = \lambda\rho \quad \text{and} \quad \sigma_Z^2 = \mu_Y + \sigma_N^2 = \lambda\rho + \sigma_N^2, \tag{5.10}$$

its pdf is given by

$$p_Z(z) = \frac{1}{\sqrt{2\pi\sigma_Z^2}} \exp(-\frac{(z-\mu_Z)^2}{2\sigma_Z^2}). \tag{5.11}$$

That is, $Z \sim N(\mu_Z, \sigma_Z^2)$. Eq. (5.10) shows that the intrinsic variance of the signal component $\sigma_Y^2 = \lambda\rho$ sets the ultimate limit on the variance σ_Z^2 of the photon measurement in X-ray CT.

The photon measurement Z has the following statistical property.

Property 5.1 The photon measurement Z of Eq. (5.9) represents the outcome of a cascaded photon emission process and photon detection process in X-ray CT. It is characterized by a Gaussian random variable with the mean and the variance of Eq. (5.10) and the pdf of Eq. (5.11).

Proof.

Using Eq. (5.1) (for X in the photon emission), Eqs. (5.5) through (5.7) (for Y in the photon detection), Eq. (5.8) (for N - the noise component), and Eqs. (5.9)-(5.11), the property is proved. ■

5.3 Statistics of Projections

5.3.1 Statistics of a Single Projection

The projection in X-ray CT was described and defined in Section 2.4.1. From Eqs. (2.16) and (2.17), it can be rewritten as[‡]

$$p(l,\theta) = -\ln \frac{N_{ad}/N_{ar}}{N_{cd}/N_{cr}} = -\ln N_{ad} + \ln N_{ar} + \ln N_{cd} - \ln N_{cr}, \tag{5.12}$$

where N_{ad} and N_{ar} are the numbers of emitted photons counted by the detector and the reference detector in the actual measurement process, and N_{cd} and N_{cr} are the numbers of emitted photons counted by the detector and the reference detector in the calibration measurement process.

As described in Section 5.2.3, each of N_{ad}, N_{ar}, N_{cd} and N_{cr} can be characterized by a Gaussian random variable as Z of Eq. (5.9) with pdf given by Eq. (5.11) and parameters given by Eq. (5.10), but has the different statistical contents. In order to see these different contents, some practical issues in photon emission and detection should be discussed.

(1) It is known that not every photon that reaches the detector is necessarily counted by the detector. The probability of a photon that reaches the detector being counted by the detector is called efficiency, denoted by ς. The overall efficiency of a detector is the product of the geometric efficiency (a ratio of the X-ray-sensitive area of the detector to the total exposed area) and the quantum efficiency (a fraction of incoming photons that are absorbed and contributed to the output signal) [1, 9, 10]. Clearly, $\varsigma < 1$, and the transmittance ρ in Section 5.2 should be replaced by $\rho\varsigma$.

(2) As shown in Figures 2.4 through 2.6, photons emitted from an X-ray source are counted in two paths: a detection path and a reference detection path. Let f_d and f_r denote the fractions of photon emission in these two paths. Then, λ in Section 5.2, which stands for the average number of photons emitted per unit time by a stable X-ray source in the detection path and the reference detection path, should be $f_d\lambda$ and $f_r\lambda$, respectively.

Combining the above two factors (1) and (2), the mean of each random variable of N_{ad}, N_{ar}, N_{cd}, and N_{cr} will have the form $f\lambda\rho\varsigma$ and their variance will have the form $f\lambda\rho\varsigma + \sigma_N^2$. More specifically, let $\mu_{Z_{ad}}$, $\mu_{Z_{ar}}$, $\mu_{Z_{cd}}$ and $\mu_{Z_{cr}}$ denote the means, $\sigma^2_{Z_{ad}}$, $\sigma^2_{Z_{ar}}$, $\sigma^2_{Z_{cd}}$, and $\sigma^2_{Z_{cr}}$ denote the variances, of N_{ad}, N_{ar}, N_{cd}, and N_{cr}, then, they are

$$\mu_{Z_{ad}} = f_d\lambda_a\rho_a\varsigma_d \quad \text{and} \quad \sigma^2_{Z_{ad}} = f_d\lambda_a\rho_a\varsigma_d + \sigma_N^2$$
$$\mu_{Z_{ar}} = f_r\lambda_a\rho_r\varsigma_r \quad \text{and} \quad \sigma^2_{Z_{ar}} = f_r\lambda_a\rho_r\varsigma_r + \sigma_N^2$$

[‡] $p(l,\theta)$ is used for the general description. Specifically, $p(l,\theta)$ and $p(\sigma,\beta)$ are used for the parallel and divergent projections, respectively.

$$\mu_{Z_{cd}} = f_d \lambda_c \rho_c \varsigma_d \quad \text{and} \quad \sigma^2_{Z_{cd}} = f_d \lambda_c \rho_c \varsigma_d + \sigma^2_N$$
$$\mu_{Z_{cr}} = f_r \lambda_c \rho_r \varsigma_r \quad \text{and} \quad \sigma^2_{Z_{cr}} = f_r \lambda_c \rho_r \varsigma_r + \sigma^2_N, \tag{5.13}$$

where the subscript a stands for the actutal measurement process, c for the calibration measurement process, d for the detection, and r for the reference detection.

The following discussions give the statistical properties of a single projection $p(l, \theta)$ defined by Eq. (5.12).

Property 5.2a Let the random variable Z defined by Eq. (5.9) denote each of N_{ad}, N_{ar}, N_{cd}, and N_{cr} of Eq. (5.12). Let the random variable W defined by

$$W = \ln Z \tag{5.14}$$

represent each of $\ln N_{ad}$, $\ln N_{ar}$, $\ln N_{cd}$, and $\ln N_{cr}$. W has the pdf $p_W(w)$ given by

$$p_W(w) = \frac{1}{\sqrt{2\pi\sigma^2_Z}} \exp(-\frac{(e^w - \mu_Z)^2}{2\sigma^2_Z}) e^w, \tag{5.15}$$

and the mean μ_W and the variance σ^2_W given by

$$\mu_W \simeq \ln \mu_Z - \frac{\sigma^2_Z}{2\mu^2_Z} \quad \text{and} \quad \sigma^2_W \simeq \frac{\sigma^2_Z}{\mu^2_Z}(1 + \frac{\sigma^2_Z}{2\mu^2_Z}), \tag{5.16}$$

where μ_Z and σ^2_Z are the mean and the variance of Z given by Eq. (5.10) but with different contents given by Eq. (5.13).

Proof.

Section 5.2 shows that Z is in the range $[0, M]$, where $M = 2\mu_Z - 1$. Because $W = \ln Z$ is not defined for $Z = 0$, a truncated random variable Z is adopted [1, 10]. That is, Z is in the range $[1, M]$.[§] Thus, W is in the range $[0, \ln M]$.

The random variable Z has a Gaussian pdf given by Eq. (5.11). Using the transformation of random variables, the pdf of the random variable W is

$$p_W(w) = \frac{c_n}{\sqrt{2\pi\sigma^2_Z}} \exp(-\frac{(e^w - \mu_Z)^2}{2\sigma^2_Z}) e^w, \tag{5.17}$$

where c_n is a normalization constant determined by

$$\int_0^{\ln M} p_W(w) = 1. \tag{5.18}$$

Appendix 5A shows that $c_n \simeq 1$. Thus, Eq. (5.17) becomes Eq. (5.15).

[§]The probability $P(Z = 1)$ of the truncated random variable is understood to be equal to the probability $P(Z = 0) + P(Z = 1)$ of the original random variable, whose meaning can be seen in the property (iv) of Section 5.2.1.1.

In Eq. (5.15), let $e^w = u$, $\frac{e^w - \mu_Z}{\sigma_Z} = v$, and use the approximation $\ln(1+x) \simeq x - \frac{x^2}{2}$ $(-1 < x \leq 1)$, after some mathematical and statistical manipulations, Appendix 5A shows that the mean μ_W and the variance σ_W^2 of W are given by Eq. (5.16). ■

Discussions. In Eq. (5.10), if $\sigma_N^2 \ll \mu_Z$, then, in Eq. (5.16), $\frac{\sigma_Z^2}{\mu_Z^2} = \frac{\mu_Z + \sigma_Z^2}{(\mu_Z)^2} \simeq \frac{1}{\mu_Z}$. Thus, the mean μ_W and the variance σ_W^2 of W, that is, Eq. (5.16), become

$$\mu_W \simeq \ln \mu_Z - \frac{1}{2\mu_Z} \quad \text{and} \quad \sigma_W^2 \simeq \frac{1}{\mu_Z}(1 + \frac{1}{2\mu_Z}). \tag{5.19}$$

In practice, the transmittance ρ defined and described in Section 2.3.1 and Section 5.2.1 is spatially variant because ρ is different at the different locations inside the object. From Eq. (5.10), the mean $\mu_Z = \lambda\rho$ and the variance $\sigma_Z^2 = \lambda\rho + \sigma_N^2$ are spatially variant. Thus, the mean μ_W and the variance σ_W^2 are also spatially variant.

Using the definition and the notation of W of Eq. (5.14), the projection $p(l, \theta)$ of Eq. (5.12) can be rewritten as

$$p(l, \theta) = -W_{ad} + W_{ar} + W_{cd} - W_{cr}. \tag{5.20}$$

Based on statistical properties of W (Property 5.2a), the compound W s of Eq. (5.20) give the following statistical property of the projection $p(l, \theta)$.

Property 5.2b The mean $\mu_{p(l,\theta)}$ and the variance $\sigma_{p(l,\theta)}^2$ of the projection $p(l, \theta)$ are

$$\mu_{p(l,\theta)} = -\ln \frac{\mu_{Z_{ad}}/\mu_{Z_{ar}}}{\mu_{Z_{cd}}/\mu_{Z_{cr}}} - \frac{1}{2}\left(-\frac{\sigma_{Z_{ad}}^2}{\mu_{Z_{ad}}^2} + \frac{\sigma_{Z_{ar}}^2}{\mu_{Z_{ar}}^2} + \frac{\sigma_{Z_{cd}}^2}{\mu_{Z_{cd}}^2} - \frac{\sigma_{Z_{cr}}^2}{\mu_{Z_{cr}}^2}\right), \tag{5.21}$$

and

$$\begin{aligned} \sigma_{p(l,\theta)}^2 = & \frac{\sigma_{Z_{ad}}^2}{\mu_{Z_{ad}}^2}\left(1 + \frac{\sigma_{Z_{ad}}^2}{2\mu_{Z_{ad}}^2}\right) + \frac{\sigma_{Z_{ar}}^2}{\mu_{Z_{ar}}^2}\left(1 + \frac{\sigma_{Z_{ar}}^2}{2\mu_{Z_{ar}}^2}\right) \\ & + \frac{\sigma_{Z_{cd}}^2}{\mu_{Z_{cd}}^2}\left(1 + \frac{\sigma_{Z_{cd}}^2}{2\mu_{Z_{cd}}^2}\right) + \frac{\sigma_{Z_{cr}}^2}{\mu_{Z_{cr}}^2}\left(1 + \frac{\sigma_{Z_{cr}}^2}{2\mu_{Z_{cr}}^2}\right). \end{aligned} \tag{5.22}$$

Proof.

The pdf of each W is given by Eq. (5.15). However, the pdf of $p(l, \theta)$ of Eq. (5.20) may be too complicated to be expressed in a closed form. Thus, instead of being based on the pdf of $p(l, \theta)$, we derive the mean $\mu_{p(l,\theta)}$ and the variance $\sigma_{p(l,\theta)}^2$ of $p(l, \theta)$ in terms of means μ_W and variances σ_W^2 of W s.

By taking the expectation of Eq. (5.20), we have

$$\mu_{p(l,\theta)} = E[p(l, \theta)] = -\mu_{W_{ad}} + \mu_{W_{ar}} + \mu_{W_{cd}} - \mu_{W_{cr}}. \tag{5.23}$$

Using Eq. (5.16) to Eq. (5.23), we obtain Eq. (5.21).

The variance $\sigma^2_{p(l,\theta)}$ of the projection $p(l,\theta)$ is

$$\begin{aligned}\sigma^2_{p(l,\theta)} &= E[(p(l,\theta) - \mu_{p(l,\theta)})^2] \\ &= E[(-(W_{ad} - \mu_{W_{ad}}) + (W_{ar} - \mu_{W_{ar}}) \\ &\quad + (W_{cd} - \mu_{W_{ac}}) - (W_{cr} - \mu_{W_{cr}}))^2]. \end{aligned} \tag{5.24}$$

Let $W_{ad} - \mu_{W_{ad}} = \delta W_{ad}$, $W_{ar} - \mu_{W_{ar}} = \delta W_{ar}$, $W_{cd} - \mu_{W_{cd}} = \delta W_{cd}$, and $W_{cr} - \mu_{W_{cr}} = \delta W_{cr}$; Eq. (5.24) can be expressed as

$$\sigma^2_{p(l,\theta)} = E[(\delta W_{ad})^2 + (\delta W_{ar})^2 + (\delta W_{cd})^2 + (\delta W_{cr})^2] \tag{5.25}$$

$$+ E[-\delta W_{ad}(\delta W_{ar} + \delta W_{cd} - \delta W_{cr})] \tag{5.26}$$

$$+ E[\delta W_{ar}(-\delta W_{ad} + \delta W_{cd} - \delta W_{cr})] \tag{5.27}$$

$$+ E[\delta W_{cd}(-\delta W_{ad} + \delta W_{ar} - \delta W_{cr})] \tag{5.28}$$

$$+ E[-\delta W_{cr}(-\delta W_{ad} + \delta W_{ar} + \delta W_{cd})]. \tag{5.29}$$

In Eq. (5.26), the co-variances $E[\delta W_{ad}\delta W_{cd}] = 0$; and $E[\delta W_{ad}\delta W_{cr}] = 0$, this is because δW_{ad} and δW_{cd}, δW_{ad} and δW_{cr}, are in different processes (actual measurement and calibration measurement); hence, they are independent. The co-variance $E[\delta W_{ad}\delta W_{ar}] = 0$; this is because, although δW_{ad} and δW_{ar} are in the same process, but they are a-dependent [12]. Here, δW_{ad} and δW_{ar} are treated as two divergent projections (Section 5.3.2). Thus, Eq. (5.26) becomes zero. Similarly, Eqs. (5.27), (5.28), and (5.29) are all equal to zero.

The remaining Eq. (5.25) is a sum of four variances that gives

$$\sigma^2_{p(l,\theta)} = \sigma^2_{W_{ad}} + \sigma^2_{W_{ar}} + \sigma^2_{W_{cd}} + \sigma^2_{W_{cr}}. \tag{5.30}$$

Using Eq. (5.16) to Eq. (5.30), we obtain Eq. (5.22). ■

Discussion. When $\sigma^2_N \ll \mu_Z$, Eq. (5.19) holds. With this approximation, Eqs. (5.21) and (5.22) become

$$\mu_{p(l,\theta)} = -\ln\frac{\mu_{Z_{ad}}/\mu_{Z_{ar}}}{\mu_{Z_{cd}}/\mu_{Z_{cr}}} - \frac{1}{2}\left(-\frac{1}{\mu_{Z_{ad}}} + \frac{1}{\mu_{Z_{ar}}} + \frac{1}{\mu_{Z_{cd}}} - \frac{1}{\mu_{Z_{cr}}}\right), \tag{5.31}$$

and

$$\sigma^2_{p(l,\theta)} = \frac{1}{\mu_{Z_{ad}}} + \frac{1}{\mu_{Z_{ar}}} + \frac{1}{\mu_{Z_{cd}}} + \frac{1}{\mu_{Z_{cr}}} + \frac{1}{2\mu^2_{Z_{ad}}} + \frac{1}{2\mu^2_{Z_{ar}}} + \frac{1}{2\mu^2_{Z_{cd}}} + \frac{1}{2\mu^2_{Z_{cr}}}. \tag{5.32}$$

The discussion of Property 5.2a indicates the mean μ_W and the variance σ^2_W of W are spatially variant. Thus, from Eqs. (5.23) and (5.30), the mean

$\mu_{p(l,\theta)}$ and the variance $\sigma^2_{p(l,\theta)}$ of the projection $p(l,\theta)$ are spatially variant. That is, projections $p(l,\theta)$ with different (l,θ) may have different means and variances, but the same type of pdf.

Property 5.2c The projection $p(l,\theta)$ defined by Eq. (5.20) can be characterized by an approximated Gaussian distribution $N(\mu_{p(l,\theta)},\sigma^2_{p(l,\theta)})$. That is, the pdf of $p(l,\theta)$ is approximated by

$$p(p(l,\theta)) = \frac{1}{\sqrt{2\pi\sigma^2_{p(l,\theta)}}}\exp(-\frac{(p(l,\theta)-\mu_{p(l,\theta)})^2}{2\sigma^2_{p(l,\theta)}}), \tag{5.33}$$

where the mean $\mu_{p(l,\theta)}$ and the variance $\sigma^2_{p(l,\theta)}$ are given by Eqs. (5.21) and (5.22), respectively.

Proof.

Each component on the right side of Eq. (5.20): W_{ad}, W_{ar}, W_{cd}, and W_{cr}, has a similar pdf: $p_{W_{ad}}(w)$, $p_{W_{ar}}(w)$, $p_{W_{cd}}(w)$, and $p_{W_{cr}}(w)$, given by Eq. (5.15), and the finite mean $\mu_{p(l,\theta)}$ and the finite variance $\sigma^2_{p(l,\theta)}$, given by Eq. (5.16) (or Eq. (5.19)). As shown in the proof of Property 5.2b, these four components are also independent. Thus, the true pdf of the projection $p(l,\theta)$ of Eq. (5.20), denoted by $\hat{p}(p(l,\theta))$, can be expressed as

$$\hat{p}(p(l,\theta)) = p_{W_{ad}}(-w) \star p_{W_{ar}}(w) \star p_{W_{cd}}(w) \star p_{W_{cr}}(-w), \tag{5.34}$$

where $\star$ denotes the convolution.

Based on the Central-Limit theorem, $\hat{p}(p(l,\theta))$ can be approximated by a Gaussian pdf $p(p(l,\theta))$ of Eq. (5.33). An error correction procedure is given in Appendix 5B. It provides the means to evaluate the accuracy of this approximation. An example is also included in Appendix 5B. It demonstrates that this approximation is quite accurate. ■

Note. Generally, we use $p(l,\theta)$ to represent a projection in X-ray CT imaging. Specifically, as shown in Chapter 2 and in the following sections, $p(l,\theta)$ is a notation for the parallel projection. It is clear that Properties 5.2a, 5.2b, and 5.2c also hold for the divergent projection $p(\sigma,\beta)$.

5.3.2 Statistics of Two Projections

This subsection describes the statistical relationship between two projections.

Property 5.3 In the parallel mode ((l,θ)-space), two projections $p(l_1,\theta_1)$ and $p(l_2,\theta_2)$ in either the same view ($\theta_1=\theta_2$) or the different views ($\theta_1\neq\theta_2$) are uncorrelated. That is, the correlation $R_p((l_1,\theta_1),(l_2,\theta_2))$ of $p(l_1,\theta_1)$ and $p(l_2,\theta_2)$ is

$$R_p((l_1,\theta_1),(l_2,\theta_2)) = E[p(l_1,\theta_1)p(l_2,\theta_2)] = \mu_{p(l_1,\theta_1)}\mu_{p(l_2,\theta_2)}$$

$$+ r_p((l_1,\theta_1),(l_2,\theta_2))\sigma_{p(l_1,\theta_1)}\sigma_{p(l_2,\theta_2)}\delta[l_1-l_2,\theta_1-\theta_2], \tag{5.35}$$

where $\mu_{p(l_1,\theta_1)}$ and $\mu_{p(l_2,\theta_2)}$ are the means, $\sigma_{p(l_1,\theta_1)}$ and $\sigma_{p(l_2,\theta_2)}$ are the standard deviations, $r_p((l_1,\theta_1),(l_2,\theta_2))$ is the correlation coefficient, of $p(l_1,\theta_1)$ and $p(l_2,\theta_2)$, $\delta[l_1-l_2,\theta_1-\theta_2]$ is 2-D Kronecker delta function given by

$$\delta[l_1-l_2,\theta_1-\theta_2] = \begin{cases} 1 & (l_1 = l_2 \text{ and } \theta_1 = \theta_2) \\ 0 & (l_1 \neq l_2 \text{ or } \quad \theta_1 \neq \theta_2) , \end{cases} \tag{5.36}$$

In the divergent mode ((σ,β)-space), two projections $p(\sigma_1,\beta_1)$ and $p(\sigma_2,\beta_2)$ in the same view ($\beta_1 = \beta_2$) are a-dependent and in the different views ($\beta_1 \neq \beta_2$) are uncorrelated. That is, the correlation $R_p((\sigma_1,\beta_1),(\sigma_2,\beta_2))$ of $p(\sigma_1,\beta_1)$ and $p(\sigma_2,\beta_2)$ is

$$R_p((\sigma_1,\beta_1),(\sigma_2,\beta_2)) = E[p(\sigma_1,\beta_1)p(\sigma_2,\beta_2)] = \mu_{p(\sigma_1,\beta_1)}\mu_{p(\sigma_2,\beta_2)}$$

$$+ r_p((\sigma_1,\beta_1),(\sigma_2,\beta_2))\sigma_{p(\sigma_1,\beta_1)}\sigma_{p(\sigma_2,\beta_2)}\delta[\beta_1-\beta_2]rect\left(\frac{\sigma_1-\sigma_2}{a}\right), \tag{5.37}$$

where $\mu_{p(\sigma_1,\beta_1)}$ and $\mu_{p(\sigma_2,\beta_2)}$ are the means, $\sigma_{p(\sigma_1,\beta_1)}$ and $\sigma_{p(\sigma_2,\beta_2)}$ are the standard deviations, and $r_p((\sigma_1,\beta_1),(\sigma_2,\beta_2))$ is the correlation coefficient, of $p(\sigma_1,\beta_1)$ and $p(\sigma_2,\beta_2)$, and $\delta[\beta_1-\beta_2]$ is 1-D Kronecker delta function, $rect(\frac{\sigma_1-\sigma_2}{a})$ is a rect function given by

$$rect(\frac{\sigma_1-\sigma_2}{a}) = \begin{cases} 1 & |\sigma_1-\sigma_2| \leq \frac{a}{2} \\ 0 & |\sigma_1-\sigma_2| > \frac{a}{2}. \end{cases} \tag{5.38}$$

Proof.

We first make some simplifications. Eq. (5.10) shows that $\mu_Z = \lambda\rho$. Because λ is normally very large, the terms $\frac{1}{2\mu_Z^2}$ of Eq. (5.32) are much less than their corresponding terms $\frac{1}{\mu_Z}$. Thus, the last four terms on the right side of Eq. (5.32) can be omitted, which gives

$$\sigma^2_{p(l,\theta)} \simeq \frac{1}{\mu_{Z_{ad}}} + \frac{1}{\mu_{Z_{ar}}} + \frac{1}{\mu_{Z_{cd}}} + \frac{1}{\mu_{Z_{cr}}}. \tag{5.39}$$

In the design of X-ray CT, the variance $\sigma^2_{p(l,\theta)}$ is to be minimized. $\mu_{Z_{cd}}$ and $\mu_{Z_{cr}}$ can be made very large such that $\frac{1}{\mu_{Z_{cd}}}$ and $\frac{1}{\mu_{Z_{cr}}}$ are negligible. Because $\mu_{Z_{cd}} = f_d\lambda_c\rho_c\varsigma_d$ and $\mu_{Z_{cr}} = f_r\lambda_c\rho_r\varsigma_r$ (Eq. (5.13)), the very large $\mu_{Z_{cd}}$ and $\mu_{Z_{cr}}$ can be achieved by making λ_c vary large.¶ λ_c is in the process of the calibration measurement, controllable, and not harmful to patients. Eq. (5.13) also shows $\mu_{Z_{ad}} = f_d\lambda_a\rho_a\varsigma_d$ and $\mu_{Z_{ar}} = f_r\lambda_a\rho_r\varsigma_r$. In $\mu_{Z_{ar}}$, the transmittance

¶The saturation of the counting capability of the photon detectors may set an up-limit to λ_c.

ρ_r is normally very large, close to 1. Thus, compared with $\frac{1}{\mu Z_{ad}}$, $\frac{1}{\mu Z_{ar}}$ can be ignored. Therefore, as a result, Eq. (5.39) can be further approximated as

$$\sigma^2_{p(l,\theta)} \simeq \frac{1}{\mu Z_{ad}} . \tag{5.40}$$

Examining Eq. (5.30), (5.39), and (5.40) shows that $\sigma^2_{W_{ar}} \simeq \frac{1}{\mu Z_{ar}} \rightarrow 0$, $\sigma^2_{W_{cd}} \simeq \frac{1}{\mu Z_{cd}} \rightarrow 0$, and $\sigma^2_{W_{cr}} \simeq \frac{1}{\mu Z_{cr}} \rightarrow 0$. These results imply that the random variables W_{ar}, W_{cd}, and W_{cr} can be considered deterministic. Thus, the randomness of the projection $p(l, \theta)$ is mainly derived from the random variable W_{ad}, that is, the photon measurements in the detection path in the process of the actual measurement.

In the parallel projection, which is a translation-rotation mode, two distinctive projections in either the same view or in different views are formed at the different, nonoverlapping time intervals. According to the statistical properties of photons described in Section 5.2.1.1, the number of photons counted by the detector in the detection path in the process of the actual measurement Z_{ad}, hence, the W_{ad} and $p(l, \theta)$ are statistically independent. Thus, the correlation of two parallel projections is given by Eq. (5.35).

In the divergent projection that is a rotation mode, the situation is more complicated than in the parallel projection. When two distinctive projections are in the different views, because they are formed in the different, nonoverlapping time intervals, which is similar to the parallel projection, they are statistically independent. When two distinctive projections are in the same view, because they are formed in the same time interval, which is different from the parallel projection, they may not be statistically independent.

As shown in Section 2.2.1, two mechanisms—fluorescence and Bremsstrahlung—produce X-ray. For most medical and industrial applications, Bremsstrahlung is dominant. In Bremsstrahlung, the high-speed electrons are scattered by the strong electric field near nuclei and the resulting decelerations emit photons. In the real X-ray tube as shown in Figure 2.3, the shape of the high-speed electron beam is not an ideal delta function. The area of the spot struck by the electron beam on the focus volume of the anode is not a single point.

At a given time interval, if X-rays are caused by the one-electron-nucleus pair ("collision"), then these rays are statistically dependent because they are produced by the same physical source; if X-rays are caused by different electron-nucleus pairs that are in the different locations of the focal spot, they are statistically independent because these X-rays are produced by the different physical sources. Thus, projections that are spatially closer are more statistically dependent than those that are more spatially separated. Therefore, it is reasonable to consider that, in the one view of divergent mode, the projections within a small angle interval (say, a, $a \ll \alpha$, α is shown in Figure 2.8) are statisticaly dependent and outside this angle interval are statistically independent. This is a so-called a-dependent process [12, 15]. Thus, the statistical relationship between two devergent projections can be expressed by

$$R_p((\sigma_1,\beta_1),(\sigma_2,\beta_2)) = E[p(\sigma_1,\beta_1)p(\sigma_2,\beta_2)] \tag{5.41}$$
$$= \begin{cases} \mu_{p(\sigma_1,\beta_1)}\mu_{p(\sigma_2,\beta_2)} & \text{(cond.1)} \\ \\ \mu_{p(\sigma_1,\beta_1)}\mu_{p(\sigma_2,\beta_2)} + r_p((\sigma_1,\beta_1),(\sigma_2,\beta_2))\sigma_{p(\sigma_1,\beta_1)}\sigma_{p(\sigma_2,\beta_2)} & \text{(cond.2)} \end{cases},$$

with

$$\begin{cases} \text{cond.1}: & \beta_1 \neq \beta_2 \quad \text{or} \quad |\sigma_1 - \sigma_2| > a/2 \\ \\ \text{cond.2}: & \beta_1 = \beta_2 \quad \text{and} \quad |\sigma_1 - \sigma_2| \leq a/2, \end{cases} \tag{5.42}$$

where $\mu_{p(\sigma_1,\beta_1)}$ and $\mu_{p(\sigma_2,\beta_2)}$ are the means, $\sigma_{p(\sigma_1,\beta_1)}$ and $\sigma_{p(\sigma_2,\beta_2)}$ are the standard deviations, and $r_p((\sigma_1,\beta_1),(\sigma_2,\beta_2))$ is the correlation coefficient, of $p(\sigma_1,\beta_1)$ and $p(\sigma_2,\beta_2)$. Eq. (5.41) can be further simplified as Eq. (5.36). ■

5.4 Statistical Interpretation of X-Ray CT Image Reconstruction

5.4.1 Signal Processing Paradigms

Let $p(n\eta, m\Delta\theta)$ $(-N \leq n \leq N,\ 0 \leq m \leq M-1)$ denote the sampled projections. When $\eta = d$, it represents the parallel projections $p(nd, m\Delta\theta)$ in (l,θ)-space; when $\eta = \delta$, it represents the divergent projections $p(n\delta, m\Delta\theta)$ in (σ,β)-space. In computational implementations of the convolution reconstruction method described by Eqs. (2.83) and (2.91), data flow in the order of the following three operations.

The first operation is a convolution of $2N+1$ projections $p(n\eta, m\Delta\theta)$ in one view $m\Delta\theta$ and a convolution function q or functions q_1, q_2, which generates $2N+1$ convolved projections $t(n\eta, m\Delta\theta)$ in that view.

The second operation is a convolution of $2N+1$ convolved projections $t(n\eta, m\Delta\theta)$ in one view $m\Delta\theta$ and an interpolation function ψ, which generates an interpolated data $s_m(r,\phi)$ in that view.

The third operation is a backprojection of M interpolated data $s_m(r,\phi)$ over M views, which produces a backprojected data $f(r,\phi)$, that is, a value of a pixel centered at (r,ϕ).

The pixel value represents the relative linear attenuation coefficient at that point. Repeat this three-step operation for a set of grids; a X-ray CT image is created. This three-step operation is illustrated by Eq. (5.43).

$$
\begin{array}{c}
\textit{for a given point } (r,\phi) \\
\Downarrow \\
M(2N+1) \ \textit{projections}: \ p(i\eta, m\Delta) \\
\overbrace{p(-N\eta,0)\cdots p(N\eta,0)}^{view\ 0}\cdots\overbrace{p(-N\eta,m\Delta)\cdots p(N\eta,m\Delta)}^{view\ m}\cdots\overbrace{p(-N\eta,(M-1)\Delta)\cdots p(N\eta,(M-1)\Delta)}^{view\ M-1} \\
\Downarrow \qquad\qquad \Downarrow \qquad\qquad \Downarrow \\
M(2N+1) \ \textit{convolved projections}: \ t(i\eta, m\Delta) = c_1\sum_{k=-N}^{N} p(k\eta, m\Delta)q^*((i-k)\eta) \\
\overbrace{t(-N\eta,0)\cdots t(N\eta,0)}^{view\ 0}\cdots\overbrace{t(-N\eta,m\Delta)\cdots t(N\eta,m\Delta)}^{view\ m}\cdots\overbrace{t(-N\eta,(M-1)\Delta)\cdots t(N\eta,(M-1)\Delta)}^{view\ M-1} \\
\Downarrow \qquad\qquad \Downarrow \qquad\qquad \Downarrow \\
M \ \textit{interpolated data}: \ s_m(r,\phi) = \sum_{k=-\infty}^{\infty} t(k\eta, m\Delta)\psi(u-k\eta) \\
\underbrace{\overbrace{s_0(r,\phi)}^{view\ 0} \ \cdots\cdots\cdots \ \overbrace{s_m(r,\phi)}^{view\ m} \ \cdots\cdots\cdots \ \overbrace{s_{M-1}(r,\phi)}^{view\ (M-1)}} \\
\Downarrow \\
1 \ \textit{backprojected data set} \\
f(r,\phi) = c_2\sum_{m=0}^{M-1} c' s_m(r,\phi) \ ,
\end{array}
\tag{5.43}
$$

where for the parallel projection shown in Eq. (2.83), $c_1 = d$, $c_2 = \Delta$, $c' = 1$, and $q^*((i-k)\eta)$ denotes $q((n-k)d)$, for the divergent projection shown in Eq. (2.91), $c_1 = \delta$, $c_2 = \Delta$, $c' = W$, and $q^*((i-k)\eta)$ denotes $q_1((n-k)\delta)\cos(k\delta) + q_2((n-k)\delta)\cos(n\delta)$.

5.4.2 Statistical Interpretations

From Sections 5.3 and 5.4.1, we observed that

1). When X-ray CT data acquisition consists of $M \times (2N+1)$ projections, and the reconstructed X-ray CT image consists of $I \times J$ pixels, then X-ray CT image reconstruction constitutes a transform from a set of $M \times (2N+1)$ random variables to another set of $I \times J$ random variables.

2). These new $I \times J$ random variables form a random process, also known as a 2-D random field. The image reconstruction is a realization of this random process. The reconstructed image is a configuration of the entire random process. Each pixel intensity in the image is a value in the state space of a corresponding random variable in this process.

5.5 Appendices

5.5.1 Appendix 5A

This appendix proves $c_n = 1$ in Eq. (5.17) and derives Eq. (5.16).

From Eqs. (5.17) and (5.18), we have

$$\frac{c_n}{\sqrt{2\pi\sigma_Z^2}}\int_0^{\ln M} \exp\left(-\frac{(e^w - \mu_Z)^2}{2\sigma_Z^2}\right)e^w dw = 1. \tag{5.44}$$

Let

$$e^w = u \ , \quad \frac{e^w - \mu_Z}{\sigma_Z} = v \ , \quad M = 2\mu_Z - 1, \tag{5.45}$$

Eq. (5.44) becomes

$$\frac{c_n}{\sqrt{2\pi}} \int_{-\frac{\mu_Z-1}{\sigma_Z}}^{\frac{\mu_Z-1}{\sigma_Z}} \exp\left(-\frac{v^2}{2}\right) dv = 1. \tag{5.46}$$

Let $\Phi(x) = \frac{1}{\sqrt{2\pi}} \int_{-\infty}^{x} e^{-\frac{y^2}{2}} dy$ denote the cumulative density function (cdf) of the standard Gaussian random variable $Y \sim N(0,1)$; Eq. (5.46) gives

$$\left(\Phi(\frac{\mu_Z - 1}{\sigma_Z}) - \Phi(-\frac{\mu_Z - 1}{\sigma_Z})\right) c_n = 1. \tag{5.47}$$

Because $\mu_Z = \lambda\rho$, hence, $\frac{\mu_Z-1}{\sigma_Z}$ is very large, $\Phi(\frac{\mu_Z-1}{\sigma_Z}) \to 1$, and $\Phi(-\frac{\mu_Z-1}{\sigma_Z}) \to 0$. Thus, $c_n \simeq 1$.

From Eq. (5.17) with $c_n \simeq 1$, the mean μ_W of the random variable $W = \ln Z$ is given by

$$\mu_W = \frac{1}{\sqrt{2\pi\sigma_Z^2}} \int_0^{\ln M} w \exp\left(-\frac{(e^w - \mu_Z)^2}{2\sigma_Z^2}\right) e^w dw. \tag{5.48}$$

Using Eq. (5.45), Eq. (5.48) becomes

$$\mu_W = \frac{1}{\sqrt{2\pi}} \int_{-\frac{\mu_Z-1}{\sigma_Z}}^{\frac{\mu_Z-1}{\sigma_Z}} \ln\left(\mu_Z\left(1 + \frac{\sigma_Z}{\mu_Z}v\right)\right) \exp\left(-\frac{v^2}{2}\right) dv. \tag{5.49}$$

Using the approximation $\ln(1+x) \simeq x - \frac{x^2}{2}$ $(-1 < x \leq 1)$, Eq. (5.49) becomes

$$\mu_W \simeq \frac{1}{\sqrt{2\pi}} \int_{-\frac{\mu_Z-1}{\sigma_Z}}^{\frac{\mu_Z-1}{\sigma_Z}} (\ln\mu_Z + \frac{\sigma_Z}{\mu_Z}v - \frac{1}{2}\left(\frac{\sigma_Z}{\mu_Z}\right)^2 v^2) \exp\left(-\frac{v^2}{2}\right) dv. \tag{5.50}$$

It has been shown that Eq. (5.50) is

$$\mu_W \simeq \ln\mu_Z - \frac{1}{2}\left(\frac{\sigma_Z}{\mu_Z}\right)^2. \tag{5.51}$$

From Eq. (5.17) with $c_n \simeq 1$, the variance σ_W^2 of the random variable $W = \ln Z$ can be computed by

$$\sigma_W^2 = \frac{1}{\sqrt{2\pi\sigma_Z^2}} \int_0^{\ln M} w^2 \exp\left(-\frac{(e^w - \mu_Z)^2}{2\sigma_Z^2}\right) e^w dw - \mu_W^2. \tag{5.52}$$

Using Eq. (5.45) and the approximation $\ln(1+x) \simeq x - \frac{x^2}{2}$ $(-1 < x \leq 1)$, the first item of the right side of Eq. (5.52) becomes

$$\frac{1}{\sqrt{2\pi}} \int_{-\frac{\mu_Z - 1}{\sigma_Z}}^{\frac{\mu_Z - 1}{\sigma_Z}} \left(\ln \mu_Z + \frac{\sigma_Z}{\mu_Z} v - \frac{1}{2} \left(\frac{\sigma_Z}{\mu_Z} \right)^2 v^2 \right)^2 \exp\left(-\frac{v^2}{2} \right) dv$$
$$= (\ln \mu_Z)^2 + \left(\frac{\sigma_Z}{\mu_Z} \right)^2 (1 - \ln \mu_Z) + \frac{3}{4} \left(\frac{\sigma_Z}{\mu_Z} \right)^4 . \quad (5.53)$$

The second item on the right side of Eq. (5.52) is

$$\mu_W^2 = (\ln \mu_Z)^2 - \ln \mu_Z \left(\frac{\sigma_Z}{\mu_Z} \right)^2 + \frac{1}{4} \left(\frac{\sigma_Z}{\mu_Z} \right)^4 . \quad (5.54)$$

By substituting Eqs. (5.53) and (5.54) into Eq. (5.52), we obtain

$$\sigma_W^2 = \left(\frac{\sigma_Z}{\mu_Z} \right)^2 + \frac{1}{2} \left(\frac{\sigma_Z}{\mu_Z} \right)^4 . \quad (5.55)$$

■

5.5.2 Appendix 5B

This appendix (1) describes an error correction procedure to evaluate the difference between the true pdf $\hat{p}(p(l,\theta))$ of Eq. (5.34) and the approximated Gaussian pdf $p(p(l,\theta))$ of Eq. (5.33), (2) gives an example to show that $p(p(l,\theta))$ is a good approximation of $\hat{p}(p(l,\theta))$, and (3) shows, by applying the error correction procedure to this example, that the procedure can lead to an accurate evaluation of the error in a Gaussian approximation.

(1) Let the difference between pdfs $\hat{p}(p(l,\theta))$ and $p(p(l,\theta))$ be defined by

$$\epsilon(p) = \hat{p}(p(l,\theta)) - p(p(l,\theta)). \quad (5.56)$$

Because calculating $\hat{p}(p(l,\theta))$ by means of the convolutions of Eq. (5.34) is difficult, we use an error correction procedure [12, 16, 17] to evaluate this difference.

The Hermite polynomials defined by

$$H_n(x) = x^n - \binom{n}{2} x^{n-2} + 1 \cdot 3 \binom{n}{4} x^{n-4} + \cdots \quad (5.57)$$

form a complete orthogonal set in the interval $(-\infty, \infty)$

$$\int_{-\infty}^{+\infty} e^{-\frac{x^2}{2}} H_n(x) H_m(x) dx = \begin{cases} n!\sqrt{2\pi} & (n = m) \\ 0 & (n \neq m). \end{cases} \quad (5.58)$$

Thus, a large class of continuous functions over this interval can be expressed in a Hermite polynomial series. In particular, the difference $\epsilon(p)$ can be expressed by

$$\epsilon(p) = \frac{1}{\sqrt{2\pi\sigma^2_{p(l,\theta)}}} \exp(-\frac{(p(l,\theta)-\mu_{p(l,\theta)})^2}{2\sigma^2_{p(l,\theta)}}) \sum_{n=0}^{\infty} C_n H_n(\frac{p(l,\theta)-\mu_{p(l,\theta)}}{\sigma_{p(l,\theta)}}), \tag{5.59}$$

where $\mu_{p(l,\theta)}$ and $\sigma^2_{p(l,\theta)}$ are given by Eq. (5.31) and Eq. (5.32), respectively.

The coefficients C_n $(n = 0, 1, 2, \cdots)$ can be determined by

$$n!\sigma^2_{p(l,\theta)} C_n + a_{n,n-2}\sigma^2_{p(l,\theta)} C_{n-2} + \cdots\cdots = B_n - \sigma^2_{p(l,\theta)} A_n, \tag{5.60}$$

where

$$C_0 = C_1 = C_2 = 0, \tag{5.61}$$

A_n $(n = 1, 2, \cdots)$ is the n-th moment of the standard Gaussian $N(0, 1)$

$$A_{2n-1} = 0 \quad \text{and} \quad A_{2n} = \frac{(2n)!}{2^n n!}, \tag{5.62}$$

B_n $(n = 1, 2, \cdots)$ is the n-th moment of $\hat{p}(p(l,\theta))$

$$B_n = \int_{-\infty}^{+\infty} (p(l,\theta) - \mu_{p(l,\theta)})^n \hat{p}(p(l,\theta)) dp(l,\theta), \tag{5.63}$$

and $a_{n,m}$ are determined by

$$A_{n+m} - \binom{m}{2} A_{n+m-2} + 1 \cdot 3 \binom{m}{4} A_{n+m-4} + \cdots\cdots, \tag{5.64}$$

with

$$a_{n,n} = n! \quad \text{and} \quad a_{n,m} = 0 \;\; (m > n \;\text{or odd}\; (n+m)). \tag{5.65}$$

Eq. (5.59) shows that in order to evaluate $\epsilon(p)$, the coefficients C_n $(n = 3, 4, \cdots)$ must be known. Because A_n are known, finding C_n is equivalent to finding B_n, the moment of $\hat{p}(p(l,\theta))$. Let the characteristic function of W (Eq. (5.14) be $h(t)$, and the characteristic functions of W_{ad}, W_{ar}, W_{cd}, and W_{cr} be $g_{W_{ad}}(t)$, $g_{W_{ar}}(t)$, $g_{W_{cd}}(t)$, and $g_{W_{cr}}(t)$. Then, the characteristic function of $\hat{p}(p(l,\theta))$ is

$$g(t) = g_{W_{ad}}(t) g_{W_{ar}}(t) g_{W_{cd}}(t) g_{W_{cr}}(t) = h^4(t), \tag{5.66}$$

where

$$g_{W_{ad}}(t) = g_{W_{ar}}(t) = g_{W_{cd}}(t) = g_{W_{cr}}(t) = h(t). \tag{5.67}$$

Thus, the n-th moment of $\hat{p}(p(l,\theta))$ is

$$B_n = (i)^{-n} g^{(n)}(0), \tag{5.68}$$

where $g^{(n)}(t)$ is the n-th derivative of $g(t)$. $g^{(n)}(t)$, can be generated iteratively through

$$g^{(n)}(t) = \Delta_n g^{(n-1)}(t), \tag{5.69}$$

with

$$\Delta_n = \Delta_{n-1} + \frac{\Delta'_{n-1}}{\Delta_{n-1}} \quad \text{and} \quad \Delta_1 = 4\frac{h'(t)}{h(t)}, \tag{5.70}$$

where $'$ denotes the first derivative.

(2) The following example shows that although W_{ar}, W_{ar}, W_{cd}, and W_{cr} are non-Gaussian (see $p_W(w)$ of Eq. (5.15)), their sum $p(l,\theta)$ (Eq. (5.20)) has quite an accurate Gaussian distribution. In this example, in order to compute the true pdf $\hat{p}(p(l,\theta))$ (Eq. (5.34)), for the purpose of simplicity, instead of using W (Eq. (5.14)) and its pdf $p_W(w)$ (Eq. (5.15)), we use four random variables $\delta\theta_i$ $(i = 1,2,3,4)$, each of them having a uniform pdf over $[-\pi, +\pi]$

$$p(\delta\theta_i) = \begin{cases} 1 & [-\pi, +\pi] \\ \\ 0 & \text{otherwise,} \end{cases} \tag{5.71}$$

with the mean and variance given by

$$\mu_{\delta\theta_i} = 0 \quad \text{and} \quad \sigma^2_{\delta\theta_i} = \frac{\pi^2}{3}. \tag{5.72}$$

It has been verified that the sum of two $\delta\theta$, $\sum_{i=1}^{2}\delta\theta_i$, has a triangle pdf $\hat{p}(\delta\theta) = p(\delta\theta) \star p(\delta\theta)$ ($\star$ - the convolution.); the sum of three $\delta\theta$, $\sum_{i=1}^{3}\delta\theta_i$, has a pdf $\hat{p}(\delta\theta) = p(\delta\theta) \star p(\delta\theta) \star p(\delta\theta)$ consisting of three pieces of parabolic curves; and the sum of four $\delta\theta$, $\sum_{i=1}^{4}\delta\theta_i$, has a pdf $\hat{p}(\delta\theta) = p(\delta\theta) \star p(\delta\theta) \star p(\delta\theta) \star p(\delta\theta)$ consisting multiple pieces of curves. These true pdfs are shown in Figures 5.1 through 5.3 by the solid lines. The corresponding approximated Gaussian pdfs are $p(\delta\theta) = \frac{1}{\sqrt{2\pi\sigma_k^2}}\exp(-\frac{(\delta\theta)^2}{2\sigma_k^2})$ $(k = 2,3,4)$, where the variances $\sigma_k^2 = \frac{1}{k}\sigma^2_{\delta\theta_i}$. The Gaussian pdfs are shown in Figures 5.1 through 5.3 by the dash lines. The virtual examination of two pdfs in each figure indicates that Gaussian approximation is quite accurate.

(3) Now, we apply the error correction procedure in (1) to the above example. Define the difference between the true pdf $\hat{p}(\delta\theta)$ and the approximated Gaussian pdf $p(\delta\theta)$ by

$$\epsilon(p) = \hat{p}(\delta\theta) - p(\delta\theta). \tag{5.73}$$

Because $p(\delta\theta)$ is an even function, the first-order expansion of Eq. (5.59) in terms of Hermite polynomials is

$$\epsilon(p) = \frac{1}{\sqrt{2\pi\sigma_k^2}}\exp(-\frac{(\delta\theta)^2}{2\sigma_k^2})\{\frac{1}{4!}(\frac{B_4}{\sigma_k^4} - 3)[(\frac{\delta\theta}{\sigma_k})^4 - 6(\frac{\delta\theta}{\sigma_k})^2 + 3]\}, \tag{5.74}$$

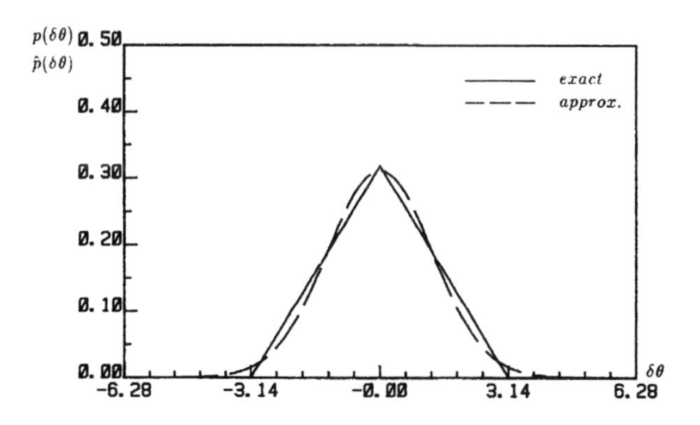

FIGURE 5.1
The true pdf $\hat{p}(\delta\theta)$ and the approximated Gaussian pdf $p(\delta\theta)$ of $\sum_{i=1}^{2} \delta\theta_i$.

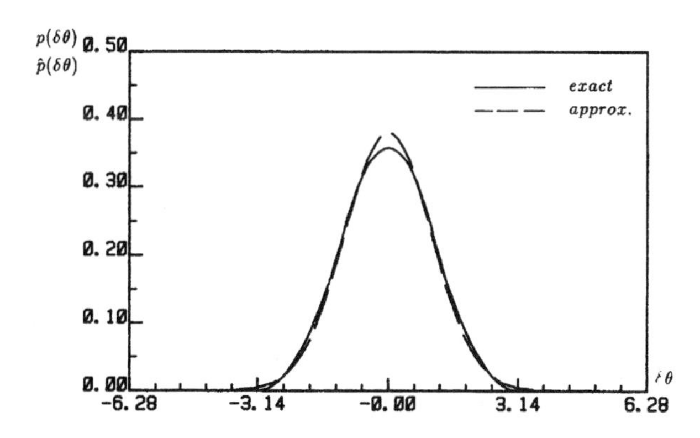

FIGURE 5.2
The true pdf $\hat{p}(\delta\theta)$ and the approximated Gaussian pdf $p(\delta\theta)$ of $\sum_{i=1}^{3} \delta\theta_i$.

where

$$\sigma_k^2 = \frac{1}{k}\sigma_{\delta\theta_i}^2 = \frac{1}{k} \cdot \frac{\pi^2}{3} \quad \text{and} \quad B_4 = g^{(4)}(0) = \frac{d^4}{dt^4}(h^k(t))_{t=0}. \tag{5.75}$$

For $k = 2, 3, 4$, due to

$$\begin{aligned} k &= 2\,, \quad \sigma_2^2 = \tfrac{\pi^2}{6}\,, \quad B_4 = \tfrac{2}{5}\pi^4, \\ k &= 3\,, \quad \sigma_3^2 = \tfrac{\pi^2}{9}\,, \quad B_4 = \tfrac{13}{405}\pi^4, \\ k &= 4\,, \quad \sigma_4^2 = \tfrac{\pi^2}{12}\,, \quad B_4 = \tfrac{13}{672}\pi^4, \end{aligned} \tag{5.76}$$

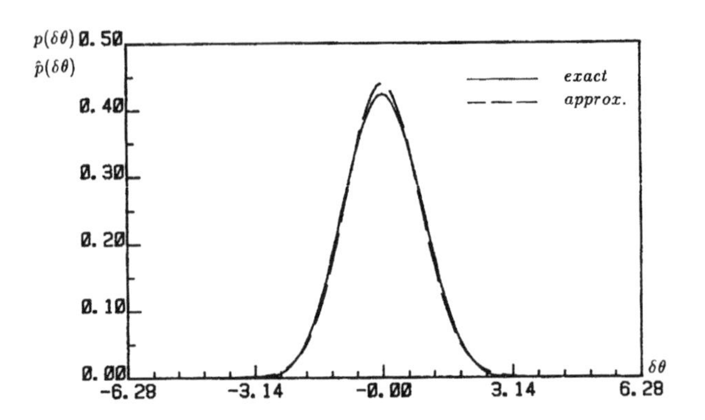

FIGURE 5.3
The true pdf $\hat{p}(\delta\theta)$ and the approximated Gaussian pdf $p(\delta\theta)$ of $\sum_{i=1}^{4}\delta\theta_i$.

we have $\epsilon(p)$ in the cases of $\sum_{i=1}^{k}\delta\theta_i$ $(k=2,3,4)$

$$\begin{aligned}
\epsilon(p) &= \frac{-\sqrt{6}}{40\pi\sqrt{2\pi}}e^{-\frac{3}{\pi^2}}\left(\tfrac{36}{\pi^4}(\delta\theta)^4 - \tfrac{36}{\pi^2}(\delta\theta)^2 + 3\right), \\
\epsilon(p) &= \frac{-\sqrt{9}}{60\pi\sqrt{2\pi}}e^{-\frac{4.5}{\pi^2}}\left(\tfrac{81}{\pi^4}(\delta\theta)^4 - \tfrac{54}{\pi^2}(\delta\theta)^2 + 3\right), \\
\epsilon(p) &= \frac{-\sqrt{12}}{112\pi\sqrt{2\pi}}e^{-\frac{6}{\pi^2}}\left(\tfrac{144}{\pi^4}(\delta\theta)^4 - \tfrac{72}{\pi^2}(\delta\theta)^2 + 3\right).
\end{aligned} \tag{5.77}$$

These $\epsilon(p)$ are ploted in Figure 5.4.

Figures 5.1 through 5.3 show the differences between the true pdf $\hat{p}(\delta\theta)$ and the approximated Gaussian pdf $p(\delta\theta)$. Figure 5.4 shows these differences evaluated by the error correction procedure. The comparison of Figure 5.4 with Figures 5.1-5.3 indicates that the differences in the two appraoches are consistent. This consistency proves that the error correction procedure can lead to an accurate evaluation of the Gaussian approximation. When the true pdf is complicated, this procedure will be very useful.

Problems

5.1. Prove $P(\lambda\rho) \longrightarrow N(\lambda\rho, \lambda\rho)$ as $\lambda\rho \to \infty$ in Section 5.2.1.3.

5.2. Validate $\frac{\sigma_Z}{\mu_Z} < 1$ in Eq. (5.49).

5.3. Prove Eq. (5.51) from Eq. (5.50).

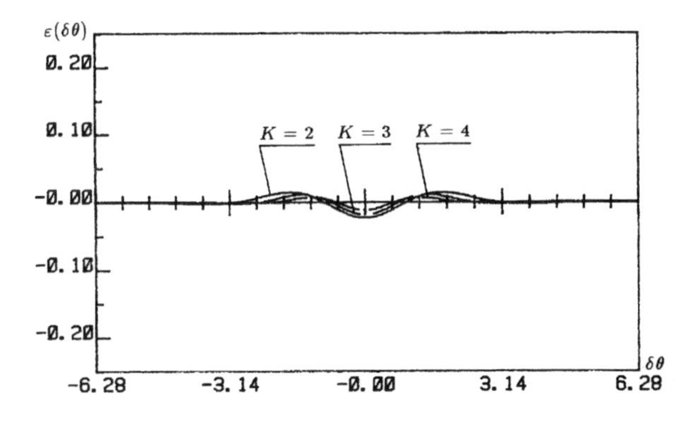

FIGURE 5.4
Differences between the true pdf $\hat{p}(\delta\theta)$ and the approximate Gaussian pdfs $p(\delta\theta)$in the example of (2).

5.4. Prove Eq. (5.53) from Eq. (5.52).

5.5. Describe the pattern of $p_W(w)$ of Eq. (5.15) and discuss the difference between $p_W(w)$ and $p_W(-w)$.

References

[1] Herman, G.: *Image Reconstruction from Projections.* Academic Press, New York (1980).

[2] Macovski, A.: *Medical Imaging Systems.* Prentice-Hall Inc., New York (1982).

[3] Lee, H., (Ed), G.W.: *Imaging Technology.* IEEE Press, New York (1986).

[4] Kak, A., Slaney, M.: *Principles of Computerized Tomographic Imaging.* IEEE Press, New York (1988).

[5] Jain, A.K.: *Fundamentals of Digital Image Processing.* Prentice Hall, Englewood Cliffs, New Jersey (1989).

[6] Cho, Z.H., Jones, J.P., Singh, M.: *Foundations of Medical Imaging.* John Wiley & Sons, Inc., New York (1993).

[7] Hsieh, J.: *Computed Tomography, Principles, Design, Artifacts and Recent Advances.* SPIE, Bellingham (2004).

[8] Kalender, W.: *Computed Tomography: Fundamentals, System Technology, Image Quality, Applications.* Publics MCD Verlag, Germany (2005).

[9] Buzug, T.: *Computed Tomography.* Springer, Berlin, Germany (2008).

[10] Herman, G.: *Fundamentals of Computerized Tomography.* Springer, London, United Kingdom (2009).

[11] Press, S.: *Applied Multivariate Analysis.* Robert E. Krieger Publishing Company, Malabar, Florida (1982).

[12] Papoulis, A.: *Probability, Random Variables and Stochastic Processes.* McGraw-Hill Book Company Inc., New York (1984).

[13] Lei, T., Sewchand, W.: Statistical approach to x-ray CT imaging and its applications in image analysis. Part 1: Statistical analysis of x-ray CT imaging. *IEEE Trans. Medical Imaging* **11**(1) (1992) 53–61.

[14] Cramer, H.: *The Elements of Probability Theory.* John Wiley & Sons, New York (1955).

[15] Papoulis, A.: *Signal Analysis.* McGraw-Hill Book Company Inc., New York (1977).

[16] Papoulis, A.: *The Fourier Integral and Its Applications.* McGraw-Hill Book Company Inc., New York (1962).

[17] Papoulis, A.: *Probability, Random Variables and Stochastic Processes.* McGraw-Hill Book Company Inc., New York (1968).

6

Statistics of X-Ray CT Image

6.1 Introduction

As shown in Section 5.4, X-ray CT image reconstruction constitutes a transform from a set of random variables (projections) to another set of random variables (pixel intensities). Statistics of X-ray CT data in the imaging domain propagate to the statistics in the image domain through image reconstruction. This chapter describes the statistics of the X-ray CT image generated by using basic CT data acquisition schemes (parallel and divergent projections) and the typical image reconstruction technique (convolution method). A statistical description of X-ray CT image is given at three levels of the image: a single pixel, any two pixels, and a group of pixels (also referred to as an image region).

The Gaussianity of the intensity of a single pixel in an X-ray CT image is proved in two ways. Then this chapter gives conditions for two pixel intensities to be independent with a probability measure for the parallel and divergent projections, respectively. These conditions essentially imply that (1) the intensities of any two pixels are correlated, (2) the degree of the correlation decreases as the distance between two pixels increases, and (3) in the limiting case of the distance approaching the infinity, the intensities of two pixels become independent. These properties are summarized as spatially asymptotically independent, abbreviated as SAI.

An X-ray CT image consists of piecewise contiguous regions. This fact reveals that each image region, that is, a group of pixels, may have some unique statistical properties. This chapter shows that, based on Gaussianity and SAI, pixel intensities in each image region are stationary in the wide sense, hence, in the strict sense; and ergodic, hence, satisfy the ergodic theorems.

Gaussianity, spatially, asymptotically independent, stationarity, and ergodicity are described in the order of a single pixel $\Longrightarrow$ any two pixels $\Longrightarrow$ a group of pixels. These properties provide a basis for creating a stochastic image model and developing new image analysis methodologies for X-ray CT image analysis, which are given in Chapters 9, 10, and 11.

6.2 Statistics of the Intensity of a Single Pixel

Statistical properties of the intensity of a single pixel in an X-ray CT image are described in Property 6.1. Similar to Chapters 2 and 5, we use $f(r, \phi)$ to denote the intensity of a pixel centered at (r, ϕ), where (r, ϕ) is the polar coordinate.

6.2.1 Gaussianity

Property 6.1 The intensity $f(r, \phi)$ of a single pixel in an X-ray CT image has a Gaussian distribution $N(\mu_{f(r,\phi)}, \sigma^2_{f(r,\phi)})$, that is, pdf of $f(r, \phi)$ is

$$p(f(r,\phi)) = \frac{1}{\sqrt{2\pi\sigma^2_{f(r,\phi)}}} \exp\left(-\frac{(f(r,\phi) - \mu_{f(r,\phi)})^2}{2\sigma^2_{f(r,\phi)}}\right), \tag{6.1}$$

where the mean $\mu_{f(r,\phi)}$ and the variance $\sigma^2_{f(r,\phi)}$ of $f(r, \phi)$ are given by

$$\mu_{f(r,\phi)} = \begin{cases} \int_0^{\pi} d\theta \int_{-\infty}^{+\infty} \mu_{p(l,\theta)} q_p(l' - l) dl & \text{(parallel projection)} \\ \int_0^{2\pi} W d\beta \int_{-\infty}^{+\infty} \mu_{p(\sigma,\beta)} q_d(\sigma' - \sigma) d\sigma & \text{(divergent projection),} \end{cases} \tag{6.2}$$

and

$$\sigma^2_{f(r,\phi)} = \begin{cases} \int_0^{\pi} d\theta \int_{-\infty}^{+\infty} \sigma^2_{p(l,\theta)} q_p^2(l' - l) dl & \text{(parallel projection)} \\ \int_0^{2\pi} W^2 d\beta \int_{-\infty}^{+\infty} \sigma^2_{p(\sigma,\beta)} q_d^2(\sigma' - \sigma) d\sigma & \text{(divergent projection),} \end{cases} \tag{6.3}$$

where $\mu_{p(l,\theta)}$, $\mu_{p(\sigma,\beta)}$, and $\sigma^2_{p(l,\theta)}$, $\sigma^2_{p(\sigma,\beta)}$ are the means and variances of the parallel and divergent projections given by Eq. (5.21) and Eq. (5.22), respectively; for the parallel projection, $q_p(l' - l)$ denotes the convolution function $q(l' - l)$ of Eq. (2.60), $l' = r\cos(\theta - \phi)$ is given by Eq. (2.15); for the divergent projection, $q_d(\sigma' - \sigma)$ denotes the convolution function $q_1(\sigma' - \sigma)\cos\sigma + q_2(\sigma' - \sigma)\cos\sigma'$ of Eq. (2.73), $\sigma' = \tan^{-1}\frac{r\cos(\theta-\phi)}{D+r\cos(\theta-\phi)}$ and $W = \frac{D}{(2\pi D')^2}$ are given by Eq. (2.64) and Eq. (2.90), respectively.

Proof.

Because (1) the projection (either parallel projection $p(l, \theta)$ or divergent projection $p(\sigma, \beta)$) has a Gaussian distribution (Property 5.2.c) and (2) the operations in the convolution reconstruction method (Eq. (2.60) for the parallel projection and Eq. (2.73) for the divergent projection) are linear, the Gaussianity of the pixel intensity $f(r, \phi)$ holds.

(1) For the parallel projection, from Eq. (2.60), the mean $\mu_{f(r,\phi)}$ of the pixel intensity $f(r,\phi)$ is

$$\mu_{f(r,\phi)} = E[f(r,\phi)] = \int_0^{\pi} d\theta \int_{-\infty}^{+\infty} \mu_{p(l,\theta)} q_p(l'-l)dl. \tag{6.4}$$

The correlation $R_f((r_1,\phi_1),(r_2,\phi_2))$ of two pixel intensities $f(r_1,\phi_1)$ and $f(r_2,\phi_2)$, in terms of Property 5.3 (Eq. (5.35)), is

$$\begin{aligned}
R_f((r_1,\phi_1),(r_2,\phi_2)) &= E[f(r_1,\phi_1)f(r_2,\phi_2)] \\
&= \int_0^{\pi}\int_0^{\pi}\int_{-\infty}^{+\infty}\int_{-\infty}^{+\infty} E[p(l_1,\theta_1)p(l_2,\theta_2)] \\
&\quad q_p(l_1'-l_1)q_p(l_2'-l_2)dl_1 dl_2 d\theta_1 d\theta_2 \\
&= \int_0^{\pi}\int_0^{\pi}\int_{-\infty}^{+\infty}\int_{-\infty}^{+\infty} \mu_{p(l_1,\theta_1)}\mu_{p(l_2,\theta_2)} \\
&\quad q_p(l_1'-l_1)q_p(l_2'-l_2)dl_1 dl_2 d\theta_1 d\theta_2 \\
&+ \int_0^{\pi}\int_{-\infty}^{+\infty} r_p((l_2,\theta_2),(l_2,\theta_2))\sigma_{p(l_2,\theta_2)}\sigma_{p(l_2,\theta_2)} \\
&\quad q_p(l_1'-l_2)q_p(l_2'-l_2)dl_2 d\theta_2 \\
&= \mu_{f(r_1,\phi_1)}\mu_{f(r_2,\phi_2)} \\
&+ \int_0^{\pi}\int_{-\infty}^{+\infty} \sigma^2_{p(l_2,\theta_2)} q_p(l_1'-l_2)q_p(l_2'-l_2)dl_2 d\theta_2.
\end{aligned} \tag{6.5}$$

In deriving the previous formula in Eq. (6.5), $r_p((l_2,\theta_2)(l_2,\theta_2)) = 1$ is used.

When $(r_1,\phi_1) = (r_2,\phi_2)$, $l_1' = l_2'$. Thus, Eq. (6.5) becomes

$$R_f((r,\phi),(r,\phi)) = \mu^2_{f(r,\phi)} + \int_0^{\pi} d\theta \int_{-\infty}^{+\infty} \sigma^2_{p(l,\theta)} q_p^2(l'-l)dl. \tag{6.6}$$

Therefore, the variance $\sigma^2_{f(r,\phi)}$ of the pixel intensity $f(r,\phi)$ is given by

$$\sigma^2_{f(r,\phi)} = R_f((r,\phi),(r,\phi)) - \mu^2_{f(r,\phi)} = \int_0^{\pi} d\theta \int_{-\infty}^{+\infty} \sigma^2_{p(l,\theta)} q_p^2(l'-l)dl. \tag{6.7}$$

(2) For the divergent projection, from Eq. (2.73), the mean $\mu_{f(r,\phi)}$ of the pixel intensity $f(r,\phi)$ is

$$\mu_{f(r,\phi)} = E[f(r,\phi)] = \int_0^{2\pi} d\beta \int_{-\infty}^{+\infty} \mu_{p(\sigma,\beta)} q_d(\sigma'-\sigma)d\sigma. \tag{6.8}$$

The correlation $R_f((r_1,\phi_1),(r_2,\phi_2))$ of two pixel intensities $f(r_1,\phi_1)$ and $f(r_2,\phi_2)$, in terms of Property 5.3 (Eq. (5.37)), is

$$R_f((r_1,\phi_1),(r_2,\phi_2)) = E[f(r_1,\phi_1)f(r_2,\phi_2)]$$

$$
\begin{aligned}
&= \int_0^{2\pi}\int_0^{2\pi}\int_{-\infty}^{+\infty}\int_{-\infty}^{+\infty} E[p(\sigma_1,\beta_1)p(\sigma_2,\beta_2)]W_1W_2 \\
&\quad q_d(\sigma_1'-\sigma_1)q_d(\sigma_2'-\sigma_2)d\sigma_1 d\sigma_2 d\beta_1 d\beta_2 \\
&= \int_0^{2\pi}\int_0^{2\pi}\int_{-\infty}^{+\infty}\int_{-\infty}^{+\infty} \mu_{p(\sigma_1,\beta_1)}\mu_{p(\sigma_2,\beta_2)}W_1W_2 \\
&\quad q_d(\sigma_1'-\sigma_1)q_d(\sigma_2'-\sigma_2)d\sigma_1 d\sigma_2 d\beta_1 d\beta_2 \\
&+ \int_0^{2\pi} W_2 d\beta_2 \int_{-\infty}^{+\infty} \sigma_{p(\sigma_2,\beta_2)} q_d(\sigma_2'-\sigma_2)d\sigma_2 \\
&\quad \int_{\sigma_2-a/2}^{\sigma_2+a/2} W_1 r_p((\sigma_1,\beta_2),(\sigma_2,\beta_2))\sigma_{p(\sigma_1,\beta_2)} q_d(\sigma_1'-\sigma_1)d\sigma_1 \\
&= \mu_{f(r_1,\phi_1)}\mu_{f(r_2,\phi_2)} \\
&+ \int_0^{2\pi}\int_{-\infty}^{+\infty} W_1W_2\sigma^2_{p(\sigma_2,\beta_2)} q_d(\sigma_1'-\sigma_2)q_d(\sigma_2'-\sigma_2)d\sigma_2 d\beta_2. \qquad (6.9)
\end{aligned}
$$

In deriving the previous formula in Eq. (6.9), the following approximations are used. For $|\sigma_1 - \sigma_2| < \frac{a}{2}$ (a is a small angle, see Property 5.3), that is, for two closer projections $p((\sigma_1,\beta_2))$ and $p((\sigma_2,\beta_2))$, $\sigma_1 \simeq \sigma_2$, $r_p((\sigma_1,\beta_2),(\sigma_2,\beta_2)) \simeq 1$.

When $(r_1,\phi_1) = (r_2,\phi_2)$, $\sigma_1' = \sigma_2'$, $W_1 = W_2$, and $q_d(\sigma_1'-\sigma_1) \simeq q_d(\sigma_2'-\sigma_2)$. Thus, Eq. (6.9) becomes

$$
R_f((r,\phi),(r,\phi)) = \mu^2_{f(r,\phi)} + \int_0^{2\pi} W^2 d\beta \int_{-\infty}^{+\infty} \sigma^2_{p(\sigma,\beta)} q_d^2(\sigma'-\sigma)d\sigma. \qquad (6.10)
$$

Therefore, the variance $\sigma^2_{f(r,\phi)}$ of the pixel intensity $f(r,\phi)$ is given by

$$
\begin{aligned}
\sigma^2_{f(r,\phi)} &= R_f((r,\phi),(r,\phi)) - \mu^2_{f(r,\phi)} \\
&= \int_0^{2\pi} W^2 d\beta \int_{-\infty}^{+\infty} \sigma^2_{p(\sigma,\beta)} q_d^2(\sigma'-\sigma)d\sigma. \qquad (6.11)
\end{aligned}
$$

■

Discussion.

(1) Figure 2.7 and Figure 2.8 show that projections, either parallel or divergent, are acquired over M views and with $(2N+1)$ samples in each view. These $M(2N+1)$ projections have the finite means and the finite variances (Property 5.2.b) and are independent for parallel projections and a-dependent for divergent projections (Property 5.3).

Signal processing paradigms of Section 5.4.1 show that the computational implementation of the convolution reconstruction method consists of two convolutions and one backprojection. As shown by Eq. (5.43), these operations are the weighted summations of the input projections $p(n\eta, m\Delta)$ $(-N \leq n \leq N,\ 0 \leq m \leq M)$ and are linear.

The number $M(2N+1)$ is usually very large. Thus, according to the Central Limit theorem [1], $f(r,\phi)$ of Eq. (5.43) generated using parallel projections has an asymptotical Gaussian distribution; according to the extended Central Limit theorem [2], $f(r,\phi)$ of Eq. (5.43) generated using divergent projections also has an asymptotical Gaussian distribution.

(2) Let $\delta(x)$ and $\delta[x]$ denote the Dirac and Kronecker delta function, respectively. Let $g(x)$ be a continuous function on $(-\infty,+\infty)$. It is known that

$$\int_{-\infty}^{+\infty} g(x)\delta(x-x_0)dx = g(x_0), \tag{6.12}$$

but

$$\int_{-\infty}^{+\infty} g(x)\delta[x-x_0]dx = g(x_0) \tag{6.13}$$

has not been proved. However, if $E > 0$ is a finite value and $-E < x_0 < E$, then

$$\int_{-E}^{+E} g(x)\delta[x-x_0]dx = g(x_0). \tag{6.14}$$

In the proof of Property 6.1, the Property 5.3 is used where $\delta[l_1-l_2,\theta_1-\theta_2]$ in Eq. (5.35) and $\delta[\beta_1-\beta_2]$ in Eq. (5.37) are Kronecker delta funtions. In Eq. (6.5), based on Eq. (6.14),

$$\int_0^{\pi}\int_{-\infty}^{+\infty} r_p((l_1,\theta_1),(l_2,\theta_2))\sigma_{p(l_1,\theta_1)}\sigma_{p(l_2,\theta_2)}$$

$$\delta[l_1-l_2,\theta_1-\theta_2]q_p(l_1'-l_2)q_p(l_2'-l_2)dl_1d\theta_1$$

$$= r_p((l_2,\theta_2),(l_2,\theta_2))\sigma_{p(l_2,\theta_2)}\sigma_{p(l_2,\theta_2)}q_p(l_1'-l_2)q_p(l_2'-l_1). \tag{6.15}$$

This is because in the practical X-ray CT image reconstruction, the integral for l_1 over $(-\infty,+\infty)$ is actually performed over a finite interval $[-E,+E]$ as shown in Figures 2.7 and 2.9.

(3) Although the parallel projetions are independent and the divergent projections are a-dependent, but this difference is not reflected in the variances of pixel intensities generated using these two types of projections, respectively. The expressions of Eqs. (6.7) and (6.11) are nearly identical except a weight W^2. The reason for this scenario is that in Eq. (6.9) when using the a-dependent, the angle a is assumed to be very small. This condition leads to an approximation $\sigma_1 \simeq \sigma_2$ that is almost equivalent to applying a Kronecker function $\delta[\sigma_1-\sigma_2]$ in Eq. (6.9), which would be the same as $\delta[l_1-l_2]$ in Eq. (6.5).

6.3 Statistics of the Intensities of Two Pixels

This section describes statistical properties of the intensities of any two pixels in the X-ray CT image generated by the convolution reconstruction method. Specifically, it gives a description of the relation between the correlation of pixel intensities and the distance between pixels. It shows that (1) the intensities of any two pixels in an X-ray CT image are correlated, (2) the degree of the correlation decreases as the distance between pixels increases, and (3) in the limiting case of the distance approaching infinity, the intensities of two pixels become independent. These properties are called spatially asymptotically independent, abbreviated as SAI. We prove it probabilistically.

6.3.1 Spatially Asymptotic Independence

Property 6.2 Pixel intensities of X-ray CT image are spatially asymptotically independent.

Proof.

For the purpose of easy reading, the three equations in Eq. (5.43) are written separately below. (i) A pixel intensity $f(r,\phi)$, that is, backprojected data are computed by

$$f(r,\phi) = c_2 \sum_{m=0}^{M-1} c' s_m(r,\phi), \tag{6.16}$$

(ii) M interpolated data $s_m(r,\phi)$ $(0 \leq m \leq M-1)$ are computed by

$$s_m(r,\phi) = \sum_{k=-\infty}^{\infty} t(k\eta, m\Delta)\psi(u - k\eta), \tag{6.17}$$

where

$$u = \begin{cases} l' = r\cos(m\Delta - \phi) & \text{(parallel projection)} \\ \sigma' = \tan^{-1} \frac{r\cos(m\Delta-\phi)}{D+r\sin(m\Delta-\phi)} & \text{(divergent projection),} \end{cases} \tag{6.18}$$

(iii) $M(2N+1)$ convolved projections $t(i\eta, m\Delta)$ $(-N \leq i \leq N,\ 0 \leq m \leq M-1)$ are computed by

$$t(i\eta, m\Delta) = c_1 \sum_{k=-N}^{N} p(k\eta, m\Delta)q^*((i-k)\eta), \tag{6.19}$$

where $p(k\eta, m\Delta)$ $(-N \leq k \leq N,\ 0 \leq m \leq M-1)$ are $M(2N+1)$ input projections; for the parallel projection shown in Eq. (2.83), $c_1 = d$, $c_2 = \Delta$, $c' = 1$, and $q^*((i-k)\eta)$ denotes $q((n-k)d)$; for the divergent projection shown in Eq. (2.91), $c_1 = \delta$, $c_2 = \Delta$, $c' = W$, and $q^*((i-k)\eta)$ denotes $q_1((n-k)\delta)\cos(k\delta) + q_2((n-k)\delta)\cos(n\delta)$.*

Let $R_f(i,j)$ denote the correlation of the intensities $f(r_i,\phi_i)$ and $f(r_j,\phi_j)$ of two pixels at (r_i,ϕ_i) and (r_j,ϕ_j). By ignoring the constants c_2 and c', $R_f(i,j)$ is given by

$$R_{\hat{f}}(i,j) = E[f(r_i,\phi_i)f(r_j,\phi_j)] = E[(\sum_{m=0}^{M-1} s_m(r_i,\phi_i))(\sum_{n=0}^{M-1} s_n(r_j,\phi_j))]$$
$$= \sum_{m=0}^{M-1} E[s_m(r_i,\phi_i)s_m(r_j,\phi_j)] + \sum_{m=0}^{M-1}\sum_{\substack{n=0 \\ n\neq m}}^{M-1} E[s_m(r_i,\phi_i)]E[s_n(r_j,\phi_j)]. \quad (6.20)$$

$s_m(r_i,\phi_i)$ and $s_n(r_j,\phi_j)$ $(m \neq n)$ of Eq. (6.20) are uncorrelated because the projections contained in $s_m(r_i,\phi_i)$ and $s_n(r_j,\phi_j)$ come from different views, and therefore are statistically independent (Property 5.3).

On the other hand, because $s_m(r_i,\phi_i)$ of Eq. (6.20) is the interpolation of two convolved projections $t(n_{m_i}\eta, m\Delta)$ and $t((n_{m_i}+1)\eta, m\Delta)$, it can be written as

$$s_m(r_i,\phi_i) = \sum_{l=0}^{1} C_l^{(m_i)} t((n_{m_i}+l)\eta, m\Delta). \quad (6.21)$$

In Eq. (6.21), n_{m_i} and $C_l^{(m_i)}$ $(l = 0,1)$ are determined by

$$n_{m_i}\eta \leq \zeta_i \leq (n_{m_i}+1)\eta \quad \text{and} \quad C_l^{(m_i)} = (-1)^{l+1}\frac{\zeta_i - (n_{m_i}+1-l)\eta}{\eta}, \quad (6.22)$$

where

$$\zeta_i = \begin{cases} r_i\cos(m\Delta - \phi_i) & \text{(parallel projection)} \quad (a) \\ \tan^{-1}\dfrac{r_i\cos(m\Delta-\phi_i)}{D+r_i\sin(m\Delta-\phi_i)} & \text{(divergent projection)} \quad (b). \end{cases} \quad (6.23)$$

It is easy to verify that the summation in Eq. (6.17) is identical to the operation in Eq. (6.21). Substituting Eq. (6.21) into the first sum on the right side of Eq. (6.20), we have

$$E[s_m(r_i,\phi_i)s_m(r_j,\phi_j)]$$
$$= E[(\sum_{l=0}^{1} C_l^{(m_i)} t((n_{m_i}+l)\eta, m\Delta))(\sum_{l=0}^{1} C_l^{(m_j)} t((n_{m_j}+l)\eta, m\Delta))]. \quad (6.24)$$

*More specifically, $W = W_1W_2$, $W_1 = \frac{D}{(2\pi)^2}$, $W_2 = \frac{1}{(D')^2}$, and $(D')^2 = (r\cos(m\Delta-\phi))^2 + (D + r\sin(m\Delta-\phi))^2$.

In the following, the proof is divided into two paths: the parallel projection and the divergent projection.

(1) *Parallel Projection.* For the parallel projection, $\eta = d$. From Eq. (6.19), we have

$$t((n_{m_i} + l)d, m\triangle) = \sum_{k=-N}^{N} p(kd, m\triangle)q^*((n_{m_i} + l - k)d) \quad (l = 0, 1). \quad (6.25)$$

The convolution function $q^*((n-k)d) = q((n-k)d)$ ($q((n-k)d)$ is defined by Eq. (2.83) and Eq. (2.61)) can be approximated by a bandlimit function, that is,

$$q^*(nd) \simeq 0 \quad (|n| > n_0), \quad (6.26)$$

where n_0 depends on the window width A of the window function $F_A(U)$ of Eq. (2.57) and $A \simeq \frac{1}{d}$ [3, 4]. Applying Eq. (6.26) to Eq. (6.25), the convolved projections $t((n_{m_i} + l)d, m\triangle)$ $(l = 0, 1)$ in Eq. (6.25) are determined by the some (not all) input projections (in the m-th view) that are located in the interval $[n_{m_i} + l - n_0, n_{m_i} + l + n_0]$ $(l = 0, 1)$. Thus the first and second sums of Eq. (6.24) are determined by some input projections $p(nd, m\triangle)$ with n that satisfies $[n_{m_i} - n_0 \leq n \leq n_{m_i} + n_0 + 1]$ and $[n_{m_j} - n_0 \leq n \leq n_{m_j} + n_0 + 1]$, respectively. For a given convolution function q^*, if the i-th and j-th pixels are separated enough such that the two intervals $[n_{m_i} - n_0, n_{m_i} + n_0 + 1]$ and $[n_{m_j} - n_0, n_{m_j} + n_0 + 1]$ are not overlapping, namely

$$|n_{m_j} - n_{m_i}| > 2n_0 + 1, \quad (6.27)$$

then Eq. (6.24) becomes

$$E[s_m(r_i, \phi_i)s_m(r_j, \phi_j)] = E[s_m(r_i, \phi_i)]E[s_m(r_j, \phi_j)], \quad (6.28)$$

because all input projections $p(nd, m\triangle)$ in the parallel projection are independent (Property 5.3). Substituting Eq. (6.28) into Eq. (6.20), we obtain

$$\begin{aligned} R_f(i, j) &= \sum_{m=0}^{M-1} E[s_m(r_i, \phi_i)] \cdot \sum_{n=0}^{M-1} E[s_n(r_j, \phi_j)] \\ &= E(f(r_i, \phi_i))E(f(r_j, \phi_j)). \end{aligned} \quad (6.29)$$

Eq. (6.29) shows that when $|n_{m_j} - n_{m_i}| > 2n_0 + 1$, $f(r_i, \phi_i)$ and $f(r_j, \phi_j)$ are statistically independent.

Clearly, the condition of Eq. (6.27) is inconvenient to use. Below we derive a formula that is simple and has obvious physical meaning, even though it is somewhat approximate. Let (r_i, ϕ_i), (x_i, y_i), (r_j, ϕ_j), and (x_j, y_j) denote the polar and rectangular coordinates of the centers of two pixels labeled by i and j, and ΔR and ψ denote the distance between these two centers and the angle of the line connecting these two centers with the positive X-axis, that is,

$$\Delta R = \sqrt{(x_i - x_j)^2 + (y_i - y_j)^2} \quad \text{and} \quad \psi = \tan^{-1}\frac{y_i - y_j}{x_i - x_j}. \quad (6.30)$$

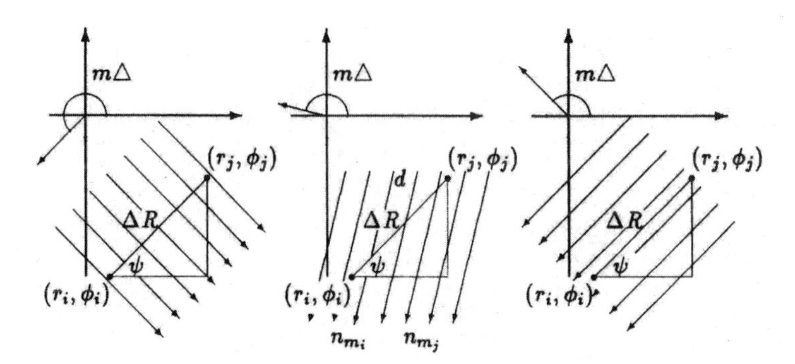

FIGURE 6.1
The geometry of projections and pixel locations.

The geometry of the projections and the locations of pixels is shown in Figure 6.1, where $m\Delta$ is assumed to have a uniform distribution on $[0, 2\pi]$, that is, its pdf is

$$p_{m\Delta} = \begin{cases} 1/2\pi & [0, 2\pi] \\ 0 & \text{elsewhere}. \end{cases} \tag{6.31}$$

Figures 6.1a and 6.1c show the two extreme situations in which

$$m\Delta = \psi + \pi \text{ (Figure 6.1.a)} \quad \text{and} \quad m\Delta = \psi + \frac{\pi}{2} \text{ (Figure 6.1c)}. \tag{6.32}$$

For the given pixels labeled by i and j, ψ is fixed; thus the probabilities

$$P(m\Delta = \psi + \pi) = P(m\Delta = \psi + \frac{\pi}{2}) = 0\,, \tag{6.33}$$

and therefore we only consider the general situation shown in Figure 6.1b. We approximate Eq. (6.23a) by using the nearest neighbor interpolation method, that is,

$$r_i \cos(m\Delta - \phi_i) \simeq n_{m_i} d \quad \text{and} \quad r_j \cos(m\Delta - \phi_j) \simeq n_{m_j} d, \tag{6.34}$$

where n_{m_i} and n_{m_j} are the nearest integers of $r_i \cos(m\Delta - \phi_i)/d$ and $r_j \cos (m\Delta - \phi_j)/d$, respectively. Thus

$$\begin{aligned}(n_{m_j} - n_{m_i})d &\simeq (x_i - x_j)\cos m\Delta + (y_i - y_j)\sin m\Delta \\ &= \Delta R \cos(m\Delta - \psi).\end{aligned} \tag{6.35}$$

Substituting Eq. (6.35) into Eq. (6.27), the statement represented by Eq. (6.27) and (6.28) becomes that when

$$|\Delta R \cos(m\Delta - \psi)| > (2n_0 + 1)d, \tag{6.36}$$

then $f(r_i, \phi_i)$ and $f(r_j, \phi_j)$ are statistically independent. $\Delta R \cos(m\triangle - \psi)$ of Eq. (6.36) is the projected distance of the distance ΔR onto the direction perpendicular to the projection direction. Thus, Eq. (6.36) states that when this projected distance is greater than the "*correlation*" distance $(2n_0 + 1)d$, then $f(r_i, \phi_i)$ and $f(r_j, \phi_j)$ are statistically independent.

It is clear from Eq. (6.36) that for some m, $\Delta R \cos(m\triangle - \psi)$ is less than $(2n_0 + 1)d$. Thus, what we have to do is find the condition such that the probability that this event occurs is less than a specified small value $(1 - P_0)$ (say 0.05). Eq. (6.31) shows that $(m\triangle - \psi)$ has a uniform distribution on $[0, 2\pi]$. It is easy to show that if γ_0 is the smallest angle among all possible $(m\triangle - \psi) < \frac{\pi}{2}$ such that when $m\triangle - \psi \geq \gamma_0$,

$$\Delta R \cos(m\triangle - \psi) \leq (2n_0 + 1)d, \tag{6.37}$$

then from the above discussion we have

$$P(\gamma_0 \leq m\triangle - \psi \leq \frac{\pi}{2}) = \frac{\frac{\pi}{2} - \gamma_0}{2\pi} \leq \frac{1 - P_0}{4}, \tag{6.38}$$

which leads to

$$\gamma_0 \geq \frac{\pi}{2} P_0. \tag{6.39}$$

γ_0 can be determined when the equality of Eq. (6.37) holds. Thus, we obtain

$$\cos^{-1}(\frac{(2n_0 + 1)d}{\Delta R}) \geq \frac{\pi}{2} P_0, \tag{6.40}$$

and therefore

$$\Delta R \geq \frac{(2n_0 + 1)d}{\cos(\frac{\pi}{2} P_0)}. \tag{6.41}$$

Eq. (6.41) gives the condition for two pixel intensities generated by the convolution reconstruction method (for the parallel projection) to be statistically independent with a probability greater than P_0. From Eq. (6.41), we have observed: (1) When the projection spacing d and the effective bandwidth n_0 of the convolution function q are fixed, the larger the probability P_0, the larger the distance ΔR is. In the limiting case of the probability P_0 approaching 1, the distance ΔR approaches infinity. (2) When d and P_0 are fixed, the smaller the n_0, the smaller ΔR is. (3)) When P_0 and n_0 are fixed, the smaller the d, the smaller ΔR is.

(2) *Divergent Projection.* For the divergent projection, $\eta = \delta$ and the convolution reconstruction method is different from that for the parallel projection in two aspects: (1) The convolution function q^* of Eq. (6.19) consists of two functions q_1 and q_2 (Eq. (2.74)) that are different from q^* for the parallel projection. (2) The original projections $p(n\delta, m\triangle)$ are a-dependent, not independent as in parallel projection. Below we show that even though these two differences exist, the conclusion similar to Eq. (6.41) for the divergent projection can still be achieved.

It has been shown that the convolution function for the divergent projection $q^*((n-k)\delta) = q_1((n-k)\delta)\cos(k\delta) + q_2((n-k)\delta)\cos(n\delta)$ (see Eq. (2.91) and Eq. (2.74)) behaves similarly to $q^*((n-k)d))$ for the parallel projection, that is, q^* can be approximated by a bandlimit function

$$q^*(n\delta) \simeq 0 \qquad (|n| > n_0), \tag{6.42}$$

where n_0 depends on the window width A of the window function $F_A(U)$ of Eq. (2.57) and $A \simeq \frac{1}{\delta}$.

The a-dependency in each view of divergent projections implies that if $|k - n| < n_0'$, where n_0' is defined by

$$n_0' = [\frac{a}{\delta}] + 1 \tag{6.43}$$

($[x]$ denotes the integer part of x), then $p(k\delta, m\triangle)$ and $p(n\delta, m\triangle)$ are correlated (Property 5.3).

Thus, when Eq. (6.27) holds for the divergent projection, then n_0 of Eq. (6.27) should be replaced by $n_0 + n_0'$. Thus, we obtain that, for the divergent projection, when

$$|n_{m_j} - n_{m_i}| > 2(n_0 + n_0') + 1, \tag{6.44}$$

then the pixel intensities $f(r_i, \phi_i)$ and $f(r_j, \phi_j)$ are statistically independent.

In order to obtain a simple formula for Eq. (6.44), we approximate Eq. (6.23b) using the nearest neighbor interpolation

$$\begin{aligned} \tan^{-1} \frac{r_i \cos(m\triangle - \phi_i)}{D + r_i \sin(m\triangle - \phi_i)} &\simeq n_{m_i}\delta \\ \tan^{-1} \frac{r_j \cos(m\triangle - \phi_j)}{D + r_j \sin(m\triangle - \phi_j)} &\simeq n_{m_j}\delta. \end{aligned} \tag{6.45}$$

After mathematical manipulations, we have

$$\tan(n_{m_j} - n_{m_i})\delta = \frac{D\Delta R\cos(m\triangle - \psi) + r_i r_j \sin(\phi_j - \phi_i)}{D^2 + D\Delta R' \sin(m\triangle - \psi') + r_i r_j \cos(\phi_j - \phi_i)}, \tag{6.46}$$

where ΔR and ψ are defined by Eq. (6.30), and $\Delta R'$ and ψ' are defined by

$$\Delta R' = \sqrt{(x_i + x_j)^2 + (y_i + y_j)^2} \quad \text{and} \quad \psi' = \tan^{-1} \frac{y_i + y_j}{x_i + x_j}. \tag{6.47}$$

In most cases, $D \gg r_i$ and r_j, thus, Eq. (6.46) can be approximated by

$$\tan(n_{m_j} - n_{m_i})\delta \simeq \frac{\Delta R\cos(m\triangle - \psi)}{D}. \tag{6.48}$$

Therefore, the statement represented by Eq. (6.44) becomes that when

$$\tan^{-1}(\frac{\Delta R\cos(m\triangle - \psi)}{D}) \geq (2(n_0 + n_0') + 1)\delta\ , \tag{6.49}$$

then $f(r_i, \phi_i)$ and $f(r_j, \phi_j)$ are statistically independent.

$\Delta R\cos(m\triangle - \psi)$ is the projected distance of ΔR onto the direction perpendicular to the central projection of the divergent projections. The left side of Eq. (6.49), $\tan^{-1}(\frac{\Delta R\cos(m\triangle - \psi)}{D})$, is the angle subtended by this projected distance at the source of a distance D. Thus, the physical meaning of Eq. (6.49) is that when the angle separation of two pixels (with respect to the source position) is greater than the "*correlation*" angle $(2(n_0+n_0')+1)\delta$, then the two pixel intensities are statistically independent.

Taking the same steps as Eq. (6.37) - Eq. (6.40), we obtain

$$\Delta R \geq \frac{D\tan(2(n_0 + n_0') + 1)\delta}{\cos(\frac{\pi}{2}P_0)}. \tag{6.50}$$

Eq. (6.50) gives the condition for two pixel intensities generated by the convolution reconstruction method (for the divergent projection) to be statistically independent with a probability greater than P_0. From Eq. (6.50), we have observed that (i) When the projection spacing δ, the effective bandwidth n_0 of the convolution function q, and a-dependency width n_0' are fixed, the larger the probability P_0, the larger the distance ΔR is. In the limiting case of the probability P_0 approaching 1, the distance ΔR approaches infinity. (ii) When δ and P_0 are fixed, the smaller the n_0 and/or n_0', the smaller the ΔR is. (iii)) When P_0 and n_0 and n_0' are fixed, the smaller δ, the smaller the ΔR is.

In a summary, $(2n_0 + 1)d$ in Eq. (6.41) and $D\tan(2(n_0 + n_0') + 1)\delta$ in Eq. (6.50) are the *correlation* distance, because if ΔR is equal to the *correlation* distance, $P_0 = 0$, that is, $1 - P_0 = 1$, which implies that two pixel intensities are correlated with probability 1. Let R_{cor} denote the *correlation* distance. Eq. (6.41) and Eq. (6.50) can be rewritten as

$$1 - P_0 \geq \frac{2}{\pi}\sin^{-1}\left(\frac{R_{cor}}{\Delta R}\right). \tag{6.51}$$

By taking the equality of Eq. (6.51), the relation between the probability $(1 - P_0)$ of two pixel intensities being correlated and their distance ΔR is plotted in Figure 6.2. The curve in Figure 6.2 shows that any two pixel intensities in an X-ray CT image generated by the convolution reconstruction method are correlated; the probability $(1 - P_0)$ of two pixel intensities being correlated is a monotonically decreasing function of the distance ΔR between pixels; when ΔR approaches infinity, $(1 - P_0)$ approaches the zero, which implies that pixel intensities are independent.

■

Let l_p be the pixel size and define

$$\Delta J = \left[\frac{\Delta R}{l_p}\right] + 1, \tag{6.52}$$

where ΔR is given by Eq. (6.41) or Eq. (6.50) by taking the equality, we obtain that when two pixels are spatially separated no less than ΔJ, then these two

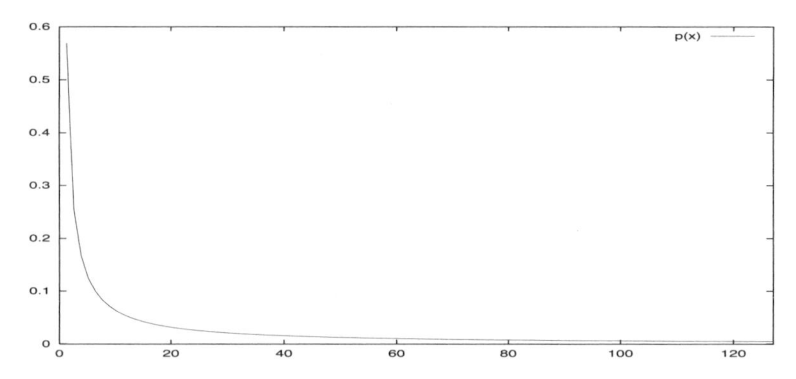

FIGURE 6.2
The relationship between the probability $(1-P_0)$ of two pixel intensities being correlated and the distance ΔR between two pixels.

pixel intensities are statistically independent with probability greater than P_0. The following example illustrates the relationship between ΔJ and the probability P_0. In this example, we assume that $n_0 = 2$ and $l_p = 10d$ for the parallel projection and $n_0 = n'_0 = 2$, $\delta = 0.05^o$, and $D = 450l_p$ for the divergent projection. The curves of ΔJ versus P_0 for the above settings are shown in Figure 6.3. It is clear from Figure 6.3, when two pixel intensities are statistically independent with probability greater than 0.7, then these two pixels must be separated by 2 for parallel or 8 for divergent projections, respectively.

6.3.2 Exponential Correlation Coefficient

Property 6.3 The magnitude of the correlation coefficient of pixel intensities of an X-ray CT image decreases exponentially with the distance between pixels.

Proof.

By taking the equality of Eq. (6.51), it becomes

$$1 - P_0 = \frac{2}{\pi}\sin^{-1}(\frac{R_{cor}}{\Delta R}) \qquad (\Delta R \geq R_{cor} > 0), \tag{6.53}$$

where ΔR is the distance between two pixels, R_{cor} is the correlation distance that is determined by the settings of the imaging system and the parameters of the image reconstruction algorith, and $(1 - P_0)$ is the probability of two pixel intensities being correlated.

Eq. (6.53) and Figure 6.2 show that any two pixel intensities in X-ray CT image generated by the convolution reconstruction method are correlated; the probability $(1-P_0)$ of two pixel intensities being correlated is a monotonically decreasing function of the distance ΔR between pixels; when ΔR approaches

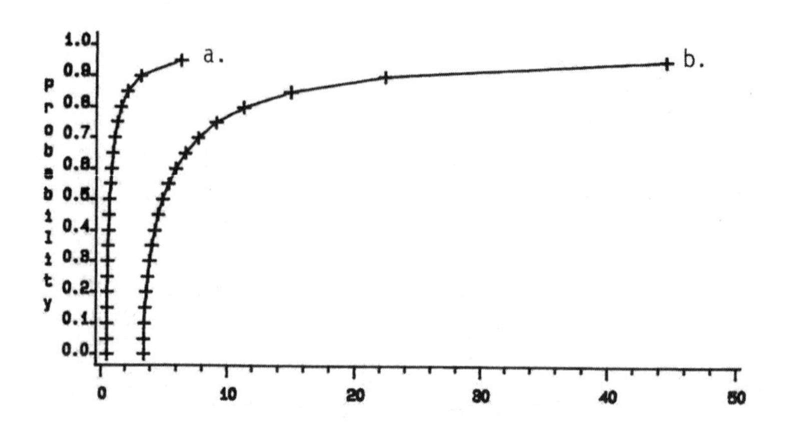

FIGURE 6.3
Relationship between the probability P_0 of two pixel intensities being independent and the pixel separation ΔJ: (a) for parallel projections, (b) for divergent projections.

infinity, $(1 - P_0)$ approaches, which implies that pixel intensities are uncorrelated, and hence independent.

These statements are equivalent to

(a) When two pixel intensities are less likely to be correlated, that is, ΔR is larger, then $(1 - P_0)$ becomes smaller. In the limiting case of $\Delta R \to \infty$, that is, two pixel intensities are uncorrelated, $(1 - P_0) = 0$.

(b) When two pixel intensities are more likely to be correlated, that is, ΔR is smaller, then $(1 - P_0)$ becomes larger. In the limiting case of $\Delta R = R_{cor}$, that is, two pixel intensities are fully correlated, then $(1 - P_0) = 1$.

Thus, the variation of $(1 - P_0)$, monotonically increasing from 0 to 1, corresponds to the change in the relationship between two pixel intensities, from uncorrelated to the fully correlated. Therefore, $(1 - P_0)$ can be used as a measure of the magnitude of the correlation coefficient, $|r_x((i,j),(k,l))|$, of two pixel intensities.

Let $(1 - P_0)$ be denoted by $\hat{r}_{|r_x((i,j),(k,l))|}$. Because Eq. (6.53) has an approximately exponential shape, we have

$$\hat{r}_{|r_x((i,j),(k,l))|} = e^{-a(\Delta R - R_{cor})} \qquad (\Delta R - R_{cor} > 0), \tag{6.54}$$

where $a > 0$ is a constant. Because $\hat{r}_{|r_x((i,j),(k,l))|}$ of Eq. (6.54) is very similar to $(1 - P_0)$ of Eq. (6.53) and also possesses several desirable analytical and computational properties, we use Eq. (6.54) to approximate Eq. (6.53). Parameter a can be determined such that $\hat{r}_{|r_x((i,j),(k,l))|}$ fits $(1 - P_0)$ according to some objective requirements. For instance, we can choose a by minimizing the sum of the square differences of $(\hat{r}_{|r_x((i,j),(k,l))|} - (1 - P_0))$ over n points of ΔR. From Eqs. (6.53) and (6.54), the dimensionality of a is the inverse of the length.

$$\hat{r}_{|r_x((i,j),(k,l))|} = e^{-a \cdot \Delta R} \qquad (\Delta R \geq 0, \ a > 0) \tag{6.55}$$

is a shifted version of Eq. (6.54). For most imaging systems and image reconstruction algorithms, the correlation distance R_{cor} is less than the one-dimensional pixel size δ_p. Thus, $\Delta R = R_{cor}$ in Eq. (6.54) and $\Delta R = 0$ of Eq. (6.55) are equivalent. Due to its simplicity, we use Eq. (6.55). By using $\Delta R = \delta_p \sqrt{(i-k)^2 + (j-l)^2} = \delta_p \sqrt{m^2 + n^2}$ and letting $a\delta_p = \alpha$, Eq. (6.55) becomes

$$\hat{r}_{|r_x((i,j),(k,l))|} = e^{-a \cdot \delta_p \sqrt{m^2+n^2}} = e^{-\alpha \sqrt{m^2+n^2}} \qquad (\alpha > 0). \tag{6.56}$$

Clearly, α has no dimensionality. ■

6.4 Statistics of the Intensities of a Group of Pixels

As shown in Section 2.3.2, the pixel intensity $f(r, \phi)$ represents the relative linear attenuation coefficient (RLAC) of an X-ray at the location (r, ϕ). The homogeneous tissue type or organ system has its unique RLAC. Pixels in some regions of the image that correspond to these homogeneous tissue type or organ system have essentially the same intensities. The small differences among them are caused by system noise. These nearly equal pixel intensities form an image region in an X-ray CT image.

In the image, a region refers to a group of pixels that are connected to each other, and their intensities have the same mean and the same variance. An image region is a union of these regions that may be or not be adjacent to each other. X-ray CT image consists of piecewise contiguous image regions. This fact reveals that each image region may possess unique statistical properties. This section proves that pixel intensities in each image region of an X-ray CT image form a spatial, stationary and ergodic random process, hence satisfying ergodic theorems.

6.4.1 Stationarity

A random process is said to be stationary in a wide sense if the mean is constant and the correlation function is shift invariant (either in time or in space) [5].

Property 6.4a Pixel intensities in an image region of an X-ray CT image form a spatial, stationary random process in the wide sense.

Proof.

Based on the concept of the image region given in this section, pixel intensities in an image region of an X-ray CT image have the same mean and the same variance, which are constant. In Eq. (6.51), $(1 - P_0)$—the probability of pixel intensities being correlated - is a measure of the correlation $R_f((r_i, \phi_i), (r_j, \phi_j))$ of pixel intensities. Eq. (6.51) shows that this measure only depends on the distance between pixels; it does not depend on pixel locations in the image. That is, the correlation of pixel intensities is spatially shift invariant. Thus, pixel intensities in an image region of an X-ray CT image are stationary in the wide sense. ■

Property 6.4b Pixel intensities in an image region of an X-ray CT image form a spatial, stationary random process in the strict sense.

Proof.

Property 6.1 shows that pixel intensities in an image region of an X-ray CT image have a Gaussian distribution. Property 6.4a shows these pixel intensities are stationary in the wide sense. Thus, pixel intensities in an image region of an X-ray CT image are stationary in the strict sense [5]. ■

Remarks on Stationarity. An X-ray CT image is a piecewise stationary random field.

6.4.2 Ergodicity

The concept of ergodicity has a precise definition for random process application. One such definition is given by [5]: *A random process is said to be ergodic if for any invariant event, F, either $m(F) = 0$ or $m(F) = 1$* (here m is the process distribution). *Thus, if an event is "closed" under time shift, then it must have all the probability or none of it.* An ergodic process is not the same as a process that satisfies an ergodic theorem. However, ergodicity is a sufficient condition of the Birkhoff-Khinchin ergodic theorem [5]. Thus, we first prove that pixel intensities in an image region of an X-ray CT image form an ergodic process, and then prove that this process satisfies the ergodic theorem.

Property 6.5a Pixel intensities in an image region of an X-ray CT image form an ergodic process.

Proof.

Property 6.4b shows that pixel intensities in an image region of an X-ray CT image form a spatial stationary random process. Property 6.2 shows that this process is spatially asymptotically independent. A stationary, asymptotically independent random process is ergodic [5]. ■

Property 6.5b Pixel intensities in an image region of an X-ray CT image satisfy the ergodic theorem, at least with probability 1.

Proof.

Because the random process formed by pixel intensities in an image region of an X-ray CT image is stationary (Property 6.4b) and ergodic (Property 6.5a), this process satisfies Birkhoff-Khinchin ergodic theorems [5]. That is, all spatial averages converge to the corresponding ensemble averages of the process, at least with probability 1. ■

Remarks on Ergodicity. An X-ray CT image is a piecewise ergodic random field.

As shown in Section 5.4.2, the reconstructed image is a configuration of the underlying random process, that is, an outcome of the image reconstruction. Each pixel intensity is a value in the state space of the corresponding random variable in this random process. Let $f(r,\phi)$ denote the underlying random process, and let $f^{(k)}(r,\phi)$ $(k = 1, 2, \cdots)$ denote the k-th configuration of $f(r,\phi)$. $f^{(k)}(r_i,\phi_i)$ represents a pixel intensity in $f^{(k)}(r,\phi)$. In order to estimate the mean or variance of $f(r_i,\phi_i)$, K independently identically distributed samples $f^{(k)}(r_i,\phi_i)$ $(k = 1, \cdots, K)$ from K images are required to generate an ensemble, for example, the sample mean or sample variance. With ergodicity, a spatial average of $f^{(k)}(r_i,\phi_i)$ over an image region inside one image $f^{(k)}(r,\phi)$ (a given k) can be performed to generate these estimates, because under ergodicity, the ensemble average equals the spatial (or time) average. Thus, only one reconstructed image is required.

6.5 Appendices

6.5.1 Appendix 6A

A signal-to-noise ratio (SNR) of an image is one of the fundamental measures of image quality. An image with a high SNR has some unique features. For an X-ray CT image, these features are given by Property 6.6. For a pixel centered at (r,ϕ), let $\mu_{f(r,\phi)}$ and $\sigma_{f(r,\phi)}$ be the mean and the standard deviation of the pixel intensity $f(r,\phi)$; its SNR is defined by

$$SNR_{f(r,\phi)} = \frac{\mu_{f(r,\phi)}}{\sigma_{f(r,\phi)}} . \tag{6.57}$$

Property 6.6 For an X-ray CT image with a high SNR, its pixel intensities tend to be statistically independent; pixel intensities in an image region are stationary and ergodic in the mean and the autocorrelation.

Proof.

The second items on the right side of Eq. (6.5) and Eq. (6.9), as shown by Eq. (6.6) and Eq. (6.10), actually represent the variance of the pixel intensity. Let them be denoted by $k_i\sigma_{f(r_i,\phi_i)}k_j\sigma_{f(r_j,\phi_j)}$ approximately, where k_1 and

k_2 are two coefficients. Thus, Eqs. (6.5) and (6.9) can be expressed by one formula:

$$R_f((r_i,\phi_i),(r_j,\phi_j))$$

$$= \mu_{f(r_i,\phi_i)}\mu_{f(r_j,\phi_j)} + k_i\sigma_{f(r_i,\phi_i)}k_j\sigma_{f(r_j,\phi_j)}$$

$$= \mu_{f(r_i,\phi_i)}\mu_{f(r_j,\phi_j)}[1 + k_ik_j/(\frac{\mu_{f(r_i,\phi_i)}\mu_{f(r_j,\phi_j)}}{\sigma_{f(r_i,\phi_i)}\sigma_{f(r_j,\phi_j)}})]$$

$$= \mu_{f(r_i,\phi_i)}\mu_{f(r_j,\phi_j)}[1 + \frac{k_ik_j}{SNR_{f(r_i,\phi_i)}SNR_{f(r_j,\phi_j)}}], \quad (6.58)$$

When the SNR is high, the second item in the bracket on the right side of Eq. (6.58) can be ignored. Thus, Eq. (6.58) becomes

$$R_f((r_i,\phi_i),(r_j,\phi_j)) \simeq \mu_{f(r_i,\phi_i)}\mu_{f(r_j,\phi_j)}, \quad (6.59)$$

which implies that pixel intensities $f(r_i,\phi_i)$ and $f(r_j,\phi_j)$ are approximately considered as statistically independent.

Under the approximated independence of pixel intensities in an X-ray CT image with high SNR, the stationarity and ergodicity of pixel intensities in an image region are proved in a different way.

In an image region, pixel intensities have the same mean μ and the same variance σ^2, that is,

$$\mu_{f(r_i,\phi_i)} = \mu \quad \text{and} \quad \sigma^2_{f(r_i,\phi_i)} = \sigma^2, \quad (6.60)$$

thus, the autocorrelation function $R_f(i,j)$ of pixel intensities is

$$R_f(i,j) = E[f(r_i,\phi_i)f(r_j,\phi_j)] = \begin{cases} \mu^2 & (i \neq j) \\ \sigma^2 + \mu^2 & (i = j). \end{cases} \quad (6.61)$$

Eqs. (6.60) and (6.61) show that $f(r,\phi)$ is a stationary process in the wide sense, and hence in the strict sense.

Because the autocovariance $C_f(i,j)$ of pixel intensities is

$$C_f(i,j) = R_f(i,j) - \mu_{f(r_i,\phi_i)}\mu_{f(r_j,\phi_j)} = \begin{cases} 0 & (i \neq j) \\ \sigma^2 & (i = j), \end{cases} \quad (6.62)$$

which means that the autovariance $C_f(i,i) < \infty$ and the autocovariance $C_f(i,j) \longrightarrow 0$ when $|j-i| \to \infty$. Thus, $f(r,\phi)$ is ergodic in the mean [1, 6].

In order to prove that $f(r,\phi)$ is ergodic in the autocorrelation, we define a new process $e_\tau(r,\phi)$:

$$e_\tau(r_i,\phi_i) = f(r_{i+\tau},\phi_{i+\tau})f(r_i,\phi_i) \quad (\tau > 0). \quad (6.63)$$

Thus, proving that $f(r,\phi)$ is ergodic in autocorrelation is equivalent to proving that $e_\tau(r,\phi)$ is ergodic in the mean. The mean of $e_\tau(r,\phi)$ is

$$\mu_{e_\tau} = E[e_\tau(r_i,\phi_i)] = E[f(r_{i+\tau},\phi_{i+\tau})f(r_i,\phi_i)] = R_f(i+\tau,i). \quad (6.64)$$

From Eq. (6.61), the mean μ_{e_τ} of $e_\tau(r,\phi)$ is

$$\mu_{e_\tau} = \mu^2. \quad (6.65)$$

The variance $\sigma^2_{e_\tau}$ of $e_\tau(r,\phi)$ is

$$\begin{aligned}\sigma^2_{e_\tau} &= E[e^2_\tau(r_i,\phi_i)] - [E(e_\tau(r_i,\phi_i))]^2 \\ &= E[f^2(r_{i+\tau},\phi_{i+\tau})f^2(r_i,\phi_i)] - \mu^2_{e_\tau} \\ &= E[f^2(r_{i+\tau},\phi_{i+\tau})]E[f^2(r_i,\phi_i)] - \mu^2_{e_\tau} \\ &= (\sigma^2+\mu^2)(\sigma^2+\mu^2) - \mu^4 \\ &= \sigma^2(\sigma^2+2\mu^2). \end{aligned} \quad (6.66)$$

The autocorrelation $R_{e_\tau}(i,j)$ of $e_\tau(r,\phi)$ (for $i \neq j$) is

$$\begin{aligned} R_{e_\tau}(i,j) &= E[e_\tau(r_i,\phi_i)e_\tau(r_j,\phi_j)] \\ &= E[f(r_{i+\tau},\phi_{i+\tau})f(r_i,\phi_i)f(r_{j+\tau},\phi_{j+\tau})f(r_j,\phi_j)] \\ &= \begin{cases} E[f(r_{i+\tau},\phi_{i+\tau})]E[f(r_i,\phi_i)]E[f(r_{j+\tau},\phi_{j+\tau})]E[f(r_j,\phi_j)] \\ E[f(r_{i+\tau},\phi_{i+\tau})]E[f(r_i,\phi_i)f(r_{j+\tau},\phi_{j+\tau})]E[f(r_j,\phi_j)] \end{cases} \\ &= \begin{cases} \mu^4 & (|i-j| \neq \tau) \\ \mu^2(\sigma^2+\mu^2) & (|i-j| = \tau) \end{cases}. \end{aligned} \quad (6.67)$$

Thus, from Eqs. (6.65) through (6.67), $e_\tau(r,\phi)$ is stationary.

Because the autocovariance $C_{e_\tau}(i,j)$ of $e_\tau(r,\phi)$ is

$$\begin{aligned} C_{e_\tau}(i,j) &= R_{e_\tau}(i,j) - E[e_\tau(r_i,\phi_i)]E[e_\tau(r_j,\phi_j)] \\ &= \begin{cases} 0 & (|i-j| \neq \tau) \\ \mu^2\sigma^2 & (|i-j| = \tau) \end{cases} \end{aligned} \quad (6.68)$$

and the autovariance $C_{e_T}(i, i)$ of $e_T(r, \phi)$ is

$$C_{e_T}(i, i) = \sigma^2(\sigma^2 + 2\mu^2) < \infty. \tag{6.69}$$

Thus, $e_T(r, \phi)$ is ergodic in the mean, and hence $f(r, \phi)$ is ergodic in the autocorrelation [1, 6]. ■

Problems

6.1. Verify that the operations represented by Eqs. (6.21) and (6.17) are identical.

6.2. Verify Eq. (6.42).

6.3. Derive Eq. (6.46).

6.4. Prove Eq. (6.50).

6.5. Derive Eq. (6.51).

References

[1] Papoulis, A.: *Probability, Random Variables and Stochastic Processes.* McGraw-Hill Book Company, New York (1984).

[2] Dvoretzky, A.: Asymptotic normality of sums of dependent random vectors. In Krishraish, P., Ed.: Multivariate analysis 4, *Proceedings of Fourth International Symposium on Multivariate Analysis.* North-Holland Publishing Company, Amsterdam, The Netherlands (1977).

[3] Herman, G.: *Image Reconstruction from Projections.* Academic Press, New York (1980).

[4] Herman, G.: *Fundamentals of Computerized Tomography.* Springer, London, United Kingdom, (2009).

[5] Gray, R., Davisson, L.: *Random Processes: A Mathematical Approach for Engineers.* Prentice-Hall, Englewood Cliffs, New Jersey, (1986).

[6] Papoulis, A.: *Probability, Random Variables and Stochastic Processes.* McGraw-Hill Book Company, New York (1968).

7

Statistics of MR Imaging

7.1 Introduction

As shown in Chapter 3, MR images are reconstructed from k-space samples, while k-space samples are formulated from analog-to-digital conversion (ADC) signals by applying adequate pulse sequences. An ADC signal is a discrete version of a phase sensitive detection (PSD) signal (sampled at the proper frequency), while a PSD signal is formed from a free induction decay (FID) signal via quadrature PSD. FID signals are induced by transverse precessing macroscopic magnetization (TPMM), while TPMM originates from thermal equilibrium macroscopic magnetization (TEMM).

Thus, in MR imaging, the term MR data means the macroscopic magnetizations (TEMM, TPMM), MR signals (FID, PSD, ADC), and k-space samples. Among them, TEMM is spatially distributed, TPMM varies with both time and the spatial location, FID, PSD and ADC are the temporal signals, and k-space samples are in the spatial frequency domain. TEMM and FID are real data; TPMM, PSD, ADC, and k-space samples are complex data. Similar to any type of realistic data, each type of MR data consists of its signal and noise components.

In some statistics studies on MR imaging, the randomness of the signal components of MR data is ignored. Therefore, the randomness of MR data is mainly derived from their noise components, particularly from the thermal noise [1–19]. MR noise studies are often associated with the signal-to-noise ratio (SNR) evaluation. This is because the SNR can provide an absolute scale for assessing imaging system performance and lead to instrumentation design goals and constraints for system optimization; also, it is one of the fundamental measures of image quality.

This chapter describes statistics of both signal and noise components of each type of MR data, and focuses on their second-order statistics. Based on the physical principles of MRI described in Chapter 3 and according to MR data acquisition procedures, the statistical description of MR imaging progresses in the following natural and logical order: macroscopic magnetizations (TEMM $\rightarrow$ TPMM) $\Longrightarrow$ MR signals (FID $\rightarrow$ PSD $\rightarrow$ ADC) $\Longrightarrow$ k-space samples. When MR data travel in the space–time–(temporal and spatial)–frequency domains, their statistics are evolving step by step.

For the typical MR data acquisition protocols (the rectilinear and the radial k-space sampling), this chapter provides signal processing paradigms for the basic image reconstruction methods (Fourier transform (FT) and projection reconstruction (PR)). Then it gives a statistical interpretation of MR image reconstruction. That is, MR image reconstruction can be viewed as a transform from a set of random variables (k-space samples) to another set of random variables (pixel intensities). These new random variables form a random process, also known as a random field. Statistics of MR data in the imaging domain propagate to the statistics in the image domain through image reconstruction.

7.2 Statistics of Macroscopic Magnetizations

Several studies on magnetization are at the microscopic scale and have established theoretical models for (a) the classical response of a single spin to a magnetic field and (b) the correlation of two individual spins in 1-D Ising model [4, 18, 20]. Macroscopic magnetization represents a vector sum of all microscopic magnetic moments of spins in a unit volume of sample. As shown in Chapter 3, because signal components of k-space samples (Eq. (3.110) and MR signals (Eqs. (3.96), (3.101), (3.102)) represent the collective behavior of a spin system, medical applications utilizing magnetic resonance for imaging objects (tissues or organs) are based on macroscopic magnetization, which is often called the bulk magnetization. Statistics of two types of bulk magnetizations—TEMM and TPMM—are analyzed in this section.

7.2.1 Statistics of Thermal Equilibrium Magnetization

As shown in Section 3.4.2, when a sample is placed in an external, static magnetic field $\vec{B}_0 = B_0\vec{k}$ (where $\vec{k}$ is the unit directional vector at the Z direction of a Cartesian coordinate system $\{U, V, Z\}$),* the magnitude of TEMM of the spin-$\frac{1}{2}$ systems such as ^1H, ^{13}C, ^{19}F, and ^{31}P, etc., is given by

$$M_z^o(\mathbf{r}) = \frac{1}{2}(n_l - n_h)\gamma\hbar = \frac{1}{2}\gamma\hbar\epsilon n, \tag{7.1}$$

where $\mathbf{r} = (u, v, z)$ denotes a location, n_l and n_h are the numbers of spins at the lower and higher energy states, $n = n_l + n_h$, γ is the gyromagnetic

*Instead of $\{X, Y, Z\}$ used in the previous chapters, starting from this chapter, $\{U, V, Z\}$ denotes a Cartesian coordinate system. $\vec{i}$, $\vec{j}$,and $\vec{k}$ are the unit directional vectors at the U, V, and Z directions. $\vec{i}$ and $\vec{j}$ define the transverse plane and $\vec{k}$ specifies the longitudinal direction.

ratio $\hbar = \frac{h}{2\pi}$ and h is Planck's constant, and ϵ is the ratio of the population difference and the total population of nuclear spins in a unit volume of sample and is given by Eq. (3.22). $M_z^o(\mathbf{r})$ is in the $\vec{k}$ direction. Statistics of TEMM are given by the following property.

Property 7.1 TEMM has a binomial distribution $B(n,p)$ on the order n with the parameter p (the probability for a single spin in a population of the size n to be in the lower energy state). Under the normal conditions and the ordinary settings of MRI, TEMM can be characterized by a spatially Gaussian random process with constant mean and almost-zero variance: $M_z^o(\mathbf{r}) \sim N(\mu_{M_z^o}, \sigma_{M_z^o} \to 0)$, where $\mu_{M_z^o(\mathbf{r})} = \frac{1}{2}\gamma\hbar\epsilon n$. That is, TEMM is spatially deterministic with probability 1. $\square$

Proof.

Because the probability that a single spin in a population of the size n is in the lower energy state is p (Section 3.4.2), the probability that n_l spins in this population are in the lower energy state and $n_h = n - n_l$ spins in this population are in the higher energy state is

$$\binom{n}{n_l} p^{n_l}(1-p)^{n-n_l}. \tag{7.2}$$

Eq. (7.2) is the probability mass function (pmf) of a binomial distribution of the order n with the parameter p, $B(n,p)$. The parameter p can be estimated by $\frac{n_l}{n}$, which, by using Eq. (3.23), leads to $p = \frac{1+\epsilon}{2}$. Let ν be a random variable with $B(n,p)$; its mean and variance are [21]

$$\begin{aligned} \mu_\nu &= E[\nu] = np \\ \sigma_\nu^2 &= Var[\nu] = np(1-p). \end{aligned} \tag{7.3}$$

In MRI, $n \gg 1$ and $np(1-p) \gg 1$. Thus, according to the Bernoulli-Poisson theorem [22] or DeMoivre-Laplace theorem [23], the binomial distribution $\nu \sim B(n,p)$ can be accurately approximated by a Gaussian distribution $\nu \sim N(\mu_\nu, \sigma_\nu)$ $(0 < n_l < n)$, that is,

$$\binom{n}{\nu} p^{\nu}(1-p)^{n-\nu} \simeq \frac{1}{\sqrt{2\pi}\sigma_\nu} \exp\left(-\frac{(\nu-\mu_\nu)^2}{2\sigma_\nu^2}\right). \tag{7.4}$$

Let ν and $(n-\nu)$ represent two populations of spins in the low and high energy states, respectively, and let Δ be the population difference; then $\Delta = \nu - (n-\nu) = 2\nu - n$. By applying the transform of random variables, Δ is characterized by a Gaussian distribution: $N(\mu_\Delta, \sigma_\Delta)$ $(0 < \Delta < n)$, where the mean, the variance, and the coefficient of variation of Δ are, respectively,

$$\begin{aligned} \mu_\Delta &= E[\Delta] = 2\mu_\nu - n = \epsilon n, \\ \sigma_\Delta^2 &= Var[\Delta] = 4\sigma_\nu^2 = (1-\epsilon^2)n, \\ (\frac{\sigma}{\mu})_\Delta &= \frac{\sqrt{1-\epsilon^2}}{\epsilon\sqrt{n}} \simeq \frac{1}{\epsilon\sqrt{n}}. \end{aligned} \tag{7.5}$$

Using Δ, Eq. (7.1) becomes $M_z^o(\mathbf{r}) = \frac{1}{2}\gamma\hbar\Delta$. By applying the transform of random variables again, TEMM $M_z^o(\mathbf{r})$ is characterized by a Gaussian distribution, $N(\mu_{M_z^o}, \sigma_{M_z^o})$, where the mean, the variance, and the coefficient of variation of $M_z^o(\mathbf{r})$ are, respectively,

$$\begin{aligned}\mu_{M_z^o(\mathbf{r})} &= E[M_z^o(\mathbf{r})] = \frac{1}{2}\gamma\hbar\mu_\Delta = \frac{1}{2}\gamma\hbar\epsilon n \triangleq \mu_{M_z^o}, \\ \sigma^2_{M_z^o(\mathbf{r})} &= Var[M_z^o(\mathbf{r})] = \frac{1}{4}\gamma^2\hbar^2\sigma^2_\Delta = \frac{1}{4}\gamma^2\hbar^2(1-\epsilon^2)n \triangleq \sigma^2_{M_z^o}, \\ (\frac{\sigma}{\mu})_{M_z^o(\mathbf{r})} &= (\frac{\sigma}{\mu})_\Delta \simeq \frac{1}{\epsilon\sqrt{n}} \triangleq (\frac{\sigma}{\mu})_{M_z^o}.\end{aligned} \tag{7.6}$$

Under normal conditions and ordinary settings of MRI as shown in Table 3.2, we have

$$\begin{aligned}\mu_{M_z^o} &\simeq 3.21 \times 10^{-12}, \\ \sigma^2_{M_z^o} &\simeq 1.33 \times 10^{-32}, \\ (\frac{\sigma}{\mu})_{M_z^o} &\simeq 3.60 \times 10^{-5}.\end{aligned} \tag{7.7}$$

The values in Eq. (7.7) show that $\sigma^2_{M_z^o}$ is almost zero and $(\frac{\sigma}{\mu})_{M_z^o}$ is very small. These two values indicate that the inherent noise of TEMM is very small and can be negligible.

Necessary and sufficient conditions for a random variable to be a constant (with probability 1) is its variance being equal to zero [23, 24]. $\sigma^2_{M_z^o} \to 0$ in Eq. (7.7) leads to that, at a given location $\mathbf{r}$, $M_z^o(\mathbf{r})$ can be approximated by a constant, or characterized by a Gaussian random variable with a constant mean $\mu_{M_z^o}$ and an almost-zero variance: $N(\mu_{M_z^o}, \sigma_{M_z^o} \to 0)$. This constant approximation of $M_z^o(\mathbf{r})$ at any given location $\mathbf{r}$ infers that TEMM can be considered a spatially deterministic process with probability 1. As a result, the correlation of TEMM at two locations is

$$E[M_z^o(\mathbf{r}_1)M_z^o(\mathbf{r}_2)] = E[M_z^o(\mathbf{r}_1)]E[M_z^o(\mathbf{r}_2)] \quad (\mathbf{r}_1 \neq \mathbf{r}_2), \tag{7.8}$$

which represents that $M_z^o(\mathbf{r}_1)$ and $M_z^o(\mathbf{r}_2)$ are uncorrelated, and therefore independent due to the Gaussianity of $M_z^o(\mathbf{r})$. ■

Remarks. The values in Eq. (7.7) are computed based on data listed in Table 3.2, where $n = 6.69 \times 10^{19}/(\text{mm})^3$ for H_2O. The voxel volume $v = 1.0\ (\text{mm})^3$ is typical for the conventional MRI. When it becomes very small, e.g., $v = 10^{-6}\ (\text{mm})^3$ such as in Micro-MRI, $(\frac{\sigma}{\mu})_{M_z^o(\mathbf{r})} \simeq 3.60 \times 10^{-2}$. In this case, the inherent noise of TEMM may not be negligible and should be considered.

7.2.1.1 Spin Noise and Its Statistics

Spin noise is inherent in magnetic resonance. It is caused by incomplete cancellation of spin moments when the external static magnetic field is absent or by their small but finite fluctuations when the magnetic field is applied. Thus, spin noise can be viewed as the variation of TEMM. In other words, TEMM $M_z^o(\mathbf{r})$ can be decomposed into a signal component $s(\mathbf{r})$ and a noise component $n(\mathbf{r})$ and expressed as

$$M_z^o(\mathbf{r}) = s(\mathbf{r}) + n(\mathbf{r}). \tag{7.9}$$

$s(\mathbf{r})$ and $n(\mathbf{r})$ are two random variables; $n(\mathbf{r})$ denotes the spin noise. Statistics of spin noise are given by the following property.

Property 7.2 Spin noise $n(\mathbf{r})$ in MRI can be characterized by a spatial Gaussian random process with the zero mean and the standard deviation σ_n: $n(\mathbf{r}) \sim N(0, \sigma_n)$, where $\sigma_n = \sigma_{M_z^o} = \frac{1}{2}\gamma\hbar\sqrt{(1-\epsilon^2)n}$ (ϵ is given by Eq. (3.22)). $n(\mathbf{r})$ and $s(\mathbf{r})$ are independent. □

Proof.

Let μ_s, μ_n and σ_s, σ_n denote the means and the standard deviations of the random variables $s(\mathbf{r})$ and $n(\mathbf{r})$, respectively.

(a) Similar to any other type of noise, spin noise $n(\mathbf{r})$ is commonly considered to have the zero mean $\mu_n = 0$. (b) From the definition of the spin noise, we have $\sigma_n = \sigma_{M_z^o}$. The arguments (a) and (b) lead to that (i) $\mu_s = \mu_{M_z^o}$, $\sigma_s = 0$; thus the random variable $s(\mathbf{r})$ is deterministic with probability 1, therefore, it can be characterized by a Gaussian distribution with the zero variance: $s(\mathbf{r}) \sim N(\mu_{M_z^o}, 0)$; and (ii) $n(\mathbf{r})$ and $s(\mathbf{r})$ are independent (because $s(\mathbf{r})$ is deterministic) and $n(\mathbf{r})$ has a Gaussian distribution: $n(\mathbf{r}) \sim N(0, \sigma_n)$ (because $n(\mathbf{r})$ and $M_z^o(\mathbf{r})$ are linear dependent—Eq. (7.9)). From Property 7.1,

$$\sigma_n = \sigma_{M_z^o} = \frac{1}{2}\gamma\hbar\sqrt{(1-\epsilon^2)n}. \tag{7.10}$$

Thus, we have

$$\begin{aligned} M_z^o(\mathbf{r}) &\sim N(\mu_{M_z^o}, \sigma_{M_z^o}), \\ s(\mathbf{r}) \sim N(\mu_s, 0) &= N(\mu_{M_z^o}, 0), \\ n(\mathbf{r}) \sim N(0, \sigma_n) &= N(0, \sigma_{M_z^o}). \end{aligned} \tag{7.11}$$

$\mu_{M_z^o}$ and $\sigma_{M_z^o}$ are given by Eq. (7.6). ■

Based on Eqs. (7.9) and (7.6), the intrinsic SNR of TEMM is defined as

$$SNR = (\frac{\mu}{\sigma})_{M_z^o} = \frac{\mu_s}{\sigma_n} = \frac{\epsilon\sqrt{n}}{\sqrt{1-\epsilon^2}}. \tag{7.12}$$

From Property 7.1 and Property 7.2, we have the following observations.

(1) At a sample size of $1 \times 1 \times 1(\text{mm})^3$ ($n = 6.69 \times 10^{19}/(\text{mm})^3$) and under the conditions: $B_o = 1\ T$ and $T = 300\text{K}$, Eqs. (3.22), (7.6), (7.11), and (7.12) give $\epsilon \simeq 3.40 \times 10^{-6}$, $\mu_s \simeq 3.21 \times 10^{-12}$ (A/m), $\sigma_n \simeq 1.16 \times 10^{-16}$ (A/m),

TABLE 7.1
Relations between B_o, T, ϵ, and SNR

B_o	1.0	1.5	3.0	4.7	9.4
T	300	300	300	300	300
$\epsilon\ 10^{-5}\times$	0.340	0.511	1.021	1.600	3.200
$\log SNR$	4.477	4.621	4.921	5.115	5.420

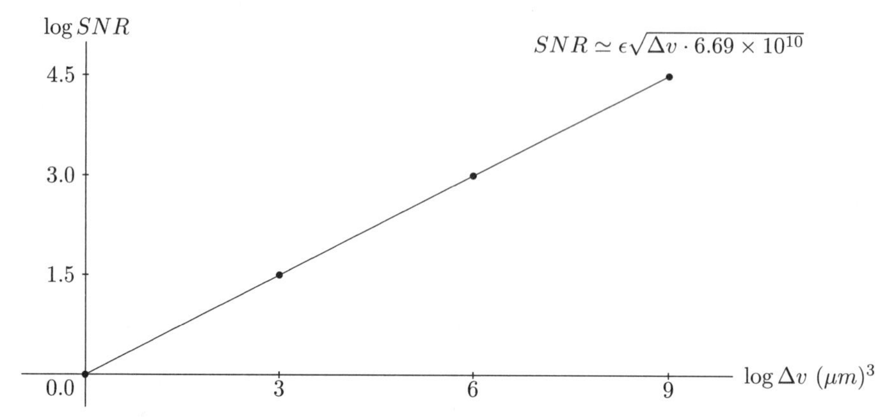

FIGURE 7.1
Relationship between spatial resolution Δv and SNR.

and $SNR \simeq 3 \times 10^4$. These numbers show that even when the population difference is just about 3 ppm and the mean of TEMM is very small, the intrinsic SNR of TEMM is very high at this sample size.

(2) For various values of B_o, Eqs. (3.22) and (7.12) give values of the ratio ϵ and SNR of TEMM shown in Table 7.1. Eq. (7.12) also links the intrinsic SNR of TEMM and the spatial resolution $\Delta v = \Delta u \cdot \Delta v \cdot \Delta z$ (via $\Delta v = n/n_o$, n_o— spin density). For MRI of H_2O ($n_o = 6.69 \times 10^{10}/(\mu\text{m})^3$), under conditions: $B_o = 1.0\ T$ and $T = 300$K, the relationship between SNR and Δv is shown in Table 7.2 and Figure 7.1, which indicate that as the sample size decreases, the intrinsic SNR decreases.

TABLE 7.2
Relations between Spatial Resolution Δv $(\mu m)^3$ and SNR

$\Delta v\ (\mu m)^3$	$1000 \cdot 1000 \cdot 1000$	$100 \cdot 100 \cdot 100$	$10 \cdot 10 \cdot 10$	$1 \cdot 1 \cdot 1$
$\log SNR$	4.477	2.944	1.444	-0.055

Table 7.2 and Figure 7.1 also show that when the voxel size decreases to some scale, the intrinsic SNR of TEMM will become small and approach 1 so that TEMM cannot be accurately measured and observed. In other words,

the spin noise may set a theoretical limit on the spatial resolution that conventional MRI can achieve.

7.2.1.2 Some Applications on Spin Noise

(1) Bloch stated [25] ("Nuclear Induction") that "*Even in the absence of any orientation by an external magnetic field one can expect in a sample with n nuclei of magnetic moment μ to find a resultant moment of the order $(n)^{\frac{1}{2}}\mu$ because of statistically incomplete cancellation.*"

Eq. (7.10) not only confirms Bloch's prediction: when $B_o = 0$, then $\epsilon = 0$ and Eq. (7.10) becomes $\sigma_n = \frac{1}{2}\gamma\hbar\sqrt{n}$, where $\frac{1}{2}\gamma\hbar = \mu$; but also gives a description of the behavior of spin noise in the case of $B_o \neq 0$: as shown by Eqs. (10) and (3.22) ($\epsilon \simeq \frac{\gamma\hbar B_o}{2\kappa T}$), σ_n decreases as B_o increases or T decreases. These observations are reasonable and consistent.

(2) Glover and Mansfield [26] have investigated the limits of the spatial resolution of magnetic resonance microscopy (MRM). They found that self-diffusion (in the liquid state); the NMR line width (in the solid state); and the diffusion, NMR line width, and few molecules per unit volume (in the gas state) limit the achievable image resolution. They concluded that "*As we have seen there are no hard and fast limits to MRM. The boundaries are flexible. Nevertheless, there are practical limits imposed by magnetic field strength and imaging time which currently limit practical resolution to about* 1 (μm). *Ostensibly this limit has been reached.*"

Table 7.2 and Figure 7.1 indicate that even in the ideal case of no thermal noise, a meaningful MR image with spatial resolution of $1 \times 1 \times 1$ $(\mu\text{m})^3$ may not be attainable, as spin noise limits the achievable SNR. This outcome provides insight from a statistical point of view to the limit of spatial resolutions of MRM.

(3) The basic idea and procedure of spin noise imaging (SNI) are as follows. The ratio between the magnitude of spin noise and TEMM is about 10^{-7}. Because the SNR of pulse-NMR experiments for a sample of the size 200 ml is in the range $10^7 - 10^9$, the spin noise signal should be measurable.

In the absence of a radiofrequency pulse and in the presence of a constant field gradient (e.g., the shim), the NMR noise is digitized and recorded as a continuous sequence of packets. The individual packets are Fourier transformed separately and co-added after calculation of the magnitude of the complex data points. In some experiments of SNI [27], 31 one-dimensional images were acquired, with the direction of the transverse magnetic field gradient being rotated by 6° to cover a 180° area. The projection reconstruction method is applied to this data set to generate final two-dimensional image.

In SNI, spin noise is treated as a kind of "signal." The peak-to-peak spin-noise-to-thermal-noise ratios σ_n/σ_m are typically 20 to 25 for the individual one-dimensional image. However, this ratio is evaluated experimentally. Eqs. (7.10) and (3.22) give a theoretical estimate of the magnitude of spin noise. It would be interesting to verify this outcome through experimental work of SNI.

7.2.2 Statistics of Transverse Precession Magnetizations

As described in Section 3.5.2, TPMM $\vec{M}_{uv}(\mathbf{r},t)$ is defined as the transverse component of free PMM $\vec{M}(\mathbf{r},t)$ in the relaxation process. Section 3.6 shows that it is given by the solution of the Block equation (Eqs. (3.46) through (3.48)) and its dynamic behavior is shown in Figure 3.4. Section 3.7 gives slice TPMM profiles for various RF excitation pulses in the excitation process. All MR signals, FID (Eq. (3.96)), PSD (Eq. (3.101)), and ADC (Eq. (3.102)) of Section 3.8, and k-space sample (Eq. (3.110)) of Section 3.9.1 are functions of TPMM. Section 3.9.2 indicates that TPMM also appears at the readout sampling.

All these equations demonstrate that the magnitude of $\vec{M}_{uv}(\mathbf{r},t)$ can be expressed by a product of its initial value $\vec{M}_{uv}(\mathbf{r},0)$ (a spatial factor of the location $\mathbf{r}$) and an exponential decay (a temporal factor of the time t). The following discussion is focused on $\vec{M}_{uv}(\mathbf{r},0)$. For simplicity and clarity, the initial value $\vec{M}_{uv}(\mathbf{r},0)$ of TPMM $\vec{M}_{uv}(\mathbf{r},t)$ is also called TPMM. Let $t=0$ denote the time instant at the start of the free precession process of spins. $t=0$ is also understood as the time instant at the end of the forced precession process of spins. Thus, $|\vec{M}(\mathbf{r},0)| = M_z^o(\mathbf{r})$. Using the complex notation, TPMM is expressed as

$$M_{uv}(\mathbf{r},0) = M_u(\mathbf{r},0) + iM_v(\mathbf{r},0) = |M_{uv}(\mathbf{r},0)|e^{i\phi_{M_{uv}}(\mathbf{r},0)}, \tag{7.13}$$

where $|M_{uv}(\mathbf{r},0)|$ and $\phi_{M_{uv}}(\mathbf{r},0)$ are the magnitude and the phase of TPMM $M_{uv}(\mathbf{r},0)$. Statistics of TPMM are given by the following property.

Property 7.3 TPMM $M_{uv}(\mathbf{r},0)$ can be characterized by a complex Gaussian random process with a constant mean and a almost-zero variance: $M_{uv}(\mathbf{r},0) \sim N(\mu_{M_{uv}}, \sigma_{M_{uv}} \rightarrow 0)$, where $\mu_{M_{uv}} = \mu_{M_z^o} \sin\alpha \; e^{i\phi_{M_{uv}}}$, $\alpha = \omega_1\tau_p = \gamma B_1\tau_p$, $\phi_{M_{uv}} = \omega_0\tau_p = \gamma B_0\tau_p$, τ_p is the duration of RF pulse. That is, $M_{uv}(\mathbf{r},0)$ is spatially deterministic with probability 1. ∎

Proof.

Section 3.7 shows that an RF pulse transforms TEMM $M_z^o(\mathbf{r})\vec{k}$ to a free precessing magnetization $\vec{M}(\mathbf{r},0) = \vec{M}_{uv}(\mathbf{r},0) + \vec{M}_z(\mathbf{r},0)$. In the $V'-Z'$ plane of the rotating reference frame,† $\vec{M}(\mathbf{r},0)$ rotates at an angle $\alpha = \gamma B_1\tau_p$ about $\vec{B}_1$. In the $U-V$ plane of the laboratory reference frame, $\vec{M}(\mathbf{r},0)$ rotates at

†Similar to the coordinate system $\{X,Y,Z\}$ in Chapter 3, the coordinate system $\{U,V,Z\}$ is called the laboratory reference frame. A coordinate system $\{U',V',Z'\}$ whose transverse plane $U'-V'$ is rotating clockwise at an angle frequency ω_0 and its longitudinal axis Z' is identical to Z is called the rotating reference frame.

an angle $\phi_{M_{uv}} = \gamma B_0 \tau_p$ about $\vec{B}_0$. From Eq. (7.13),

$$\begin{aligned} M_{uv}(\mathbf{r},0) &= |M_{uv}(\mathbf{r},0)|e^{i\phi_{M_{uv}}(\mathbf{r},0)} \\ &= |\vec{M}_{uv}(\mathbf{r},0)|e^{i\phi_{M_{uv}}(\mathbf{r},0)} \\ &= |\vec{M}(\mathbf{r},0)|\sin\alpha\; e^{i\phi_{M_{uv}}(\mathbf{r},0)} \\ &= M_z^o(\mathbf{r})\sin\alpha\; e^{i\phi_{M_{uv}}(\mathbf{r},0)}. \end{aligned} \tag{7.14}$$

For the magnitude of $M_{uv}(\mathbf{r},0)$, because (i) $|M_{uv}(\mathbf{r},0)| = M_z^o(\mathbf{r})\sin\alpha$ (α is constant), and (ii) $M_z^o(\mathbf{r}) \sim N(\mu_{M_z^o}, \sigma_{M_z^o} \to 0)$, then using the transform of random variables, $|M_{uv}(\mathbf{r},0)|$ is characterized by a Gaussian distribution with the mean $\mu_{|M_{uv}|} = \mu_{M_z^o}\sin\alpha$ and the variance $\sigma^2_{|M_{uv}|} = \sigma^2_{M_z^o}\sin^2\alpha \to 0$, that is, $|M_{uv}(\mathbf{r},0)| \sim N(\mu_{|M_{uv}|}, \sigma_{|M_{uv}|} \to 0)$.

For the phase of $M_{uv}(\mathbf{r},0)$, because at the end of the RF excitation or at the beginning of the relaxation, $\phi_{M_{uv}}(\mathbf{r},0)$ are the same for all $\mathbf{r}$, otherwise the incoherence of the phase $\phi_{M_{uv}}(\mathbf{r},0)$ would make $M_{uv}(\mathbf{r},0)$ vanish. Thus, $\phi_{M_{uv}}(\mathbf{r},0) = \gamma B_0 \tau_p \stackrel{\Delta}{=} \phi_{M_{uv}}$. That is, $\phi_{M_{uv}}(\mathbf{r},0)$ is constant for all $\mathbf{r}$.

Thus, from Eq. (7.13) and using the transform of random variables, $M_{uv}(\mathbf{r},0) = |M_{uv}(\mathbf{r},0)|e^{i\phi_{M_{uv}}}$ can be characterized by a Gaussian distribution: $M_{uv}(\mathbf{r},0) \sim N(\mu_{M_{uv}}, \sigma_{M_{uv}})$, where the mean and the variance are

$$\mu_{M_{uv}} = E[|M_{uv}(\mathbf{r},0)|e^{i\phi_{M_{uv}}}] = \mu_{|M_{uv}|}e^{i\phi_{M_{uv}}} = \mu_{M_z^o}\sin\alpha\; e^{i\phi_{M_{uv}}}, \tag{7.15}$$
$$\sigma^2_{M_{uv}} = Var[|M_{uv}(\mathbf{r},0)|e^{i\phi_{M_{uv}}}] = Var[|M_{uv}(\mathbf{r},0)|]|e^{i\phi_{M_{uv}}}|^2 = \sigma^2_{M_z^o}\sin^2\alpha \to 0.$$

This constant approximation of $M_{uv}(\mathbf{r},0)$ at any given location $\mathbf{r}$ leads to that TPMM can be considered a spatially deterministic process with probability one. ■

Remarks. Discussions so far were for a homogeneous spin system. $0 \leq \phi_{M_{uv}} < 2\pi$. $\phi_{M_{uv}} = 0$ and $\frac{\pi}{2}$ indicate that $\vec{M}_{uv}(\mathbf{r},0)$ is in the direction $\vec{i}$ and $\vec{j}$, respectively.

7.2.2.1 Observations and Discussions

Let $\delta(x)$ be a Dirac delta function and $g(x|\mu,\sigma)$ be a Gaussian pdf for the random variable $\mathbf{x}$. We observed that

(a) The Dirac delta function $\delta(x-\mu)$ sometimes is defined as [23, 28]

$$\delta(x-\mu) \geq 0 \quad \text{and} \quad \int_{-\infty}^{+\infty} \delta(x-\mu)dx = 1. \tag{7.16}$$

By comparing Eq. (7.16) with the definition of the pdf of a random variable, $\delta(x-\mu)$ can be considered a pdf of a random variable in some sense.

(b) [19, 28] show that

$$\delta(x-\mu) = \lim_{\sigma\to 0} \frac{1}{\sqrt{2\pi}\sigma} e^{-\frac{(x-\mu)^2}{2\sigma^2}} = \lim_{\sigma\to 0} g(x|\mu,\sigma). \tag{7.17}$$

Thus, Eq. (7.17) implies that a Dirac delta function can be used as an approximation of the pdf of a Gaussian random variable with the fixed mean and the almost-zero variance.

The $\lim_{\sigma \to 0} g(x|\mu, \sigma)$ in Eq. (7.17) can be written as $g(x|\mu, \sigma \to 0)$ and is known as the degenerate Gaussian pdf. Eqs. (7.16) and (7.17) show that a delta function and a degenerate Gaussian pdf are equivalent and can be expressed by

$$\delta(x - \mu) = g(x|\mu, \sigma \to 0). \tag{7.18}$$

Therefore, the properties of the delta function can be used to simplify statistical derivation in cases with degenerate Gaussians. For TPMM, we have

$$\begin{aligned} g(|M_{uv}(\mathbf{r},0)| \mid \mu_{|M_{uv}|}, \sigma_{|M_{uv}|} \to 0) &= \delta(|M_{uv}(\mathbf{r},0)| - \mu_{|M_{uv}|}), \\ g(M_{uv}(\mathbf{r},0) \mid \mu_{M_{uv}}, \sigma_{M_{uv}} \to 0) &= \delta(M_{uv}(\mathbf{r},0) - \mu_{M_{uv}}). \end{aligned} \tag{7.19}$$

7.3 Statistics of MR Signals

At the different stages of the signal detection module, MR signals take different forms and can be classified in different ways. For example, they may be classified as free induction decay, RF echos, gradient echos, etc. [18, 19]. These signals are generated using various pulse sequences and their amplitudes are functions of MR intrinsic parameters (e.g., spin density, T_1, T_2, T_I) and the operative parameters (e.g., T_E, T_R), etc. Because only the basic excitation/reception pulse sequences are considered in this statistical study, a simple classification—free induction decay (FID) signal, phase sensitive detection (PSD) signal, analog-to-digital conversion (ADC) signal—is adopted. This section analyzes statistics of FID, PSD, and ADC signals in terms of statistics of their signal and noise components.

(1) *FID signal.* The signal component $s_r(t)$ an of FID signal is given by Eq. (3.96). For the static gradient $\vec{G}(\mathbf{r},\tau) = \vec{G} = G_u\vec{i} + G_v\vec{j} + G_z\vec{k}$ (Section 3.6.3), the free precession frequency $\Delta\omega(\mathbf{r},t)$ in Eqs. (3.45) through (3.48) becomes spatially dependent only,that is, $\Delta\omega(\mathbf{r},t) \triangleq \gamma\vec{G}(\mathbf{r},t)\cdot\mathbf{r} = \gamma\vec{G}(\mathbf{r})\cdot\mathbf{r} \triangleq \Delta\omega(\mathbf{r})$. With minor manipulations, Eq. (3.96) can be rewritten as

$$\begin{aligned} s_r(t) = \omega_o \int_V &|B_{r_{uv}}(\mathbf{r})|\,|M_{uv}(\mathbf{r},0)|e^{-\frac{t}{T_2(\mathbf{r})}} \\ &\sin((\omega_o + \Delta\omega(\mathbf{r}))t + \phi_{B_{r_{uv}}}(\mathbf{r}) - \phi_{M_{uv}}(\mathbf{r},0))d\mathbf{r}, \end{aligned} \tag{7.20}$$

where V represents the sample volume, $\omega_o = \gamma B_0$ is the Larmor frequency, $\Delta\omega(\mathbf{r})$ is given by

$$\Delta\omega(\mathbf{r}) = \gamma\vec{G}(\mathbf{r})\cdot\mathbf{r} = \gamma(G_u u + G_v v + G_z z), \tag{7.21}$$

$|B_{r_{uv}}(\mathbf{r})|$ and $\phi_{B_{r_{uv}}}(\mathbf{r})$ are the magnitude and phase of the transverse component indicative of the sensitivity of the receiver coil, $\vec{B}_r(\mathbf{r})$, $|M_{uv}(\mathbf{r},0)|$ and $\phi_{M_{uv}}(\mathbf{r},0)$ are the magnitude and phase of TPMM $M_{uv}(\mathbf{r},0)$, and $T_2(\mathbf{r})$ is the transverse relaxation time constant at $\mathbf{r}$.

Eq. (7.20) is derived under several conditions: (i) the contribution of LPMM $M_z(\mathbf{r},t)$ to $s_r(t)$ is ignored because it varies slowly compared to TPMM $M_{uv}(\mathbf{r},t)$, and (ii) approximations such as $\omega_o \gg \Delta\omega(\mathbf{r},t)$ and $\omega_o \gg \frac{1}{T_2(\mathbf{r})}$ are adopted because they are valid in practice.

Randomly fluctuating noise currents in the sample induced mainly by the time-varying magnetic field are also picked up by the receiver coil, which constitutes a counterpart of $s_r(t)$, denoted by $n_r(t)$. Thus, a real-valued physical signal detected by the receiver coil can be expressed as

$$V_r(t) = s_r(t) + n_r(t). \tag{7.22}$$

$V_r(t)$ is the FID signal; $s_r(t)$ and $n_r(t)$ are its signal and noise components.

(2) *PSD signal.* As shown in Figure 3.8, with $s_r(t)$ of Eq. (7.20) as the input, the quadrature PSD results in a frequency shift of $s_r(t)$ by ω_0 and produces a complex-valued baseband signal component $s_c(t)$, which is given by Eq. (3.100). With minor manipulations, $s_c(t)$ is expressed by

$$s_c(t) = \frac{\omega_o}{2}\int_V B^*_{r_{uv}}(\mathbf{r})M_{uv}(\mathbf{r},0)e^{-\frac{t}{T_2(\mathbf{r})}}e^{-i\Delta\omega(\mathbf{r})t}d\mathbf{r}. \tag{7.23}$$

where $*$ represents the complex conjugate, $B_{r_{uv}}(\mathbf{r}) = |B_{r_{uv}}(\mathbf{r})|e^{i\phi_{B_{r_{uv}}}(\mathbf{r})}$, $M_{uv}(\mathbf{r},0) = |M_{uv}(\mathbf{r},0)|e^{i\phi_{M_{uv}}(\mathbf{r},0)}$, and $\Delta\omega(\mathbf{r})$ is given by Eq. (7.21).

$s_c(t)$ consists of a real part $s_I(t)$ (Eq. (3.97)) in the in-phase channel I and an imaginary part $s_Q(t)$ (Eq. (3.98)) in the quadrature channel Q. Similarly, $n_c(t)$—the noise counterpart of $s_c(t)$—also consists of a real part $n_I(t)$ in the channel I and an imaginary part $n_Q(t)$ in the channel Q. Thus, the signal component $s_c(t)$, the noise component $n_c(t)$, the in-phase signal $I(t)$, and the quadrature signal $Q(t)$ are, respectively,

$$\begin{aligned} s_c(t) &= s_I(t) + is_Q(t)\,, \qquad n_c(t) = n_I(t) + in_Q(t), \\ I(t) &= s_I(t) + n_I(t)\,, \qquad Q(t) = s_Q(t) + n_Q(t). \end{aligned} \tag{7.24}$$

Therefore, the complex-valued baseband MR signal $V_c(t)$ from the quadrature PSD can be expressed in two ways:

$$V_c(t) = I(t) + iQ(t) = (s_I(t) + n_I(t)) + i(s_Q(t) + n_Q(t))$$

or (7.25)

$$V_c(t) = s_c(t) + n_c(t) = (s_I(t) + is_Q(t)) + (n_I(t) + in_Q(t)).$$

$V_c(t)$ is the PSD signal; $s_c(t)$ and $n_c(t)$ are its signal and noise components.

(3) *ADC signal.* $V_c(t)$ is sampled at the Nyquist frequency [29] (for the given bandwidth of the anti-aliasing filter) to yield a discrete, complex-valued

baseband signal $\hat{V}_c(j)$, which is given by Eq. (3.102), where j denotes the time instant. With minor manipulations, $\hat{s}_c(j)$ is expressed by

$$\hat{s}_c(j) = \frac{\omega_o}{2}\int_V B^*_{r_{uv}}(\mathbf{r})M_{uv}(\mathbf{r},0)e^{-\frac{j}{T_2(\mathbf{r})}}e^{-i\Delta\omega(\mathbf{r})j}d\mathbf{r}. \tag{7.26}$$

Let $\hat{n}_c(j)$, $\hat{I}(j)$, $\hat{Q}(j)$, $\hat{s}_I(j)$, $\hat{s}_Q(j)$, $\hat{n}_I(j)$, and $\hat{n}_Q(j)$ be the discrete versions of $n_c(t)$, $I(t)$, $Q(t)$, $s_I(t)$, $s_Q(t)$, $n_I(t)$, and $n_Q(t)$. Similar to Eq. (7.25), $\hat{V}_c(j)$ can be expressed in two ways:

$$\hat{V}_c(j) = \hat{I}(j) + i\hat{Q}(j) = (\hat{s}_I(j) + \hat{n}_I(j)) + i(\hat{s}_Q(j) + \hat{n}_Q(j))$$

or (7.27)

$$\hat{V}_c(j) = \hat{s}_c(j) + \hat{n}_c(j) = (\hat{s}_I(j) + i\hat{s}_Q(j)) + (\hat{n}_I(j) + i\hat{n}_Q(j)).$$

The sampling is carried out by ADC. $\hat{V}_c(j)$ is the ADC signal; $\hat{s}_c(j)$ and $\hat{n}_c(j)$ are its signal and noise components.

Eqs. (7.22), (7.25) and (7.27) indicate that the randomness of MR signals results from both their signal and noise components. Statistics of signal and noise components of these signals are analyzed separately.

7.3.1 Statistics of Signal Components of MR Signals

Statistics of signal components of MR signals are described in the following property.

Property 7.4 (1) The signal component $s_r(t)$ of the real-valued physical FID signal $V_r(t)$ (Eq. (7.22)) can be characterized by a Gaussian random process with the mean $\mu_{s_r(t)}$, the variance $\sigma^2_{s_r(t)}$, and the autocorrelation function $R_{s_r(t)}(t_1, t_2)$ given by

$$\begin{gathered}\mu_{s_r(t)} = c_r e^{-\frac{t}{T_2}} sinc(\bar{\gamma}G_u u_0 t, \bar{\gamma}G_v v_0 t)\sin((\omega_0 + \gamma G_z z_0)t + \phi_{B_{r_{uv}}} - \phi_{M_{uv}}) \,, \\ \sigma^2_{s_r(t)} \to 0 \,, \\ R_{s_r(t)}(t_1,t_2) = \mu_{s_r(t_1)}\mu_{s_r(t_2)} + \sigma^2_{s_r(t)}\delta[t_2 - t_1],\end{gathered} \tag{7.28}$$

where $c_r = \omega_0|B_{r_{uv}}|\mu_{M_z^o}\sin\alpha V_0$ is a constant, $\mu_{M_z^o}$ is the mean of TEMM, α is the flip angle, V_0 is the volume of the sample,‡ $\bar{\gamma} = \frac{\gamma}{2\pi}$ and γ is the gyromagnetic ratio, $sinc(\bar{\gamma}G_u u_0 t, \bar{\gamma}G_v v_0 t) = sinc\,(\bar{\gamma}G_u u_0 t) sinc(\bar{\gamma}G_v v_0 t)$ is a 2-D sinc function, z_0 denotes the location of the selected slice, and $\delta[t]$ is the Kronecker delta function.§ That is, $s_r(t) \sim N(\mu_{s_r(t)}, \sigma^2_{s_r(t)} \to 0)$ is an

‡In the 2-D case, it represents an area. When it is specified by $[-\frac{u_0}{2} \le u \le \frac{u_0}{2}, -\frac{v_0}{2} \le v \le \frac{v_0}{2}]$, $V_0 = u_0 v_0$.

§Kronecker delta function $\delta[t]$—a discrete counterpart of the Dirac delta function $\delta(t)$—is mathematically defined as $\delta[t] = 1$ $(t = 0)$ and 0 $(t \neq 0)$ [15, 19, 30].

independent random process. It is also independent of the noise component $n_r(t)$ of $V_r(t)$.

(2) The real and imaginary parts $s_I(t)$ and $s_Q(t)$ of the signal component $s_c(t) = s_I(t) + is_Q(t)$ of the complex-valued baseband PSD signal $V_c(t)$ (Eq. (7.25)) can be characterized by two real Gaussian random processes: $s_I(t) \sim N(\mu_{s_I(t)}, \sigma^2_{s_I} \to 0)$ and $s_Q(t) \sim N(\mu_{s_Q(t)}, \sigma^2_{s_Q} \to 0)$, where $\mu_{s_I(t)} = \Re\{\mu_{s_c(t)}\}$, $\mu_{s_Q(t)} = \Im\{\mu_{s_c(t)}\}$ ($\Re$ and $\Im$—the real and imaginary parts of a complex-valued quantity, $\mu_{s_c(t)}$ - the mean of $s_c(t)$).

$s_I(t)$ and $s_Q(t)$ are independent random processes. They are independent of each other and also independent of their corresponding noise components $n_I(t)$ and $n_Q(t)$, respectively.

$s_c(t)$ is characterized by a bivariate Gaussian random process with the mean $\mu_{s_c(t)}$, the variance $\sigma^2_{s_c(t)}$, and the autocorrelation function $R_{s_c(t)}(t_1, t_2)$ given by

$$\begin{aligned} \mu_{s_c(t)} &= c_c e^{-\frac{t}{T_2}} sinc(\bar{\gamma} G_u u_0 t, \bar{\gamma} G_v v_0 t) e^{-i\gamma G_z z_0 t}, \\ \sigma^2_{s_c(t)} &\to 0, \\ R_{s_c(t)}(t_1, t_2) &= \mu_{s_c(t_1)} \mu^*_{s_c(t_2)} + \sigma^2_{s_c(t)} \delta[t_2 - t_1], \end{aligned} \tag{7.29}$$

where $c_c = \frac{\omega_0}{2} B^*_{r_{uv}} \mu_{M_z^o} \sin\alpha V_0$ is a constant, and $*$ denotes the complex conjugate. That is, $s_c(t) \sim N(\mu_{s_c(t)}, \sigma^2_{s_c(t)} \to 0)$ is an independent random process. It is also independent of the noise component $n_c(t)$ of $V_c(t)$.

(3) The real and imaginary parts $\hat{s}_I(j)$ and $\hat{s}_Q(j)$ of the signal component $\hat{s}_c(j) = \hat{s}_I(j) + i\hat{s}_Q(j)$ of the discrete complex-valued baseband ADC signal $\hat{V}_c(j)$ (Eq. (7.27)) can be characterized by two real Gaussian random processes: $\hat{s}_I(j) \sim N(\mu_{\hat{s}_I(j)}, \sigma^2_{\hat{s}_I} \to 0)$ and $\hat{s}_Q(j) \sim N(\mu_{\hat{s}_Q(j)}, \sigma^2_{\hat{s}_Q} \to 0)$, where $\mu_{\hat{s}_I(j)} = \Re\{\mu_{\hat{s}_c(j)}\}$ and $\mu_{\hat{Q}_I(j)} = \Im\{\mu_{\hat{s}_c(j)}\}$ ($\mu_{\hat{s}_c(j)}$—the mean of $\hat{s}_c(j)$).

$\hat{s}_I(j)$ and $\hat{s}_Q(j)$ are independent random processes. They are independent of each other and also independent of their corresponding noise components $\hat{n}_I(j)$ and $\hat{n}_Q(j)$, respectively.

$\hat{s}_c(j)$ is characterized by a bivariate Gaussian random process with the mean $\mu_{\hat{s}_c(j)}$, the variance $\sigma^2_{\hat{s}_c(j)}$, and the autocorrelation function $R_{\hat{s}_c(j)}(j_1, j_2)$ given by

$$\begin{aligned} \mu_{\hat{s}_c(j)} &= \hat{c}_c e^{-\frac{j}{T_2}} sinc(\bar{\gamma} G_u u_0 j, \bar{\gamma} G_v v_0 j) e^{-i\gamma G_z z_0 j}, \\ \sigma^2_{\hat{s}_c(j)} &\to 0, \\ R_{\hat{s}_c(j)}(j_1, j_2) &= \mu_{\hat{s}_c(j_1)} \mu^*_{\hat{s}_c(j_2)} + \sigma^2_{\hat{s}_c(j)} \delta[j_2 - j_1], \end{aligned} \tag{7.30}$$

where $\hat{c}_c = c_c$, $sinc(\bar{\gamma} G_u u_0 j, \bar{\gamma} G_v v_0 j) = sinc(\bar{\gamma} G_u u_0 j) sinc(\bar{\gamma} G_v v_0 j)$ is a 2-D sinc function. That is, $\hat{s}_c(j) \sim N(\mu_{\hat{s}_c(j)}, \sigma^2_{\hat{s}_c(j)} \to 0)$ is an independent random process. It is also independent of the noise component $\hat{n}_c(j)$ of $\hat{V}_c(j)$. □

This property is proved for the homogeneous and inhomogeneous spin systems, respectively. In the proof, the reception field $\vec{B}_{r,uv}(\mathbf{r})$ is assumed to be uniform over the sample; thus, its magnitude $|B_{r_{uv}}(\mathbf{r})|$ and phase $\phi_{B_{r_{uv}}}(\mathbf{r})$ are constant and denoted by $|B_{r_{uv}}|$ and $\phi_{B_{r_{uv}}}$, respectively.

Proof. (*for homogeneous spin systems*)

(1) For $s_r(t)$ of $V_r(t)$. In the case of homogeneous spin systems, $T_2(\mathbf{r}) = T_2$. Due to $\phi_{M_{uv}}(\mathbf{r}, 0) = \phi_{M_{uv}}$ [Property 7.3], Eq. (7.20) becomes

$$s_r(t) = \omega_0 |B_{r_{uv}}| e^{-\frac{t}{T_2}} \int_V |M_{uv}(\mathbf{r}, 0)| \sin((\omega_0 + \Delta\omega(\mathbf{r}))t + \phi_{B_{r_{uv}}} - \phi_{M_{uv}}) d\mathbf{r}. \tag{7.31}$$

In Eq. (7.31), because $|M_{uv}(\mathbf{r}, 0)|$ is characterized by a Gaussian random process [Property 7.3] and the integration is a linear operation, the Gaussianity is preserved for $s_r(t)$. Thus, $s_r(t)$ can be characterized by a Gaussian random process. Appendix 7A shows that at a given time t, the mean and the variance of $s_r(t)$ are given by Eq. (7.28).

The almost-zero variance $\sigma^2_{s_r(t)} \to 0$ implies that $s_r(t)$ is a temporally deterministic process with probability 1. Thus, for any $t_1 \neq t_2$, $R_{s_r(t)}(t_1, t_2) = \mu_{s_r(t_1)} \mu_{s_r(t_2)}$, i.e., $s_r(t)$ is an uncorrelated, and hence an independent random process. As a result, its autocorrelation function can be expressed as Eq. (7.28). The deterministic nature of $s_r(t)$ also infers that $s_r(t)$ is independent of the noise component $n_r(t)$ of $V_r(t)$.

(2) For $s_c(t)$ of $V_c(t)$. The statistical property of $s_c(t)$ is proved first, because the statistical properties of $s_I(t)$ and $s_Q(t)$ can be proved in a similar but simpler manner.

In the case of homogeneous spin systems, $T_2(\mathbf{r}) = T_2$. Eq. (7.23) becomes

$$s_c(t) = \frac{\omega_0}{2} B^*_{r_{uv}} e^{-\frac{t}{T_2}} \int_V M_{uv}(\mathbf{r}, 0) e^{-i\Delta\omega(\mathbf{r})t} d\mathbf{r}. \tag{7.32}$$

In Eq. (7.32), because $M_{uv}(\mathbf{r}, 0)$ is characterized by a Gaussian random process [Property 7.3] and the integration is a linear operation, the Gaussianity is preserved for $s_c(t)$. Thus, $s_c(t)$ can be characterized by a complex Gaussian random process. Appendix 7A shows that at a given time t, the mean and the variance of $s_c(t)$ are given by Eq. (7.29).

The almost-zero variance $\sigma^2_{s_c(t)} \to 0$ implies that $s_c(t)$ is a temporally deterministic process with probability 1. Thus, for any $t_1 \neq t_2$, $R_{s_c(t)}(t_1, t_2) = \mu_{s_c(t_1)} \mu^*_{s_c(t_2)}$, i.e., $s_c(t)$ is an uncorrelated, hence, an independent random process. As a result, its autocorrelation function can be expressed as Eq. (7.29). The deterministic nature of $s_c(t)$ also infers that $s_c(t)$ is independent of the noise component $n_c(t)$ of $V_c(t)$.

$s_I(t)$ and $s_Q(t)$ are given by Eqs. (3.97) and (3.98), respectively. Similarly to the above proof for $s_c(t)$, it is easy to verify that $s_I(t)$ and $s_Q(t)$ can be characterized by two Gaussian random processes with almost-zero variances:

$s_I(t) \sim N(\mu_{s_I(t)}, \sigma^2_{s_I} \to 0)$ and $s_Q(t) \sim N(\mu_{s_Q(t)}, \sigma^2_{s_Q} \to 0)$, where $\mu_{s_I(t)} = \Re\{\mu_{s_c(t)}\}$ and $\mu_{s_Q(t)} = \Im\{\mu_{s_c(t)}\}$.

The almost-zero variances, $\sigma^2_{s_I} \to 0$ and $\sigma^2_{s_Q} \to 0$, imply that $s_I(t)$ and $s_Q(t)$ are the temporally deterministic processes with probability 1. Therefore, $s_I(t)$ and $s_Q(t)$ are (a) independent random processes, (b) independent of each other, and (c) independent of their corresponding noise components $n_I(t)$ and $n_Q(t)$, respectively.

(3) For $\hat{s}_c(j)$ of $\hat{V}_c(j)$. Because $\hat{s}_c(j)$ is a discrete version of $s_c(t)$, its statistics are essentially the same as those of $s_c(t)$. Therefore, the proof of its statistical properties is basically the same as that for $s_c(t)$. As a result, $\hat{s}_c(j)$ can be characterized by a discrete, complex Gaussian random process, and its mean, the variance, and the autocorrelation function are given by Eq. (7.30).

Similar to $s_I(t)$ and $s_Q(t)$ of $s_c(t)$, the real and imaginary parts, $\hat{s}_I(j)$ and $\hat{s}_Q(j)$, of $\hat{s}_c(j)$ are characterized by two Gaussian random processes: $\hat{s}_I(j) \sim N(\mu_{\hat{s}_I(j)}, \sigma^2_{\hat{s}_I} \to 0)$ and $\hat{s}_Q(j) \sim N(\mu_{\hat{s}_Q(j)}, \sigma^2_{\hat{s}_Q} \to 0)$, respectively. Also, they are (a) independent random processes, (b) independent of each other, and (c) independent of their corresponding noise counterparts $\hat{n}_I(j)$ and $\hat{n}_Q(j)$, respectively.

■

Remarks. Eq. (7.28) shows a real-valued signal at the frequency $(\omega_0 + \gamma G_z z_0)$ in the laboratory reference frame. Eqs. (7.29),(7.30) show two complex-valued baseband signals at frequency $\gamma G_z z_0$ in the rotating reference frame. All three signals are modulated by a 2-D sinc function. This is because $M_{uv}(\mathbf{r}, 0)$ is constant in the area $[-\frac{u_0}{2} \le u \le \frac{u_0}{2}, -\frac{v_0}{2} \le v \le \frac{v_0}{2}]$ with probability 1 [Property 7.3] for a homogeneous spin system; therefore, its Fourier counterpart must be a 2-D sinc function $sinc(\bar{\gamma} G_u u_0 t, \bar{\gamma} G_v v_0 t)$.

In Eqs. (7.31) and (7.32), if the constant parameters such as ω_0, $\vec{B}_{r_{uv}}$ are omitted, as they often were in some literature, the physical unit of the quantity represented by the resulting formula would be different. For this reason, these constants are kept in the derivations and proofs.

Proof. *(for inhomogeneous spin systems)*

In this proof, an inhomogeneous spin system is considered to consist of several homogeneous sub-spin systems. That is, a sample can be partitioned into several sub-samples, and each of them corresponds to a homogeneous sub-spin system with its own spin density, T_1 and T_2 constants. For example, in the uniform reception field and for a single spectral component, the brain can be considered to consist of the cerebrospinal fluid (CSF), the gray matter (GM), and the white matter (WM) with the different spin densities, T_1 and T_2 constants [18].

This proof does not consider (i) the multiple isochromats¶ mainly caused by the existence of inhomogeneities in the $\vec{B}_0$ field and the chemical shift

¶A group of nuclear spins that share the same resonance frequency is called an isochromat.

effect, (ii) the off-resonance excitations and frequency selectivity of the RF pulse, and (iii) the partial volume effect. The following proof is for $s_c(t)$ only, because $s_r(t)$ can be proved in the same way and $\hat{s}_c(j)$ is a discrete version of $s_c(t)$.

Let the sample S be partitioned into N sub-samples S_n ($n = 1, \cdots, N$), that is, $S = \bigcup_{n=1}^{N} S_n$. Each S_n corresponds to a homogeneous sub-spin system that has TPMM $M_{uv,n}(\mathbf{r}, 0)$. The subscript n denotes both the n-th sub-sample and the n-th sub-spin system. For the n-th sub-sample S_n, its $s_c(t)$, denoted by $s_{c,n}(t)$, is given by Eq. (7.32) as

$$s_{c,n}(t) = \frac{\omega_0}{2} B^*_{r_{uv}} e^{-\frac{t}{T_{2,n}}} \int_V M_{uv,n}(\mathbf{r}, 0) e^{-i\Delta\omega(\mathbf{r})t} d\mathbf{r}. \tag{7.33}$$

For $n = 1, \cdots, N$, $M_{uv,n}(\mathbf{r}, 0) \neq 0$ ($\mathbf{r} \in S_n$) and $M_{uv,n}(\mathbf{r}, 0) = 0$ ($\mathbf{r} \bar{\in} S_n$), that is, $M_{uv,n}(\mathbf{r}, 0)$ is mutually exclusive. Thus, TPMM of this inhomogeneous spin system can be expressed as

$$M_{uv}(\mathbf{r}, 0) = \sum_{n=1}^{N} M_{uv,n}(\mathbf{r}, 0), \tag{7.34}$$

therefore

$$s_c(t) = \sum_{n=1}^{N} s_{c,n}(t). \tag{7.35}$$

$s_{c,n}(t)$ ($n = 1, \cdots, N$) are characterized by Gaussian random processes: $s_{c,n}(t) \sim N(\mu_{s_{c,n}(t)}, \sigma^2_{s_{c,n}(t)} \to 0)$ [Property 7.4 for the homogeneous spin system]. $\sigma^2_{s_{c,n}(t)} \to 0$ implies that$s_{c,n}(t)$ is a temporally deterministic process with probability 1. This implies that $s_{c,n}(t)$ ($n = 1, \cdots, N$) are independent. Therefore, $s_c(t)$ of Eq. (7.35)—a sum of N independent Gaussian random processes—can be characterized by a Gaussian random process. From Eq. (7.29), at a given time t, its mean and variance are

$$\mu_{s_c(t)} = \sum_{n=1}^{N} \mu_{s_{c,n}(t)} = (\sum_{n=1}^{N} c_{c,n} e^{-\frac{t}{T_{2,n}}}) sinc(\bar{\gamma} G_u u_0 t, \bar{\gamma} G_v v_0 t) e^{-i\gamma G_z z_0 t},$$

$$\sigma^2_{s_c(t)} = \sum_{n=1}^{N} \sigma^2_{s_{c,n}(t)} \to 0, \tag{7.36}$$

where the constant $c_{c,n} = \frac{\omega_0}{2} B^*_{r_{uv}} \mu_{M^o_{z,n}} \sin\alpha V_{0,n}$, and $\mu_{M^o_{z,n}}$, α, and $V_{0,n}$ are the mean of TEMM, the flip angle, and the volume of the n-th homogeneous sub-spin system. $\sigma^2_{s_c(t)} \to 0$ implies that $s_c(t)$ of this inhomogeneous spin system is temporally deterministic with probability 1. This implies that (a) $s_c(t)$ is an independent random process; that is, its autocorrelation function is $R_{s_c(t)}(t_1, t_2) = \mu_{s_c(t_1)} \mu^*_{s_c(t_2)}$ ($t_1 \neq t_2$), and (b) $s_c(t)$ is independent with the noise component $n_c(t)$ of of $V_c(t)$.

■

7.3.2 Statistics of the Noise Components of MR Signals

The noise in MRI is thermal in origin, arising from the Brownian motion of electrons in the conductive medium. The main noise contributors are associated with the resistance of the receiver coil and the effective resistance of the sample (observed by the receiver coil). Except for some cases such as low-field imaging and small-volume imaging, the resistance inherent in the sample is dominant. Let R denote the resistance or the effective resistance, the power spectral density of the thermal noise is $4\kappa RT$, where κ is Boltzmann's constant and T is absolute temperature [2, 6, 14]. Statistics of the noise components—$n_r(t)$ of FID signal, $n_c(t)$ of PSD signal, and $\hat{n}_c(j)$ of ADC signals—are investigated based on the theory of statistical mechanics [31, 32] and the theory of statistical communication [33], and given by the following property.

Property 7.5 (1) The noise component $n_r(t)$ of the real-valued FID signal $V_r(t)$ (Eq. (7.22)) can be characterized by a Gaussian random process, $n_r(t) \sim N(0, \sigma^2_{n_r})$, with the zero mean, the power spectral density $P_{n_r}(f)$, the autocorrelation function $R_{n_r}(\tau)$, and the variance $\sigma^2_{n_r}$ given by

$$\begin{aligned} P_{n_r}(f) &= 4\kappa RT \\ R_{n_r}(\tau) &= 4\kappa RT\delta(\tau) \\ \sigma^2_{n_r} &= 4\kappa RT, \end{aligned} \tag{7.37}$$

where R is the effective resistance of the sample, $\tau = t_2 - t_1$, and $\delta(\tau)$ is a Dirac delta function.

After anti-aliasing filtering

$$\begin{aligned} P_{n_r}(f) &= 4\kappa RT\ rect(\frac{f}{\Delta f}) \\ R_{n_r}(\tau) &= 4\kappa RT\Delta f\ sinc(\tau\Delta f) \\ \sigma^2_{n_r} &= 4\kappa RT\Delta f, \end{aligned} \tag{7.38}$$

where Δf is the bandwidth of the anti-aliasing filter, and $rect(\frac{f}{\Delta f})$ is a rectangular function defined by

$$rect(\frac{f}{\Delta f}) = \begin{cases} 1 & |f| \leq \frac{\Delta f}{2}, \\ 0 & |f| > \frac{\Delta f}{2}. \end{cases} \tag{7.39}$$

The front-end amplification further degrades $\sigma^2_{n_r}$ by the noise figure F to $4\kappa RT\Delta fF$.

$n_r(t)$ is stationary, additive, and independent of the signal component $s_r(t)$.

(2) The real and imaginary parts, $n_I(t)$ and $n_Q(t)$, of the noise component $n_c(t) = n_I(t) + in_Q(t)$ of the complex-valued baseband PSD signal $V_c(t)$ (Eq. (7.25)) can be characterized by two real Gaussian random processes with

the zero mean, the power spectral densities $P_{n_I}(f)$, $P_{n_Q}(f)$, the autocorrelation functions $R_{n_I}(\tau)$, $R_{n_Q}(\tau)$, and the variances $\sigma^2_{n_I}$, $\sigma^2_{n_Q}$ given by

$$\begin{aligned} P_{n_I}(f) = P_{n_Q}(f) = P_{n_r}(f) \\ R_{n_I}(\tau) = R_{n_Q}(\tau) = \sigma^2_{n_r}\, sinc(\tau \Delta f) \\ \sigma^2_{n_I} = \sigma^2_{n_Q} = \sigma^2_{n_r}, \end{aligned} \tag{7.40}$$

where $\tau = t_2 - t_1$, Δf is the bandwidth of the lowpass filter (LPF) $H(f)$ in the quadrature PSD and is equal to that of the anti-aliasing filter.

$n_I(t)$ and $n_Q(t)$ are independent of each other. They are stationary, additive, and also independent of their corresponding signal components $s_I(t)$ and $s_Q(t)$, respectively.

$n_c(t)$ can be characterized by a bivariate Gaussian random process with zero mean and variance $\sigma^2_{n_c}$: $n_c(t) \sim N(0, \sigma^2_{n_c})$, where $\sigma^2_{n_c} = 2\sigma^2_{n_r}$. It is stationary, additive, and independent of its signal component $s_c(t)$.

(3) The real and imaginary parts, $\hat{n}_I(j)$ and $\hat{n}_Q(j)$, of the noise component $\hat{n}_c(j) = \hat{n}_I(j) + i\hat{n}_Q(j)$ of the discrete complex-valued baseband ADC signal $\hat{V}_c(j)$ (Eq. (7.27), $j = \Delta t,\ 2\Delta t,\ \cdots,\ (\Delta t = 1/\Delta f)$) can be characterized by two real Gaussian random processes with zero mean; the power spectral densities $P_{\hat{n}_I}(f)$, $P_{\hat{n}_Q}(f)$; the autocorrelation functions $R_{\hat{n}_I}(\tau)$, $R_{\hat{n}_Q}(\tau)$; and the variances $\sigma^2_{\hat{n}_I}$, $\sigma^2_{\hat{n}_Q}$ given by

$$\begin{aligned} P_{\hat{n}_I}(f) = P_{\hat{n}_Q}(f) = P_{n_r}(f) \\ R_{\hat{n}_I}(\tau) = R_{\hat{n}_Q}(\tau) = \sigma^2_{n_r} \delta[\tau] \\ \sigma^2_{\hat{n}_I} = \sigma^2_{\hat{n}_Q} = \sigma^2_{n_r}, \end{aligned} \tag{7.41}$$

where $\tau = j_2 - j_1$, $\delta[\tau]$ is a Kronecker delta function.

$\hat{n}_I(j)$ and $\hat{n}_Q(j)$ are independent random processes. They are independent of each other. They are also stationary, additive, and independent of their corresponding signal components $\hat{s}_I(j)$ and $\hat{s}_Q(j)$, respectively.

$\hat{n}_c(j) = \hat{n}_I(j) + i\hat{n}_Q(j)$ can be characterized by a bivariate Gaussian random process with the zero mean and the variance $\sigma^2_{\hat{n}_c}$: $\hat{n}_c(j) \sim N(0, \sigma^2_{\hat{n}_c})$, where $\sigma^2_{\hat{n}_c} = 2\sigma^2_{n_r}$. It is an independent random process. It is stationary, additive, and independent of the signal component $\hat{s}_c(j)$. □

Proof.

(1) For $n_r(t)$ of $V_r(t)$. It is known that a noise voltage always develops across the ends of a resistor. In fact, this noise arises in any conductor medium because of the Brownian motion of electrons from thermal agitation of the electrons in the structure. Brownian motion of (free) particles is described by the Wiener-Lévy process, which is the limiting case of the random walk and is characterized by a Gaussian distribution. [23, 29]

In MRI, randomly fluctuating noise currents through the effective resistance of the sample are picked up by the receiver coil and becomes the noise component $n_r(t)$ of FID signal $V_r(t)$. Thus, $n_r(t)$ can be characterized by a Gaussian random process. It has two characteristics: the zero mean and the constant power spectral density:

$$\mu_{n_r} = E[n_r(t)] = 0 \quad \text{and} \quad P_{n_r}(f) = 4\kappa RT. \tag{7.42}$$

From Eq. (7.42), the autocorrelation function and variance of $n_r(t)$ are $R_{n_r}(\tau) = \mathcal{F}^{-1}\{P_{n_r}(f)\} = 4\kappa RT\delta(\tau)$ and $\sigma_{n_r}^2 = Var[n_r(t)] = R_{n_r}(0) = 4\kappa RT$, respectively. Therefore, $n_r(t) \sim N(0, \sigma_{n_r}^2)$.

In Eq. (7.42), f is implicitly assumed to be in $(-\infty, +\infty)$. After anti-aliasing filtering with the bandwidth Δf, the power spectral density of $n_r(t)$ is given by the first formula of Eq. (7.38); thus, the autocorrelation function and variance of $n_r(t)$ are $R_{n_r}(\tau) = \mathcal{F}^{-1}\{4\kappa\ RTrect(\frac{f}{\Delta f})\} = 4\kappa RTsinc(\tau\Delta f)$ and $\sigma_{n_r}^2 = Var[n_r(t)] = R_{n_r}(0) = 4\kappa RT\Delta f$, respectively. Note, after anti-aliasing filtering, $n_r(t)$ can be expressed as $\alpha_n(t)\cos(\omega_o t + \phi_n(t))$.

Because the autocorrelation function $R_{n_r}(\tau)$ of $n_r(t)$ only depends on the time lags τ (not the time instants t) and $n_r(t)$ is characterized by a Gaussian random process, $n_r(t)$ is stationary in both the wide and the strict sense.

Because $s_r(t)$ is induced by TPMM of spins in the sample and $n_r(t)$ is due to thermal noise in the sample (which are two different physical sources), they are additive. And because $s_r(t)$ is a temporally deterministic process with probability 1 [Property 7.4], $n_r(t)$ and $s_r(t)$ are independent.

(2) For $n_c(t)$ of $V_c(t)$. $n_r(t)$ is split into the in-phase (I) and the quadrature (Q) channels, passes the ideal lowpass filter (LPF) $h(t)$ (Figure 3.8), and becomes $n_I(t)$ and $n_Q(t)$. Because LPF is a linear operation, the properties of $n_r(t)$ - Gaussianity, additive, and independent of its signal component $s_r(t)$—are also preserved for $n_I(t)$ and $n_Q(t)$, that is, $n_I(t) \sim N(0, \sigma_{n_I}^2)$, $n_Q(t) \sim N(0, \sigma_{n_Q}^2)$, and are additive and independent of their corresponding signal components $s_I(t)$ and $s_Q(t)$.

For an ideal LPF $h(t)$, $|H(f)| = |\mathcal{F}\{h(t)\}| = 1$. Let the bandwidth of $H(f)$ be Δf—the same as the anti-aliasing filter. Then, the power spectral densities of $n_I(t)$ and $n_Q(t)$ are $P_{n_I}(f) = P_{n_Q}(f) = |H(f)|^2 P_{n_r}(f) = 4\kappa RTrect(\frac{f}{\Delta f})$. Thus, the autocorrelation functions and variances of $n_I(t)$ and $n_Q(t)$ are $R_{n_I}(\tau) = R_{n_Q}(\tau) = \sigma_{n_r}^2 sinc(\tau\Delta f)$ and $\sigma_{n_I}^2 = \sigma_{n_Q}^2 = \sigma_{n_r}^2$, respectively.

Because $R_{n_I}(\tau)$ and $R_{n_Q}(\tau)$ only depend on the time lags τ (not the time instants t) and $n_I(t)$ and $n_Q(t)$ are characterized by Gaussian random processes, $n_I(t)$ and $n_Q(t)$ are stationary in both the wide and the strict sense.

$n_c(t) = n_I(t) + in_Q(t)$ is viewed as the thermal noise resulting from 2-D Brownian motion; thus, $n_c(t)$ can be characterized by a 1-D complex Gaussian random process and its pdf is [34]

$$\begin{aligned} p(n_c(t)) &= \pi^{-1}(\sigma_{n_c}^2)^{-1}\exp(-(n_c^*(t)(\sigma_{n_c}^2)^{-1}n_c(t)) \\ &= \pi^{-1}(2\sigma_{n_r}^2)^{-1}\exp(-(n_I(t) + in_Q(t))^*(2\sigma_{n_r}^2)^{-1}(n_I(t) + in_Q(t))) \\ &= (2\pi\sigma_{n_r}^2)^{-1}\exp(-(n_I^2(t) + n_Q^2(t))/(2\sigma_{n_r}^2)) \end{aligned} \tag{7.43}$$

$$= (2\pi\sigma_{n_I}^2)^{-1/2}\exp(-(n_I^2(t))/(2\sigma_{n_I}^2)) \cdot (2\pi\sigma_{n_Q}^2)^{-1/2}\exp(-(n_Q^2(t))/(2\sigma_{n_Q}^2))$$
$$= p(n_I(t)) \cdot p(n_Q(t)).$$

$p(n_c(t))$ is the joint pdf of $n_I(t)$ and $n_Q(t)$, that is, $p(n_I(t), n_Q(t)) = p(n_c(t))$. Thus, Eq. (7.43) shows that $p(n_I(t), n_Q(t)) = p(n_I(t))p(n_Q(t))$. Therefore, $n_I(t)$ and $n_Q(t)$ are independent.

$n_c(t)$ has the following two characteristics: the zero mean and the constant power spectral density:

$$\mu_{n_c} = E[n_c(t)] = E[n_I(t)] + iE[n_Q(t)] = 0,$$
$$P_{n_c}(f) = P_{n_I}(f) + P_{n_Q}(f) = 8\kappa RTrect\left(\frac{f}{\Delta f}\right). \tag{7.44}$$

From Eq. (7.44), the autocorrelation function and variance of $n_c(t)$ are

$$R_{n_c}(\tau) = \mathcal{F}^{-1}\{P_{n_c}(f)\} = 8\kappa RTsinc(\tau\Delta f)$$
$$\sigma_{n_c}^2 = R_{n_c}(0) = 8\kappa RT = 2\sigma_{n_r}^2. \tag{7.45}$$

Thus, $n_c(t) \sim N(0, \sigma_{n_c}^2)$, where $\sigma_{n_c}^2 = 2\sigma_{n_r}^2$. Because $n_I(t)$ and $n_Q(t)$ are stationary, additive, and independent of their signal components $s_I(t)$ and $s_Q(t)$, $n_c(t)$ is stationary, additive, and independent of its signal component $s_c(t)$.

It can be verified that $n_I(t)$ and $n_Q(t)$ can be expressed as $\frac{1}{2}\alpha_n(t)\cos\phi_n(t)$ and $-\frac{1}{2}\alpha_n(t)\sin\phi_n(t)$. Thus, $n_c(t)$ can be expressed as $\frac{1}{2}\alpha_n(t)e^{-i\phi_n(t)}$.

(3) For $\hat{n}_c(j)$ of $\hat{V}_c(j)$. Because $\hat{n}_c(j)$ is a discrete version of $n_c(t)$, its statistics are essentially the same as those of $n_c(t)$. Therefore, the proof of its statistical properties is basically same as that for $n_c(t)$. As a result, $\hat{n}_c(j)$ is characterized by a bivariate Gaussian random process with zero mean and variance $\sigma_{\hat{n}_c}^2$: $\hat{n}_c(j) \sim N(0, \sigma_{\hat{n}_c}^2)$, where $\sigma_{\hat{n}_c}^2 = 2\sigma_{n_r}^2$. It is stationary, additive, and independent of the signal component $\hat{s}_c(j)$.

Similar to $n_I(t)$ and $n_Q(t)$ of $n_c(t)$, the real and imaginary parts, $\hat{n}_I(j)$ and $\hat{n}_Q(j)$, of $\hat{n}_c(j)$ are (a) characterized by two Gaussian random processes: $\hat{n}_I(j) \sim N(0, \sigma_{\hat{n}_I}^2)$ and $\hat{n}_Q(j) \sim N(0, \sigma_{\hat{n}_Q}^2)$; (b) independent of each other; and (c) stationary, additive, and independent of their corresponding signal components $\hat{s}_I(j)$ and $\hat{s}_Q(j)$, respectively.

However, sampling makes the differences in the autocorrelation functions $R_{\hat{n}_I}(\tau)$, $R_{\hat{n}_Q}(\tau)$, and $R_{\hat{n}_c}(\tau)$.

According to the Sampling theorem, in order to avoid aliasing, the time interval Δt of the readout samples should be less than or equal to the reciprocal of the bandwidth Δf of the baseband signal component $s_c(t)$, which is restricted by the anti-aliasing filter. Appendix 7B shows that when $\Delta t = \frac{1}{\Delta f}$ and the samples are taken at $j = l\Delta t$ (l - integers), for $j_1 \neq j_2$, $R_{\hat{n}_I}(j_2 - j_1) = R_{\hat{n}_Q}(j_2 - j_1) = 0$; because $\hat{n}_I(j)$ and $\hat{n}_Q(j)$ have the zero means $\mu_{\hat{n}_I(j)} = \mu_{\hat{n}_Q(j)} = 0$, we have $R_{\hat{n}_I}(j_2 - j_1) = \mu_{\hat{n}_I(j_1)}\mu_{\hat{n}_I(j_2)}^*$ and

$R_{\hat{n}_Q}(j_2 - j_1) = \mu_{\hat{n}_Q(j_1)}\mu^*_{\hat{n}_Q(j_2)}$. This implies that both $\hat{n}_I(j)$ and $\hat{n}_Q(j)$ are the uncorrelated, hence the independent random processes. For $j_1 = j_2$, $R_{\hat{n}_I}(j_2 - j_1) = R_{\hat{n}_I}(0) = \sigma^2_{n_r}$ and $R_{\hat{n}_Q}(j_2 - j_1) = R_{\hat{n}_Q}(0) = \sigma^2_{n_r}$. Thus, letting $j_2 - j_1 = \tau$, we obtain $R_{\hat{n}_I}(\tau) = R_{\hat{n}_Q}(\tau) = \sigma^2_{n_r}\delta[\tau]$.

For $\hat{n}_c(j)$, because $\hat{n}_I(j)$ and $\hat{n}_Q(j)$ are independent random processes with zero means, it is easy to verify that for $j_1 \neq j_2$,

$$E[\hat{n}_c(j_1)\hat{n}^*_c(j_2)] = E[\hat{n}_c(j_1)]E[\hat{n}^*_c(j_2)] = 0,$$

which implies that $\hat{n}_c(j)$ is an independent random process. ∎

Remarks. Eq. (7.37) shows that before anti-aliasing filtering, the noise component $n_r(t)$ of FID signal $V_r(t)$ is an uncorrelated, hence independent random process. Eq. (7.38) shows that after anti-aliasing filtering, $n_r(t)$ is not an independent random process, except for some time points t_i and t_j that satisfy $(t_i - t_j)\Delta f = \tau\Delta f = n\pi$ $(n = 1, 2, \cdots)$. Eq. (7.40) shows that the real and imaginary parts, $n_I(t)$ and $n_Q(t)$, of the noise component $n_c(t)$ of PSD signal $V_c(t)$ are not independent random processes. Its exception is the same as $n_r(t)$ (after anti-aliasing filtering). However, as shown by Eq. (7.41), the real and imaginary parts, $\hat{n}_I(t)$ and $\hat{n}_Q(t)$, of the noise component $\hat{n}_c(t)$ of ADC signal $\hat{V}_c(t)$ are the uncorrelated, hence, independent random processes.

Remarks. In the above proof, $n_r(t)$ and $n_c(t)$ are expressed as $\alpha_n(t)\cos(\omega_o t + \phi_n(t))$ and $\frac{1}{2}\alpha_n(t)e^{-i\phi_n(t)}$, respectively. It is worth comparing them with Eqs. (7.20) and (7.23), where $s_r(t)$ is a narrowband signal with frequencies $\omega_o + \Delta\omega(\mathbf{r})$ and $s_c(t)$ is a baseband signal with frequencies $\Delta\omega(\mathbf{r})$.

7.3.3 Statistics of MR Signals

In terms of the statistics of their signal and noise components, the statistics of MR signals are given by the following property.

Property 7.6 (1) The real-valued physical FID signal $V_r(t)$ (Eq. (7.22)), the complex-valued baseband PSD signal $V_c(t)$ (Eq. (7.25)), and the discrete complex-valued baseband ADC signal $\hat{V}_c(j)$ (Eq. (7.27)) can be characterized by Gaussian random processes:

$$\begin{aligned} V_r(t) &\sim N(\mu_{V_r(t)}, \sigma^2_{V_r}), \\ V_c(t) &\sim N(\mu_{V_c(t)}, \sigma^2_{V_c}), \\ \hat{V}_c(j) &\sim N(\mu_{\hat{V}_c(j)}, \sigma^2_{\hat{V}_c}), \end{aligned} \tag{7.46}$$

where the means $\mu_{V_r(t)} = \mu_{s_r(t)}$, $\mu_{V_c(t)} = \mu_{s_c(t)}$; and $\mu_{\hat{V}_c(j)} = \mu_{\hat{s}_c(j)}$, ($\mu_{s_r(t)}$, $\mu_{s_c(t)}$; and $\mu_{\hat{s}_c(j)}$ are given by Property 7.4); the variances $\sigma^2_{V_r} = \sigma^2_{n_r}$, $\sigma^2_{V_c} = \sigma^2_{n_c}$, and $\sigma^2_{\hat{V}_c} = \sigma^2_{\hat{n}_c}$, ($\sigma^2_{n_r}$, $\sigma^2_{n_c}$, and $\sigma^2_{\hat{n}_c}$ are given by Property 7.5).

$\hat{V}_c(j)$ is an independent random process, that is, its autocorrelation function $R_{\hat{V}_c(j)}(j_1, j_2)$ is given by

$$R_{\hat{V}_c(j)}(j_1, j_2) = \mu_{\hat{V}_c(j_1)}\mu^*_{\hat{V}_c(j_2)} + \sigma^2_{\hat{V}_c}\delta[j_2 - j_1], \tag{7.47}$$

where $\delta[j]$ is the Kronecker delta function.

(2) The signal components—$s_r(t)$ of $V_r(t)$, $s_c(t)$ of $V_c(t)$, and $\hat{s}_c(j)$ of $\hat{V}_c(j)$—can be characterized by Gaussian random processes with almost-zero variances: $s_r(t) \sim N(\mu_{s_r(t)}, \sigma^2_{s_r} \to 0)$, $s_c(t) \sim N(\mu_{s_c(t)}, \sigma^2_{s_c} \to 0)$, and $\hat{s}_c(j) \sim N(\mu_{\hat{s}_c(j)}, \sigma^2_{\hat{s}_c} \to 0)$, where the means $\mu_{s_r(t)}$, $\mu_{s_c(t)}$, and $\mu_{\hat{s}_c(j)}$ are given by Property 7.4.

The noise components—$n_r(t)$ of $V_r(t)$, $n_c(t)$ of $V_c(t)$, and $\hat{n}_c(j)$ of $\hat{V}_c(j)$—can be characterized by Gaussian random processes with zero means $n_r(t) \sim N(0, \sigma^2_{n_r})$, $n_c(t) \sim N(0, \sigma^2_{n_c})$, and $\hat{n}_c(j) \sim N(0, \sigma^2_{\hat{n}_c})$, where the variances $\sigma^2_{n_r}$, $\sigma^2_{n_c}$, and $\sigma^2_{\hat{n}_c}$ are given by Property 7.5. $\hat{n}_c(j)$ is an independent random process.

The noise components $n_r(t)$, $n_c(t)$, and $\hat{n}_c(j)$ are stationary, additive, and independent of their corresponding signal components $s_r(t)$, $s_c(t)$, and $\hat{s}_c(j)$.

In detail, for PSD signal $V(t)$, $s_I(t)$ and $s_Q(t)$ of $s_c(t) = s_I(t) + is_Q(t)$ are characterized by two Gaussian random processes: $s_I(t) \sim N(\mu_{s_I(t)}, \sigma^2_{s_I} \to 0)$ and $s_Q(t) \sim N(\mu_{s_Q(t)}, \sigma^2_{s_Q} \to 0)$, where $\mu_{s_I(t)} = \Re\{\mu_{s_c(t)}\}$ and $\mu_{s_Q(t)} = \Im\{\mu_{s_c(t)}\}$. They are independent random processes and are independent of each other.

$n_I(t)$ and $n_Q(t)$ of $n_c(t) = n_I(t) + in_Q(t)$ are characterized by two Gaussian random processes: $n_I(t) \sim N(0, \sigma^2_{n_I})$ and $n_Q(t) \sim N(0, \sigma^2_{n_Q})$, where $\sigma^2_{n_I} = \sigma^2_{n_Q} = \sigma^2_{n_r}$. They are independent random processes and are independent of each other.

$n_I(t)$ and $n_Q(t)$ are stationary, additive, and independent of their corresponding signal components $s_I(t)$ and $s_Q(t)$, respectively.

For ADC signal $\hat{V}(j)$, $\hat{s}_I(j)$ and $\hat{s}_Q(j)$ of $\hat{s}_c(j) = \hat{s}_I(j) + i\hat{s}_Q(j)$ are characterized by two Gaussian random processes: $\hat{s}_I(j) \sim N(\mu_{\hat{s}_I(j)}, \sigma^2_{\hat{s}_I} \to 0)$ and $\hat{s}_Q(j) \sim N(\mu_{\hat{s}_Q(j)}, \sigma^2_{\hat{s}_Q} \to 0)$, where $\mu_{\hat{s}_I(j)} = \Re\{\mu_{\hat{s}_c(j)}\}$ and $\mu_{\hat{s}_Q(j)} = \Im\{\mu_{\hat{s}_c(j)}\}$. They are independent random processes and are independent of each other.

$\hat{n}_I(j)$ and $\hat{n}_Q(j)$ of $\hat{n}_c(j) = \hat{n}_I(j) + i\hat{n}_Q(j)$ are characterized by two Gaussian random processes: $\hat{n}_I(j) \sim N(0, \sigma^2_{\hat{n}_I})$ and $\hat{n}_Q(j) \sim N(0, \sigma^2_{\hat{n}_Q})$, where $\sigma^2_{\hat{n}_I} = \sigma^2_{\hat{n}_Q} = \sigma^2_{n_r}$. They are independent random processes and are independent of each other.

$\hat{n}_I(j)$ and $\hat{n}_Q(j)$ are stationary, additive, and independent of their corresponding signal components $\hat{s}_I(j)$ and $\hat{s}_Q(j)$, respectively.

(3) The real and imaginary parts, $I(t)$ and $Q(t)$, of $V_c(t)$ can be characterized by Gaussian random processes: $I(t) \sim N(\mu_{I(t)}, \sigma^2_I)$ and $Q(t) \sim N(\mu_{Q(t)}, \sigma^2_Q)$, where the means $\mu_{I(t)} = \Re\{\mu_{s_c(t)}\} = \mu_{s_I(t)}$ and $\mu_{Q(t)} = \Im\{\mu_{s_c(t)}\} = \mu_{s_Q(t)}$, and the variances $\sigma^2_I = \sigma^2_{n_I}$ and $\sigma^2_Q = \sigma^2_{n_Q}$ $(\sigma^2_{n_I} = \sigma^2_{n_Q} = \sigma^2_{n_r})$. They are independent: $E[I(t)Q(t)] = \mu_{I(t)}\mu_{Q(t)}$.

The real and imaginary parts, $\hat{I}(j)$ and $\hat{Q}(j)$, of $\hat{V}_c(j)$ can be characterized by Gaussian random processes: $\hat{I}(j) \sim N(\mu_{\hat{I}(j)}, \sigma^2_{\hat{I}})$ and $\hat{Q}(j) \sim N(\mu_{\hat{Q}(j)}, \sigma^2_{\hat{Q}})$, where the means $\mu_{\hat{I}(j)} = \Re\{\mu_{\hat{s}_c(j)}\} = \mu_{\hat{s}_I(t)}$ and $\mu_{\hat{Q}(j)} = \Im\{\mu_{\hat{s}_c(j)}\} = \mu_{\hat{s}_Q(t)}$, and the variances $\sigma^2_{\hat{I}} = \sigma^2_{\hat{n}_I}$ and $\sigma^2_{\hat{Q}} = \sigma^2_{\hat{n}_Q}$ $(\sigma^2_{\hat{n}_I} = \sigma^2_{\hat{n}_Q} = \sigma^2_{n_r})$. They are independent: $E[\hat{I}(j)\hat{Q}(j)] = \mu_{\hat{I}(j)}\mu_{\hat{Q}(j)}$. Furthermore, $\hat{I}(j)$ and $\hat{Q}(j)$ are independent random processes. That is, for $j_1 \neq j_2$,

$$E[\hat{I}(j_1)\hat{I}(j_2)] = \mu_{\hat{I}(j_1)}\mu_{\hat{I}(j_2)} \quad \text{and} \quad E[\hat{Q}(j_1)\hat{Q}(j_2)] = \mu_{\hat{Q}(j_1)}\mu_{\hat{Q}(j_2)} \,. \tag{7.48}$$

□

Proof.

(1) Signal components of $s_r(t)$ of $V_r(t)$, $s_c(t)$ of $V_c(t)$, and $\hat{s}_c(j)$ of $\hat{V}_c(j)$ are characterized by Gaussian random processes with the almost-zero variances: $s_r(t) \sim N(\mu_{s_r(t)}, \sigma_{s_r} \to 0)$, $s_c(t) \sim N(\mu_{s_c(t)}, \sigma_{s_c(t)} \to 0)$, and $\hat{s}_c(j) \sim N(\mu_{\hat{s}_c(j)}, \sigma_{\hat{s}_c(j)} \to 0)$ [Property 7.4]. Noise components of $n_r(t)$ of $V_r(t)$, $n_c(t)$ of $V_c(t)$, and $\hat{n}_c(j)$ of $\hat{V}_c(j)$ are characterized by Gaussian random processes with zero means: $n_r(t) \sim N(0, \sigma^2_{n_r})$, $n_c(t) \sim N(0, \sigma^2_{n_c})$, $\hat{n}_c(j) \sim N(0, \sigma^2_{\hat{n}_c})$ [Property 7.5]. $s_r(t)$ and $n_r(t)$, $s_c(t)$ and $n_c(t)$, and $\hat{s}_c(j)$ and $\hat{n}_c(j)$ are independent [[Property 7.4 and Property 7.5]. Thus, each of $V_r(t) = s_r(t) + n_r(t)$, $V_c(t) = s_c(t) + n_c(t)$, and $\hat{V}_c(j) = \hat{s}_c(j) + \hat{n}_c(j)$ represents a sum of two independent Gaussian random processes. Therefore, $V_r(t)$, $V_c(t)$, and $\hat{V}_c(j)$ can be characterized by Gaussian random processes with the means of their signal components as the means and with the variances of their noise components as the variances. That is, $V_r(t) \sim N(\mu_{s_r(t)}, \sigma^2_{n_r})$, $V_c(t) \sim N(\mu_{s_c(t)}, \sigma^2_{n_c})$, and $\hat{V}_c(j) \sim N(\mu_{\hat{s}_c(j)}, \sigma^2_{\hat{n}_c})$.

The signal component $\hat{s}_c(j)$ of ADC signal $\hat{V}_c(j)$ is a temporally deterministic process with probability 1 [Property 7.4]. The noise component $\hat{n}_c(j)$ of ADC signal $\hat{V}_c(j)$ is an independent random process with zero mean, and $\hat{n}_c(j)$ and $\hat{s}_c(j)$ are independent [Property 7.5]. Thus, it is easy to verify that $\hat{V}_c(j) = \hat{s}_c(j) + \hat{n}_c(j)$ is an independent random process. Therefore, its autocorrelation function can be expressed by Eq. (7.46).

(2) The proofs are given in the Proofs of Property 7.4 and Property 7.5.

(3) The signal components $s_I(t)$ and $s_Q(t)$ are characterized by two Gaussian random processes with almost-zero variance: $s_I(t) \sim N(\mu_{s_I(t)}, \sigma^2_{s_I} \to 0)$ and $s_Q(t) \sim N(\mu_{s_Q(t)}, \sigma^2_{s_Q} \to 0)$ [Property 7.4]. The noise components $n_I(t)$ and $n_Q(t)$ are characterized by two Gaussian random processes with zero means: $n_I(t) \sim N(0, \sigma^2_{n_I})$ and $n_Q(t) \sim N(0, \sigma^2_{n_Q})$ [Property 7.5]. $s_I(t)$ and $n_I(t)$, $s_Q(t)$ and $n_Q(t)$ are independent [Property 7.4 and Property 7.5]. Thus, each of $I(t) = s_I(t) + n_I(t)$ and $Q(t) = s_Q(t) + n_Q(t)$ represents a sum of two independent Gaussian random processes. Therefore, $I(t)$ and $Q(t)$ can be characterized by Gaussian random processes with the means of their signal components as the mean and with the variances of their noise components as the variances. That is, $I(t) \sim N(\mu_{s_I(t)}, \sigma^2_{n_I})$ and $Q(t) \sim N(\mu_{s_Q(t)}, \sigma^2_{n_Q})$.

Signal components $s_I(t)$ and $s_Q(t)$ are the temporally deterministic processes with probability 1 [Property 7.4]. Noise components $n_I(t)$ and $n_Q(t)$ are the random processes with zero means, and they are independent [Property 7.5]. Thus, $I(t)$ and $Q(t)$ are independent, that is, $E[I(t)Q(t)] = \mu_{I(t)}\mu_{Q(t)}$.

The Gaussianity of $\hat{I}(j)$ and $\hat{Q}(j)$, that is, $\hat{I}(j) \sim N(\mu_{\hat{I}(j)}, \sigma^2_{\hat{n}_I})$ and $\hat{Q}(j) \sim N(\mu_{\hat{Q}(j)}, \sigma^2_{\hat{n}_Q})$, as well as the independence of $\hat{I}(j)$ and $\hat{Q}(j)$, that is, $E[\hat{I}(j)\hat{Q}(j)] = \mu_{\hat{I}(j)}\mu_{\hat{Q}(j)}$, can be proved in the same way as for $I(t)$ and $Q(t)$ shown above.

Signal components $\hat{s}_I(j)$ and $\hat{s}_Q(j)$ are the temporally deterministic processes with probability 1 [Property 7.4]. Noise components $\hat{n}_I(j)$ and $\hat{n}_Q(j)$ are the independent random processes with zero means [Property 7.5]. Thus, $\hat{I}(j)$ and $\hat{Q}(j)$ are independent random processes, that is, Eq. (7.48) is proved.

■

7.4 Statistics of *k*-Space Samples

By defining the spatial frequency, k-space sample, and revealing the underlying mechanism that transforms the ADC signal to a k-space sample; Section 3.9 shows that a k-space sample is an alternative representation of ADC signal in terms of the spatial frequency. The one-to-one relation between the ADC signal and k-space sample can be seen from Eq. (3.110), a k-space version of the MR signal equation.

As indicated in Section 3.9.1, Eq. (3.110) actually shows the unique relation between $\hat{s}_c(j)$—the *signal component* of the ADC signal $\hat{V}_c(j)$—and $\mathcal{M}_s(k_u(j), k_v(j))$—the *signal component* of the rectilinear k-space sample $\mathcal{M}(k_u(j), k_v(j))$.[‖] Letting the noise component of k-space sample $\mathcal{M}(k_u(j), k_v(j))$ be denoted by $\mathcal{M}_n(k_u(j), k_v(j))$, which is a counterpart of the noise component $\hat{n}_c(j)$ of the ADC signal $\hat{V}_c(j)$, we have $\mathcal{M}(k_u(j), k_v(j)) = \mathcal{M}_s(k_u(j), k_v(j)) + \mathcal{M}_n(k_u(j), k_v(j))$. The k-space sample is a complex-valued quantity. Letting $\mathcal{M}_R(k_u(j), k_v(j))$ and $\mathcal{M}_I(k_u(j), k_v(j))$ denote its real and imaginary parts, we have $\mathcal{M}(k_u(j), k_v(j)) = \mathcal{M}_R(k_u(j), k_v(j)) + i\mathcal{M}_I(k_u(j), k_v(j))$. Similar to Eq. (7.27), The k-space sample can be expressed in two ways:

$$\mathcal{M}(k_u(j), k_v(j)) = \mathcal{M}_R(k_u(j), k_v(j)) + i\mathcal{M}_I(k_u(j), k_v(j))$$

or (7.49)

[‖] As stated in Section 7.2.1, in this and the following chapters, the Cartesian coordinate system $\{X, Y, Z\}$ is denoted by $\{U, V, Z\}$. That is, the coordinate (x, y) is replaced by (u, v).

$$\mathcal{M}(k_u(j), k_v(j)) = \mathcal{M}_s(k_u(j), k_v(j)) + \mathcal{M}_n(k_u(j), k_v(j)),$$

with

$$\begin{aligned}
\mathcal{M}_s(k_u(j), k_v(j)) &= \mathcal{M}_{s_R}(k_u(j), k_v(j)) + i\mathcal{M}_{s_I}(k_u(j), k_v(j)) \\
\mathcal{M}_n(k_u(j), k_v(j)) &= \mathcal{M}_{n_R}(k_u(j), k_v(j)) + i\mathcal{M}_{n_I}(k_u(j), k_v(j)) \\
\mathcal{M}_R(k_u(j), k_v(j)) &= \mathcal{M}_{s_R}(k_u(j), k_v(j)) + \mathcal{M}_{n_R}(k_u(j), k_v(j)) \\
\mathcal{M}_I(k_u(j), k_v(j)) &= \mathcal{M}_{s_I}(k_u(j), k_v(j)) + \mathcal{M}_{n_I}(k_u(j), k_v(j)),
\end{aligned}$$

where $\mathcal{M}_{s_R}(k_u(j), k_v(j))$ and $\mathcal{M}_{n_R}(k_u(j), k_v(j))$ are the signal and noise components of the real part $\mathcal{M}_R(k_u(j), k_v(j))$ of the k-space sample $\mathcal{M}(k_u(j), k_v(j))$, while $\mathcal{M}_{s_I}(k_u(j), k_v(j))$ and $\mathcal{M}_{n_I}(k_u(j), k_v(j))$ are the signal and noise components of the imaginary part $\mathcal{M}_I(k_u(j), k_v(j))$ of the k-space sample $\mathcal{M}(k_u(j),\ k_v(j))$. The above expressions can also be applied to the radial k-space sample $\mathcal{M}(k(j),\ \theta(j))$ defined in Section 3.9.

In the image reconstruction, $\mathcal{M}(k_u(t), k_v(t))$ and $\mathcal{M}(k_u(j), k_v(j))$ are used to denote the continuous and discrete rectilinear k-space *data* and abbreviated as $\mathcal{M}(k_u, k_v)$; $\mathcal{M}(m\Delta k_u, n\Delta k_v)$ is used to denote the rectilinear k-space *sample* and abbreviated as $\mathcal{M}(m, n)$. Similarly, $\mathcal{M}(k(t), \theta(t))$ and $\mathcal{M}(k(j), \theta(j))$ are used to denote the continuous and discrete radial k-space *data* and abbreviated as $\mathcal{M}(k, \theta)$; $\mathcal{M}(n\Delta k, m\Delta\theta)$ is used to denote the radial k-space *sample* and abbreviated as $\mathcal{M}(n, m)$. In the following, unless specified, the general term "k-space sample" is commonly used. Statistics of k-space samples are given by the following property.

Property 7.7 (1) k-space samples $\mathcal{M}(k_u, k_v)$ and $\mathcal{M}(k, \theta)$ can be characterized by a complex Gaussian random process $N(\mu_{\mathcal{M}}, \sigma^2_{\mathcal{M}})$ with the mean $\mu_{\mathcal{M}} = \mu_{\hat{V}_c(j)}$ and the variance $\sigma^2_{\mathcal{M}} = \sigma^2_{\hat{V}_c}$. It is an independent random process; that is, its autocorrelation function $R_{\mathcal{M}}$ is given by

$$\begin{aligned}
R_{\mathcal{M}}((k_{u_1}, k_{v_1}), (k_{u_2}, k_{v_2})) &= \mu_{\mathcal{M}(k_{u_1}, k_{v_1})}\mu^*_{\mathcal{M}(k_{u_2}, k_{v_2})} + \sigma^2_{\mathcal{M}}\delta[k_{u_2} - k_{u_1}, k_{v_2} - k_{v_1}] \\
R_{\mathcal{M}}((k_1, \theta_1), (k_2, \theta_2)) &= \mu_{\mathcal{M}(k_1, \theta_1)}\mu^*_{\mathcal{M}(k_2, \theta_2)} + \sigma^2_{\mathcal{M}}\delta[k_2 - k_1, \theta_2 - \theta_1], \quad (7.50)
\end{aligned}$$

where $*$ represents the complex conjugate and $\delta[m, n]$ is a 2-D Kronecker delta function.

(2) The signal component of the k-space sample, $\mathcal{M}_s(k_u, k_v)$ and $\mathcal{M}_s(k, \theta)$, can be characterized by complex Gaussian random processes $N(\mu_{\mathcal{M}_s}, \sigma^2_{\mathcal{M}_s})$ with the mean $\mu_{\mathcal{M}_s} = \mu_{\hat{V}_c(j)}$ and the variance $\sigma^2_{\mathcal{M}_s} \to 0$. The noise component of the k-space sample, $\mathcal{M}_n(k_u, k_v)$ and $\mathcal{M}_n(k, \theta)$, can be characterized by complex Gaussian random processes $N(\mu_{\mathcal{M}_n}, \sigma^2_{\mathcal{M}_n})$ with mean $\mu_{\mathcal{M}_n} = 0$ and the variance $\sigma^2_{\mathcal{M}_n} = \sigma^2_{\hat{V}_c}$.

$\mathcal{M}_n(k_u, k_v)$ and $\mathcal{M}_n(k, \theta)$ are stationary, additive, and independent of $\mathcal{M}_s(k_u, k_v)$ and $\mathcal{M}_s(k, \theta)$, respectively.

(a) Rectlinear samples. The real and imaginary parts of the signal component $\mathcal{M}_s(k_u, k_v)$, that is, $\mathcal{M}_{s_R}(k_u, k_v)$ and $\mathcal{M}_{s_I}(k_u, k_v)$, are characterized by two real Gaussian random processes: $\mathcal{M}_{s_R}(k_u, k_v) \sim N(\mu_{\mathcal{M}_{s_R}}, \sigma^2_{\mathcal{M}_{s_R}} \to 0)$ and $\mathcal{M}_{s_I}(k_u, k_v) \sim N(\mu_{\mathcal{M}_{s_I}}, \sigma^2_{\mathcal{M}_{s_I}} \to 0)$, where $\mu_{\mathcal{M}_{s_R}} = \Re\{\mu_{\hat{s}_c(j)}\} = \mu_{\hat{s}_I(j)}$ and $\mu_{\mathcal{M}_{s_I}} = \Im\{\mu_{\hat{s}_c(j)}\} = \mu_{\hat{s}_Q(j)}$. They are independent random processes and are independent of each other.

The real and imaginary parts of the noise component $\mathcal{M}_n(k_u, k_v)$, that is, $\mathcal{M}_{n_R}(k_u, k_v)$ and $\mathcal{M}_{n_I}(k_u, k_v)$, are characterized by two real Gaussian random processes: $\mathcal{M}_{n_R}(k_u, k_v) \sim N(0, \sigma^2_{\mathcal{M}_{n_R}})$ and $\mathcal{M}_{n_I}(k_u, k_v) \sim N(0, \sigma^2_{\mathcal{M}_{n_I}})$, where $\sigma^2_{\mathcal{M}_{n_R}} = \sigma^2_{\hat{n}_I}$ and $\sigma^2_{\mathcal{M}_{n_I}} = \sigma^2_{\hat{n}_Q}$ $(\sigma^2_{\hat{n}_I} = \sigma^2_{\hat{n}_Q} = \sigma^2_{n_r})$. They are independent random processes and are independent of each other.

$\mathcal{M}_{n_R}(k_u, k_v)$ and $\mathcal{M}_{n_I}(k_u, k_v)$ are stationary, additive, and independent of their corresponding $\mathcal{M}_{s_R}(k_u, k_v)$ and $\mathcal{M}_{s_I}(k_u, k_v)$.

(b) Radial samples. The real and imaginary parts of the signal component $\mathcal{M}_s(k, \theta)$, that is, $\mathcal{M}_{s_R}(k, \theta)$ and $\mathcal{M}_{s_I}(k, \theta)$, are characterized by two real Gaussian random processes: $\mathcal{M}_{s_R}(k, \theta) \sim N(\mu_{\mathcal{M}_{s_R}}, \sigma^2_{\mathcal{M}_{s_R}} \to 0)$ and $\mathcal{M}_{s_I}(k, \theta) \sim N(\mu_{\mathcal{M}_{s_I}}, \sigma^2_{\mathcal{M}_{s_I}} \to 0)$, where $\mu_{\mathcal{M}_{s_R}} = \Re\{\mu_{\hat{s}_c(j)}\} = \mu_{\hat{s}_I(j)}$ and $\mu_{\mathcal{M}_{s_I}} = \Im\{\mu_{\hat{s}_c(j)}\} = \mu_{\hat{s}_Q(j)}$. They are independent random processes and are independent of each other.

The real and imaginary parts of the noise component $\mathcal{M}_n(k, \theta)$, that is, $\mathcal{M}_{n_R}(k,\ \theta)$ and $\mathcal{M}_{n_I}(k, \theta)$, are characterized by two real Gaussian random processes: $\mathcal{M}_{n_R}(k, \theta) \sim N(0,\ \sigma^2_{\mathcal{M}_{n_R}})$ and $\mathcal{M}_{n_I}(k, \theta) \sim N(0, \sigma^2_{\mathcal{M}_{n_I}})$, where $\sigma^2_{\mathcal{M}_{n_R}} = \sigma^2_{\hat{n}_I}$ and $\sigma^2_{\mathcal{M}_{n_I}} = \sigma^2_{\hat{n}_Q}$ $(\sigma^2_{\hat{n}_I} = \sigma^2_{\hat{n}_Q} = \sigma^2_{n_r})$. They are independent random processes and are independent of each other.

$\mathcal{M}_{n_R}(k, \theta)$ and $\mathcal{M}_{n_I}(k, \theta)$ are stationary, additive, and independent of their corresponding $\mathcal{M}_{s_R}(k, \theta)$ and $\mathcal{M}_{s_I}(k, \theta)$.

(3) The real part of the k-space sample, $\mathcal{M}_R(k_u, k_v)$ and $\mathcal{M}_R(k, \theta)$, can be characterized by a real Gaussian random process $N(\mu_{\mathcal{M}_R}, \sigma^2_{\mathcal{M}_R})$ with mean $\mu_{\mathcal{M}_R} = \Re\{\mu_{\hat{s}_c(j)}\} = \mu_{\hat{s}_I(j)}$ and variance $\sigma^2_{\mathcal{M}_R} = \sigma^2_{\hat{n}_I}$. It is an independent random process.

The imaginary part of the k-space sample, $\mathcal{M}_I(k_u, k_v)$ and $\mathcal{M}_I(k, \theta)$, can be characterized by a real Gaussian random process $N(\mu_{\mathcal{M}_I}, \sigma^2_{\mathcal{M}_I})$ with mean $\mu_{\mathcal{M}_I} = \Im\{\mu_{\hat{s}_c(j)}\} = \mu_{\hat{s}_Q(j)}$ and variance $\sigma^2_{\mathcal{M}_I} = \sigma^2_{\hat{n}_Q}$. It is an independent random process.

The real part and the imaginary part of the k-space sample are independent. In the above, $\sigma^2_{\mathcal{M}_R} = \sigma^2_{\mathcal{M}_I}$ $(\sigma^2_{\hat{n}_I} = \sigma^2_{\hat{n}_Q} = \sigma^2_{n_r})$. □

Proof.

Eq. (3.110) maps the signal component $\hat{s}_c(j)$ of ADC signal $\hat{V}_c(j)$ (in the time domain) to the signal component $\mathcal{M}_s(k_x(j), k_y(j))$ of the k-space sample $\mathcal{M}(k_x(j), k_y(j))$ (in the spatial frequency domain). Similarly, the noise component $\hat{n}_c(j)$ of ADC signal $\hat{V}_c(j)$ is mapped to the noise component $\mathcal{M}_n(k_x(j), k_y(j))$ of the k-space sample $\mathcal{M}(k_x(j), k_y(j))$. Thus, the ADC signal $\hat{V}_c(j) = \hat{s}_c(j) + \hat{n}_c(j)$ and k-space sample $\mathcal{M}(k_u(j), k_v(j)) = \mathcal{M}_s(k_x(j),$

$k_y(j)) + \mathcal{M}_n(k_x(j), k_y(j))$ has a one-to-one relationship. In other words, the k-space sample is an alternative representation of the ADC signal in terms of spatial frequency. Therefore, statistics of k-space sample are the same as those of $\hat{V}_c(j)$, which are given by Eqs. (7.46), (7.47), and (7.48). This unique one-to-one relation also holds between ADC signal $\hat{V}_c(j)$ and the radial k-space sample $\mathcal{M}(k(j),\ \theta(j))$. ∎

Remarks on the second formula of Eq. (7.50). It seems to be that $\mathcal{M}(0, \theta_1)$ and $\mathcal{M}(0, \theta_2)$ should be correlated, because the location $k = 0$ belongs to all views. This is a misconception. As shown in the above proof, the k-space sample is an alternative representation of the ADC signal. The k-space sample and ADC signal have a one-to-one relationship. $\mathcal{M}(0, \theta_1)$ and $\mathcal{M}(0, \theta_2)$ correspond to two ADC signals $\hat{V}_c(j_1)$ and $\hat{V}_c(j_2)$, which are acquired at the different time instants j_1 and j_2 ($j_1 \neq j_2$). See Section 3.9.2.2 and Figure 3.12. Property 7.6 (Eq. (7.47)) shows that $\hat{V}_c(j_1)$ and $\hat{V}_c(j_2)$ are independent. Therefore, $\mathcal{M}(0, \theta_1)$ and $\mathcal{M}(0, \theta_2)$ are independent.

We can prove the independence of k-space samples, that is, Eq. (7.50), in a different way—from the viewpoint of data sampling and image reconstruction. A Fourier transform has the following property: if $s(t) \stackrel{\mathcal{F}}{\Longleftrightarrow} S(f)$ (here $\mathcal{F}$ represents Fourier transform), then

$$s(t) \sum_{k=-\infty}^{\infty} \delta(t - k\Delta t) \stackrel{\mathcal{F}}{\Longleftrightarrow} \frac{1}{\Delta t} \sum_{k=-\infty}^{\infty} S(f - k\frac{1}{\Delta t}),$$

where Δt is the sampling period. This property shows that equal sampling in one domain (image or k-space) amounts to replication in the corresponding transform domain. Applications of this property to MR image reconstruction can be found in [12, 15, 19, 35].

[23, 33] show that Fourier series expansion of a *periodic random process*** has two properties: (i) it converges mean square to the process and (ii) its coefficients are uncorrelated. In the rectilinear sampling as shown in Figure 3.10, when $\mathcal{M}(k_u(j), k_v(j))$ are acquired at Cartesian grids with the spacings Δk_u and Δk_v in k-space, then in the image domain the resulting images are replicas with periods equal to the reciprocals of the sample spacings $\frac{1}{\Delta k_u}$ and $\frac{1}{\Delta k_v}$, and consequently constitute a 2-D spatial, periodic process. The correlation of pixel intensities in this periodic process is periodic with those periods. Thus, this periodic process is a periodic random process; therefore, Fourier series coefficients of this periodic random process—k-space samples $\mathcal{M}(k_u(j), k_v(j))$—are uncorrelated, and hence statistically independent.

This reasoning can be extended to radial sampling, which is shown in Figure 3.12. When $\mathcal{M}(k(j), \theta(j))$ are acquired at polar grids with the spacings

**A random process $\mathbf{x}(t)$ is periodic in the sense, if its autocorrelation function $R(\tau)$ is periodic, that is, if $R(\tau + T) = R(\tau)$ for every τ. The period of the process is the smallest T satisfying the above equation.

Δk and $\Delta\theta$ in k-space, then theoretically we can form Cartesian grids with sufficiently small spacings Δk_u and Δk_v such that each point in the polar grid is surrounded by four different points in a Cartesian grid, that is, the eight points in Cartesian grids that surround two adjacent points in polar grids are different. Suppose that $\mathcal{M}(k_u(j), k_v(j))$ are sampled with these fine spacings, then each $\mathcal{M}(k(j), \theta(j))$ which is sampled at a point in polar grids will have an interpolated value of four surroundings $\mathcal{M}(k_u(j), k_v(j))$, that is, a weighted average. Due to the independence of $\mathcal{M}(k_u(j), k_v(j))$, any two $\mathcal{M}(k(j), \theta(j))$ are uncorrelated, and therefore statistically independent. In this extension, we assume that there is no aliasing artifact in image reconstruction caused by reduced sampling periods Δk_u and Δk_v.

[33] also shows that Fourier series coefficients are approximately uncorrelated for any process whose spectral density is approximately constant over intervals large compared to the sample spacing. In the radial sampling of PR, Eq. (3.128) shows that the sample spacings Δk and $\Delta\theta$ must satisfy the requirements: $\Delta k \leq \frac{1}{FOV_r}$ and $\Delta\theta \leq \frac{2\Delta p}{FOV_r}$. This means that Δk and $\Delta\theta$ must be sufficiently small; it follows that the spectral density over intervals large compared to these small Δk and $\Delta\theta$ can be considered approximately constant. With this approximation, Fourier series coefficients of this random process (i.e., the reconstructed images)—k-space samples $\mathcal{M}(k(j), \theta(j))$—are uncorrelated, and therefore independent due to Gaussianity.

7.5 Statistical Interpretation of MR Image Reconstruction

Two k-space sampling schemes (rectilinear and radial) lead to two basic MR image reconstruction methods: Fourier transform (FT) and projection reconstruction (PR). In the 2-D case, they are abbreviated as 2DFT and 2DPR. Based on mathematical descriptions and computational implementations of these two methods given in Section 3.10, this section provides signal processing paradigms for 2DFT and 2DPR and presents a statistical interpretation of MR image reconstruction.

7.5.1 Signal Processing Paradigms

For the 2DFT shown in Section 3.10.1, by properly choosing the sampling parameters (the sampling spacing: $\Delta k_u, \Delta k_u$) and the image parameters (the image size: FOV_u, FOV_v; the pixel size: $\delta u, \delta v$), an $I \times J$ MR image can be reconstructed using 2-D DFT on $I \times J$ samples directly acquired from k-space in the Cartesian grids. The signal processing paradigm for 2DFT is shown in

Eq. (7.51),

$$\begin{array}{c} I \times J \; rectilinear \; k-space \; samples \; \mathcal{M}(m\Delta k_u, n\Delta k_v) \\ (m = 1, \cdots, I, \;\; n = 1, \cdots, J) \\ \Downarrow \\ for \; a \; given \; point \; (u, v) \\ \Downarrow \\ x(u, v) = \Delta k_u \Delta k_v \sum_{m=-I/2}^{I/2-1} \sum_{n=-J/2}^{J/2-1} \\ \mathcal{M}(m\Delta k_u, n\Delta k_v)\Phi(m\Delta k_u, n\Delta k_v) e^{i2\pi(um\Delta k_u + vn\Delta k_v)} \\ \Downarrow \\ an \; I \times J \; MR \; image \; X(u, v), \end{array} \tag{7.51}$$

where Φ is the filter function.

For the transform-based 2DPR method shown in Section 3.10.2, by properly selecting a filter function, an $I \times J$ MR image can be reconstructed on $M \times N$ samples acquired from k-space in the polar grids. FBP is numerically implemented by a 1-D Fourier transform and a backprojection described in Section 3.10.2. The signal processing paradigm given by Eq. (7.52) depicts these operations and the associated signal flow in 2DPR:

$$\begin{array}{c} M \times N \; radial \; k-space \; samples \; \mathcal{M}(n\Delta k, m\Delta\theta) \\ (m = 0, \cdots, M-1, \;\; n = 1, \cdots, N) \\ \Downarrow \\ for \; a \; given \; point \; (u, v) \\ \Downarrow \\ M \;\; filtered \;\; projections \\ t(u', m\Delta\theta) = \Delta k \sum_{n=-N/2}^{N/2-1} \mathcal{M}(n\Delta k, m\Delta\theta)|n\Delta k| e^{i2\pi u' n\Delta k} \\ {\scriptstyle at \; the \; m-th \; view \; (m=0,\cdots,M-1)} \\ \overbrace{t(u', m\Delta\theta)} \\ \Downarrow \\ 1 \;\; backprojected \; data \\ x(u, v) = \Delta\theta \sum_{m=0}^{M-1} t(u', m\Delta\theta) \\ \Downarrow \\ an \; I \times J \; MR \; image \; X(u, v), \end{array} \tag{7.52}$$

where Δk and $\Delta\theta$ are the radial and angular sample periods, and $u' = u \cos m\Delta\theta + v \sin m\Delta\theta$.

For the convolution-based 2DPR method shown in Section 3.10.2, by properly selecting a convolution function and an interpolation function, an $I \times J$ MR image can be reconstructed on $M \times N$ samples acquired from the k-space in the polar grids. FBP is numerically implemented by two convolutions and a backprojection (in terms of the measured projections) described in Section 3.10.2. The signal processing paradigm given by Eq. (7.53) depicts these operations and the associated signal flow in 2DPR.

$$\begin{array}{c} M \times N \ radial\ k-space\ samples\ \mathcal{M}(n\Delta k, m\Delta\theta) \\ (m=0,\cdots,M-1,\ \ n=1,\cdots,N) \\ \Downarrow \\ for\ a\ given\ point\ (u,v) \\ \Downarrow \\ M \times N\ \ measured\ \ projections \\ \overbrace{\underset{at\ \ the\ \ m-th\ \ view\ \ (m=0,\cdots,M-1)}{p(l\Delta p, m\Delta\theta) = \Delta k \sum_{k=-N/2}^{N/2-1} \mathcal{M}(k\Delta k, m\Delta\theta) e^{i2\pi kl\Delta k\Delta p}}} \\ p(-N/2\Delta p, m\Delta\theta)\cdots p((N/2-1)\Delta p, m\Delta\theta) \\ \Downarrow \\ M \times N\ \ convolved\ \ projections \\ \overbrace{\underset{at\ \ the\ \ m-th\ \ view\ \ (m=0,\cdots,M-1)}{t(n\Delta p, m\Delta\theta) = \Delta p \sum_{l=-N/2}^{N/2-1} p(l\Delta p, m\Delta\theta) q((n-l)\Delta p)}} \\ t(-N/2\Delta p, m\Delta\theta)\cdots t((N/2-1)\Delta p, m\Delta\theta) \\ \Downarrow \\ M\ \ interpolated\ data \\ \overbrace{\underset{at\ \ the\ \ m-th\ \ view\ \ (m=0,\cdots,M-1)}{s_m(u,v) = \Delta\theta \sum_{n=-\infty}^{\infty} t(n\Delta p, m\Delta\theta)\psi(u^{'}-n\Delta p)}} \\ s_m(u,v) \\ \Downarrow \\ 1\ \ backprojected\ data \\ x(u,v) = \sum_{m=0}^{M-1} s_m(u,v) \\ \Downarrow \\ an\ I \times J\ MR\ image\ X(u,v) \end{array} \tag{7.53}$$

where Δk and $\Delta\theta$ are the radial and angular sample periods, Δp is the spacing between two measured projections in one view,†† q and ψ are the filter and interpolation functions, and $u' = u\cos m\Delta\theta + v\sin m\Delta\theta$.

Although 2DFT and 2DPR are two different MR image reconstruction methods, a common feature exists. That is the Fourier transform. In the signal processing paradigm of 2DFT, Fourier transform is performed on $I \times J$ rectilinear k-space samples. In the signal processing paradigm of 2DPR, both the transform-based and the convolution-based methods perform 1-D Fourier transform on N radial k-space samples in each view.

In 2DPR, in addition to 1-D Fourier transform in each view, the transform-based and the convolution-based methods share the backprojection over all views. Furthermore, there is another unrevealed common feature in these two methods. That is the use of the measured projection. This is clear in the convolution-based method, because FBP is performed on $M \times N$ measured projections. It is unclear in the transform-based method, because FBP is

††Statistics of the measured projections are given in Appendix 7C.

directly performed on k-space samples. To get insight into this, we can revisit Eq. (3.143), in which the $|k|$ filter implicitly utilizes Eq. (3.141)—the definition of the measured projection. As a new variable in MR image reconstruction, the measured projection and its statistics are given in Appendix 7C.

7.5.2 Statistical Interpretations

Section 7.4 shows that k-space samples come from an independent Gaussian random process and are the input to the image reconstruction algorithms; thus, Section 7.5.1 actually demonstrates that

1) When MR data acquisition consists of $I \times J$ k-space samples $\mathcal{M}(m\Delta k_u, n\Delta k_v)$ $(1 \leq m \leq I,\ 1 \leq n \leq J)$ in Cartesian grids (for 2DFT) or of $M \times N$ k-space samples $\mathcal{M}(n\Delta k, m\Delta\theta)$ $(0 \leq m \leq M-1,\ 1 \leq n \leq N)$ in radial grids (for 2DPR) and the reconstructed MR image $X(k\delta x, l\delta y)$ $(1 \leq k \leq I,\ 1 \leq l \leq J)$ consists of $I \times J$ pixels, then MR image reconstruction (via 2DFT or 2DPR) constitutes a transform from a set of $I \times J$ or $M \times N$ random variables to another set of $I \times J$ random variables.

2) These new $I \times J$ random variables $x(k,l)$ $(1 \leq k \leq I,\ 1 \leq l \leq J)$ form a new random process, sometimes referred to as a 2-D random field. Each pixel value $x(k,l)$ is a value in the state space of a corresponding random variable in this process and the whole image $X(k,l)$ is a configuration (i.e., a realization) of the entire random process.

3) Statistics of the MR data in the imaging domain are propagated to the statistics of the MR image in the image domain through image reconstruction. Chapter 8 discusses this transform process and the statistics of an MR image.

7.6 Appendices

7.6.1 Appendix 7A

This appendix proves Eqs. (7.28) and (7.29).

Proof.

(1) For Eq. (7.28). Let $p(|M_{uv}|(\mathbf{r},0))$ denote pdf of the magnitude of TPMM $|M_{uv}|(\mathbf{r},0)$. Applying Property 7.3 and Eq. (7.19) to Eq. (7.31), the mean of $s_r(t)$ is given by

$$\begin{aligned}\mu_{s_r(t)} &= E[s_r(t)] \\ &= \int_{|M_{uv}|} s_r(t) p(|M_{uv}|(\mathbf{r},0)) d|M_{uv}(\mathbf{r},0)| \\ &= \int_{|M_{uv}|} \delta(|M_{uv}(\mathbf{r},0)| - \mu_{|M_{uv}|}) d|M_{uv}(\mathbf{r},0)| \qquad (7.54)\end{aligned}$$

$$
(\int_V \omega_0 |B_{r_{uv}}| \, |M_{uv}(\mathbf{r},0)| e^{-\frac{t}{T_2}} \sin((\omega_0 + \Delta\omega(\mathbf{r}))t + \phi_{B_{r_{uv}}} - \phi_{M_{uv}}) d\mathbf{r})
$$

$$
= \int_V \omega_0 |B_{r_{uv}}| \mu_{|M_{uv}|} e^{-\frac{t}{T_2}} \sin((\omega_0 + \Delta\omega(\mathbf{r}))t + \phi_{B_{r_{uv}}} - \phi_{M_{uv}}) d\mathbf{r}.
$$

For the slice at z_0 with unity thinkness ($dz = 1$), $d\mathbf{r} = dudv$. In the 2-D case, the sample volume V becomes an area; let it be specified by $[-\frac{u_0}{2} \le u \le \frac{u_0}{2}, -\frac{v_0}{2} \le v \le \frac{v_0}{2}]$. For a constant gradient (Eq. (7.21)), Eq. (7.54) becomes

$$
\begin{aligned}
\mu_{s_r(t)} = & \int_{-u_0/2}^{u_0/2} \int_{-v_0/2}^{v_0/2} \omega_0 |B_{r_{uv}}| \mu_{|M_{uv}|} e^{-\frac{t}{T_2}} \\
& \sin((\omega_0 + \gamma G_z z_0)t + \gamma(G_u u + G_v v)t + \phi_{B_{r_{uv}}} - \phi_{M_{uv}}) dudv,
\end{aligned} \tag{7.55}
$$

where G_u, G_v and G_z are gradients. After some trigonometric manipulations (see Eq. (7.59)) and using Property 7.3, Eq. (7.55) becomes

$$
\begin{aligned}
\mu_{s_r(t)} = & \, \omega_0 |B_{r_{uv}}| \mu_{M_z^o} \sin\alpha V_0 \, e^{-\frac{t}{T_2}} \\
& sinc(\bar{\gamma} G_u u_0 t, \bar{\gamma} G_v v_0 t) \sin((\omega_0 + \gamma G_z z_0)t + \phi_{B_{r_{uv}}} - \phi_{M_{uv}}) \,,
\end{aligned} \tag{7.56}
$$

where $V_0 = u_0 v_0$, $\bar{\gamma} = \frac{\gamma}{2\pi}$, $sinc(\bar{\gamma} G_u u_0 t, \bar{\gamma} G_v v_0 t) = sinc(\bar{\gamma} G_u u_0 t) sinc(\bar{\gamma} G_v v_0 t)$ is a 2-D sinc function.

The variance of $s_r(t)$ is given by

$$
\begin{aligned}
\sigma^2_{s_r(t)} &= Var[s_r(t)] \\
&= \int_{|M_{uv}|} (s_r(t) - \mu_{s_r(t)})^2 p(|M_{uv}|(\mathbf{r},0)) d|M_{uv}(\mathbf{r},0)| \\
&= \int_{|M_{uv}|} (s_r(t))^2 \delta(|M_{uv}(\mathbf{r},0)| - \mu_{|M_{uv}|}) d|M_{uv}(\mathbf{r},0)| - \mu^2_{s_r(t)} = 0.
\end{aligned} \tag{7.57}
$$

$\sigma^2_{s_r(t)} = 0$ is understood as $\sigma^2_{s_r(t)} \to 0$ because it is the result of directly applying $\delta(|M_{uv}| - \mu_{|M_{uv}|})$—a limiting expression for the Gaussian pdf $g(|M_{uv}||\mu_{|M_{uv}|}, \sigma_{|M_{uv}|} \to 0)$.

(2) For Eq. (7.29). Let $p(M_{uv}(\mathbf{r},0))$ denote pdf of TPMM $M_{uv}(\mathbf{r},0)$. Applying Property 7.3 and Eq. (7.19) to Eq. (7.32), the mean of $s_c(t)$ is given by

$$
\begin{aligned}
\mu_{s_c(t)} &= E[s_c(t)] \\
&= \int_{M_{uv}} s_c(t) p(M_{uv}(\mathbf{r},0)) dM_{uv}(\mathbf{r},0) \\
&= \frac{\omega_0}{2} B^*_{r_{uv}} e^{-\frac{t}{T_2}} \int_{M_{uv}} \delta(M_{uv}(\mathbf{r},0) - \mu_{M_{uv}}) dM_{uv}(\mathbf{r},0) \\
&\quad (\int_V M_{uv}(\mathbf{r},0) e^{-i\Delta\omega(\mathbf{r})t} d\mathbf{r}) \\
&= \frac{\omega_0}{2} B^*_{r_{uv}} \mu_{M_{uv}} e^{-\frac{t}{T_2}} \int_V e^{-i\Delta\omega(\mathbf{r})t} d\mathbf{r}.
\end{aligned} \tag{7.58}
$$

Similar to $s_r(t)$, in the 2-D case, $d\mathbf{r} = dudv$, the sample volume V is replaced by the area specified by $[-\frac{u_0}{2} \leq u \leq \frac{u_0}{2}, -\frac{v_0}{2} \leq v \leq \frac{v_0}{2}]$. Thus, using Eq. (7.21)

$$\int_V e^{-i\Delta\omega(\mathbf{r})t} d\mathbf{r} = \int_V (\cos(\Delta\omega(\mathbf{r})t) - i\sin(\Delta\omega(\mathbf{r})t))d\mathbf{r}$$
$$= \int_{-u_0/2}^{u_0/2} \int_{-v_0/2}^{v_0/2} (\cos\gamma(G_u u + G_v v + G_z z_0)t - i\sin\gamma(G_u u + G_v v + G_z z_0)t)dudv$$
$$= \int_{-u_0/2}^{u_0/2} \int_{-v_0/2}^{v_0/2} [\cos\gamma(G_u u + G_v v)t\cos\gamma G_z z_0 t - \sin\gamma(G_u u + G_v v)t\sin\gamma G_z z_0 t$$
$$-i\sin\gamma(G_u u + G_v v)t\cos\gamma G_z z_0 t - i\cos\gamma(G_u u + G_v v)t\sin\gamma G_z z_0 t]dudv$$
$$= V_0 sinc(\bar{\gamma}G_u u_0 t, \bar{\gamma}G_v v_0 t)\cos\gamma G_z z_0 t - iV_0 sinc(\bar{\gamma}G_u u_0 t, \bar{\gamma}G_v v_0 t)\sin\gamma G_z z_0 t$$
$$= V_0 sinc(\bar{\gamma}G_u u_0 t, \bar{\gamma}G_v v_0 t)e^{-i\gamma G_z z_0 t}. \tag{7.59}$$

Therefore, Eq. (7.58) becomes

$$\mu_{s_c(t)} = \frac{\omega_0}{2} B^*_{r,uv} \mu_{M_{uv}} V_0 \, e^{-\frac{t}{T_2}} \, sinc(\bar{\gamma}G_u u_0 t, \bar{\gamma}G_v v_0 t)e^{-i\gamma G_z z_0 t} \, . \tag{7.60}$$

The variance of $s_c(t)$ is given by

$$\sigma^2_{s_c(t)} = Var[s_c(t)]$$
$$= \int_{M_{uv}} |s_c(t) - \mu_{s_c(t)}|^2 p(M_{uv}(\mathbf{r},0))dM_{uv}(\mathbf{r},0) \tag{7.61}$$
$$= \int_{M_{uv}} |s_c(t)|^2 \delta(M_{uv}(\mathbf{r},0) - \mu_{M_{uv}})dM_{uv}(\mathbf{r},0) - |\mu_{s_c(t)}|^2 = 0.$$

Similar to the interpretation of Eq. (7.57), $\sigma^2_{s_c(t)} = Var[s_c(t)] = 0$ is understood as $\sigma^2_{s_c(t)} = Var[s_c(t)] \to 0$. ■

7.6.2 Appendix 7B

This appendix proves the second formula in Eq. (7.41).

Proof.

For the rectilinear k-space sampling shown in Figure 3.10 and described in Section 3.9.3.1, let FOV_u and FOV_v be the dimensions of a rectangular field of view, Δk_u and Δk_v be the sample periods at k_u and k_v directions of k-space. According to the Sampling theorem [15, 18, 19, 29], in order to avoid to aliasing, Δk_u and Δk_v should satisfy the conditions

$$\Delta k_u \leq \frac{1}{FOV_u} \quad \text{and} \quad \Delta k_v \leq \frac{1}{FOV_v}. \tag{7.62}$$

When the frequency encoding is used in the U direction, Eq. (3.115) shows that

$$\Delta k_u = \frac{1}{2\pi}\gamma G_u \Delta t, \tag{7.63}$$

where γ is the gyromagnetic ratio, G_u is the amplitude of the frequency encoding gradient, Δt is the time interval between the readout samples. Eqs. (7.62) and (7.63) give

$$\Delta t \leq \frac{1}{\frac{1}{2\pi}\gamma G_u \cdot FOV_u}. \tag{7.64}$$

In Eq. (7.64), $G_u \cdot FOV_u$ is the dynamic range of magnetic field strength over the dimension FOV_u caused by the frequency encoding gradient G_u. Thus, $\frac{1}{2\pi}\gamma G_u \cdot FOV_u$ is the corresponding (temporal) frequency range and should be equal to the bandwidth Δf of the baseband signal $\hat{s}_c(j)$. Taking $\Delta t = \frac{1}{\Delta f}$ and substituting it into $R_{n_I}(\tau)$ and $R_{n_Q}(\tau)$ of Eq. (7.40), $\tau = j_2 - j_1 = l\Delta t$ (l is an integer), we have

$$R_{\hat{n}_I}(l\Delta t) = R_{\hat{n}_Q}(l\Delta t) = \sigma^2_{n_r} sinc(l\Delta t\Delta f) = 0. \tag{7.65}$$

This is also known as orthogonal [23]. Due to $E[\hat{n}_I(t)] = E[\hat{n}_Q(t)] = 0$, samples of $\hat{n}_I(t)$ and $\hat{n}_Q(t)$ taken at $j = \Delta t,\ 2\Delta t,\ \cdots,\ n\Delta t$ are uncorrelated, and therefore independent.

For the radial k-space sampling shown in Figure 3.12 and Section 3.9.3.2, let FOV_r be the dimensions of a circular field of view and Δk be the sample periods. According to the Sampling theorem, in order to avoid aliasing, Δk should satisfy the condition

$$\Delta k \leq \frac{1}{FOV_r}. \tag{7.66}$$

When frequency encoding is used in the radial direction, Eq. (7.63) leads to

$$\Delta k = \frac{1}{2\pi}\gamma G\Delta t, \tag{7.67}$$

where $G = \sqrt{G_u^2 + G_v^2}$, Δt is the time interval between the readout samples. Eqs. (7.66) and (7.67) give

$$\Delta t \leq \frac{1}{\frac{1}{2\pi}\gamma G \cdot FOV_r}. \tag{7.68}$$

Similar to the above rectilinear k-space sampling, $\frac{1}{2\pi}\gamma G \cdot FOV_r$ should be equal to the bandwidth Δf of the baseband signal $\hat{s}_c(j)$. By taking $\Delta t = \frac{1}{\Delta f}$, we have that samples $\hat{n}_I(t)$ and $\hat{n}_Q(t)$ at $j = \Delta t, 2\Delta t, \cdots, n\Delta t$ are independent.

Thus, for the real and imaginary parts $\hat{n}_I(j)$ and $\hat{n}_Q(j)$ of the noise component $\hat{n}_c(j)$ of the sampled complex-valued baseband signal $\hat{V}_c(j)$, the samples $\hat{n}_I(j)$ and $\hat{n}_Q(j)$ at $j = \Delta t, 2\Delta t, \cdots$ ($\Delta t = \frac{1}{\Delta f}$) are independent. Therefore, due to the zero mean, the autocorrelation functions of $\hat{n}_I(j)$ and $\hat{n}_Q(j)$ can be expressed as $\sigma^2_{n_r}\delta[j_2 - j_1]$.

■

7.6.3 Appendix 7C

This appendix provides further discussion on the measured projection and shows its statistics.

7.6.3.1 The Measured Projection

The theoretical representation. When the constant gradients G_u and G_v are applied, by rotating the rectangular U–V coordinate system with an angle $\theta = \tan^{-1}(\frac{G_v}{G_u})$, the resulting U'–V' coordinate system and U–V coordinate system are related by

$$u' = u\cos\theta + v\sin\theta \quad \text{and} \quad v' = -u\sin\theta + v\cos\theta. \tag{7.69}$$

A gradient G oriented at the angle θ and the gradients G_u, G_v are related by

$$G_u = G\cos\theta \quad \text{and} \quad G_v = G\sin\theta. \tag{7.70}$$

Using Eq. (3.108) and Eqs. (7.69) and (7.70), the exponential item in Eq. (3.110) becomes

$$k_u(j)u + k_v(j)v = \frac{\gamma}{2\pi}(uG_u + vG_v)j = \frac{\gamma}{2\pi}G(u\cos\theta + v\sin\theta)j = \frac{\gamma}{2\pi}Gu'j. \tag{7.71}$$

Thus, by ignoring the constant c', the k-space version of MR signal equation (3.110) in the U'–V' coordinate system is

$$\mathcal{M}(k_{u'}(j), k_{v'}(j)) = \int_{u'}\int_{v'} M_{uv}(u', v', 0)e^{-i2\pi(\frac{\gamma}{2\pi}Gu'j)}du'dv'. \tag{7.72}$$

When θ varies, let the *rotating rectangular coordinate* (u', v') be replaced by the polar coordinate (k, θ). In k-space, let

$$\mathcal{M}(k_{u'}(j), k_{v'}(j)) = \mathcal{M}(k(j), \theta(j)) \quad \text{and} \quad k(j) = \frac{\gamma}{2\pi}Gj. \tag{7.73}$$

In Eq. (7.72), by defining

$$\int_{v'} M_{uv}(u', v', 0)dv' = \int_{v'} M_{uv}(u', v')dv' = p(u', \theta(j)), \tag{7.74}$$

it becomes

$$\mathcal{M}(k(j), \theta(j)) = \int_{u'} p(u', \theta(j))e^{-i2\pi k(j)u'}du' = \mathcal{F}_{u'}\{p(u', \theta)\}. \tag{7.75}$$

That is,

$$p(u', \theta(j)) = \int_{k} \mathcal{M}(k(j), \theta(j))e^{i2\pi k(j)u'}dk = \mathcal{F}_k^{-1}\{\mathcal{M}(k(j), \theta(j))\}, \tag{7.76}$$

where $\mathcal{F}_k^{-1}$ represents Inverse FT with respect to k. Eq. (7.75) is the MR signal equation in a polar (k, θ) coordinate system.

Eq. (7.74) shows that $p(u', \theta)$ is the line integral of the *object function* $M_{uv}(u', v')$ over v'. This expression is exactly the same as the *projection* used in X-ray CT. Therefore, $p(u', \theta)$ defined by Eq. (7.74) is a kind of *projection*. Eqs. (7.75) and (7.76) provide an operational definition of $p(u', \theta)$. They directly relate $p(u', \theta)$ to k-space *measurement*. Thus, $p(u', \theta)$ is called the *measured projection*. Eqs. (7.75), (7.73) and (7.72) show that, at a given view, 1-D FT of the projection equals 2-D FT of the object function. This is the version of the Fourier Slice theorem in MR 2DPR [15, 19, 36, 37].

Remarks. In mapping the temporally sampled ADC signal $\hat{V}_c(j)$ to the radial k-space sample $\mathcal{M}(k(j), \theta(j))$, although a time index j corresponds to a location $(k(j), \theta(j))$ in k-space, the relation between the time index j and the angle $\theta(j)$, however, is characterized by a staircase function (as indicated in Section 3.9 and shown by Figure 3.12). That is, for a given view θ_m ($m = 0, 1, \cdots, M-1$) and a *set* of time indices $j_1, j_2, \cdots\cdots, j_n$, $\theta \sim j$ dependence is $\theta(j_1) = \theta(j_2) = \cdots\cdots = \theta(j_n) = \theta_m$. This $\theta \sim j$ relationship that results from the radial k-space sampling scheme is the basis for defining and formulating the measured projection.

A practical representation. Using a typical and conventional notation adopted in computed tomography, that is, letting l take a value of u', Eq. (7.76) can be written as

$$p(l, \theta) = \int_{-\infty}^{+\infty} \mathcal{M}(k, \theta) e^{i2\pi kl} dk. \qquad (7.77)$$

In practice, the measured projection $p(l, \theta)$ is computed by

$$p(l, \theta) = \int_{-\infty}^{+\infty} (\mathcal{M}(k, \theta) \cdot \frac{1}{\Delta k} comb(\frac{k}{\Delta k}) \cdot rect(\frac{k}{W})) e^{i2\pi kl} dk. \qquad (7.78)$$

where Δk is the radial k-space sample period (see Section 3.9.3, Figure 3.12, and Eq. (3.123)), $comb(\frac{k}{\Delta k})$ is a comb function defined by

$$comb(\frac{k}{\Delta k}) = \Delta k \sum_{n=-\infty}^{+\infty} \delta(k - n\Delta k), \qquad (7.79)$$

$rect(\frac{k}{W})$ is a rect function defined by

$$rect(\frac{k}{W}) = \begin{cases} 1 & (-\frac{W}{2} \leq k \leq \frac{W}{2}) \\ 0 & \text{(otherwise)}, \end{cases} \qquad (7.80)$$

and W is the window width.

In Eq. (7.78), $\mathcal{M}(k,\theta)$ represents the continuous radial k-space data in the view θ, and k ranges from $-\infty$ to $+\infty$;

$$\mathcal{M}(k,\theta) \cdot \frac{1}{\Delta k} comb(\frac{k}{\Delta k}) = \sum_{n=-\infty}^{+\infty} \mathcal{M}(n\Delta k,\theta)\delta(k - n\Delta k)$$

represents the discrete radial k-space data in the view θ, k ranges from $-\infty$ to $+\infty$;

$$\mathcal{M}(k,\theta) \cdot \frac{1}{\Delta k} comb(\frac{k}{\Delta k}) \cdot rect(\frac{k}{W}) = \sum_{n=-N/2}^{+N/2} \mathcal{M}(n\Delta k,\theta)\delta(k - n\Delta k)$$

represents the discrete radial k-space data in the view θ, k ranges from $-\frac{N}{2}$ to $+\frac{N}{2}$, where $N = [\frac{W}{\Delta k}]$ ($[x]$ denotes the integer part of x). Thus, the integral of Eq. (7.78) represents an Inverse Fourier transform of N radial k-space samples (in the radial direction) in the view θ. In practice, θ is also discretized as $m\Delta\theta$, where $m = 0, 1, \cdots, M-1$ and $(M-1)\Delta\theta = \pi$.

7.6.3.2 Statistics of the Measured Projections

Statistics of the measured projection $p(l,\theta)$ are given by the following property.

Property 7.8 The measured projection $p(l,\theta)$ defined by Eq. (7.78) can be characterized by a Gaussian random process: $p(l,\theta) \sim N(\mu_{p(l,\theta)}, \sigma_p^2)$, with the mean $\mu_{p(l,\theta)}$ and the variance σ_p^2 given by

$$\mu_{p(l,\theta)} = \mathcal{F}_k^{-1}\{\mu_{\mathcal{M}(k,\theta)}\} \quad \text{and} \quad \sigma_p^2 = \frac{\sigma_{\mathcal{M}}^2}{\Delta k} W, \tag{7.81}$$

where $\mu_{\mathcal{M}(k,\theta)}$ and $\sigma_{\mathcal{M}}^2$ are the mean and the variance of the radial k-space sample $\mathcal{M}(k,\theta)$ of Eq. (7.50).

The correlation of the measured projections $p(l_1,\theta_1)$ and $p(l_2,\theta_2)$ is given by

$$R_p((l_1,\theta_1),(l_2,\theta_2)) = \mu_{p(l_1,\theta_1)}\mu^*_{p(l_2,\theta_2)} + \frac{\sigma_{\mathcal{M}}^2}{\Delta k} W\delta[l_1 - l_2, \theta_1 - \theta_2], \tag{7.82}$$

where $*$ denotes the complex conjugate, and $\delta[\Delta l, \Delta\theta]$ is a 2-D Kronecker delta function. That is, $p(l,\theta)$ is an independent random process. □

Proof.

In Eq. (7.78), because $\mathcal{M}(k,\theta) \cdot \frac{1}{\Delta k} comb(\frac{k}{\Delta k}) \cdot rect(\frac{k}{W}) = \mathcal{M}(n\Delta k, m\Delta\theta)$ represents a (truncated) Gaussian random process [Property 7.7] and the

Fourier transform is a linear operation, $p(l,\theta)$ can be characterized by a Gaussian random process. Eqs. (7.81) and (7.82) are proved in the following three steps.

S.1. The continuous radial k-space data over an infinite radial extent in a view θ, $\mathcal{M}(k,\theta)$, are used; the corresponding projection is given by

$$\tilde{p}(l,\theta) = \int_{-\infty}^{+\infty} \mathcal{M}(k,\theta)e^{i2\pi kl}dk. \tag{7.83}$$

This is a hypothetical case. The mean of $\tilde{p}(l,\theta)$ is

$$\mu_{\tilde{p}(l,\theta)} = E[\tilde{p}(l,\theta)] = \int E[\mathcal{M}(k,\theta)]e^{i2\pi kl}dk = \int \mu_{\mathcal{M}(k,\theta)}e^{i2\pi kl}dk, \tag{7.84}$$

where $\mu_{\mathcal{M}(k,\theta)}$ is the mean of the k-space data $\mathcal{M}(k,\theta)$ given by Eq. (7.50).

The correlation $R_{\tilde{p}}((l_1,\theta_1),(l_2,\theta_2))$ of $\tilde{p}(l_1,\theta_1)$ and $\tilde{p}(l_2,\theta_2)$, using Property 7.7, is

$$\begin{aligned}
&R_{\tilde{p}}((l_1,\theta_1),(l_2,\theta_2)) = E[\tilde{p}((l_1,\theta_1)\tilde{p}^*(l_2,\theta_2)] \\
&= E[\int_{-\infty}^{+\infty} \mathcal{M}(k_1,\theta_1)e^{i2\pi k_1 l_1}dk_1 \int_{-\infty}^{+\infty} \mathcal{M}^*(k_2,\theta_2)e^{-i2\pi k_2 l_2}dk_2] \\
&= \int_{-\infty}^{+\infty}\int_{-\infty}^{+\infty} E[\mathcal{M}(k_1,\theta_1)\mathcal{M}^*(k_2,\theta_2)]e^{i2\pi(k_1l_1-k_2l_2)}dk_1dk_2 \\
&= \int_{-\infty}^{+\infty}\int_{-\infty}^{+\infty} \mu_{\mathcal{M}(k_1,\theta_1)}\mu^*_{\mathcal{M}(k_2,\theta_2)}e^{i2\pi(k_1l_1-k_2l_2)}dk_1dk_2 \\
&\quad +\sigma^2_{\mathcal{M}}\int_{-\infty}^{+\infty}\int_{-\infty}^{+\infty} \delta[k_1-k_2,\theta_1-\theta_2]e^{i2\pi(k_1l_1-k_2l_2)}dk_1dk_2.
\end{aligned} \tag{7.85}$$

The first item on the right side of Eq. (7.85) equals $\mu_{\tilde{p}(l_1,\theta_1)}\mu^*_{\tilde{p}(l_2,\theta_2)}$. The second item is the covariance of $\tilde{p}(l_1,\theta_1)$ and $\tilde{p}(l_2,\theta_2)$:

$$C_{\tilde{p}}((l_1,\theta_1),(l_2,\theta_2)) = \sigma^2_{\mathcal{M}}\int_{-\infty}^{+\infty}\int_{-\infty}^{+\infty} \delta[k_1-k_2,\theta_1-\theta_2]e^{i2\pi(k_1l_1-k_2l_2)}dk_1dk_2. \tag{7.86}$$

When $\tilde{p}(l_1,\theta_1)$ and $\tilde{p}(l_2,\theta_2)$ are in the different views, that is, $\theta_1 \neq \theta_2$,

$$C_{\tilde{p}}((l_1,\theta_1),(l_2,\theta_2)) = 0\ . \tag{7.87}$$

When $\tilde{p}(l_1,\theta_1)$ and $\tilde{p}(l_2,\theta_2)$ are in the same view, that is, $\theta_1 = \theta_2$,

$$\begin{aligned}
&C_{\tilde{p}}((l_1,\theta_1),(l_2,\theta_2)) \\
&= \sigma^2_{\mathcal{M}}\int_{-\infty}^{+\infty}\int_{-\infty\ (k_1\neq k_2)}^{+\infty} \delta[k_1-k_2,0]e^{i2\pi(k_1l_1-k_2l_2)}dk_1dk_2
\end{aligned}$$

$$+\sigma^2_{\mathcal{M}} \int_{-\infty}^{+\infty} \int_{-\infty\ (k_1=k_2)}^{+\infty} \delta[k_1 - k_2, 0] e^{i2\pi(k_1 l_1 - k_2 l_2)} dk_1 dk_2$$
$$= \sigma^2_{\mathcal{M}} \int_{-\infty}^{+\infty} e^{i2\pi k(l_1 - l_2)} dk$$
$$= \sigma^2_{\mathcal{M}} \delta(l_1 - l_2)\ . \tag{7.88}$$

Thus, from Eqs. (7.87) and (7.88), $C_{\tilde{p}}((l_1, \theta_1), (l_2, \theta_2))$ can be expressed as

$$C_{\tilde{p}}((l_1, \theta_1), (l_2, \theta_2)) = \sigma^2_{\mathcal{M}} \delta(l_1 - l_2) \delta[\theta_1 - \theta_2]. \tag{7.89}$$

Therefore, Eq. (7.85) becomes

$$R_{\tilde{p}}((l_1, \theta_1), (l_2, \theta_2)) = \mu_{\tilde{p}(l_1, \theta_1)} \mu^*_{\tilde{p}(l_2, \theta_2)} + \sigma^2_{\mathcal{M}} \delta(l_1 - l_2) \delta[\theta_1 - \theta_2]. \tag{7.90}$$

Eq. (7.90) shows that $\tilde{p}(l, \theta)$ is an independent random process. The variance of $\tilde{p}(l, \theta)$, from Eq. (7.89), is

$$\sigma^2_{\tilde{p}} = C_{\tilde{p}}((l, \theta), (l, \theta)) = \sigma^2_{\mathcal{M}} \delta(0). \tag{7.91}$$

S.2. The discrete radial k-space data over an infinite radial extent in a view θ, $\mathcal{M}(k, \theta) \cdot \frac{1}{\Delta k} comb(\frac{k}{\Delta k})$, are used; the corresponding projection is given by

$$\hat{p}(l, \theta) = \int_{-\infty}^{+\infty} \left(\mathcal{M}(k, \theta) \cdot \frac{1}{\Delta k} comb\left(\frac{k}{\Delta k}\right)\right) e^{i2\pi kl} dk\ . \tag{7.92}$$

This is also a hypothetical case. From Eq. (7.92), $\hat{p}(l, \theta)$ can be expressed as

$$\hat{p}(l, \theta) = \int_{-\infty}^{+\infty} \mathcal{M}(k, \theta) \sum_{n=-\infty}^{+\infty} \delta(k - n\Delta k) e^{i2\pi kl} dk$$
$$= \sum_{n=-\infty}^{+\infty} \mathcal{M}(n\Delta k, m\Delta\theta) e^{i2\pi n\Delta k\, l}. \tag{7.93}$$

The mean of $\hat{p}(l, \theta)$ is

$$\mu_{\hat{p}(l,\theta)} = E[\hat{p}(l, \theta)] = \sum_{n=-\infty}^{+\infty} \mu_{\mathcal{M}(n\Delta k, m\Delta\theta)} e^{i2\pi n\Delta k\, l}, \tag{7.94}$$

where $\mu_{\mathcal{M}(n\Delta k, m\Delta\theta)}$ is the mean of the k-space data $\mathcal{M}(n\Delta k, m\Delta\theta)$ given by Eq. (7.50).

From Eqs. (7.83) and (7.92), we have

$$\hat{p}(l, \theta) = \tilde{p}(l, \theta) \star comb(\Delta k\, l), \tag{7.95}$$

where $\star$ denotes the convolution and $comb(\Delta k\ l) = \mathcal{F}_k^{-1}\{\frac{1}{\Delta k}comb(\frac{k}{\Delta k})\}$. That is,

$$\hat{p}(l,\theta) = \frac{1}{\Delta k}\sum_{n=-\infty}^{+\infty} \tilde{p}\left(\frac{n}{\Delta k}\right)\delta\left(l - \frac{n}{\Delta k}\right). \tag{7.96}$$

Thus, the correlation $R_{\hat{p}}((l_1,\theta_1),(l_2,\theta_2))$ of $\hat{p}(l_1,\theta_1)$ and $\hat{p}(l_2,\theta_2)$, using Eq. (7.90), is

$$\begin{aligned}
&R_{\hat{p}}((l_1,\theta_1),(l_2,\theta_2)) = E[\hat{p}((l_1,\theta_1)\hat{p}^*(l_2,\theta_2)] \\
&= E\left[\left(\frac{1}{\Delta k}\sum_{n_1=-\infty}^{+\infty} \tilde{p}\left(\frac{n_1}{\Delta k}\right)\delta\left(l_1 - \frac{n_1}{\Delta k}\right)\right)\left(\frac{1}{\Delta k}\sum_{n_2=-\infty}^{+\infty} \tilde{p}\left(\frac{n_2}{\Delta k}\right)\delta\left(l_2 - \frac{n_2}{\Delta k}\right)\right)^*\right] \\
&= \frac{1}{(\Delta k)^2}\sum_{n_1=-\infty}^{+\infty}\sum_{n_2=-\infty}^{+\infty} E\left[\tilde{p}\left(\frac{n_1}{\Delta k}\right)\tilde{p}^*\left(\frac{n_2}{\Delta k}\right)\right]\delta\left(l_1 - \frac{n_1}{\Delta k}\right)\delta\left(l_2 - \frac{n_2}{\Delta k}\right) \\
&= \frac{1}{(\Delta k)^2}\sum_{n_1=-\infty}^{+\infty}\sum_{n_2=-\infty}^{+\infty} \mu_{\tilde{p}}\left(\frac{n_1}{\Delta k},\theta_1\right)\mu_{\tilde{p}}^*\left(\frac{n_2}{\Delta k},\theta_2\right)\delta\left(l_1 - \frac{n_1}{\Delta k}\right)\delta\left(l_2 - \frac{n_2}{\Delta k}\right) \\
&+ \frac{\sigma_{\mathcal{M}}^2}{(\Delta k)^2}\sum_{n_1=-\infty}^{+\infty}\sum_{n_2=-\infty}^{+\infty} \delta\left(\frac{n_1}{\Delta k} - \frac{n_2}{\Delta k}\right)\delta[\theta_1-\theta_2]\delta\left(l_1 - \frac{n_1}{\Delta k}\right)\delta\left(l_2 - \frac{n_2}{\Delta k}\right). \quad (7.97)
\end{aligned}$$

The first item on the right side of Eq. (7.97) equals $\mu_{\hat{p}}(l_1,\theta_1)\mu_{\hat{p}}^*(l_2,\theta_2)$. The second item is the covariance of $\hat{p}(l_1,\theta_1)$ and $\hat{p}(l_2,\theta_2)$,

$$\begin{aligned}
&C_{\hat{p}}((l_1,\theta_1),(l_2,\theta_2)) \\
&= \frac{\sigma_{\mathcal{M}}^2}{(\Delta k)^2}\delta[\theta_1-\theta_2]\sum_{n_1=-\infty}^{+\infty}\sum_{n_2=-\infty}^{+\infty} \delta\left(\frac{n_1}{\Delta k} - \frac{n_2}{\Delta k}\right)\delta(l_1 - \frac{n_1}{\Delta k})\delta\left(l_2 - \frac{n_2}{\Delta k}\right). \quad (7.98)
\end{aligned}$$

When $\hat{p}(l_1,\theta_1)$ and $\hat{p}(l_2,\theta_2)$ are in different views, that is, $\theta_1 \neq \theta_2$,

$$C_{\hat{p}}((l_1,\theta_1),(l_2,\theta_2)) = 0. \tag{7.99}$$

When $\hat{p}(l_1,\theta_1)$ and $\hat{p}(l_2,\theta_2)$ are in the same view, that is, $\theta_1 = \theta_2$,

$$\begin{aligned}
&C_{\hat{p}}((l_1,\theta_1),(l_2,\theta_2)) \\
&= \frac{\sigma_{\mathcal{M}}^2}{(\Delta k)^2}\sum_{n_1=-\infty}^{+\infty}\sum_{n_2=-\infty}^{+\infty} \delta\left(\frac{n_1}{\Delta k} - \frac{n_2}{\Delta k}\right)\delta\left(l_1 - \frac{n_1}{\Delta k}\right)\delta\left(l_2 - \frac{n_2}{\Delta k}\right) \\
&= \sigma_{\mathcal{M}}^2\Delta k\sum_{n_1=-\infty}^{+\infty}\sum_{n_2=-\infty}^{+\infty} \delta(n_1-n_2)\delta(n_1 - l_1\Delta k)\delta(n_2 - l_2\Delta k) \\
&= \sigma_{\mathcal{M}}^2\Delta k\sum_{n_2=-\infty}^{+\infty}\left(\sum_{n_1=-\infty}^{+\infty} \delta(n_1-n_2)\delta(n_1 - l_1\Delta k)\right)\delta(n_2 - l_2\Delta k)
\end{aligned}$$

$$= \sigma_{\mathcal{M}}^2 \Delta k \sum_{n_2=-\infty}^{+\infty} \delta(n_2 - l_1\Delta k)\delta(n_2 - l_2\Delta k)$$
$$= \sigma_{\mathcal{M}}^2 \delta((l_1 - l_2)\Delta k)$$
$$= \frac{\sigma_{\mathcal{M}}^2}{\Delta k}\delta(l_1 - l_2). \tag{7.100}$$

Thus, from Eqs. (7.99) and (7.100), $C_{\hat{p}}((l_1, \theta_1), (l_2, \theta_2))$ can be expressed as

$$C_{\hat{p}}((l_1, \theta_1), (l_2, \theta_2)) = \frac{\sigma_{\mathcal{M}}^2}{\Delta k}\delta(l_1 - l_2)\delta[\theta_1 - \theta_2]. \tag{7.101}$$

Therefore, Eq. (7.97) becomes

$$R_{\hat{p}}((l_1, \theta_1), (l_2, \theta_2)) = \mu_{\hat{p}(l_1,\theta_1)}\mu^*_{\hat{p}(l_2,\theta_2)} + \frac{\sigma_{\mathcal{M}}^2}{\Delta k}\delta(l_1 - l_2)\delta[\theta_1 - \theta_2]. \tag{7.102}$$

Eq. (7.102) shows that $\hat{p}(l, \theta)$ is an independent random process. The variance of $\hat{p}(l, \theta)$, from Eq. (7.101), is

$$\sigma_{\hat{p}}^2 = C_{\hat{p}}((l, \theta), (l, \theta)) = \frac{\sigma_{\mathcal{M}}^2}{\Delta k}\delta(0). \tag{7.103}$$

S.3. The discrete radial k-space data over a finite radial extent in a view θ, $\mathcal{M}(k, \theta) \cdot \frac{1}{\Delta k} comb(\frac{k}{\Delta k} \cdot rect(\frac{k}{W})$, are used; the measured projection is given by

$$\bar{p}(l, \theta) = \int_{-\infty}^{+\infty} (\mathcal{M}(k, \theta) \cdot \frac{1}{\Delta k} comb(\frac{k}{\Delta k} \cdot rect(\frac{k}{W}))e^{i2\pi kl}dk. \tag{7.104}$$

This is a practical case. From Eq. (7.104), $\bar{p}(l, \theta)$ can be expressed as

$$\bar{p}(l, \theta) = \int_{-W/2}^{+W/2} \mathcal{M}(k, \theta) \sum_{n=-\infty}^{+\infty} \delta(k - n\Delta k)e^{i2\pi kl}dk$$
$$= \sum_{n=-N/2}^{+N/2} \mathcal{M}(n\Delta k, m\Delta\theta)e^{i2\pi n\Delta k\, l}. \tag{7.105}$$

The mean of $\bar{p}(l, \theta)$ is

$$\mu_{\bar{p}(l,\theta)} = E[\bar{p}(l, \theta)] = \sum_{n=-N/2}^{+N/2} \mu_{\mathcal{M}(n\Delta k, m\Delta\theta)}e^{i2\pi n\Delta k\, l}. \tag{7.106}$$

Eq. (7.106) shows that the mean of $\bar{p}(l, \theta)$ is the Fourier transform (with respect to k) of the mean of the k-space data $\mathcal{M}(n\Delta k, m\Delta\theta)$ given by Eq. (7.50).

From Eqs. (7.92) and (7.104), we have

$$\bar{p}(l, \theta) = \hat{p}(l, \theta) \star W sinc(Wl), \tag{7.107}$$

where $\star$ denotes the convolution and $sinc(Wl) = \mathcal{F}^{-1}\{rect(\frac{k}{W})\}$ is a sinc function. That is,

$$\bar{p}(l,\theta) = \int \hat{p}(v,\theta) W sinc(W(l-v))dv. \quad (7.108)$$

Thus, the correlation $R_{\bar{p}}((l_1,\theta_1),(l_2,\theta_2)$ of $\bar{p}(l_1,\theta_1)$ and $\bar{p}(l_2,\theta_2)$, using Eq. (7.102), is

$$\begin{aligned}
&R_{\bar{p}}((l_1,\theta_1),(l_2,\theta_2)) = E[\bar{p}((l_1,\theta_1)\bar{p}^*(l_2,\theta_2)] \\
&= E[(\int \hat{p}(v_1,\theta) W sinc(W(l_1-v_1))dv_1)(\int \hat{p}(v_2,\theta) W sinc(W(l_2-v_2))dv_2)^*] \\
&= \int\int E[\hat{p}(v_1,\theta)\hat{p}^*(v_2,\theta)] W sinc(W(l_1-v_1)) W sinc(W(l_2-v_2))dv_1dv_2 \\
&= \int\int \mu_{\hat{p}(l_1,\theta_1)}\mu^*_{\hat{p}(l_2,\theta_2)} W sinc(W(l_1-v_1)) W sinc(W(l_2-v_2))dv_1dv_2 \\
&+ \frac{\sigma^2_{\mathcal{M}}}{\Delta k} W^2 \delta[\theta_1-\theta_2] \\
&\int\int \delta(v_1-v_2) sinc(W(l_1-v_1)) sinc(W(l_2-v_2))dv_1dv_2. \quad (7.109)
\end{aligned}$$

The first item on the right side of Eq. (7.109) equals $\mu_{\bar{p}}(l_1,\theta_1)\mu^*_{\bar{p}}(l_2,\theta_2)$. The second item is the covariance of $\bar{p}(l_1,\theta_1)$ and $\bar{p}(l_2,\theta_2)$

$$\begin{aligned}
C_{\bar{p}}((l_1,\theta_1),(l_2,\theta_2)) &= \frac{\sigma^2_{\mathcal{M}}}{\Delta k} W^2 \delta[\theta_1-\theta_2] \\
&\int\int \delta(v_1-v_2) sinc(W(l_1-v_1)) sinc(W(l_2-v_2))dv_1dv_2. \quad (7.110)
\end{aligned}$$

When $\bar{p}(l_1,\theta_1)$ and $\bar{p}(l_2,\theta_2)$ are in the different views, that is, $\theta_1 \neq \theta_2$,

$$C_{\bar{p}}((l_1,\theta_1),(l_2,\theta_2)) = 0\ . \quad (7.111)$$

When $\bar{p}(l_1,\theta_1)$ and $\bar{p}(l_2,\theta_2)$ are in the same view, that is, $\theta_1 = \theta_2$,

$$\begin{aligned}
&C_{\bar{p}}((l_1,\theta_1),(l_2,\theta_2)) \\
&= \frac{\sigma^2_{\mathcal{M}}}{\Delta k} W^2 \delta[\theta_1-\theta_2] \int sinc(W(l_1-v_2)) sinc(W(l_2-v_2))dv_2 \\
&= \frac{\sigma^2_{\mathcal{M}}}{\Delta k} W sinc(W(l_1-l_2))\delta[\theta_1-\theta_2]. \quad (7.112)
\end{aligned}$$

$\bar{p}(l,\theta)$ and $\mathcal{M}(k,\theta)$ are discretized as $\bar{p}(n\Delta p, m\Delta\theta)$ and $\mathcal{M}(n\Delta k, m\Delta\theta)$. Section 3.9.3.2 shows that $\frac{1}{N\Delta p} < \Delta k$. By assuming $\frac{1}{N\Delta p} = \frac{1}{i}\Delta k$ ($i > 1$—an integer), we have $\Delta p \Delta k = \frac{i}{N}$, which leads to $W(l_1-l_2) = N\Delta k \cdot (n_1-n_2)\Delta p = i(n_1-n_2)$. Thus, Eq. (7.112) becomes

$$C_{\bar{p}}((l_1,\theta_1),(l_2,\theta_2)) = \frac{\sigma^2_{\mathcal{M}}}{\Delta k} W \frac{\sin i(n_1-n_2)\pi}{i(n_1-n_2)\pi} = \begin{cases} 0 & (l_1 \neq l_2) \\ \frac{\sigma^2_{\mathcal{M}}}{\Delta k} W & (l_1 = l_2) \end{cases}. \quad (7.113)$$

Thus, from Eqs. (7.111) and (7.113), $C_{\bar{p}}((l_1,\theta_1),(l_2,\theta_2))$ can be expressed as

$$C_{\bar{p}}((l_1,\theta_1),(l_2,\theta_2)) = \frac{\sigma_{\mathcal{M}}^2}{\Delta k} W\delta[l_1-l_2,\theta_1-\theta_2]. \tag{7.114}$$

Therefore, Eq. (7.109) becomes

$$R_{\bar{p}}((l_1,\theta_1),(l_2,\theta_2)) = \mu_{\bar{p}(l_1,\theta_1)}\mu^*_{\bar{p}(l_2,\theta_2)} + \frac{\sigma_{\mathcal{M}}^2}{\Delta k} W\delta[l_1-l_2,\theta_1-\theta_2]. \tag{7.115}$$

Eq. (7.115) shows that $\bar{p}(l,\theta)$ is an independent random process. The variance of $\bar{p}(l,\theta)$, from Eq. (7.114), is

$$\sigma_{\bar{p}}^2 = C_{\bar{p}}((l,\theta),(l,\theta)) = \frac{\sigma_{\mathcal{M}}^2}{\Delta k} W. \tag{7.116}$$

■

Problems

7.1. In the Proof of Property 7.2 (Section 7.2.1.1), one outcome of the arguments of (a) and (b) is $\sigma_s = 0$. Prove it.

7.2. In the Proof of Property 7.5 (Section 7.3.2), $n_r(t)$ and $n_c(t)$ are expressed as $\alpha_n(t)\cos(\omega_o t + \phi_n(t))$ and $\frac{1}{2}\alpha_n(t)e^{-i\phi_n(t)}$, respectively. Prove it.

7.3. In the proof of Property 7.5, we state that

$$E[\hat{n}_c(j_1)\hat{n}_c^*(j_2)] = E[\hat{n}_c(j_1)]E[\hat{n}_c^*(j_2)] = 0.$$

Prove it.

7.4. Prove Eq. (7.59).

References

[1] Hoult, D., Richards, R.: The SNR of the NMR experiment. *J. Magn. Reson.* **24**(2) (1979) 71–85.

[2] Hoult, D., Lauterbur, P.: Sensitivity of the zeugmatographic experiment involving human samples. *J. Magn. Reson.* **34**(2) (1979) 425–433.

[3] Mansfield, P., Morris, P.: *NMR Imaging in Biomedicine.* Academic Press, Paris, France, (1982).

[4] Abragam, A.: *Principles of Nuclear Magnetism.* Oxford University Press, New York (1983).

[5] Nalcioglu, O., Cho, Z.: Limits to signal-to-noise improvement by FID averaging in NMR imaging. *Phys. Med. Biol.* **29**(8) (1984) 969–978.

[6] Wagner, R., Brown, D.: Unified SNR analysis of medical imaging systems. *Phys. Med. Biol.* **30**(6) (1985). 489–518

[7] Edelstein, W., Glover, G., Hardy, C., Redington, R.: The intrinsic SNR in NMR imaging. *Magn. Reson. Med.* **3** (1986) 604–618.

[8] Fuderer, M.: The information content of MR images. *IEEE Trans. Med. Imag.* **7**(4) (1988) 368–380.

[9] Chen, C.N., Hoult, D.: *Biomedical Magnetic Resonance Technology.* IOP Publishing Ltd., New York (1989).

[10] Parker, D., Bullberg, G.: Signal-to-noise efficiency in magnetic resonance imaging. *Med. Phys.* **17** (1990) 250–257.

[11] Murphy, B., Carson, P., Ellis, J., Zhang, Y., Hyde, R., Chenevert, T.: SNR measures for MR images. *Magn. Reson. Imaging* **11** (1993) 425–428.

[12] Cho, Z.H., Jones, J.P., Singh, M.: *Foundations of Medical Imaging.* John Wiley & Sons, Inc., New York (1993).

[13] Sijbers, J., Scheunders, P., Bonnet, N., van Dyck, D., Raman, E.: Quantification and improvement of the SNR in an MR image acquisition procedure. *Magn. Reson. Imaging* **14**(10) (1996) 1157–1167.

[14] Macovski, A.: Noise in MRI. *Magn. Reson. Med.* **36** (1996) 494–497.

[15] Nishimura, D.: *Principles of Magnetic Resonance Imaging.* Stanford University, Palo Alto, California, (1996).

[16] Andersen, A., Kirsch, J.: Analysis of noise in phase contrast MR imaging. *Med. Phys.* **23 (6)** (1996) 857–869.

[17] Firbank, M., Coulthard, A., Harrison, R., Williams, E.: A comparison of two methods for measuring the signal to noise ratio on MR images. *Phys. Med. Biol.* **44**(12) (1999) 261–264.

[18] Haacke, E., Brown, R., Thompson, M., Venkatesan, R.: *Magnetic Resonance Imaging: Physical Principles and Sequence Design.* John Wiley & Sons Inc., New York (1999).

[19] Liang, Z.P., Lauterbur, P.: *Principles of Magnetic Resonance Imaging, A Signal Processing Perspective.* IEEE Press, New York (2000).

[20] Tejero, C.: A method for evaluating two-spin correlations of a one-dimensional ising model. *Am. J. Phys.* **56**(2) (1988) 169–171.

[21] Fisz, M.: *Probability Theory and Mathematical Statistics.* John Wiley & Sons, New York (1965).

[22] Cramer, H.: *The Elements of Probability Theory.* John Wiley & Sons, New York (1955).

[23] Papoulis, A.: *Probability, Random Variables and Stochastic Processes.* McGraw-Hill Book Company Inc., New York (1984).

[24] Gray, R., Davisson, L.: *Random Processes: A Mathematical Approach for Engineers.* Prentice-Hall, Inc., Englewood Cliffs, New Jersey, 1986).

[25] Bloch, F.: Nuclear induction. *Phys. Rev.* **70**(7 and 8) (1946) 460–474.

[26] Glover, P., Mansfield, S.P.: Limits on magnetic resonance microscopy. *Rep. Progr. Phys* **65** (2002) 1489–1511.

[27] Muller, N., Jerschow, A.: Nuclear spin noise imaging. *PNAS* **103**(18) (2006) 6790–6792.

[28] Poularikas, A., Seely, S.: *Signals and Systems.* PWS-KENT Publishing Company, Boston, Massachusetts, (1991).

[29] Papoulis, A.: *Signal Analysis.* McGraw Hill Book Company Inc., New York (1977).

[30] Bracewell, R.N.: *The Fourier Transform and Its Applications.* McGraw-Hill Book Company, New York (1978).

[31] Callen, H., Welton, T.: Irreversibility and generalized noise. *Phys. Rev.* **83 (1)** (1951) 34–40.

[32] Callen, H., Greene, R.: On a theorem of irreversible thermodynamics. *Phys. Rev.* **86 (5)** (1952) 702–710.

[33] Thomas, J.: *An Introduction to Statistical Communication Theory.* John Wiley & Sons, New York (1969).

[34] Anderson, T.W.: *An Introduction to Multivariate Statistical Analysis.* John Wiley & Sons, New York (1984).

[35] Peters, T., Williams, J.: *The Fourier Transform in Biomedical Engineering.* Birkhauser, Boston, Massachusetts, (1998).

[36] Herman, G.: *Image Reconstruction from Projections.* Academic Press, New York (1980).

[37] Kak, A., Slaney, M.: *Principles of Computerized Tomographic Imaging.* IEEE Press, New York (1988).

8

Statistics of MR Image

8.1 Introduction

Chapter 7 indicated that the statistics of MR imaging in its data domain propagate to its image domain through image reconstruction. Following this insight, an investigation into the statistics of MR image is conducted for the images generated using typical MR data acquisition schemes and basic image reconstruction methods: rectilinear k-space sampling/Fourier transform (FT), and radial sampling/projection reconstruction (PR). This approach is performed at three levels of the image: a single pixel, any two pixels, and a group of pixels (i.e., an image region).

In MR image analysis and other applications, pixel intensity is always assumed to have a Gaussian distribution. Because an MR image is complex valued, its magnitude image is widely utilized, and its phase image is used in some cases, Gaussianity should be elaborated in a more detailed fashion. This chapter shows that (1) pixel intensity of the complex-valued MR image has a complex Gaussian distribution, (2) its real and imaginary parts are Gaussian distributed and independent, and (3) its magnitude and phase components have non-Gaussian distributions but can be approximated by independent Gaussians when the signal-to-noise ratio (SNR) of the image is moderate or large.

Characterizing spatial relationships of pixel intensities in MR image is an important issue for MR image analysis/processing and other applications. Although theoretical and experimental studies indicate that the pixel intensities of an MR image are correlated, the explicit statements and/or the analytic formulae on the correlation have not been given. This chapter shows that (1) pixel intensities of an MR image are statistically correlated, (2) the degree of the correlation decreases as the distance between pixels increases, and (3) pixel intensities become statistically independent when the distance between pixels approaches infinity. These properties are summarized as spatially asymptotic independence (SAI). This chapter also gives a quantitative measure of the correlations between pixel intensities, that is, the correlation coefficient of the pixel intensities of an MR image decreases exponentially with the distance between pixels. This property is referred to as the Exponential correlation coefficient (ECC).

An MR image appears piecewise contiguous. This scenario suggests that each image region (i.e., a group of pixels) may possess some unique statistics. This chapter proves that each image region is stationary and ergodic (hence satisfies ergodic theorems). Thus, an MR image is a piecewise stationary and ergodic random field. Furthermore, the autocorrelation function (acf) and the spectral density function (sdf) of pixel intensities in an image region of an MR image are derived and expressed in analytic formulae.

Six statistical properties of an MR image—Gaussianity, spatially asymptotic independence, exponential correlation coefficient, stationarity, ergodicity, autocorrelation function, and spectral density function—are described in the order of a single pixel $\Longrightarrow$ any two pixels $\Longrightarrow$ a group of pixels. In addition to theoretical derivations and proofs, experimental results obtained using real MR images are also included. Theoretical and experimental results are in good agreement. These statistics provide the basis for creating stochastic image models and developing new image analysis methodologies for MR image analysis, which are given Chapters 9, 10, and 11.

8.2 Statistics of the Intensity of a Single Pixel

This section analyzes the statistics of the intensity of a single pixel in an MR image. It first derives probability density functions (pdfs) of (1) the complex-valued pixel intensity, (2) its real and imaginary parts, and (3) its magnitude and phase components, and the associated statistical parameters. Then it proves and interprets the Gaussianity of the pixel intensity of an MR image. For the convenience of description, all proofs and derivations of these statistics are given in Appendix 8A.

In general, $x(i,j)$ and $x_{i,j}$ are used to represent a pixel intensity at the location (i,j) in a 2-D image. When the pixel location (i,j) is not required, it is convenient to change notations slightly by suppressing the location index (i,j), that is, simply to use x. Let x be a pixel intensity of the complex-valued MR image, x_R and x_I be its real and imaginary parts, and x_s and x_n be its (underlying) signal and noise components. Similar to Eq. (7.49), we have

$$x = x_R + i x_I = (x_{s_R} + x_{n_R}) + i(x_{s_I} + x_{n_I})$$

or

$$x = x_s + x_n = (x_{s_R} + i x_{s_I}) + (x_{n_R} + i x_{n_I}), \tag{8.1}$$

where x_{s_R} and x_{n_R} are the signal and noise components of the real part x_R of the pixel intensity x, and x_{s_I} and x_{n_I} are the signal and noise components of the imaginary part x_I of the pixel intensity x. x_R and x_I are real-valued quantities, x_s and x_n are complex-valued quantities.

8.2.1 Gaussianity

Property 8.1a The real and imaginary parts, x_R and x_I, of any single pixel intensity x in a complex-valued MR image are characterized by two real Gaussian variables: $x_R \sim N(\mu_{x_R}, \sigma^2_{x_R})$, and $x_I \sim N(\mu_{x_I}, \sigma^2_{x_I})$, where the means $\mu_{x_R} = x_{s_R}$ and $\mu_{x_I} = x_{s_I}$, and the variances $\sigma^2_{x_R} = \sigma^2_{x_{n_R}}$ and $\sigma^2_{x_I} = \sigma^2_{x_{n_I}}$ $(\sigma^2_{x_{n_R}} = \sigma^2_{x_{n_I}} \stackrel{\Delta}{=} \sigma^2)$. x_R and x_I are independent: $p(x_R, x_I) = p(x_R)p(x_I)$.

The pixel intensity x is characterized by a complex Gaussian random variable: $x \sim N(\mu_x, \sigma^2_x)$, and its pdf is given by

$$p(x) = \pi^{-1}(\sigma^2_x)^{-1} \exp(-(x - \mu_x)^*(\sigma^2_x)^{-1}(x - \mu_x)), \tag{8.2}$$

where $*$ represents the complex conjugate, and the mean μ_x and the variance σ^2_x are*

$$\begin{aligned} \mu_x &= E[x] = x_s = x_{s_R} + i x_{s_I}, \\ \sigma^2_x &= E[(x - \mu_x)(x - \mu_x)^*] = E[x_n x^*_n] = 2\sigma^2. \end{aligned} \tag{8.3}$$

Property 8.1b Let ρ and θ denote the magnitude and the phase of the pixel intensity x of a complex-valued MR image: $\rho = |x| = \sqrt{x_R^2 + x_I^2}$ and $\theta = \angle x = \tan^{-1}(x_I/x_R)$. The joint pdf of ρ and θ is given by

$$p(\rho, \theta|\phi) = \frac{\rho}{2\pi\sigma^2} \exp(-\frac{\rho^2 + \nu^2 - 2\nu\rho\cos(\theta - \phi)}{2\sigma^2}) \quad (\rho > 0, \ -\pi \le \theta < \pi), \tag{8.4}$$

where

$$\nu = |\mu_x| = \sqrt{x^2_{s_R} + x^2_{s_I}} \quad \text{and} \quad \phi = \angle\mu_x = \tan^{-1}(x_{s_I}/x_{s_R}). \tag{8.5}$$

Property 8.1c The pdf of the magnitude ρ of the pixel intensity x of a complex-valued MR image is given by

$$p(\rho) = \frac{\rho}{\sigma^2} \exp(-\frac{\rho^2 + \nu^2}{2\sigma^2}) I_0(\frac{\nu\rho}{\sigma^2}) \quad (\rho > 0), \tag{8.6}$$

where $I_0(x)$ is the modified Bessel function of the first kind of zero order [1] defined by

$$I_0(x) = \frac{1}{2\pi} \int_0^{2\pi} e^{x\cos(\theta - \phi)} d\theta = \sum_{n=0}^{\infty} \frac{x^{2n}}{2^{2n}(n!)^2} . \tag{8.7}$$

---

*The relationship between σ^2_x and the variance $\sigma^2_{\mathcal{M}}$ of k-space samples depends on image reconstruction methods. For the filtered 2DFT, Section 8.3.1.1 shows that $\sigma^2_x = \frac{1}{N}\sigma^2_{\mathcal{M}} f(0)$, where N is the number of k-space samples used in the image reconstruction, $f(0)$ is a factor determined by the filter function, for example, for the Hanning filter, $f(0) = 0.375$.

Eq. (8.6) is known as a Rician distribution. The moments of ρ are given by

$$E[\rho^n] = (2\sigma^2)^{\frac{n}{2}}\Gamma(1+\frac{n}{2})\,{}_1F_1(-\frac{n}{2};1;-\frac{\nu^2}{2\sigma^2}), \tag{8.8}$$

where $\Gamma(x)$ is the Gamma function [1] defined by

$$\Gamma(\alpha) = \int_0^\infty x^{\alpha-1}e^{-x}dx = (\alpha-1)\Gamma(\alpha-1), \tag{8.9}$$

and ${}_1F_1[\alpha;\beta;y]$ is the confluent hypergeometric function of the first kind [1, 2] defined by

$$ {}_1F_1[\alpha;\beta;y] = 1+\frac{\alpha}{\beta}\frac{(y)}{1!}+\frac{\alpha(\alpha+1)}{\beta(\beta+1)}\frac{(y^2)}{2!}+\cdots+\frac{\alpha(\alpha+1)\cdots(\alpha+n-1)}{\beta(\beta+1)\cdots(\beta+n-1)}\frac{(y^n)}{n!}\cdots. \tag{8.10}$$

An extension of Property 8.1c Let the signal-to-noise ratio (SNR) of the pixel intensity x of a complex-valued MR image be defined by $\varrho = \frac{\nu}{\sigma}$. When $\varrho = 0$, pdf of ρ becomes

$$p(\rho) = \frac{\rho}{\sigma^2}\exp(-\frac{\rho^2}{2\sigma^2}), \tag{8.11}$$

which is a Rayleigh distribution. The moments of ρ in this limiting case are

$$E[\rho^n] = (2\sigma^2)^{\frac{n}{2}}\Gamma(1+\frac{n}{2}). \tag{8.12}$$

When $\varrho \to \infty$, pdf of ρ becomes

$$p(\rho) = \frac{1}{\sqrt{2\pi}\sigma}\exp(-\frac{(\rho-\nu)^2}{2\sigma^2}), \tag{8.13}$$

which is a Gaussian distribution. The central moments of ρ in this limiting case are

$$E[(\rho-\nu)^{2k-1}] = 0 \quad \text{and} \quad E[(\rho-\nu)^{2k}] = 1\cdot 3\cdots(2k-1)\sigma^{2k}, \tag{8.14}$$

where $k = 1, 2, \cdots$. Generally, when $\varrho \geq 2\sqrt{2}$, Eq. (8.6) can be approximated by a Gaussian distribution Eq. (8.13).

The Rician distribution Eq. (8.6), Rayleigh distribution Eq. (8.11), and approximate Gaussian distribution Eq. (8.13) with various SNR are shown in Figure 8.1.

Property 8.1d The pdf of the phase deviation $\delta\theta = \theta - \phi$ of the pixel intensity x of a complex-valued MR image is given by

$$p(\delta\theta) = \frac{1}{2\pi}\exp(-\frac{\nu^2}{2\sigma^2}) + \frac{\nu\cos(\delta\theta)}{\sqrt{2\pi}\sigma}\exp\left(-\frac{\nu^2\sin^2(\delta\theta)}{2\sigma^2}\right)\Phi\left(\frac{\nu\cos(\delta\theta)}{\sigma}\right), \tag{8.15}$$

where $-\pi \leq \theta < \pi$, $\Phi(x)$ is the cumulative distribution function (cdf) of $N(0,1)$, defined by

$$\Phi(x) = \frac{1}{\sqrt{2\pi}} \int_{-\infty}^{x} e^{-\frac{t^2}{2}} dt. \tag{8.16}$$

An extension of Property 8.1d Let the SNR of the pixel intensity x of a complex-valued MR image be defined by $\varrho = \frac{\nu}{\sigma}$. When $\varrho = 0$, pdf of $\delta\theta$ becomes

$$p(\delta\theta) = \begin{cases} \frac{1}{2\pi} & (-\pi \leq \delta\theta < \pi) \\ 0 & \text{otherwise,} \end{cases} \tag{8.17}$$

which is a uniform distribution. The moments of $\delta\theta$ in this limiting case are

$$E[(\delta\theta)^{2k-1}] = 0 \quad \text{and} \quad E[(\delta\theta)^{2k}] = \frac{\pi^{2k}}{2k+1}, \tag{8.18}$$

where $k = 1, 2, \cdots$. When $\varrho \rightarrow \infty$, the pdf of $\delta\theta$ becomes

$$p(\delta\theta) = \begin{cases} \frac{1}{\sqrt{2\pi}(\sigma/\nu)} \exp(-\frac{(\delta\theta)^2}{2(\sigma/\nu)^2}) & (-\pi \leq \delta\theta < \pi) \\ 0 & \text{otherwise,} \end{cases} \tag{8.19}$$

which is a Gaussian distribution. The moments of $\delta\theta$ in this limiting case are

$$E[(\delta\theta)^{2k-1}] = 0 \quad \text{and} \quad E[(\delta\theta)^{2k}] = 1 \cdot 3 \cdots (2k-1)(\sigma/\nu)^{2k}, \tag{8.20}$$

where $k = 1, 2, \cdots$. Generally, when SNR $\varrho \geq 1$, Eq. (8.15) can be approximated by a Gaussian distribution (8.19).

pdf Eq. (8.15), the uniform distribution Eq. (8.17) and the approximate Gaussian distribution Eq. (8.19) with various SNR are shown in Figure 8.2.

Property 8.1e When SNR ϱ is moderate or large (as specified in the Extensions of Property 8.1c and Property 8.1d), the magnitude ρ and the phase θ of the pixel intensity x of a complex-valued MR image are approximately independent, that is.,

$$p(\rho, \theta|\phi) \simeq p(\rho)p(\theta|\phi). \tag{8.21}$$

Proof.

Proofs for Property 8.1a—Property 8.1e and their extensions are given in Appendix 8A. ■

Remarks on Gaussianity. This section shows that (1) the single pixel intensity of a complex-valued MR image has a complex *Gaussian* distribution, (2) its real and imaginary parts are *Gaussian* distributed and independent, and (3) its magnitude and phase deviation are approximately *Gaussian* distributed and independent when the signal-to-noise ratio of image is moderate or large. *Gaussianity* of MR image is interpreted and understood in this sense.

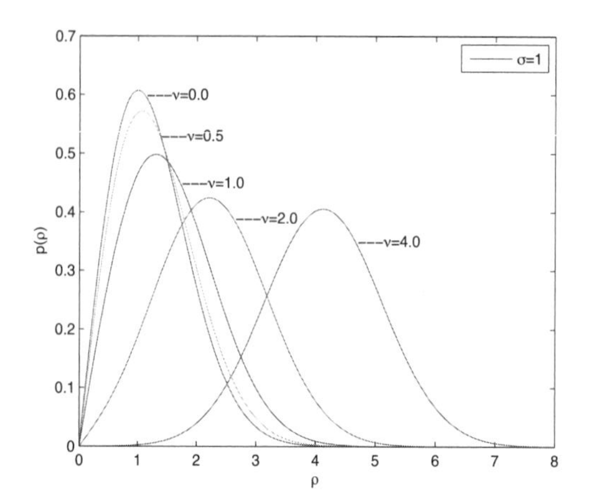

FIGURE 8.1
Rician pdfs (Eq. (8.6)) with various SNR. Among them, Rayleigh pdf (Eq. (8.11)) with the zero SNR and the approximate Gaussian pdf (Eq. (8.13)) with the moderate or larger SNR.

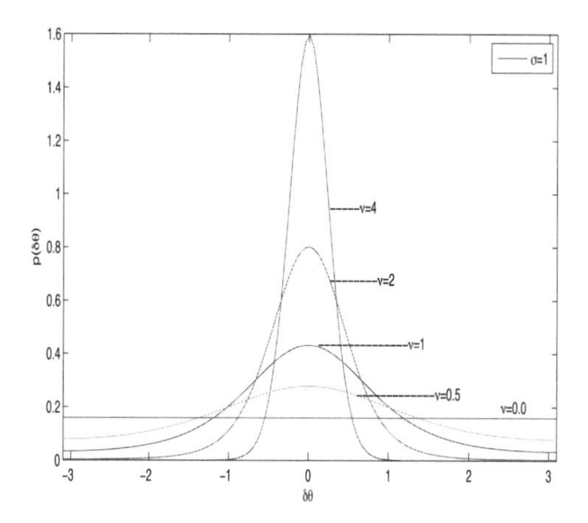

FIGURE 8.2
pdfs Eq. (8.15) with various SNR. Among them, the Uniform pdf (Eq. (8.17)) with the zero SNR, and the approximate Gaussian pdf (Eq. (8.19)) with the moderate or larger SNR.

8.3 Statistics of the Intensities of Two Pixels

This section analyzes the statistics of the intensities of any two pixels in an MR image. It first gives a qualitative description: (1) the intensities of any two pixels in MR image are statistically correlated, (2) the degree of the correlation decreases as the distance between the pixels increases, and (3) pixel intensities become statistically independent when the distance between pixels approaches infinity. These properties are summarized as spatially asymptotic independence, abbreviated as SAI. Then it gives a quantitative description: the correlation coefficient of pixel intensities decreases exponentially with the distance between pixels. This property is referred to as the exponential correlation coefficient, abbreviated as ECC.

This section shows that the correlation and the correlation coefficient are the functions of (1) MR imaging parameters, (2) statistics of MR signals and noise, (3) pixel intensities, and (4) the distance between pixels. For the convenience of description, the major steps of the proofs of SAI and ECC are outlined in this section, and the details of these proofs are given in Appendices 8B. The more extended discussions and results are reported in [3–5].

8.3.1 Spatially Asymptotic Independence

Property 8.2 Pixel intensities of MR image are spatially asymptotically independent.

The SAI of an MR image reconstructed by 2DFT and 2DPR methods is proved *progressively and probabilistically*, respectively, in the following two subsections.

8.3.1.1 The SAI of an MR Image by FT

Proof.

Because 2-D FT and other functions used in the rectilinear k-space data acquisition are separable, for simplicity, 1-D notation is used in this proof. k and u denote the locations in k and x spaces, and $\mathcal{M}(k)$ and $x(u)$ denote the k-space data and the pixel intensity. Sections 3.9 and 3.10 give detailed descriptions of rectilinear k-space sampling and the 2DFT image reconstruction method. Eq. (3.130) shows that pixel intensity $x(u)$ generated by 2DFT method can be expressed by

$$x(u) = \mathcal{F}^{-1}\{\mathcal{M}(k) \cdot \frac{1}{\Delta k} comb(\frac{k}{\Delta k}) \cdot rect(\frac{k}{W}) \cdot filt(\frac{k}{W})\}, \tag{8.22}$$

where $\mathcal{F}^{-1}$ denotes the inverse FT, Δk and W are the sampling period and window width in k-space, and *comb* (the sampling function), *rect* (the truncating function), and $filt$ (the filter function) are defined by

$$comb\left(\frac{k}{\Delta k}\right) = \Delta k \sum_{m=-\infty}^{\infty} \delta(k - m\Delta k)$$
$$rect\left(\frac{k}{W}\right) = \begin{cases} 1 & |k| \leq \frac{W}{2} \\ 0 & \text{otherwise} \end{cases} \tag{8.23}$$
$$filt\left(\frac{k}{W}\right) = \Phi\left(\frac{k}{W}\right) rect\left(\frac{k}{W}\right),$$

where $\delta(k)$ is the Dirac Delta function; $filt(\frac{k}{W})$ is a real, even, and normalized; and $\Phi(\frac{k}{W})$ can take different forms, for example, Hanning, Hamming functions, etc [2, 6–16].

The different combinations of the four factors on the right side of Eq. (8.22) constitute four different k-space data acquisition schemes:

1) $\mathcal{M}(k)$ represents the continuous k-space data over the infinity extent of k-space;
2) $\mathcal{M}(k) \cdot \frac{1}{\Delta k} comb(\frac{k}{\Delta k})$ represents the discrete k-space data over the infinity extent of k-space;
3) $\mathcal{M}(k) \cdot \frac{1}{\Delta k} comb(\frac{k}{\Delta k}) \cdot rect(\frac{k}{W})$ represents the discrete k-space data over a finite extent of k-space;
4) $\mathcal{M}(k) \cdot \frac{1}{\Delta k} comb(\frac{k}{\Delta k}) \cdot rect(\frac{k}{W}) \cdot filt(\frac{k}{W})$ represents the filtered, discrete k-space data over a finite extent of k-space.

1) and 2) represent two hypothetical cases; 3) is a practical case in which a Gibbs ringing artifact occurs [3, 7, 9, 10, 17]; and 4) is a commonly used case in which Gibbs ringing artifact is reduced [3, 7, 9, 10, 17]. SAI is proved through these four cases, *progressively*.

Case 1) The continuous k-space data over the infinity extent of k-space, $\mathcal{M}(k)$, were used. $\tilde{x}(u)$ is adopted to denote the corresponding pixel intensity:

$$\tilde{x}(u) = \mathcal{F}^{-1}\{\mathcal{M}(k)\}. \tag{8.24}$$

Appendix 8.B shows that the correlation $R_{\tilde{x}}(u_1, u_2)$ between $\tilde{x}(u_1)$ and $\tilde{x}(u_2)$ is given by

$$R_{\tilde{x}}(u_1, u_2) = \mu_{\tilde{x}(u_1)}\mu^*_{\tilde{x}(u_2)} + \sigma^2_{\mathcal{M}}\delta(\Delta u), \tag{8.25}$$

where $\mu_{\tilde{x}(u)}$ denotes the mean of $\tilde{x}(u)$, $*$ represents the complex conjugate, $\Delta u = u_1 - u_2$ is the distance between two pixels at u_1 and u_2, and $\sigma^2_{\mathcal{M}}$ is the variance of $\mathcal{M}(k)$.

1a) When $u_1 \neq u_2$, Eq. (8.25) becomes

$$R_{\tilde{x}_1,\tilde{x}_2} = \mu_{\tilde{x}(u_1)}\mu^*_{\tilde{x}(u_2)}, \tag{8.26}$$

which shows that $\tilde{x}(u_1)$ and $\tilde{x}(u_2)$ are uncorrelated.

1b) When $u_1 = u_2$, Eq. (8.25) gives the variance $\sigma_{\tilde{x}}^2$ of $\tilde{x}(u)$

$$\sigma_{\tilde{x}}^2 = \sigma_{\mathcal{M}}^2 \delta(0), \tag{8.27}$$

where $\delta(0)$ is used as a *notation* here as well as in Eq. (8.31). Because $\delta(t)$ is undefined for $t = 0$, its interpretation is given at the end of Section 8.3.1.1: *Remarks on the variance of FT MR image.*

Case 2) The discrete (i.e., the sampled) k-space data over the infinity extent of k-space, $\mathcal{M}(k) \cdot \frac{1}{\Delta k} comb(\frac{k}{\Delta k})$, were used. $\hat{x}(u)$ is adopted to denote the corresponding pixel intensity:

$$\hat{x}(u) = \mathcal{F}^{-1}\{\mathcal{M}(k) \cdot \frac{1}{\Delta k} comb(\frac{k}{\Delta k})\} = \tilde{x}(u) \star comb(\Delta k\, u), \tag{8.28}$$

where $\star$ denotes the convolution. Appendix 8B shows that the correlation $R_{\hat{x}}(u_1, u_2)$ between $\hat{x}(u_1)$ and $\hat{x}(u_2)$ is given by

$$R_{\hat{x}}(u_1, u_2) = \mu_{\hat{x}(u_1)} \mu_{\hat{x}(u_2)}^* + \frac{\sigma_{\mathcal{M}}^2}{\Delta k} \delta(\Delta u), \tag{8.29}$$

where $\mu_{\hat{x}(u)}$ denotes the mean of $\hat{x}(u)$.

2a) When $u_1 \neq u_2$, Eq. (8.29) becomes

$$R_{\hat{x}} = \mu_{\hat{x}(u_1)} \mu_{\hat{x}(u_2)}^*, \tag{8.30}$$

which shows that $\hat{x}(u_1)$ and $\hat{x}(u_2)$ are uncorrelated.

2b) When $u_1 = u_2$, Eq. (8.29) gives the variance $\sigma_{\hat{x}}^2$ of $\hat{x}(u)$

$$\sigma_{\hat{x}}^2 = \frac{\sigma_{\mathcal{M}}^2}{\Delta k} \delta(0), \tag{8.31}$$

where the factor $\frac{1}{\Delta k}$, by comparing with Eq. (8.27), indicates the difference between the discrete and the continuous data.

Case 3) The discrete k-space data over a finite extent of k-space, $\mathcal{M}(k) \cdot \frac{1}{\Delta k} comb(\frac{k}{\Delta k}) \cdot rect(\frac{k}{W})$, are used. FT reconstruction with this type of data acquisition is known as the basic FT. $\bar{x}(u)$ is adopted to denote the corresponding pixel intensity:

$$\bar{x}(u) = \mathcal{F}^{-1}\{\mathcal{M}(k) \cdot \frac{1}{\Delta k} comb(\frac{k}{\Delta k}) \cdot rect(\frac{k}{W})\} = \hat{x}(u) \star W\, sinc(Wu), \tag{8.32}$$

where $sinc(Wu) = sin(\pi W u)/(\pi W u)$ is a *sinc* function. Appendix 8B shows that the correlation $R_{\bar{x}}(u_1, u_2)$ between $\bar{x}(u_1)$ and $\bar{x}(u_2)$ is given by

$$R_{\bar{x}}(u_1, u_2) = \mu_{\bar{x}(u_1)} \mu_{\bar{x}(u_2)}^* + \frac{\sigma_{\mathcal{M}}^2}{\Delta k} W\, sinc(W \Delta u). \tag{8.33}$$

3a) When $u_1 \neq u_2$, Eq. (8.33) shows that the degree of correlation of $\bar{x}(u_1)$ and $\bar{x}(u_2)$ is characterized by

$$\frac{\sigma_{\mathcal{M}}^2}{\Delta k} W sinc(W\Delta u). \tag{8.34}$$

Appendix 8B shows that $\frac{\sigma_{\mathcal{M}}^2}{\Delta k} W sinc(W\Delta u)$ is confined to the main lobe of $sinc(W\Delta u)$. Thus, $|\frac{\sigma_{\mathcal{M}}^2}{\Delta k} W sinc(W\Delta u)|$ monotonically decreases from 1 to 0 (approaching, but not equalizing) as $|\Delta u|$ increases from 0 to its maximum. This finding shows that $\bar{x}(u_1)$ and $\bar{x}(u_2)$ are correlated and the degree of correlation decreases as the distance between pixels increases. In the limiting case of $|\Delta u| \to \infty$, due to $lim\ sinc(W\Delta u) = 0$, we have

$$\lim_{\Delta u \to \infty} R_{\bar{x}}(u_1, u_2) = \mu_{\bar{x}(u_1)} \mu^*_{\bar{x}(u_2)}, \tag{8.35}$$

which shows that $\bar{x}(u_1)$ and $\bar{x}(u_2)$ are uncorrelated, and therefore independent [Property 8.1]. Thus, pixel intensities of MR image reconstructed by the basic FT are SAI, which are also demonstrated by the curves of Eq. (8.34) in Figure 8.8 of Appendix 8B.

3b) When $u_1 = u_2$, due to $sinc(0) = 1$, Eq. (8.33) gives the variance $\sigma_{\bar{x}}^2$ of $\bar{x}(u)$

$$\sigma_{\bar{x}}^2 = \sigma_{\mathcal{M}}^2 \frac{W}{\Delta k}. \tag{8.36}$$

Case 4) The filtered, discrete k-space data over a finite extent of k-space, $\mathcal{M}(k) \cdot \frac{1}{\Delta k} comb(\frac{k}{\Delta k}) \cdot rect(\frac{k}{W}) \cdot filt(\frac{k}{W})$, are used. FT reconstruction with this type of data acquisition is known as filtered FT. $x(u)$ is adopted to denote the corresponding pixel intensity:

$$x(u) = \mathcal{F}^{-1}\{\mathcal{M}(k) \cdot \frac{1}{\Delta k} comb(\frac{k}{\Delta k}) \cdot filt(\frac{k}{W})\} = \hat{x}(u) \star \phi(Wu), \tag{8.37}$$

where

$$\phi(Wu) = \mathcal{F}^{-1}\{filt(\frac{k}{W})\}. \tag{8.38}$$

Eq. (8.37) is obtained by merging $rect(\frac{k}{W})$ of Eq. (8.22) into $filt(\frac{k}{W})$, which also contains a factor $rect(\frac{k}{W})$ (see Eq. (8.23)). Appendix 8B shows that the correlation $R_x(u_1, u_2)$ between $x(u_1)$ and $x(u_2)$ for Hanning and Hamming filtering is given by

$$R_x(u_1, u_2) = \mu_{x(u_1)} \mu^*_{x(u_2)} + \frac{\sigma_{\mathcal{M}}^2}{\Delta k} W sinc(W\Delta u) f(W\Delta u). \tag{8.39}$$

where

$$f(W\Delta u) = \begin{cases} \frac{1.5}{(W\Delta u)^4 - 5(W\Delta u)^2 + 4} & \text{(a. Hanning)} \\ 0.0064 \frac{(W\Delta u)^4 - 16.5(W\Delta u)^2 + 248.375}{(W\Delta u)^4 - 5(W\Delta u)^2 + 4} & \text{(b. Hamming).} \end{cases} \tag{8.40}$$

Other filter functions have been examined. For example, the second term in Eq. (8.39) for the Bartlett filter is $\frac{\sigma_{\mathcal{M}}^2}{\Delta k} W \frac{2(1-sinc(W\Delta u))}{(\pi W \Delta u)^2}$. Because (1) their expressions cannot be factorized as the product of a $sinc(W\Delta u)$ and a $f(W\Delta u)$ as shown in Eq. (8.39), and (2) Hamming and Hanning filters possess the desirable property of being compact in the k-space domain and having low sidelobes in the image domain, those filter functions are not included in this book. The performances of Hanning and Hamming filtering in terms of harmonic and spectral analysis are given in Appendix 8C.

4a) When $u_1 \neq u_2$, Eq. (8.39) shows that the degree of the correlation of $x(u_1)$ and $x(u_2)$ is characterized by

$$\frac{\sigma_{\mathcal{M}}^2}{\Delta k} W sinc(W\Delta u) f(W\Delta u). \tag{8.41}$$

Appendix 8B shows that $|\frac{\sigma_{\mathcal{M}}^2}{\Delta k} W sinc(W\Delta u) f(W\Delta u)|$ monotonically decreases from 1 to 0 (approaching, but not equalizing) as $|\Delta u|$ increases from 0 to its maximum. This finding shows that $x(u_1)$ and $x(u_2)$ are correlated and the degree of correlation decreases as the distance between pixels increases. In the limiting case of $|\Delta u| \to \infty$, due to (1) $lim\ sinc(W\Delta u) = 0$ and (2) $f(W\Delta u) = 0,\ 0.0064$ (Hanning, Hamming filter), Eq. (8.39) becomes

$$\lim_{\Delta u \to \infty} R_x(u_1, u_2) = \mu_{x(u_1)} \mu_{x(u_2)}^*, \tag{8.42}$$

which shows that $x(u_1)$ and $x(u_2)$ are uncorrelated, and therefore independent [Property 8.1]. Thus, pixel intensities of an MR image reconstructed by filtered FT are SAI, which are also demonstrated by the curves of Eq. (8.41) in Figures 8.9 and 8.10 of Appendix 8B, for Hanning and Hamming filters, respectively.

4b) When $u_1 = u_2$, due to $sinc(0) = 1$, Eq. (8.39) gives the variance σ_x^2 of x

$$\sigma_x^2 = \sigma_{\mathcal{M}}^2 \frac{W}{\Delta k} f(0), \tag{8.43}$$

where $f(0) = 0.375,\ 0.3974$ for Hanning, Hamming filter, respectively. ∎

Remarks on the variances of FT MR image. The variances of pixel intensities of FT MR images in the four cases of the above proof—Eqs. (8.27), (8.31), (8.36), and (8.43)—are summarized below.

Case 1):	continuous, infinite k-space sampling,	$\tilde{x}(u)$:	$\sigma_{\mathcal{M}}^2 \delta(0)$
Case 2):	discrete, infinite k-space sampling,	$\hat{x}(u)$:	$\sigma_{\mathcal{M}}^2 \delta(0) \frac{1}{\Delta k}$
Case 3):	discrete, finite k-space sampling,	$\bar{x}(u)$:	$\sigma_{\mathcal{M}}^2 \frac{W}{\Delta k}$
Case 4):	filtered, discrete, finite k-space sampling,	$x(u)$:	$\sigma_{\mathcal{M}}^2 \frac{W}{\Delta k} f(0)$

The Continuous Fourier Transform (CFT) is used in the proof. FTs in Eqs. (8.24), (8.28), (8.32), and (8.37) do not use a scale coefficient $\frac{1}{N}$ as N-points Discrete Fourier Transform (DFT) does. Here, N is the number of the samples in k-space truncated by a window with width W, which is specified by $(N-1)\Delta k < W < N\Delta k$. If this coefficient is applied in Eqs. (8.24), (8.28), (8.32), and (8.37), the following observations can be obtained. In the case 1, the variance in Eq. (8.27) should be $\sigma^2_{\tilde{x}} = \lim_{N\to\infty} \frac{1}{N^2}(\sigma^2_{\mathcal{M}}\delta(0)) = \sigma^2_{\mathcal{M}} \lim_{N\to\infty} \frac{1}{N}(\frac{1}{N}\delta(0)) = \sigma^2_{\mathcal{M}} \lim_{N\to\infty} \frac{1}{N} \longrightarrow 0.$† Similarly, the variance in Eq. (8.31) of case 2 should be $\sigma^2_{\hat{x}} = \lim_{N\to\infty} \frac{1}{N^2}(\frac{\sigma^2_{\mathcal{M}}}{\Delta k}\delta(0)) = \frac{\sigma^2_{\mathcal{M}}}{\Delta k} \lim_{N\to\infty}\frac{1}{N} \longrightarrow 0$. In case 3, the variance in Eq. (8.36) should be $\sigma^2_{\bar{x}} = \frac{1}{N^2}(\sigma^2_{\mathcal{M}}\frac{W}{\Delta k})$. Let $\frac{W}{\Delta k} \simeq [\frac{W}{\Delta k}] + 1 = N$ ($[x]$ is the integer part of x), we have $\sigma^2_{\bar{x}} = \frac{1}{N}\sigma^2_{\mathcal{M}}$, which is identical to that in the Direct FFT MR image [7, 9, 10]. Also, the variance in Eq. (8.43) of case 4 should be $\sigma^2_x = \frac{1}{N^2}(\sigma^2_{\mathcal{M}}\frac{W}{\Delta k}f(0)) = \frac{1}{N}\sigma^2_{\mathcal{M}}f(0)$ ($f(0) = 0.375$, 0.3974 for Hanning and Hamming filter).

Cases 1 and 2 are hypothetical. In case 3, the variance $\sigma^2_{\bar{x}} = \sigma^2_{\mathcal{M}}(\frac{W}{\Delta k})^2$ shows the effect of the finite samples in MR data acquisition on the reconstructed (basic FT) image, which is inevitably linked with the Gibbs ringing artifact. In case 4, the variance $\sigma^2_x = \sigma^2_{\mathcal{M}}(\frac{W}{\Delta k})^2 f(0)$ shows the impact of filtering (apodization) in MR data acquisition on the reconstructed (filtered FT) image, which, due to $f(0) < 1$, reduces the variance caused by reducing Gibbs ringing artifact. The above discussion shows that (1) CFT and DFT analysis can provide consistent results, and (2) CFT analysis is simpler than DFT and may lead to some new important outcomes.

The correlations shown by Eqs. (8.25), (8.29), (8.33), and (8.39) and the variances shown by Eqs. (8.27), (8.31), (8.36), and (8.43) are in 1-D notation. Using 2-D notation, the correlations and variances in the above four cases are, respectively,

$$
\begin{aligned}
R_{\tilde{x}}((u_1,v_1),(u_2,v_2)) &= \mu_{\tilde{x}_1}\mu^*_{\tilde{x}_2} + \sigma^2_{\mathcal{M}}\,\delta(\Delta u, \Delta v), \\
R_{\hat{x}}((u_1,v_1),(u_2,v_2)) &= \mu_{\hat{x}_1}\mu^*_{\hat{x}_2} + \frac{1}{(\Delta k})^2\sigma^2_{\mathcal{M}}\,\delta(\Delta u, \Delta v), \\
R_{\bar{x}}((u_1,v_1),(u_2,v_2)) &= \mu_{\bar{x}_1}\mu^*_{\bar{x}_2} + (\frac{W}{\Delta k})^2\sigma^2_{\mathcal{M}}\, sinc(W\Delta u, W\Delta v), \\
R_{x}((u_1,v_1),(u_2,v_2)) &= \mu_{x_1}\mu^*_{x_2} + (\frac{W}{\Delta k})^2\sigma^2_{\mathcal{M}} sinc(W\Delta u, W\Delta v) f(W\Delta u, W\Delta v), \\
\sigma^2_{\tilde{x}} &= \sigma^2_{\mathcal{M}}\delta(0,0), \\
\sigma^2_{\hat{x}} &= \sigma^2_{\mathcal{M}}\delta(0,0)\frac{1}{(\Delta k)^2}, \\
\sigma^2_{\bar{x}} &= \sigma^2_{\mathcal{M}}(\frac{W}{\Delta k})^2, \\
\sigma^2_{x} &= \sigma^2_{\mathcal{M}}(\frac{W}{\Delta k})^2 f(0,0)\,. \qquad (8.44)
\end{aligned}
$$

† Here we consider $\lim_{N\to\infty}\frac{1}{N}\delta(0) = 1$, because $1 = \int_{-1/2N}^{1/2N}\delta(t)dt \simeq \frac{1}{N}\delta(0)$.

8.3.1.2 The SAI of an MR Image by PR

Proof. (for *Filtering by Fourier transform*)

The principle of *Filtering by Fourier transform* in the PR of an MR image is mathematically described by Eqs. (3.146) and (3.147). It is a two-step approach:

A) Computing the filtered projections in each view θ by

$$t(u',\theta) = \mathcal{F}_k^{-1}\{\mathcal{M}(k,\theta) \cdot \frac{1}{\Delta k}comb(\frac{k}{\Delta k}) \cdot |k| \cdot rect(\frac{k}{W_k})\}, \tag{8.45}$$

B) Computing the backprojection over all views by

$$x(u,v) = \int_0^{\pi} t(u',\theta)d\theta = \Delta\theta \sum_{m=0}^{M-1} t(u', m\Delta\theta), \tag{8.46}$$

where Δk and $\Delta\theta$ are the radial and angular sampling periods as shown in Figure 3.12, $k = n\Delta k$ and $\theta = m\Delta\theta$ specify the locations of the samples in k-space, (u,v) is the coordinate of the pixel center in the image,

$$u' = u\cos\theta + v\sin\theta \tag{8.47}$$

is the projection of the vector $u\vec{i} + v\vec{j}$ onto U'-axis specified by the view angle θ, $\frac{1}{\Delta k}comb(\frac{k}{\Delta k})$ is a comb function (defined in Eq. (8.23)), $|k|$ is the filter function, and $rect(\frac{k}{W_k})$ is a rect function with the width W_k (defined in Eq. (8.23)). Thus, the SAI of an MR image reconstructed via *Filtering by Fourier transform* is proved at two stages: the filtered projection and the backprojection.

A) The SAI of the filtered projections $t(u',\theta)$. The different combinations of the four factors on the right side of Eq. (8.45) form different k-space data acquisition schemes. Correspondingly, they generate four versions of the filtered projections. Their statistics are discussed below.

A-1) The continuous k-space data over the infinity radial extent, $\mathcal{M}(k,\theta)$, were used; $\tilde{t}(u',\theta)$ is adopted to denote the projection in the view θ

$$\tilde{t}(u',\theta) = \mathcal{F}_k^{-1}\{\mathcal{M}(k,\theta)\}. \tag{8.48}$$

Appendix 8D shows that the correlation $R_{\tilde{t}}((u_1',\theta_1),(u_2',\theta_2))$ of $\tilde{t}(u_1', \theta_1)$ and $\tilde{t}(u_2',\theta_2)$ is given by

$$R_{\tilde{t}}((u_1',\theta_1),(u_2',\theta_2)) = \mu_{\tilde{t}(u_1',\theta_1)}\mu^*_{\tilde{t}(u_2',\theta_2)} + \sigma^2_{\mathcal{M}}\delta(u_1'-u_2')\delta[\theta_1-\theta_2], \tag{8.49}$$

where $\mu_{\tilde{t}(u',\theta)}$ is the mean of $\tilde{t}(u',\theta)$, $*$ denotes the complex conjugate, $\sigma^2_{\mathcal{M}}$ is the variance of $\mathcal{M}(k,\theta)$, and $\delta(u_1'-u_2')$ and $\delta[\theta_1-\theta_2]$ are the Dirac delta and Kronecker delta functions, respectively. From the notations of Eq. (8.46), $\delta[\theta_1 - \theta_2]$ is equivalent to $\delta[m_1 - m_2]$.

A-2) The discrete k-space data over the infinity radial extent, $\mathcal{M}(k,\theta) \cdot \frac{1}{\Delta k} comb(\frac{k}{\Delta k})$, were used; $\hat{t}(u',\theta)$ is adopted to denote the projection in the view θ

$$\hat{t}(u',\theta) = \mathcal{F}_k^{-1}\{\mathcal{M}(k,\theta)\frac{1}{\Delta k} comb(\frac{k}{\Delta k})\} = \tilde{t}(u',\theta) \star comb(\Delta k\ u'), \tag{8.50}$$

where $\star$ denotes the convolution.

Appendix 8D shows that the correlation $R_{\hat{t}}((u_1',\theta_1),(u_2',\theta_2))$ of $\hat{t}(u_1',\ \theta_1)$ and $\hat{t}(u_2',\theta_2)$ is given by

$$R_{\hat{t}}((u_1',\theta_1),(u_2',\theta_2)) = \mu_{\hat{t}(u_1',\theta_1)}\mu^*_{\hat{t}(u_2',\theta_2)} + \frac{\sigma^2_{\mathcal{M}}}{\Delta k}\delta(u_1' - u_2')\delta[\theta_1 - \theta_2], \tag{8.51}$$

where $\mu_{\hat{t}(u',\theta)}$ is the mean of $\hat{t}(u',\theta)$.

A-3) The discrete k-space data over a finite radial extent, $\mathcal{M}(k,\theta)\frac{1}{\Delta k}comb$ $(\frac{k}{\Delta k})rect(\frac{k}{W_k})$, are used; $\bar{t}(u',\theta)$ is adopted to denote the projection in the view θ

$$\bar{t}(u',\theta) = \mathcal{F}_k^{-1}\{\mathcal{M}(k,\theta)\frac{1}{\Delta k} comb(\frac{k}{\Delta k})rect(\frac{k}{W_k})\} = \hat{t}(u',\theta) \star W_k sinc(W_k u')). \tag{8.52}$$

Appendix 8D shows that the correlation $R_{\bar{t}}((u_1',\theta_1),(u_2',\theta_2))$ of $\bar{t}(u_1',\ \theta_1)$ and $\bar{t}(u_2',\theta_2)$ is given by

$$R_{\bar{t}}((u_1',\theta_1),(u_2',\theta_2)) = \mu_{\bar{t}(u_1',\theta_1)}\mu^*_{\bar{t}(u_2',\theta_2)} + \frac{\sigma^2_{\mathcal{M}}}{\Delta k} W_k sinc(W_k(u_1' - u_2'))\delta[\theta_1 - \theta_2], \tag{8.53}$$

where $\mu_{\bar{t}(u',\theta)}$ is the mean of $\bar{t}(u',\theta)$ and $sinc(W_k(u_1' - u_2'))$ is a sinc function.

A-4) The filtered, discrete k-space data over a finite radial extent, $\mathcal{M}(k,\theta)$ $\frac{1}{\Delta k}comb(\frac{k}{\Delta k})|k|rect(\frac{k}{W_k})$, are used; $t(u',\theta)$ is adopted to denote the projection in the view θ

$$t(u',\theta) = \mathcal{F}_k^{-1}\{\mathcal{M}(k,\theta)\frac{1}{\Delta k} comb(\frac{k}{\Delta k})|k|rect(\frac{k}{W_k})\} = \hat{t}(u',\theta) \star \phi(W_k\ u'), \tag{8.54}$$

where

$$\phi(W_k u') = \mathcal{F}_k^{-1}\{|k| \cdot rect\left(\frac{k}{W_k}\right)\} = \frac{W_k^2}{2} sinc(W_k u') - \frac{W_k^2}{4} sinc^2\left(\frac{W_k u'}{2}\right). \tag{8.55}$$

Appendix 8D shows that the correlation $R_t((u_1',\theta_1),(u_2',\theta_2))$ of $t(u_1',\ \theta_1)$ and $t(u_2',\theta_2)$ is given by

$$R_t((u_1',\theta_1),(u_2',\theta_2)) = \mu_{t(u_1',\theta_1)}\mu^*_{t(u_2',\theta_2)} + \frac{\sigma^2_{\mathcal{M}}}{\Delta k} f(W_k(u_1' - u_2'))\delta[\theta_1 - \theta_2], \tag{8.56}$$

where $\mu_{t(u',\theta)}$ is the mean of $t(u', \theta)$, and

$$f(W_k(u'_1 - u'_2)) = \mathcal{F}_k^{-1}\{(|k|rect(\frac{k}{W_k}))^2\}|_{(u'_1-u'_2)}$$
$$= \frac{W_k^3}{2}\left(\frac{1}{2}sinc(W_k(u'_1 - u'_2))\right.$$
$$+ \left.\frac{\cos(\pi W_k(u'_1 - u'_2)) - sinc(W_k(u'_1 - u'_2))}{(\pi W_k(u'_1 - u'_2))^2}\right). \quad (8.57)$$

The second item on the right side of Eq. (8.56)

$$\frac{\sigma_{\mathcal{M}}^2}{\Delta k} f(W_k(u'_1 - u'_2))\delta[\theta_1 - \theta_2] \quad (8.58)$$

is the covariance of $t(u'_1, \theta_1)$ and $t(u'_2, \theta_2)$ and, hence, provides a measure of the correlation between them. For the filtered projections in the different views $\theta_1 \neq \theta_2$, due to $\delta[\theta_1 - \theta_2] = 0$, they are uncorrelated. For the filtered projections in the same view $\theta_1 = \theta_2$, we have the following.

a) When $u'_1 \neq u'_2$, Appendix 8D shows that $|f(W_k(u'_1 - u'_2))|$ of Eq. (8.57) *almost* monotonically decreases as $|u'_1 - u'_2|$ increases from 0 to its maximum. This result indicates that $t(u'_1, \theta_1)$ and $t(u'_2, \theta_2)$ are correlated and the degree of the correlation decreases as the distance between them increases. In the limiting case of $|u'_1 - u'_2| \to \infty$, due to $sinc(W_k(u'_1 - u'_2)) \to 0$ and $\frac{1}{(\pi W_k(u'_1 - u'_2))^2} \to 0$, $|f(W_k(u'_1 - u'_2))| \to 0$; thus, Eq. (8.56) becomes

$$\lim_{|u'_1 - u'_2| \to \infty} R_t((u'_1, \theta_1), (u'_2, \theta_2)) = \mu_{t(u'_1,\theta_1)}\mu^*_{t(u'_2,\theta_2)}, \quad (8.59)$$

which implies that $t(u'_1, \theta_1)$ and $t(u'_2, \theta_2)$ are uncorrelated.

b) When $u'_1 = u'_2$, Eq. (8.58) gives the variance σ_t^2 of the filtered projection $t(u', \theta)$. Appendix 8D shows that

$$\sigma_t^2 = \sigma_{\mathcal{M}}^2 \frac{W_k}{\Delta k}(\frac{1}{12}W_k^2). \quad (8.60)$$

B) SAI of the pixel intensities $x(u, v)$. Appendix 8D shows that after the backprojection over all views, the correlation $R_x((u_1, v_1), (u_2, v_2))$ of the pixel intensities $x(u_1, v_1)$ and $x(u_2, v_2)$ is given by

$$R_x((u_1, v_1), (u_2, v_2))$$
$$= \mu_{x(u_1,v_1)}\mu^*_{x(u_2,v_2)} + \frac{\sigma_{\mathcal{M}}^2}{\Delta k}\int f(W_k \Delta r \cos(\theta - \Delta\phi))d\theta$$
$$= \mu_{x(u_1,v_1)}\mu^*_{x(u_2,v_2)} + \Delta\theta \frac{\sigma_{\mathcal{M}}^2}{\Delta k}\sum_{m=0}^{M-1} f(W_k \Delta r \cos(m\Delta\theta - \Delta\phi)). \quad (8.61)$$

where

$$\Delta r \cos(\theta - \Delta\phi) = \Delta r \cos(m\Delta\theta - \Delta\phi) = u_1' - u_2' , \tag{8.62}$$

and Δr and $\Delta\phi$ are given by

$$\Delta r = \sqrt{(u_1 - u_2)^2 + (v_1 - v_2)^2} \quad \text{and} \quad \Delta\phi = \tan^{-1}(\frac{v_1 - v_2}{u_1 - u_2}). \tag{8.63}$$

Using Eqs. (8.62) and (8.57), $f(W_k \Delta r \cos(m\Delta\theta - \Delta\phi))$ of Eq. (8.61) becomes

$$f(W_k\Delta r\cos(m\Delta\theta - \Delta\phi)) = \frac{W_k^3}{2}\left(\frac{1}{2}sinc(W_k\Delta r\cos(m\Delta\theta - \Delta\phi)\right.$$
$$\left. + \frac{\cos(\pi W_k\Delta r\cos(m\Delta\theta - \Delta\phi)) - sinc(W_k\Delta r\cos(m\Delta\theta - \Delta\phi))}{(\pi W_k\Delta r\cos(m\Delta\theta - \Delta\phi))^2}\right). \tag{8.64}$$

Eqs. (8.62) and (8.63) show that $(u_1' - u_2')$ is the projection of the distance Δr (i.e., the vector $(u_1 - u_2)\vec{i} + (v_1 - v_2)\vec{j}$) onto the U'-axis (i.e., the view direction) specified by the view angle $m\Delta\theta$. Thus, Eq. (8.61) indicates that the correlation $R_x((u_1, v_1), (u_2, v_2))$ between two pixel intensities $x(u_1, v_1)$ and $x(u_2, v_2)$ is determined by the projections $\Delta r \cos(m\Delta\theta - \Delta\phi)$ of the distance $\Delta r = ||(u_1, v_1), (u_2, v_2)||$ on each U'-axis specified by the view angle $m\Delta\theta$.

The second item on the right side of Eq. (8.61)

$$\Delta\theta\frac{\sigma_{\mathcal{M}}^2}{\Delta k}\sum_{m=0}^{M-1} f(W_k\Delta r\cos(m\Delta\theta - \Delta\phi)) \tag{8.65}$$

is the covariance of $x(u_1, v_1)$ and $x(u_2, v_2)$ and hence provides a measure of the correlation between them.

a) When $(u_1, v_1) \neq (u_2, v_2)$, Appendix 8D shows that excluding a few points, $\sum_{m=0}^{M-1} f(W_k \; \Delta r \cos(m\Delta\theta - \Delta\phi))$ of Eq. (8.65) monotonically decreases as Δr increases from 0 to its maximum. This result indicates that $x(u_1, v_1)$ and $x(u_2, v_2)$ are correlated and the degree of the correlation decreases as the distance between them increases. In the limiting case of $\Delta r \to \infty$, due to $sinc(W_k\Delta r \cos(m\Delta\theta - \Delta\phi)) \to 0$ and $\frac{1}{(\pi W_k \Delta r \cos(m\Delta\theta - \Delta\phi))^2} \to 0$ for each $m = 1, \cdots, M-1$, $\sum_{m=0}^{M-1} f(W_k \; \Delta r \cos(m\Delta\theta - \Delta\phi)) \to 0$; thus, Eq. (8.61) becomes

$$\lim_{\Delta r \to \infty} R_x((u_1, v_1), (u_2, v_2)) = \mu_{x(u_1,v_1)}\mu_{x(u_2,v_2)}^*, \tag{8.66}$$

which implies that $x(u_1, v_1)$ and $x(u_2, v_2)$ are uncorrelated.

b) When $(u_1, v_1) = (u_2, v_2)$, Eq. (8.65) gives the variance σ_x^2 of the pixel intensity $x(u, v)$. Appendix 8D shows that σ_x^2 is given by

$$\sigma_x^2 = \sigma_{\mathcal{M}}^2\frac{W_k}{\Delta k}(\pi\frac{1}{12}W_k^2) = \frac{1}{3}\sigma_{\mathcal{M}}^2\frac{W_k}{\Delta k}S_{W_k}, \tag{8.67}$$

where $S_{W_k} = \frac{1}{4}\pi W_k^2$ is the area of a circular region with W_k as its diameter in k-space of the radial sampling.

In the radial k-space sampling as shown in Figure 3.12, let $\frac{W_k}{\Delta k} \simeq N$ be the number of k-space samples in each view. If N is viewed as a *line density*, then $\frac{W_k}{\Delta k} S_{W_k}$ of Eq. (8.67) can be thought as the total number of k-space samples in the circular region specified by the $|k|$ filter. Thus, Eq. (8.67) essentially represents a scaled, total variance of all k-space samples used in the image reconstruction via *Filtering by Fourier transform.* ■

Proof. (for *Filtering by Convolution*)

The principle of *Filtering by Convolution* in PR of MR image is described by Eqs. (3.148) through (3.151). Based on the measured projections, it is a three-step approach:

a) Computing the convolution $t(u', \theta)$ in each view θ by

$$t(u', \theta) = p(u', \theta) \star q(u') , \tag{8.68}$$

where $u' = u\cos\theta + v\sin\theta$ (Eq. (8.47)), $p(u', \theta)$ is the measured projection (Sections 3.9.3.2 and 7.6.3), $q(u')$ is a properly selected convolution function, and $\star$ denotes the convolution.

b) Computing the interpolation $s_\theta(u, v)$ in each view θ by

$$s_\theta(u, v) = t(u', \theta) \star \varphi(u'), \tag{8.69}$$

where $\varphi(u')$ is a properly selected interpolation function.

c) Computing the backprojection $x(u, v)$ over all views by

$$x(u, v) = \int_{m=0}^{\pi} s_\theta(u, v) d\theta = \Delta\theta \sum_{0}^{M-1} s_m(u, v), \tag{8.70}$$

where $M\Delta\theta = \pi$.

In the computation, θ and u are discretized: $\theta = m\Delta\theta$ ($m = 0, 1, \cdots, M-1$), $u' = u\cos(m\Delta\theta) + v\sin(m\Delta\theta)$ and is approximated by $n\Delta p$, Δp is the spacing between two adjacent measured projections $p(n\Delta p, m\Delta\theta)$.

Unlike the progressive proof of SAI for *Filtering by Fourier transform*, the proof of SAI for *Filtering by Convolution* is probabilistical through the above three steps.

1) Let $x(u_i, v_i)$ be the intensity of the pixel at (u_i, v_i) in the image. The correlation $R_x((u_1, v_1), (u_2, v_2))$ between $x(u_1, v_1)$ and $x(u_2, v_2)$ is given by

$$\begin{aligned}
&R_x((u_1, v_1), (u_2, v_2)) \\
&= E[x(u_1, v_1) x^*(u_2, v_2)] \\
&= E[(\sum_{m_1=0}^{M-1} s_{m_1}(u_1, v_1))(\sum_{m_2=0}^{M-1} s^*_{m_2}(u_2, v_2))] \\
&= \sum_{m=0}^{M-1} E[s_m(u_1, v_1) s^*_m(u_2, v_2)] \\
&\quad + \sum_{m_1=0}^{M-1} \sum_{m_2=0(\neq m_1)}^{M-1} E[s_{m_1}(u_1, v_1)] E[s^*_{m_2}(u_2, v_2)].
\end{aligned} \tag{8.71}$$

This is because all measured projections $p(n\Delta p, m\Delta\theta)$ are independent (Section 7.6.3, Property 7.8); therefore, in different views $m\Delta\theta$, the resultant convolved projections $t(n\Delta p, m\Delta\theta)$ in each view, and hence the interpolated data $s_m(u_i, v_i)$ in each view, are uncorrelated.

When the nearest-neighbor interpolation (Section 3.10.2.1) is used, the interpolated data in the m-th view is given by

$$s_m(u_i, v_i) = t(n_{m_i}\Delta p, m\Delta\theta), \tag{8.72}$$

where $n_{m_i} = Arg\ min_n\{|u'_i - n\Delta p|\}$ $(u'_i = u_i\cos(m\Delta\theta) + v_i\sin(m\Delta\theta))$. Thus, the first term on the right side of Eq. (8.71) becomes

$$\sum_{m=0}^{M-1} E[s_m(u_1, v_1)s_m^*(u_2, v_2)]$$
$$= \sum_{m=0}^{M-1} E[t(n_{m_1}\Delta p, m\Delta\theta)t^*(n_{m_2}\Delta p, m\Delta\theta)]. \tag{8.73}$$

Because the convolution function q is bandlimited: $q(n) \simeq 0$ $(|n| > n_0)$, the convolved projections $t(n\Delta p, m\Delta\theta)$ are determined by some (not all) measured projections $p(l\Delta p, m\Delta\theta)$ $(n - n_0 \leq l \leq n + n_0)$ in each view $(m\Delta\theta)$. Thus, when two pixels (u_1, v_1) and (u_2, v_2) are spatially separated sufficiently such that two intervals $[n_{m_1} - n_0, n_{m_1} + n_0]$ and $[n_{m_2} - n_0, n_{m_2} + n_0]$ do not overlap, that is,

$$|n_{m_2} - n_{m_1}| > 2n_0, \tag{8.74}$$

then Eq. (8.73) becomes

$$\sum_{m=0}^{M-1} E[s_m(u_1, v_1)s_m^*(u_2, v_2)]$$
$$= \sum_{m=0}^{M-1} E[t(n_{m_1}\Delta p, m\Delta\theta)]E[t^*(n_{m_2}\Delta p, m\Delta\theta)], \tag{8.75}$$

This is because under the condition (8.74), the measured projections contributed to $t(n_{m_1}\Delta p, m\Delta\theta)$ and those contributed to $t(n_{m_2}\Delta p, m\Delta\theta)$ are no overlap and therefore become uncorrelated. Thus, by substituting Eq. (8.75) into Eq. (8.71), we have

$$R_x((u_1, v_1), (u_2, v_2)) = E[x(u_1, v_1)]E[x^*(u_2, v_2)]. \tag{8.76}$$

2) Due to the use of nearest-neighbor interpolation,

$$u'_i = u_i\cos(m\Delta\theta) + v_i\sin(m\Delta\theta) \simeq n_{m_i}\Delta p \qquad (i = 1, 2), \tag{8.77}$$

we have

$$(n_{m_2} - n_{m_1})\Delta p = \Delta r\cos(m\Delta\theta - \Delta\phi), \tag{8.78}$$

where

$$\Delta r = \sqrt{(u_1 - u_2)^2 + (v_1 - v_2)^2} \ \ and \ \ \Delta\phi = \tan^{-1}\left(\frac{v_1 - v_2}{u_1 - u_2}\right). \qquad (8.79)$$

Thus, through Eqs. (8.74), (8.78), and (8.76), we have

$$\begin{array}{ll} \text{if} & |\Delta r \cos(m\Delta\theta - \Delta\phi)| > 2n_0\Delta p, \\ \text{then} & R_x((u_1, v_1), (u_2, v_2)) = E[x(u_1, v_1)]E[x^*(u_2, v_2)]. \end{array} \qquad (8.80)$$

3) For some values of angles $(m\Delta\theta - \Delta\phi)$, $|\Delta r \cos(m\Delta\theta - \Delta\phi)|$ may not be greater than $2n_0\Delta p$. Because (u_i, v_i) is arbitrary in the image and m uniformly takes a value on $[0, M-1]$, it is reasonable to assume that $(m\Delta\theta - \Delta\phi) \triangleq \vartheta$ is uniformly distributed on $[0, 2\pi]$.

$$p(\vartheta) = \begin{cases} \frac{1}{2\pi} & [0, 2\pi] \\ 0 & \text{elsewhere.} \end{cases} \qquad (8.81)$$

Let γ_0 be the smallest ϑ in $[0, \frac{\pi}{2}]$, that is, $0 < \gamma_0 \leq \vartheta \leq \frac{\pi}{2}$ such that $\Delta r \cos\vartheta < 2n_0\Delta p$. In order to limit the probability of the occurrence of the event $\Delta r \cos\vartheta < 2n_0\Delta p$ (i.e., the probability for two pixel intensities to be correlated) to be a small value $(1 - P_0)$ (say, 0.05) for $0 < \vartheta < 2\pi$, we have to make

$$P(\gamma_0 \leq \vartheta \leq \frac{\pi}{2}) = \int_{\gamma_0}^{\pi/2} p(\vartheta)d\vartheta = \frac{1}{2\pi}(\frac{\pi}{2} - \gamma_0) < \frac{1}{4}(1 - P_0), \qquad (8.82)$$

which leads to $\gamma_0 \geq \frac{\pi}{2}P_0$. Thus, Eq. (8.80) can be stated as

$$\begin{array}{ll} \text{if} & \Delta r > \frac{2n_0\Delta p}{\cos(\frac{\pi}{2}P_0)} \triangleq dr, \\ \text{then} & R_x((u_1, v_1), (u_2, v_2)) = E[x(u_1, v_1)]E[x^*(u_2, v_2)], \end{array} \qquad (8.83)$$

with probability greater than P_0. It is important to note that P_0 is the probability for two pixel intensities to be uncorrelated.

Eq. (8.83) implies that for the fixed parameters (Δp and n_0), when the distance dr (hence Δr) becomes larger, the probability P_0 will be larger, that is, pixel intensities will be more likely to be uncorrelated and hence independent. In the limiting case of $dr \to \infty$, P_0 must approach 1, that is, pixel intensities approach statistically independent with $P_0 = 1$ when the distance between pixels approaches infinity. ■

Eq. (8.83) can be rewritten as $(1 - P_0) = \frac{2}{\pi}\sin^{-1}(\frac{2n_0\Delta p}{dr})$. Figure 8.4 shows the curve of $(1 - P_0)$ versus dr, that is, the probability for two pixel intensities to be correlated versus the distance between these two pixels. It

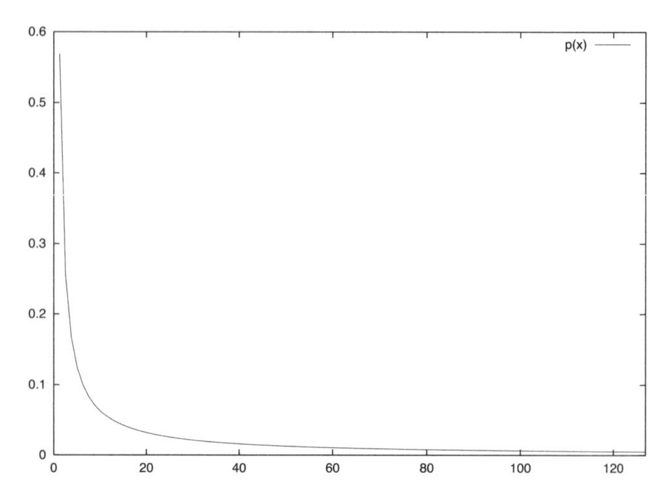

FIGURE 8.3
The SAI in a PR MR image. For given parameters n_0 and Δp, the probability for two pixel intensities to be correlated, $1 - P_0$ (the vertical axis), is a monotonically decreasing function of the distance between the pixels, dr (the horizontal axis): $(1 - P_0) = \frac{2}{\pi}\sin^{-1}(\frac{2n_0\Delta p}{dr})$.

demonstrates that (a) when dr decreases and equals $2n_0\Delta p^{\ddagger}$, $(1 - P_0) = 1$; (b) when dr increases, $(1 - P_0)$ monotonically decreases; and (c) when dr approaches infinity, $(1 - P_0)$ approaches zero, that is, $P_0 \longrightarrow 1$.

8.3.2 Exponential Correlation Coefficient

Property 8.3 The magnitude of the correlation coefficient of pixel intensities of the MR image decreases exponentially with the distance between pixels.
Proof.

This property is proved in the following six (6) steps.

1) Property 8.2 (SAI) implicitly shows that the correlations of pixel intensities of the MR image are spatially shift invariant. This is because the correlations of pixel intensities are determined only by the distances between pixels, not by their locations. Property 8.1 (Gaussianity) shows that the pixel intensities of an MR image have Gaussian distributions. Thus, the pixel intensities of an MR image form a spatial, stationary (in the both wide and strict sense), Gaussian random process (also called a random field).

‡In order to obtain accurate interpolated data $s_m(u, v)$ in each view (m) through the nearest-neighbor interpolation, the spacing Δp between two adjacent convolved projections $t(n\Delta p, m\Delta\theta)$ in each view $(m\Delta\theta)$ must be sufficiently small (see Eq. (8.72)). Assume it be a one-tenth pixel size. n_o is the bandwidth of the convolution function q. For the parallel projection, $n_o \simeq 2$ [18, 49]. Thus, $2n_0\Delta p$ is about a fraction of the pixel size and is called the correlation distance.

Property 8.2 (SAI) explicitly indicates the degree of correlations of pixel intensities of MR image decreases as the distance between pixels increases. This property leads to another property of MR image, Markovianity, which implies that the MR image is embedded in a Markov random field (MRF) with the proper neighborhood system. Markovianity is proved based on SAI and is given in Section 9.3.1. Therefore, pixel intensities of MR image form a spatial, stationary, Gaussian, Markov random process (field).

2) Let $x_{i_1}, \cdots, x_{i_n}$ represent the intensities of n pixels at $i_1, \cdots, i_n$ along a straight line in an MR image§. By defining

$$y_i = \frac{x_i - \mu_{x_i}}{\sigma_{x_i}} \quad (i = i_1, \cdots, i_n), \tag{8.84}$$

where μ_{x_i} and $\sigma^2_{x_i}$ are the mean and variance of x_i, and letting $R_y(i,j)$ and $r_x(i,j)$ be the correlation of y and the correlation coefficient of x, respectively, we have

$$R_y(i,j) = E[y_i y_j^*] = E[\frac{(x_i - \mu_{x_i})(x_j - \mu_{x_j})^*}{\sigma_{x_i}\sigma_{x_j}}] = r_x(i,j). \tag{8.85}$$

Eqs. (8.84) and (8.85) show that y_i is the normalized x_i and the correlation $R_y(i,j)$ of y is the correlation coefficient $r_x(i,j)$ of x. From 1). It is easy to verify that $\{y_{i_1}, \cdots, y_{i_n}\}$ is also a spatial, stationary, Gaussian, Markov random process, but with zero mean.

3) Due to the Gaussianity of y_i and $E[y_i] = 0$, the nonlinear and linear m.s. estimations of y_{i_n} in terms of y_{i_1}, $\cdots$, $y_{i_{n-1}}$ result in the identical solution [20]. That is

$$E[y_{i_n} | y_{i_{n-1}}, \cdots, y_{i_1}] = \sum_{i=i_1}^{i_{n-1}} a_i y_i, \tag{8.86}$$

where coefficients a_i $(i = i_1, \cdots, i_{n-1})$ are determined by the orthogonality principle [20]

$$E[(y_{i_n} - \sum_{i=i_1}^{i_{n-1}} a_i y_i) y_j] = 0 \quad (j = i_1, \cdots, i_{n-1}). \tag{8.87}$$

4) Because $\{y_{i_1}, \cdots, y_{i_n}\}$ is Markov process with the proper order neighborhood systems,¶ say the k-th order, Eqs. (8.86) and (8.87) can be expressed as

$$E[y_{i_n} | y_{i_{n-1}}, \cdots, y_{i_1}] = E[y_{i_n} | y_{i_{n-1}}, \cdots, x_{i_{n-k}}] = \sum_{i=i_{n-k}}^{i_{n-1}} a_i y_i, \tag{8.88}$$

§This straight line can be in different orientations in the image: horizontal, vertical, tilted, and passes through the centers of pixels.

¶Markovianity of the MR image and its proof are given in Chapter 9.

and

$$E[(y_{i_n} - \sum_{i=i_{n-k}}^{i_{n-1}} a_i y_i) y_j] = 0 \qquad (j = i_{n-k}, \cdots, i_{n-1}). \tag{8.89}$$

respectively. Using the vector and matrix forms, Eq. (8.89) becomes

$$\Re_{\mathbf{y}} A = R_{\mathbf{y}}, \tag{8.90}$$

where

$$\Re_{\mathbf{y}} = \begin{pmatrix} R_y(i_{n-k}, i_{n-k}) & R_y(i_{n-k+1}, i_{n-k}) & \cdots\cdots & R_y(i_{n-2}, i_{n-k}) & R_y(i_{n-1}, i_{n-k}) \\ R_y(i_{n-k}, i_{n-k+1}) & R_y(i_{n-k+1}, i_{n-k+1}) & \cdots\cdots & R_y(i_{n-2}, i_{n-k+1}) & R_y(i_{n-1}, i_{n-k+1}) \\ \cdots & \cdots & \cdots\cdots & \cdots & \cdots \\ \cdots & \cdots & \cdots\cdots & \cdots & \cdots \\ R_y(i_{n-k}, i_{n-2}) & R_y(i_{n-k+1}, i_{n-2}) & \cdots\cdots & R_y(i_{n-2}, i_{n-2}) & R_y(i_{n-1}, i_{n-2}) \\ R_y(i_{n-k}, i_{n-1}) & R_y(i_{n-k+1}, i_{n-1}) & \cdots\cdots & R_y(i_{n-2}, i_{n-1}) & R_y(i_{n-1}, i_{n-1}) \end{pmatrix}, \tag{8.91}$$

$$A = \begin{pmatrix} a_{i_{n-k}} \\ a_{i_{n-k+1}} \\ \cdots \\ \cdots \\ a_{i_{n-2}} \\ a_{i_{n-1}} \end{pmatrix}, \tag{8.92}$$

$$R_{\mathbf{y}} = \begin{pmatrix} R_y(i_n, i_{n-k}) \\ R_y(i_n, i_{n-k+1}) \\ \cdots \\ \cdots \\ R_y(i_n, i_{n-2}) \\ R_y(i_n, i_{n-1}) \end{pmatrix}. \tag{8.93}$$

5) For every $i_n > i_{n-1} > \cdots\cdots > i_2 > i_1$, let $i_n - i_{n-1} = s_{i_1}$, $i_{n-1} - i_{n-2} = s_{i_2}$, $\cdots$, $i_{n-k+2} - i_{n-k+1} = s_{i_{k-1}}$, $i_{n-k+1} - i_{n-k} = s_{i_k}$. Due to the stationarity of $\{y_{i_1}, \cdots, y_{i_n}\}$, $\Re_{\mathbf{y}}$ and $R_{\mathbf{y}}$ can be represented as

$$\Re_{\mathbf{y}} = \begin{pmatrix} R_y(0) & R_y(s_{i_k}) & \cdots\cdots \\ R_y(s_{i_k}) & R_y(0) & \cdots\cdots \\ \cdots & \cdots & \cdots\cdots \\ \cdots & \cdots & \cdots\cdots \\ R_y(s_{i_3} + \cdots + s_{i_k}) & R_y(s_{i_3} + \cdots + s_{i_{k-1}}) & \cdots\cdots \\ R_y(s_{i_2} + s_{i_3} + \cdots + s_{i_k}) & R_y(s_{i_2} + s_{i_3} + \cdots + s_{i_{k-1}}) & \cdots\cdots \end{pmatrix}$$

$$\begin{matrix} \cdots\cdots R_y(s_{i_3} + \cdots + s_{i_{k-1}} + s_{i_k}) & R_y(s_{i_2} + \cdots + s_{i_{k-1}} + s_{i_k}) \\ \cdots\cdots R_y(s_{i_3} + \cdots + s_{i_{k-1}}) & R_y(s_{i_2} + \cdots + s_{i_{k-1}}) \\ \cdots\cdots \ \cdots & \cdots \\ \cdots\cdots \ \cdots & \cdots \\ \cdots\cdots R_y(0) & R_y(s_{i_2}) \\ \cdots\cdots R_y(s_{i_2}) & R_y(0) \end{matrix} \Bigg), \tag{8.94}$$

and

$$R_{\mathbf{y}} = \begin{pmatrix} R_y(s_{i_1} + s_{i_2} + \cdots + s_{i_{k-1}} + s_{i_k}) \\ R_y(s_{i_1} + s_{i_2} + \cdots + s_{i_{k-1}}) \\ \cdots \\ \cdots \\ R_y(s_{i_1} + s_{i_2}) \\ R_y(s_{i_1}) \end{pmatrix}, \tag{8.95}$$

respectively. It is easy to verify that $\Re_{\mathbf{y}}^T = \Re_{\mathbf{y}}$ and $R_y(0) = Var[y_i] = 1$.

6) By substituting $\Re_{\mathbf{y}}$ Eq. (8.94) and $R_{\mathbf{y}}$ Eq. (8.95) into Eq. (8.90) and using the mathematical induction method, we obtain the solution for the coefficient vector A

$$A = \begin{pmatrix} a_{i_{n-k}} \\ a_{i_{n-k+1}} \\ \cdots \\ \cdots \\ a_{i_{n-2}} \\ a_{i_{n-1}} \end{pmatrix} = \begin{pmatrix} 0 \\ 0 \\ \cdots \\ \cdots \\ 0 \\ e^{c \cdot s_{i_1}} \end{pmatrix}, \tag{8.96}$$

where c is a constant.

By substituting Eqs. (8.94) through (8.96) into Eq. (8.90), we obtain a meaningful insight into this solution. That is;

$$\begin{pmatrix} R_y(s_{i_2} + s_{i_3} + \cdots + s_{i_{k-1}} + s_{i_k}) \\ R_y(s_{i_2} + s_{i_3} + \cdots + s_{i_{k-1}}) \\ \cdots \\ \cdots \\ R_y(s_{i_2}) \\ 1 \end{pmatrix} \cdot e^{c \cdot s_{i_1}} =$$

$$\begin{pmatrix} R_y(s_{i_1} + s_{i_2} + \cdots + s_{i_{k-1}} + s_{i_k}) \\ R_y(s_{i_1} + s_{i_2} + \cdots + s_{i_{k-1}}) \\ \cdots \\ \cdots \\ R_y(s_{i_1} + s_{i_2}) \\ R_y(s_{i_1}) \end{pmatrix}. \tag{8.97}$$

Because $|R_y(s_{i_1})| \leq 1$ and $s_{i_1} > 0$, c must be negative. Thus, letting $c = -\alpha$ $(\alpha > 0)$, we have

$$R_y(s) = e^{-\alpha \cdot s}. \tag{8.98}$$

■

Remarks on ECC. (i) In general, the coefficient vector A of Eq. (8.96) should be $A = (0,\ 0, \cdots\cdots, 0,\ b^{c \cdot s_{i_1}})^T$ $(b > 0,\ c < 0)$. T denotes the transpose. Because $b^{c \cdot s_{i_1}} = e^{c \cdot s_{i_1} \ln b}$ and $\ln b$ can be merged into the constant c, we use $e^{c \cdot s_{i_1}}$. Because $-1 \leq R_y(s_{i_1}) \leq 1$, we should have $R_y(s_{i_1}) = \pm e^{c \cdot s_{i_1}}$, that is, $|R_y(s_{i_1})| = e^{c \cdot s_{i_1}}$. (ii) In the proof of ECC, the single index i or j are used to represent the location of the pixel x, $s = ||i, j||$ is used to represent the distance between two pixels along a straight line with any orientations in the image. Generally, let (i_1, j_1) and (i_2, j_2) be the locations of two pixels and $m = i_2 - i_1$, $n = j_2 - j_1$, we have $s = s(m,n) = \sqrt{m^2 + n^2}$. (iii) For a given coordinate system, i, j, m, n can be either positive or negative. However, $s(m,n)$ is always positive and symmetric with respect to either m, or n, or both of them. (iv) Eqs. (8.149) and (8.154) of Appendix 8B show that the correlation coefficient is determined by the parameters in data acquisition and image reconstruction. Thus, the constant α is determined by these parameters. Because m and n are the numbers of pixels, α is dimensionalityless. Thus, from Eq. (8.98), we have

$$|r_y(m,n)| = e^{-\alpha\sqrt{m^2+n^2}} \quad (\alpha > 0). \tag{8.99}$$

Experimental results of ECC. For a 2-D random process $\mathbf{x}$, the correlation coefficient is defined by [20–22]

$$r_x(k,l) = \frac{E[(x(i,j) - \mu_{x(i,j)})(x(i+k,j+l) - \mu_{x(i+k,j+l)})^*]}{\sigma_{x(i,j)}\sigma_{x(i+k,j+l)}}, \tag{8.100}$$

where μ_x and σ_x are the mean and the standard deviation of x. $r_x(k,l)$ is estimated by [21, 22]

$$\hat{r}_x(k,l) = \frac{1}{N_i N_j} \sum_{i=1}^{N_i-k} \sum_{j=1}^{N_j-l} \frac{(x(i,j) - \hat{\mu}_{x(i,j)})(x(i+k,j+l) - \hat{\mu}_{x(i+k,j+l)})^*}{\hat{\sigma}_{x(i,j)}\hat{\sigma}_{x(i+k,j+l)}}, \tag{8.101}$$

where $\hat{\mu}_x$, $\hat{\sigma}_x^2$, and $\hat{r}_x(k,l)$ are the sample mean, sample variance, and sample correlation coefficient of x, and N_i and N_j are the numbers of the samples of x at i and j directions. For a real-valued MR image, Eq. (8.101) is simplified as

$$\hat{r}_x(k,l) = \frac{1}{N_{\bar{\text{ß}}}} \sum_{(i,j),(i+k,j+l) \in \bar{\text{ß}}} \frac{(x(i,j) - \hat{\mu}_{x(i,j)})(x(i+k,j+l) - \hat{\mu}_{x(i+k,j+l)})}{\hat{\sigma}_{x(i,j)}\hat{\sigma}_{x(i+k,j+l)}}, \tag{8.102}$$

where $\bar{\text{ß}}$ denotes the nonbackground of the image, $N_{\bar{\text{ß}}}$ is the number of nonbackground pixels in the image, and

$$\begin{aligned} \hat{\mu}_{x(i,j)} &= \tfrac{1}{N_{\mathcal{R}}} \textstyle\sum_{(i,j)\in\mathcal{R}} x(i,j), \\ \hat{\sigma}^2_{x(i,j)} &= \tfrac{1}{N_{\mathcal{R}}} \textstyle\sum_{(i,j)\in\mathcal{R}} (x(i,j) - \hat{\mu}_{x(i,j)})^2, \end{aligned} \tag{8.103}$$

where $\mathcal{R}$ is an image region that the pixel (i, j) belongs to, $N_{\mathcal{R}}$ is the number of pixels in $\mathcal{R}$, $\bar{\text{ß}} = \cup\mathcal{R}$, and $N_{\bar{\text{ß}}} = \sum N_{\mathcal{R}}$.

An example is shown in Figure 8.4 to demonstrate the ECC of an MR image. A gray matter (GM) regional MR image ($N_{\bar{\text{ß}}} = 7806$, $\hat{\mu}_x = 152.24$, $\hat{\sigma}_x^2 = 65.30$) is shown in Figure 8.4a. A 2-D (128×128) surface plot of its sample correlation coefficients $|\hat{r}_x(k,l)|$ calculated using Eq. (8.102) is shown in Figure 8.4b. Because (i) $r_x(k,l)$ is symmetric over k and/or l, and (ii) $\hat{r}_x(k,l) \simeq 0$ ($k > 128$ and/or $l > 128$), we only display $\hat{r}_x(k,l)$ for $0 \leq k, l < 128$. In order to demonstrate the details of the decreasing pattern of $|r_x(k,l)|$, Figure 8.4c shows a 2-D (16×16) surface plot of $|\hat{r}_x(k,l)|$ for $0 \leq k, l < 16$. On the basis of this plot, the contours (from the origin $(0,0)$ outward) give the regions of $|\hat{r}_x(k,l)| < 0.8$, 0.6, 0.4, 0.2, subsequently. Figure 8.4.d shows a curve of $|\hat{r}_x(0,l)|$ ($0 \leq l < 128$), that is, a line of profile. In order to demonstrate the details of the decreasing pattern of $|\hat{r}_x(0,l)|$, Figure 8.4e shows its first 16 values ($0 \leq l < 16$) on the solid line, and also an exponential function $e^{-0.8 \cdot l}$ in the dash line. The plots and curves in Figures 8.4b to 8.4e demonstrate ECC. The absolute value of correlation coefficients $|\hat{r}_x(k,l)|$ damp out very quickly, become very smaller after 10 lags, and are very well fitted by an exponential function.

The surface plots and the curves in Figure 8.5 and Figure 8.6 demonstrate ECC for the regional MR images of white matter (WM) and the cerebrospinal fluid (CSF), respectively. In Figure 8.5, $N_{\bar{\text{ß}}} = 6595$, $\hat{\mu}_x = 132.35$, $\hat{\sigma}_x^2 = 63.43$, and the fitting exponential function is $e^{-1.0 \cdot l}$. In Figure 8.6, $N_{\bar{\text{ß}}} = 2912$, $\hat{\mu}_x = 211.45$, $\hat{\sigma}_x^2 = 464.93$, and the fitting exponential function is $e^{-0.9 \cdot l}$.

The difference between the curve of the sample correlation coefficient and the fitting exponential function in Figure 8.6e (CSF) is greater than those in Figure 8.4e (GM) and Figure 8.5e (WM), this may be caused by (i) the lesser number of samples in the estimation: 2912 pixels (CSF) versus 7806 (GM) and 6595 (WM), and (ii) the larger variance of samples: 464.93 (CSF) versus 65.30 (GM) and 63.43 (WM). It is worth noting that the contours of the equal values of the sample correlation coefficient in Figure 8.5c (WM) are more "directional" than those in Figure 8.4c (GM) and Figure 8.6c (CSF). The reason for this may be that GM and CSF are more homogeneous than WM, which may include white tracks and some white lesions, for example, multiple sclerosis as in this example. However, in all these three cases (GM, WM, and CSF), the exponential fittings are quite good. Figures 8.4 through 8.6 demonstrate that the theoretical and experimental results are in good agreement.

Remarks on SAI and ECC. Although both SAI and ECC characterize the spatial, statistical relationship of pixel intensities of an MR image, some differences between them are worth discussing. Based on the statistics of k-space samples (Property 7.7), SAI is derived using the Gaussianity of pixel intensities only (Property 8.1). It shows that the degree of correlation of pixel intensities monotonically decreases as the distance between pixels increases. It can be seen from Figures 8.10 through 8.11. However, this monotonically

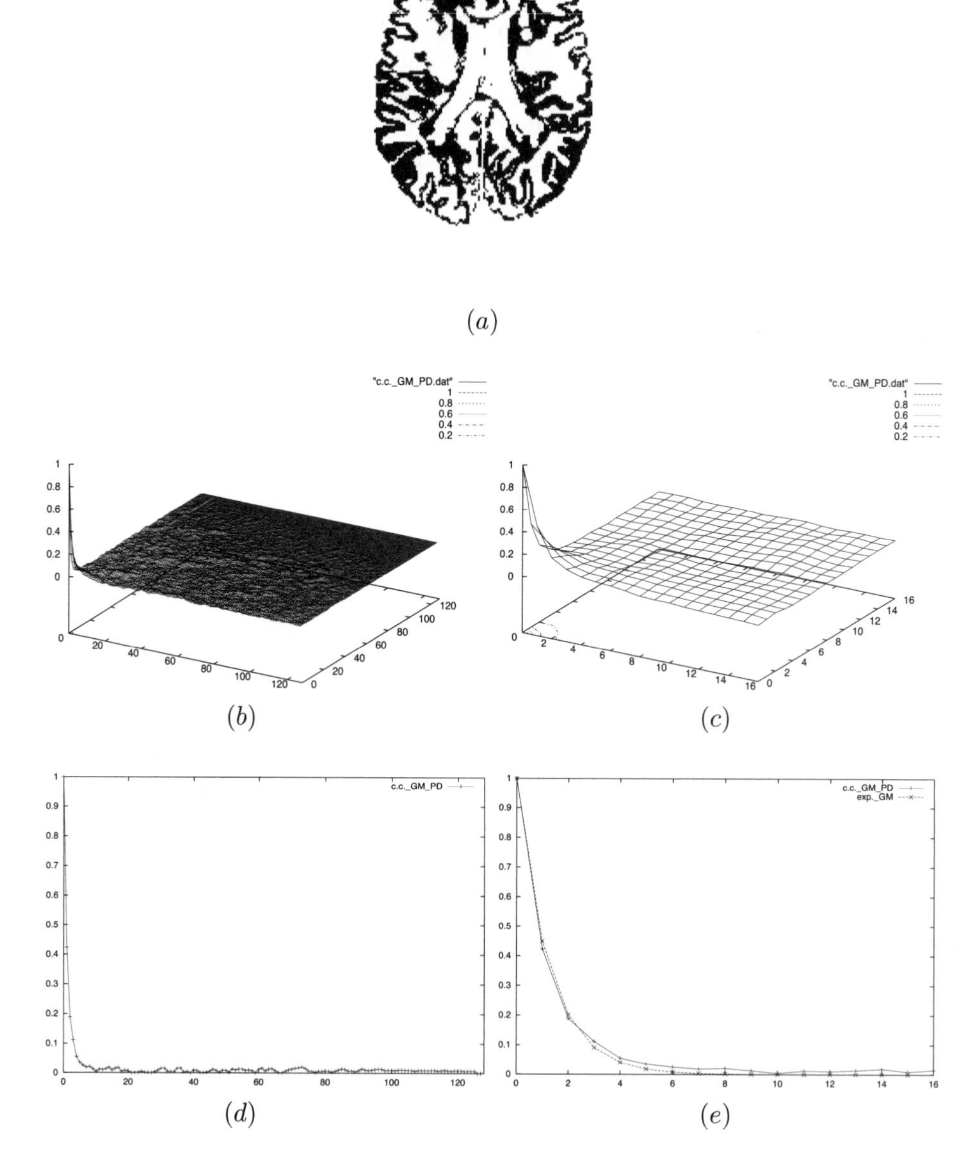

FIGURE 8.4
ECC of MR image: (a) a regional gray matter (GM) image, (b) 2-D plot of $|\hat{r}_x(k,l)|$ $(0 \leq k,l \leq 128)$, (c) 2-D plot of $|\hat{r}_x(k,l)|$ $(0 \leq k,l \leq 16)$, (d) 1-D curve of $|\hat{r}_x(0,l)|$ $(0 \leq l < 128)$, (e) 1-D curves of $|\hat{r}_x(0,l)|$ $(0 \leq l < 16)$ (the solid line) and $e^{-0.8 \cdot l}$ (the dash line).

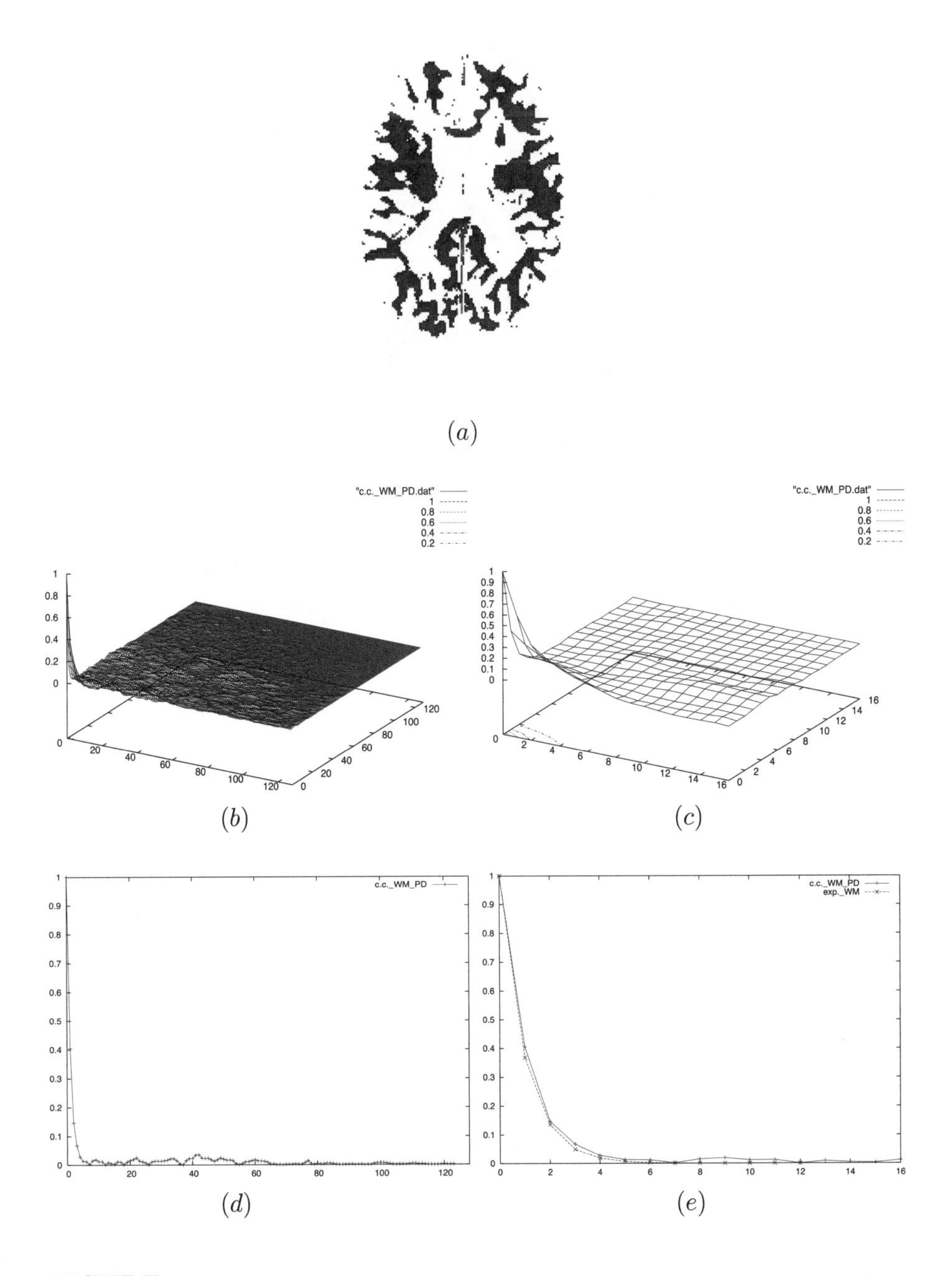

FIGURE 8.5

ECC of MR image: (a) a regional white matter (WM) image, (b) 2-D plot of $|\hat{r}_x(k,l)|$ $(0 \leq k,l \leq 128)$, (c) 2-D plot of $|\hat{r}_x(k,l)|$ $(0 \leq k,l \leq 16)$, (d) 1-D curve of $|\hat{r}_x(0,l)|$ $(0 \leq l < 128)$, (e) 1-D curves of $|\hat{r}_x(0,l)|$ $(0 \leq l < 16)$ (the solid line) and $e^{-1.0 \cdot l}$ (the dash line).

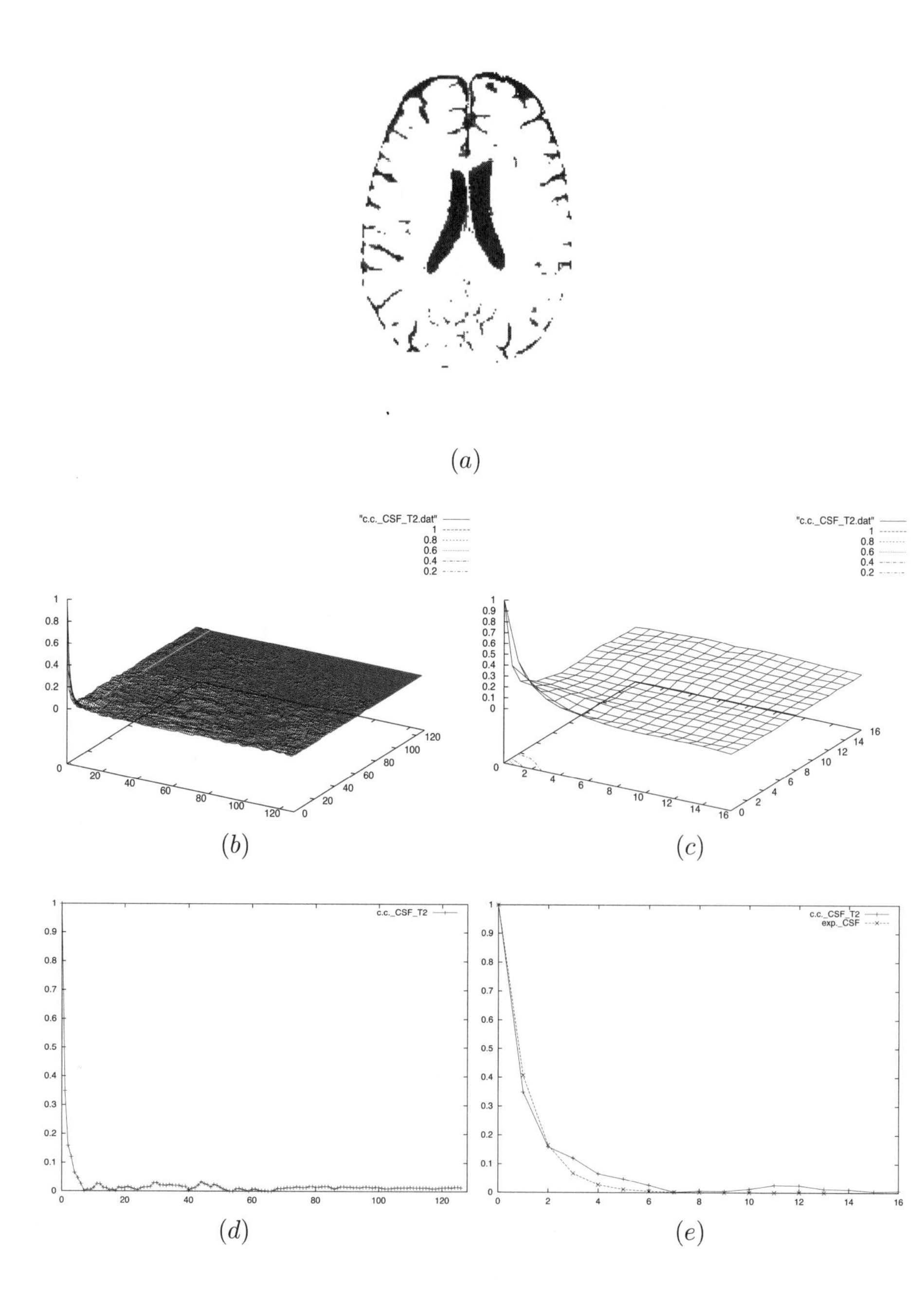

FIGURE 8.6
ECC of MR image: (a) regional cerebrospinal fluid (CSF) image, (b) 2-D plot of $|\hat{r}_x(k,l)|$ $(0 \leq k,l \leq 128)$, (c) 2-D plot of $|\hat{r}_x(k,l)|$ $(0 \leq k,l \leq 16)$, (d) 1-D curve of $|\hat{r}_x(0,l)|$ $(0 \leq l < 128)$, (e) 1-D curves of $|\hat{r}_x(0,l)|$ $(0 \leq l < 16)$ (the solid line) and $e^{-0.9 \cdot l}$ (the dash line).

decreasing pattern is exponential has not been assessed.

As shown in the proof, ECC is derived using Gaussianity (Property 8.1), SAI (Property 8.2, which further leads to Stationarity, Property 8.4), and the Markovianity of an MR image (Property 9.2). That is, more constraints are imposed. As a result, the correlation a coefficient of pixel intensities exponentially decreases with the distance between pixels. This theoretically derived conclusion has been verified by two experimental approaches. The first one directly estimates correlation coefficients using real MR images; the computed sample correlation coefficients are well fitted by exponential functions, which are shown in Figures 8.4 through 8.6. The second approach is to use large amounts of real k-space data to build (by observation) a model for the spectral density function (sdf) and then derive an autocorrelation function (acf) by FT of sdf, which is an exponential function [23]. More details on this second approach are shown in Property 8.6 and its discussion.

8.4 Statistics of the Intensities of a Group of Pixels

By a commonly accepted concept, a region refers to a group of pixels that are connected to each other and their intensities have the same mean and the same variance. An image region is a union of these regions, which may be or not be adjacent to each other. An MR image appears piecewise contiguous. This scenario suggests that each image region may possess some unique statistical properties. This section proves that pixel intensities in each image region form a spatial, stationary and ergodic random process, and hence satisfy ergodic theorems, and then derives the autocorrelation function (acf) and spectral density function (sdf) of pixel intensities in each image region.

8.4.1 Stationarity

A definition of the weak stationary random process is given in [24]: *Let $[X_t; \ , t \in \mathbf{I}]$ be a random process (continuous or discrete time). The random process is said to be weakly stationary (or wide-sense stationary or stationary in the weak (or wide) sense) if the mean*

$$E[X_t] = E[X] \ ; \quad all \ \ t \in \mathbf{I},$$

that is, the left-hand side does not depend on t, and the correlation

$$R_X(t, t+\tau) = R_X(\tau) \ ; \quad all \ \ t, \tau : \ t, t+\tau \in \mathbf{I},$$

that is, the left-hand side does not depend on t. Thus, we have

Property 8.4a Pixel intensities in an image region of an MR image form a spatial, stationary random process in the wide sense.

Proof.

By the concept of the image region, pixel intensities in an image region have the same mean and the same variance, which are constant. From SAI Property 8.2, the correlation of pixel intensities, $R_x((i_1, j_1), (i_2, j_2))$, only depends on the distance between two pixels in the image; that is, it does not depend on their spatial locations in the image. Thus, pixel intensities in each image region form a spatial, stationary random process in the wide sense. ■

A precise definition of the strict stationary random process is also given in [24]. Its simple version can be stated as: *a random process* $[X_t; \ , t \in \mathbf{I}]$ *is said to be strict stationary if the joint cdfs for all possible finite selections of random variables in the random process are unchanged by time shift.* Thus, we have

Property 8.4b Pixel intensities in an image region of an MR image form a spatial, stationary random process in the strict sense.

Proof.

Gaussianity (Property 8.1) shows that pixel intensities in an image region of an MR image form a Gaussian random process with a constant mean and variance. Property 8.4a shows that this random process is weak stationary; that is, the correlations of pixel intensities are unchanged by (pixel) location shift. These parameters (the mean and the correlation (via the variance)) completely describe all joint pdfs and hence cdfs of the Gaussian process. Thus, the resulting joint cdfs are unchanged by (pixel) location shift. Therefore, pixel intensities in each image region of the MR image form a spatial, stationary random process in the strict sense. ■

Remarks on Stationarity. An MR image is a piecewise stationary random field.

8.4.2 Ergodicity

Section 7.5.2 indicates that (i) MR image reconstruction is a transform from a set of $M \times N$ random variables (k-space samples) to another set of $I \times J$ random variables (pixel intensities), and (ii) these $I \times J$ random variables form a random process and the resultant image is a (one) realization of this random process. Image processing and analysis techniques often require estimates of some parameters of this random process (e.g., the mean, variance, etc.) from the given image. Thus, it is necessary to assess if the spatial averages of pixel intensities (e.g., the sample mean, the sample variance, etc.) over a spatial extent of the image (the one realization of the underlying random process) converge to their corresponding ensemble averages, which normally require many realizations of this random process, that is, if these spatial averages satisfy ergodic theorems.

There are several convergences, for example, (i) in the mean square, ($l.i.m.$),

(ii) with probability 1 (*a.s.* - the strong law of large number), (iii) with probability (the weak law of large number), and (iv) in the distribution. Based on the types of convergence, several mean ergodic theorems have been proved under the different conditions and with various applications [24].

However, an ergodic process is not the same as a process that satisfies an ergodic theorem. *A discrete random process* $[X_n; \ , n \in \mathbf{I}]$ *is said to satisfy an ergodic theorem if there exists a random variable* $\hat{X}$ *such that in some sense*

$$\lim_{n\to\infty} \frac{1}{n} \sum_{i=1}^{n} X_i = \hat{X}.$$

The concept of ergodicity has a precise definition for random process application. One such definition is given in [24]: *A random process is said to be ergodic if for any invariant event,* F*, either* $m(F) = 0$ *or* $m(F) = 1$ (here m is the process distribution). *Thus, if an event is "closed" under time shift, then it must have all the probability or none of it.* Because ergodicity is a sufficient condition of the well-known Birkhoff-Khinchin ergodic theorem [24], we first prove that pixels intensities in an image region of MR image form an ergodic process, and then prove that this process satisfies an ergodic theorem.

Property 8.5a Pixel intensities in an image region of MR image form an ergodic process.
Proof.

Stationarity (Property 8.4b) shows that pixel intensities in an image region of an MR image form a stationary random process. SAI (Property 8.2) shows that this process is spatially asymptotically independent. A stationary, asymptotically independent random process is ergodic [24]. ■

Property 8.5b Pixel intensities in an image region of an MR image satisfy ergodic theorem, at least with probability 1.
Proof.

Because the random process formed by pixel intensities in an image region of an MR image is stationary (Property 8.4b) and ergodic (Property 8.5a), this process satisfies Birkhoff-Khinchin ergodic theorems [24]. That is, all spatial averages converge to the corresponding ensemble averages of the process, at least with probability 1. ■

Remarks on ergodicity. An MR image is a piecewise ergodic random field.

8.4.3 Autocorrelaton and Spectral Density

Property 8.6 The autocorrelation function (acf) of pixel intensities in an image region is

$$\gamma_{m,n} = e^{-\alpha\sqrt{m^2+n^2}}, \tag{8.104}$$

where α is a dimensionalityless constant given in Property 8.3. The spectral density function (sdf) of pixel intensities in an image region is

$$p_{k,l} = \frac{\alpha}{(2\pi)^2}(k^2 + l^2 + (\frac{\alpha}{2\pi})^2)^{-\frac{3}{2}}. \tag{8.105}$$

Proof.

Let $\gamma_{m,n} = r_x(m,n)$; here $|r_x(m,n)| = e^{-\alpha\sqrt{m^2+n^2}}$ is given by Eq. (8.99). It is easy to verify that $\gamma_{m,n}$ possesses the following properties:

$$\begin{cases} \gamma_{0,0} = 1 \\ \\ \gamma_{\infty,n} = \gamma_{m,\infty} = \gamma_{\infty,\infty} = 0 \\ \\ |\gamma_{m,n}| \leq 1. \end{cases} \tag{8.106}$$

Thus, for the pixels in an image region that form a stationary random process (Property 8.4), $\gamma_{m,n}$ can serve as the acf of this stationary random process [21, 25].

The sdf of a stationary random process is the Fourier transform of its acf [20, 21, 25]. Let $p_{k,l}$ be this sdf; then it is

$$p_{k,l} = \mathcal{F}_2\{\gamma_{m,n}\} = \int_{-\infty}^{\infty}\int_{-\infty}^{\infty} \gamma_{m,n} e^{-i2\pi(km+ln)} dmdn, \tag{8.107}$$

where $\mathcal{F}_2$ denotes the 2-D Fourier transform, and k, l, m, n are temporally treated as the continuous variables. Using the polar coordinates (r, θ) and (ρ, ϕ)

$$\begin{cases} m = r\cos\theta \\ \\ n = r\sin\theta \end{cases} \quad \text{and} \quad \begin{cases} k = \rho\cos\phi \\ \\ l = \rho\sin\phi, \end{cases} \tag{8.108}$$

and note that $\gamma_{m,n}$ and $p_{k,l}$ are circularly symmetric around the origin, $\gamma_{m,n}$ and $p_{k,l}$ can be rewritten as

$$\gamma_{m,n} = \gamma_{\sqrt{m^2+n^2}} = \gamma_r \quad \text{and} \quad p_{k,l} = p_{\sqrt{k^2+l^2}} = p_\rho. \tag{8.109}$$

Substituting Eq. (8.109) into Eq. (8.107), we have

$$p_\rho = \int_0^\infty r\gamma_r \left(\int_0^{2\pi} e^{-i2\pi\rho r\cos(\theta-\phi)} d\theta\right) dr. \tag{8.110}$$

The inner integral is the Bessel function of the first kind of zero order, that is,

$$\int_0^{2\pi} e^{-i2\pi\rho r\cos(\theta-\phi)} d\theta = 2\pi J_0(2\pi\rho r)\ , \tag{8.111}$$

thus, Eq. (8.110) becomes

$$p_\rho = 2\pi \int_0^\infty r\gamma_r J_0(2\pi\rho r) dr. \tag{8.112}$$

The relation between p_ρ and γ_r in Eq. (8.112) is known as the zero-order Hankel transform [1, 16, 26].

Because $r^2 = m^2 + n^2$, Eq. (8.104) can be written as

$$\gamma_r = e^{-\alpha r} \qquad (r \geq 0). \tag{8.113}$$

Substituting Eq. (8.113) into Eq. (8.112), we have [1, 16, 26]

$$p_\rho = 2\pi \int_0^\infty r e^{-\alpha r} J_0(2\pi\rho r)dr = \frac{1}{2\pi}(\frac{\alpha}{2\pi})(\rho^2 + (\frac{\alpha}{2\pi})^2)^{-\frac{3}{2}}. \tag{8.114}$$

Because $\rho^2 = k^2 + l^2$, Eq. (8.105) is proved. ■

An Experimental verification of Property 8.6. Under the four assumptions ((a)–(d)) on MR k-space data and the two assumptions ((e) and (f)) on MR image:

(a) The signal distribution must be Gaussian (with zero mean),
(b) The noise distribution must be Gaussian (white),
(c) The noise must be additive and uncorrelated to the signal,
(d) The data samples (signal + noise) must be mutually independent,
(e) The image (the Fourier transform of the data) must be stationary,
(f) The images must be ergodic,

and based on the observations of a large number of real experimental MR images, [23] shows a model for the spectral density function of MR image. In the k-space domain, this model is

$$P_s(k_u, k_v) = c \cdot (k_u^2 + k_v^2 + (\frac{\epsilon}{2\pi})^2)^{-3/2}, \tag{8.115}$$

where $p_s(k_u, k_v)$ is an ensemble average signal power at position (k_u, k_v), ϵ is a parameter depending on the texture of the experimental object, and the coefficient c is given by

$$c = CNR^2 \frac{\epsilon n}{2(2\pi)^2} p_n, \tag{8.116}$$

where p_n is the average power of the noise, n is the number of pixels in the image, and CNR is the contrast-to-noise ratio of the image defined by

$$CNR^2 = \frac{\sum_{i,j}(x_{i,j} - \bar{x})^2}{(n-1)\sigma_n^2} - 1 \simeq \frac{\int\int p_s(k_u, k_v)dk_u dk_v}{np_n/2}, \tag{8.117}$$

where $\bar{x}$ is the sample mean of the image and σ_n^2 is the variance of the real (or the imaginary) part of the noise. From Eq. (8.115), [23] shows that autocorrelation function of MR image should be

$$\rho_{m,n} = e^{-\frac{\epsilon}{res}\sqrt{m^2+n^2}}, \tag{8.118}$$

where res denotes the resolution of the image.

By directly comparing Eq. (8.115) with (8.105), Eq. (8.118) with (8.104), we find that they are almost identical. Eq. (8.115) and Eq. (8.105) show $\alpha = \epsilon$. With $\alpha = \epsilon$, Eq. (8.118) and Eq. (8.104) only differ by a scale $\frac{1}{res}$. [23] shows that $\frac{res}{\epsilon}$ represents the average size of the distinct details of objects in the image; ϵ is about 25, a typical value for MR images. ϵ is dimensionless. Property 8.6 shows that α is also a dimensionless constant, determined by the parameters in data acquisition and image reconstruction algorithms.

Eqs. (8.116) and (8.117) show $c = (CNR^2 \frac{np_n}{2}) \frac{\epsilon}{(2\pi)^2} = (\int\int p_s(k_u, k_v) dk_u dk_v) \frac{\epsilon}{(2\pi)^2}$. Because the integration of a signal spectrum is over the entire k-space, $\int\int p_s(k_u, k_v) dk_u dk_v$ is a constant for a given MR image. Let it be denoted by P_k. We have $c = P_k \frac{\epsilon}{(2\pi)^2}$. In Eq. (8.105), the coefficient is $\frac{\alpha}{(2\pi)^2}$. Thus, Eq. (8.115) and Eq. (8.105) only differ by a scale P_k.

We have theoretically derived a formula Eq. (8.104) for the acf in the image domain and further derived a formula Eq. (8.105) for the sdf in the spectrum domain. [23] has experimentally formed a formula Eq. (8.115) for the sdf and further given a formula Eq. (8.118) for the acf. Two completely different approaches produce near-identical results.

Remarks on sdf. The experimental approach of [23] encountered a problem as it noted "*For one particular detail,* Eq. (8.115) (i.e., Eq. (11) in [23]) *does not fit the measurements: in the case of the modulus images, the power of the central point is much larger than predicted by* Eq. (8.115)." Therefore, "*The aim is to determine a value for c which will make the modeled* P_s *as close as possible to the actual spectral power* $P_{s,act}$ *for all data points except for the central point where* $k_u = k_v = 0$." The following explanation may reveal the underlying reason for this problem.

In our approach, the correlation coefficient of pixel intensities, $r_x(m,n)$, is used as the acf that possesses the properties given by Eq. (8.106) [21, 22]. In the approach of [23], the correlation of pixel intensities, $R_x(m,n)$[‖], is used as the acf [2, 27, 28]. For a stationary random process $\mathbf{x}$, $R_x(m,n)$ and $r_x(m,n)$ are related by

$$R_x(m,n) = |\mu_x|^2 + \sigma_x^2 \cdot r_x(m,n), \tag{8.119}$$

where μ_x and σ_x^2 are the mean and variance of this process.

If $\mu_x = 0$, then $R_x(m,n) = \sigma_x^2 \cdot r_x(m,n)$, which leads to

$$\mathcal{F}_2\{R_x(m,n)\} = \sigma_x^2 \cdot \mathcal{F}_2\{r_x(m,n)\}. \tag{8.120}$$

$\mathcal{F}_2\{R_x(m,n)\}$ and $\mathcal{F}_2\{r_x(m,n)\}$ give Eq. (8.115) and Eq. (8.105), respectively, which are almost identical and only different by a scale of σ_x^2.

If $\mu_x \neq 0$, then

$$\mathcal{F}_2\{R_x(m,n)\} = \mathcal{F}_2\{|\mu_x|^2\} + \sigma_x^2 \cdot \mathcal{F}_2\{r_y(m,n)\}. \tag{8.121}$$

[‖] In this case, acf and sdf are related by the Wiener-Khinchin theorem and can be computed by Image-FFT2 [NIH *Image*], a public domain image processing and analysis software package.

$\mathcal{F}_2\{|\mu_x|^2\} = |\mu_x|^2\delta(0,0)$ is the spectrum at the single point $k_u = k_v = 0$, where the Delta function $\delta(0,0)$ is not theoretically defined. However, for a given MR image, the numerical calculation of the spectrum using its k-space data can generate a value for $k_u = k_v = 0$ that is very large. For all k_u and k_v other than $k_u = k_v = 0$, $\mathcal{F}_2\{R_x(m,n)\} = \sigma_x^2 \cdot \mathcal{F}_2\{r_x(m,n)\}$, which is the same as that in the case of $\mu_x = 0$. Thus, the problem reported in [23] is not surprising. It is caused by using $R_x(m,n)$, not by $r_x(m,n)$.

8.5 Discussion and Remarks

8.5.1 Discussion

Although six statistical properties of an MR image are described in a natural order of "a single pixel $\Longrightarrow$ any two pixels $\Longrightarrow$ a group of pixels," the proofs of these properties as shown in this chapter are somewhat interlaced, and is illustrated in Figure 8.7.

The Gaussianity of the pixel intensity of an MR image was addressed in complete fashion: (i) pdfs and the moments of the complex-valued pixel intensity, its real and imaginary parts, its magnitude and phase components, and (ii) the Gaussian approximation of pdfs of the magnitude and phase have been proved.

SAI provides a qualitative measure of the statistical relationship of pixel intensities. Its proof reveals that using finite extent of k-space data for image reconstruction and the filtering for reducing Gibbs ringing artifact introduces and enhances the correlation of pixel intensities in the resultant image. SAI is also referred as strongly mixing [24]. It leads to another important property of MR image, Markovianity, which is discussed in Chapter 9.

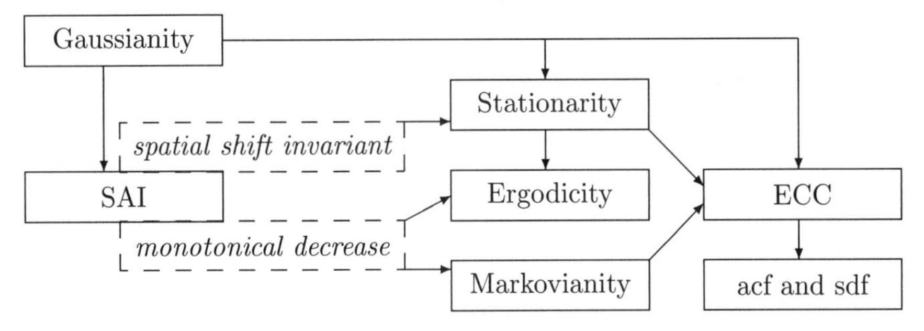

FIGURE 8.7
The interrelationships in proving six statistical properties of an MR image. Markovianity and its proof are given in Chapter 9.

ECC offers a quantitative measure of the statistical relationship of pixel intensities. Its proof actually indicates that autocorrelation of a Gaussian, stationary, Markov random process with zero mean should be exponential. ECC (hence SAI) has been verified by real MR images.

The piecewise contiguous appearance of an MR image is characterized by stationarity, which is sometimes known as spatial shift invariant. Because pixel intensities of an MR image are correlated, the independent, identical distribution (iid) is no longer validated for an MR image, especially for its parameter estimation. Ergodicity provides an alternative way to estimate ensemble averages via spatial averages.

Theoretically derived formulae of the acf and sdf and their experimentally observed counterparts are almost identical. The derivation of the sdf also reveals the underlying reason for a problem encountered in the experimental computation of sdf.

8.5.2 Remarks

A framework of statistics of MR image was presented in this chapter. Six intrinsic properties of MR image were proved and verified by the experimental results obtained using real MR images. These statistics will be used as the basis for establishing stochastic models and developing new image analysis methodologies for MR image, and will be presented in Chapters 9, 10 and 11.

CFT-based analysis is used in the derivations and proofs. It shows that CFT-based analysis (i) gives the identical results as DFT-based analysis, (ii) is simpler than DFT-based analysis, and (iii) leads to some new important outcomes. Six statistical properties are confined for the conventional MR images generated by typical MR data acquisition schemes and image reconstruction methods (the rectilinear k-space sampling/FT and the radial sampling/PR). The strategy and methods used in the derivations and proofs can be extended to the more advanced MR imaging protocols.

8.6 Appendices

8.6.1 Appendix 8A

This Appendix proves Property 8.1a through Property 8.1e and their extensions.

Proof.

1) Proof of Property 8.1a. Signal processing paradigms of Section 7.5.1 show that all operations in the numerical implementation of MR image reconstruction (2DFT and 2DPR) are linear. Let $\mathcal{L}$ represent these linear operations.

By applying $\mathcal{L}$ to k-space samples and using Eqs. (7.49) and (8.1), we have

$$x = x_R + ix_I = \mathcal{L}\{\mathcal{M}\} = \mathcal{L}\{\mathcal{M}_R\} + i\mathcal{L}\{\mathcal{M}_I\}$$

or (8.122)

$$x = x_s + x_n = \mathcal{L}\{\mathcal{M}\} = \mathcal{L}\{\mathcal{M}_s\} + \mathcal{L}\{\mathcal{M}_n\}.$$

Because $\mathcal{L}$ is a linear operator and $\mathcal{M}$ is characterized by a Gaussian random process, the Gaussianity of $x = \mathcal{L}\{\mathcal{M}\}$ can be generally assessed. However, for the purpose of deriving a more precise description of the statistics of single pixel intensity, the following proof is given.

Eq. (8.122) can be rewritten in more detail as

$$(x_{s_R} + x_{n_R}) + i(x_{s_I} + x_{n_I}) = \mathcal{L}\{(\mathcal{M}_{s_R} + \mathcal{M}_{n_R})\} + i\mathcal{L}\{(\mathcal{M}_{s_I} + \mathcal{M}_{n_I})\}$$

or (8.123)

$$(x_{s_R} + ix_{s_I}) + (x_{n_R} + ix_{n_I}) = \mathcal{L}\{(\mathcal{M}_{s_R} + i\mathcal{M}_{s_I})\} + \mathcal{L}\{(\mathcal{M}_{n_R} + i\mathcal{M}_{n_I})\}.$$

Thus, the statistical properties of $\mathcal{M}_{s_R}$, $\mathcal{M}_{n_R}$, $\mathcal{M}_{s_I}$, and $\mathcal{M}_{n_I}$ in the MR imaging domain are preserved in the corresponding x_{s_R}, x_{n_R}, x_{s_I}, and x_{n_I} in the MR image domain. Based on Property 7.7, that is, (i) the signal components, x_{s_R} and x_{s_I}, of the pixel intensity x are characterized by two Gaussian random variables: $x_{s_R} \sim N(\mu_{x_{s_R}}, \sigma^2_{x_{s_R}} \to 0)$ and $x_{s_I} \sim N(\mu_{x_{s_I}}, \sigma^2_{x_{s_I}} \to 0)$; (ii) the noise components, x_{n_R} and x_{n_I}, of the pixel intensity x are characterized by two Gaussian random variables: $x_{n_R} \sim N(0, \sigma^2_{x_{n_R}})$ and $x_{n_I} \sim N(0, \sigma^2_{x_{n_I}})$ with $\sigma^2_{x_{n_R}} = \sigma^2_{x_{n_I}} \stackrel{\Delta}{=} \sigma^2$; and (iii) x_{n_R} and x_{n_I} are stationary, additive, and independent of the corresponding x_{s_R} and x_{s_I}.

Therefore, the real and imaginary parts, $x_R = x_{s_R} + x_{n_R}$ and $x_I = x_{s_I} + x_{n_I}$, of the pixel intensity x (each of them is a sum of two independent Gaussian random variables) are characterized by two Gaussian random variables: $x_R \sim N(\mu_{x_{s_R}}, \sigma^2_{x_{n_R}})$ and $x_I \sim N(\mu_{x_{s_I}}, \sigma^2_{x_{n_I}})$. $\sigma^2_{x_{s_R}} \to 0$ and $\sigma^2_{x_{s_I}} \to 0$ imply that x_{s_R} and x_{s_I} are constant with probability 1, that is, $\mu_{x_{s_R}} = x_{s_R}$ and $\mu_{x_{s_I}} = x_{s_I}$. Also due to $\sigma^2_{x_{n_R}} = \sigma^2_{x_{n_I}} = \sigma^2$, we have $x_R \sim N(x_{s_R}, \sigma^2)$ and $x_I \sim N(x_{s_I}, \sigma^2)$.

Because x_{s_R} and x_{s_I} are constant with probability 1, x_{n_R} and x_{n_I} are independent with zero mean, it is easy to verify that

$$E[x_R x_I] = E[(x_{s_R} + x_{n_R})(x_{s_I} + x_{n_I})] = E[[x_R]E[x_I], \tag{8.124}$$

which implies that x_R and x_I are uncorrelated, and hence independent. Thus, the joint pdf of x_R and x_I is simply the product of the pdfs of x_R and x_I

$$p(x_R, x_I) = p(x_R)p(x_I) = \frac{1}{2\pi\sigma^2} \exp\left(-\frac{(x_R - x_{s_R})^2 + (x_I - x_{s_I})^2}{2\sigma^2}\right). \tag{8.125}$$

The mean and variance of the pixel intensity x are

$$\mu_x = E[x] = E[(x_{s_R} + x_{n_R}) + i(x_{s_I} + x_{n_I})] = x_{s_R} + ix_{s_I},$$
$$\sigma_x^2 = E[(x - \mu_x)(x - \mu_x)^*] = E[(x_{n_R} + ix_{n_I})(x_{n_R} + ix_{n_I})^*] = 2\sigma^2. \tag{8.126}$$

Due to $(x - \mu_x)(x - \mu_x)^* = ((x_R + ix_I) - (x_{s_R} + ix_{s_I}))((x_R + ix_I) - (x_{s_R} + ix_{s_I}))^* = (x_R - x_{s_R})^2 + (x_I - x_{s_I})^2$, Eq. (8.125) can be rewritten as

$$p(x_R, x_I) = \pi^{-1}(\sigma_x^2)^{-1} \exp(-(x - \mu_x)^*(\sigma_x^2)^{-1}(x - \mu_x)), \tag{8.127}$$

which is a standard expression of a 1-D complex Gaussian pdf, $p(x)$, of the random variable $x = x_R + ix_I$ [29, 30].

■

Remarks. A Gaussian distribution of the complex-valued pixel intensity can also be assessed from another point of view. Property 7.7 indicates that k-space samples are independent Gaussian samples with finite 1st and 2nd moments. Section 7.5.1 shows that all operations in MR image reconstruction are linear. The number of k-space samples used in image reconstruction are often very large. Thus, according to the Central Limit theorem [20], the pixel intensity of the resultant complex-valued MR image has an asymptotic Gaussian distribution. This is consistent with the above proof.

2) Proof of Property 8.1b. The complex-valued pixel intensity x can be expressed by a rectangular version $x_R + ix_I$ or a polar version $\rho e^{i\theta}$, that is,

$$x = x_R + ix_I = \rho e^{i\theta}. \tag{8.128}$$

The relationship between (x_R, x_I) and (ρ, θ) is given by

$$x_R = \rho \cos\theta \quad \text{and} \quad x_I = \rho \sin\theta, \tag{8.129}$$

where $\rho > 0$, $-\pi \leq \theta < \pi$. Similarly, the mean of x can also be expressed in a rectangular version or a polar version, that is, $\mu_x = x_{s_R} + ix_{s_I} = \nu e^{i\phi}$. Thus, we have

$$x_{s_R} = \nu \cos\phi \quad \text{and} \quad x_{s_I} = \nu \sin\phi, \tag{8.130}$$

where $\nu > 0$, $-\pi \leq \phi < \pi$.

By substituting Eqs. (8.129) and (8.130) into Eq. (8.125) and noticing that the Jacobi factor $J = |\frac{\partial(x_R, x_I)}{\partial(\rho, \theta)}| = \rho$, we obtain

$$p(\rho, \theta|\phi) = \frac{\rho}{2\pi\sigma^2} \exp\left(-\frac{\rho^2 + \nu^2 - 2\nu\rho\cos(\theta - \phi)}{2\sigma^2}\right). \tag{8.131}$$

■

3) Proof of Property 8.1c. By integrating Eq. (8.4) with respect to θ over $[-\pi, \pi]$, we have

$$p(\rho) = \int_{-\pi}^{\pi} p(\rho, \theta|\phi) d\theta = \frac{\rho}{\sigma^2} \exp\left(-\frac{\rho^2 + \nu^2}{2\sigma^2}\right) \cdot \frac{1}{2\pi} \int_{-\pi}^{\pi} e^{\frac{\nu\rho}{\sigma^2}\cos(\theta - \phi)} d\theta. \tag{8.132}$$

By substituting Eq. (8.7) into Eq. (8.132), we obtain Eq. (8.6). Furthermore, the moments of ρ are

$$\begin{aligned}
E[\rho^n] &= \int_0^{\infty} \rho^n p(\rho) d\rho = \frac{1}{\sigma^2} \int_0^{\infty} \rho^{n+1} \exp\left(-\frac{\rho^2 + \nu^2}{2\sigma^2}\right) I_0\left(\frac{\rho\nu}{\sigma^2}\right) d\rho \\
&= \frac{1}{\sigma^2} \int_0^{\infty} \rho^{n+1} \exp\left(-\frac{\rho^2 + \nu^2}{2\sigma^2}\right) \sum_{k=0}^{\infty} \frac{(\nu\rho/\sigma^2)^{2k}}{2^{2k}(k!)^2} d\rho \\
&= \frac{1}{\sigma^2} \exp\left(-\frac{\nu^2}{2\sigma^2}\right) \sum_{k=0}^{\infty} \frac{(\nu)^{2k}}{(2\sigma^2)^{2k}(k!)^2} \int_0^{\infty} \rho^{2k+n+1} \exp\left(-\frac{\rho^2}{2\sigma^2}\right) d\rho \\
&= (2\sigma^2)^{\frac{n}{2}} \exp\left(-\frac{\nu^2}{2\sigma^2}\right) \sum_{k=0}^{\infty} \frac{1}{(k!)^2} \left(\frac{\nu^2}{2\sigma^2}\right)^k \Gamma\left(k + \frac{n}{2} + 1\right) \\
&= (2\sigma^2)^{\frac{n}{2}} \exp\left(-\frac{\nu^2}{2\sigma^2}\right) (1 + (\frac{n}{2} + 1)(\frac{\nu^2}{2\sigma^2}) \\
&\quad + \frac{1}{(2!)^2} \left(\frac{\nu^2}{2\sigma^2}\right)^2 \left(\frac{n}{2} + 2\right) \left(\frac{n}{2} + 1\right) + \cdots) \Gamma\left(1 + \frac{n}{2}\right) \\
&= (2\sigma^2)^{\frac{n}{2}} \exp\left(-\frac{\nu^2}{2\sigma^2}\right) \Gamma\left(1 + \frac{n}{2}\right) {}_1F_1\left(\frac{n}{2} + 1; 1; \frac{\nu^2}{2\sigma^2}\right) \\
&= (2\sigma^2)^{\frac{n}{2}} \Gamma(1 + \frac{n}{2}) \, {}_1F_1\left(-\frac{n}{2}; 1; -\frac{\nu^2}{2\sigma^2}\right),
\end{aligned} \tag{8.133}$$

where $I_0(x)$ is the modified Bessel function of the first kind of zero order given by Eq. (8.7), $\Gamma(x)$ is the Gamma function given by Eq. (8.9), and ${}_1F_1(a; b; x)$ is the confluent hypergeometric function given by Eq. (8.10) [1, 2, 30]. The last step in Eq. (8.133) is based on ${}_1F_1(a; b; -x) = {}_1F_1(1 - a; b; x)$ [31, 32].

■

Proof of the extension of Property 8.1c. The zero SNR $\varrho = 0$ with the finite-valued variance σ^2 implies $\nu = 0$. Thus, $I_0(\frac{\nu\rho}{\sigma^2}) = 1$ [2] and Eq. (8.6) becomes Eq. (8.11). In this limiting case, the moments of ρ is

$$E[\rho^n] = \frac{1}{\sigma^2} \int_0^{\infty} \rho^{n+1} \exp\left(-\frac{\rho^2}{2\sigma^2}\right) d\rho = (2\sigma^2)^{\frac{n}{2}} \Gamma\left(1 + \frac{n}{2}\right). \tag{8.134}$$

It is known [2, 33, 34] that $I_0(x) \simeq \frac{e^x}{\sqrt{2\pi x}}$ $(x \gg 1)$. When SNR ϱ is very large, $I_0(\frac{\nu\rho}{\sigma^2}) \simeq \frac{\exp(\nu\rho/\sigma^2)}{\sqrt{2\pi\nu\rho/\sigma^2}}$. Substituting this approximation into Eq. (8.6) yields

$$p(\rho) = \gamma_n \sqrt{\frac{\rho}{2\pi\nu\sigma^2}} \exp\left(-\frac{(\rho-\nu)^2}{2\sigma^2}\right), \qquad (8.135)$$

where γ_n is a constant to ensure that $\int_0^\infty p(\rho)d\rho = 1$. $p(\rho)$ of Eq. (8.135) has a maximum value of $\frac{1}{\sqrt{2\pi\sigma^2}}$ at $\rho = \nu$. In the region around ν, $p(\rho)$ changes very slowly; and in the region $|\rho - \nu| >> 0$, the exponential factor $\exp(-\frac{(\rho-\nu)^2}{2\sigma^2})$ is dominant. Thus, by replacing ρ with ν in the first factor of Eq. (8.135) and normalizing Eq. (8.135) by $\gamma_n = 1$, we obtain Eq. (8.13). In this limiting case, the central moments of ρ is given by

$$E[(\rho-\nu)^n] = \frac{1}{\sqrt{2\pi}\sigma} \int_0^\infty (\rho-\nu)^n \exp\left(-\frac{(\rho-\nu)^2}{2\sigma^2}\right) d\rho. \qquad (8.136)$$

The SNR $\varrho \to \infty$ is caused by very large ν, or very small σ, or both, in all three situations; the integral $\int_0^\infty$ can be approximated by $\int_{-\infty}^\infty$. Thus, Eq. (8.136) can be rewritten as

$$E[(\rho-\nu)^n] \simeq \frac{1}{\sqrt{2\pi}\sigma} \int_{-\infty}^\infty (\rho-\nu)^n \exp\left(-\frac{(\rho-\nu)^2}{2\sigma^2}\right) d\rho. \qquad (8.137)$$

It is easy to verify that $E[(\rho-\nu)^{2k-1}] = 0$ and $E[(\rho-\nu)^{2k}] = 1\cdot 3\cdot(2k-1)\sigma^2$ for $k = 1, 2, 3, \cdots$.

It has been shown [30, 35, 36] that when SNR $\varrho \geq 2\sqrt{2}$, Eq. (8.6) is well approximated by Eq. (8.13). The quantitative evaluation and curve illustration of Eqs. (8.6), (8.11), and (8.13) can be found in [11, 35].

■

4) Proof of Property 8.1d. By integrating Eq. (8.4) with respect to ρ over $(0, \infty)$, we have

$$\begin{aligned} p(\delta\theta) &= \int_0^\infty p(\rho, \theta|\phi)d\rho \\ &= \frac{1}{2\pi\sigma^2} \int_0^\infty \rho \exp\left(-\frac{\rho^2 + \nu^2 - 2\nu\rho\cos\delta\theta}{2\sigma^2}\right) d\rho \\ &= \frac{1}{2\pi\sigma^2} \exp\left(-\frac{\nu^2 \sin^2\delta\theta}{2\sigma^2}\right) \int_0^\infty \rho \exp\left(-\frac{(\rho - \nu\cos\delta\theta)^2}{2\sigma^2}\right) d\rho \\ &= \frac{1}{2\pi\sigma^2} \exp\left(-\frac{\nu^2 \sin^2\delta\theta}{2\sigma^2}\right) \int_{-\frac{\nu\cos\delta\theta}{\sigma}}^\infty \sigma(\sigma t + \nu\cos\delta\theta) \exp\left(-\frac{t^2}{2}\right) dt \\ &= \frac{1}{2\pi} \exp(-\frac{\nu^2}{2\sigma^2}) + \frac{\nu\cos\delta\theta}{\sqrt{2\pi}\sigma} \exp\left(-\frac{\nu^2\sin^2\delta\theta}{2\sigma^2}\right) \Phi\left(\frac{\nu\cos\delta\theta}{\sigma}\right), \end{aligned} \qquad (8.138)$$

which is Eq. (8.15), and $\Phi(\frac{\nu\cos\delta\theta}{\sigma})$ is cdf of $N(0,1)$ given by Eq. (8.16).

■

Proof of the extension of Property 8.1d. The zero SNR $\varrho = 0$ with the finite-valued variance σ^2 implies that $\nu = 0$. Thus, $p(\delta\theta)$ of Eq. (8.15) equals $\frac{1}{2\pi}$, which gives Eq. (8.17). In this limiting case, it is easy to verify Eq. (8.18).

When SNR ϱ is very large, $\delta\theta$ is very small, which leads to $\sin\delta\theta \simeq \delta\theta$, $\cos\delta\theta \simeq 1$, $\Phi(\frac{\nu\cos\delta}{\sigma}) \simeq 1$; $\exp(-\frac{\nu^2}{2\sigma^2}) \simeq 0$. Thus, Eq. (8.15) becomes Eq. (8.19). Similar to the proof of Eq. (8.14), it is easy to verify Eq. (8.20).

It has been shown [35] that when SNR $\varrho \geq 1$, Eq. (8.15) is well approximated by Eq. (8.19). The quantitative evaluation and curve illustration of Eqs. (8.15), (8.17), and (8.19) can be found in [11, 35].

■

5) Proof of Property 8.1e. Eqs. (8.13) and (8.19) show that when SNR $\varrho \geq 2\sqrt{2}$,

$$
\begin{aligned}
p(\rho)p(\delta\theta) &\simeq \frac{\nu}{2\pi\sigma^2}\exp\left(-\frac{(\rho-\nu)^2+\nu^2(\delta\theta)^2}{2\pi\sigma^2}\right) \\
&= \frac{\nu}{2\pi\sigma^2}\exp\left(-\frac{\rho^2-2\nu\rho(1-\frac{\nu}{2\rho}(\delta\theta)^2)+\nu^2}{2\pi\sigma^2}\right).
\end{aligned} \tag{8.139}
$$

Similar to the approximations used in the proof of Eq. (8.13) (that is, in the region around $\rho = \nu$, Eq. (8.139) changes very slowly, and in the region $|\rho - \nu| >> 0$, the exponential is a dominated factor of Eq. (8.139)), ν in the first factor of Eq. (8.139) is replaced by ρ, and the fraction $\frac{\nu}{2\rho}$ in the numerator of the exponential factor of Eq. (8.139) becomes $\frac{1}{2}$, which leads to $1-\frac{\nu}{2\rho}(\delta\theta)^2 \simeq 1-\frac{1}{2}(\delta\theta)^2 \simeq \cos(\delta\theta)$ (using the first two items of a Taylor series expansion of $\cos(\delta\theta)$). Thus, Eq. (8.139) becomes

$$
p(\rho)p(\delta\theta) \simeq \frac{\rho}{2\pi\sigma^2}\exp\left(-\frac{\rho^2+\nu^2-2\nu\rho\cos(\delta\theta)}{2\pi\sigma^2}\right). \tag{8.140}
$$

With $\delta\theta = \theta - \phi$, Eq. (8.140) is identical to Eq. (8.4). Thus, we obtain $p(\rho)p(\theta|\phi) \simeq p(\rho,\theta|\phi)$, that is, when SNR ϱ is moderate or large (as described in the Extensions of Properties 8.1c and 8.1d), the magnitude ρ and the phase θ of the pixel intensity x of a complex-valued MR image are approximately statistically independent.

■

8.6.2 Appendix 8B

This appendix proves SAI (Property 8.2) for FT MR image.

Proof.

1) Proof of Eq. (8.25) in **Case 1**). Let $R_{\mathcal{M}}(k_1, k_2)$ be the correlation between

k-space samples $\mathcal{M}(k_1)$ and $\mathcal{M}(k_2)$; then from Eq. (7.50), we have

$$R_{\mathcal{M}}(k_1, k_2) = \mu_{\mathcal{M}(k_1)}\mu^*_{\mathcal{M}(k_2)} + \sigma^2_{\mathcal{M}}\delta[k_1 - k_2], \tag{8.141}$$

where $\mu_{\mathcal{M}(k)}$ and $\sigma^2_{\mathcal{M}}$ are the mean and the variance of $\mathcal{M}(k)$, and $\delta[x]$ is the Kronecker Delta function [6, 8, 16]. Thus, from Eq. (8.24) and using Eq. (8.141), the correlation of $\tilde{x}_1$ and $\tilde{x}_2$ is

$$\begin{aligned}
R_{\tilde{x}}(u_1, u_2) &= E[\tilde{x}(u_1)\tilde{x}^*(u_2)] \\
&= E[\int_{-\infty}^{\infty} \mathcal{M}(k_1)e^{i2\pi k_1 u_1}dk_1 \int_{-\infty}^{\infty} \mathcal{M}^*(k_2)e^{-i2\pi k_2 u_2}dk_2] \\
&= \int_{-\infty}^{\infty}\int_{-\infty}^{\infty} E[\mathcal{M}(k_1)\mathcal{M}^*(k_2)]e^{i2\pi(k_1u_1 - k_2u_2)}dk_1dk_2 \\
&= \int_{-\infty}^{\infty}\int_{-\infty}^{\infty} \mu_{\mathcal{M}(k_1)}\mu^*_{\mathcal{M}(k_2)}e^{i2\pi(k_1u_1 - k_2u_2)}dk_1dk_2 \\
&\quad +\sigma^2_{\mathcal{M}}\int_{-\infty}^{\infty}\int_{-\infty}^{\infty} \delta[k_1 - k_2]e^{i2\pi(k_1u_1 - k_2u_2)}dk_1dk_2 \\
&= \mu_{\tilde{x}(u_1)}\mu^*_{\tilde{x}(u_2)} + \sigma^2_{\mathcal{M}}\delta(u_1 - u_2),
\end{aligned} \tag{8.142}$$

where $\delta(x)$ is the Dirac Delta function. In Eq. (8.142), $\int_{-\infty}^{\infty} e^{i2\pi x(x_1 - x_2)}dx = \delta(x_1 - x_2)$ is used [2, 12, 27].

2) Proof of Eq. (8.29) in **Case 2**). From Eq. (8.28) and using Eq. (8.23), we have

$$\hat{x}(u) = \frac{1}{\Delta k}\sum_{m=-\infty}^{\infty} \tilde{x}\left(\frac{m}{\Delta k}\right)\delta\left(u - \frac{m}{\Delta k}\right). \tag{8.143}$$

From Eq. (8.143) and using Eq. (8.25), the correlation of $\hat{x}_1$ and $\hat{x}_2$ is

$$\begin{aligned}
R_{\hat{x}}(u_1, u_2) &= E[\hat{x}(u_1)\hat{x}^*(u_2)] \\
&= \frac{1}{(\Delta k)^2}\sum_{m=-\infty}^{\infty}\sum_{n=-\infty}^{\infty} E[\tilde{x}\left(\frac{m}{\Delta k}\right)\tilde{x}^*\left(\frac{n}{\Delta k}\right)]\delta\left(u_1 - \frac{m}{\Delta k}\right)\delta\left(u_2 - \frac{n}{\Delta k}\right) \\
&= \frac{1}{(\Delta k)^2}\sum_{m=-\infty}^{\infty}\sum_{n=-\infty}^{\infty} \mu_{\tilde{x}(\frac{m}{\Delta k})}\mu^*_{\tilde{x}(\frac{n}{\Delta k})}\delta\left(u_1 - \frac{m}{\Delta k}\right)\delta\left(u_2 - \frac{n}{\Delta k}\right) \\
&\quad +\frac{\sigma^2_{\mathcal{M}}}{(\Delta k)^2}\sum_{m=-\infty}^{\infty}\sum_{n=-\infty}^{\infty} \delta\left(\frac{m}{\Delta k} - \frac{n}{\Delta k}\right)\delta\left(u_1 - \frac{m}{\Delta k}\right)\delta\left(u_2 - \frac{n}{\Delta k}\right) \\
&= \mu_{\hat{x}(u_1)}\mu^*_{\hat{x}(u_2)} + \frac{\sigma^2_{\mathcal{M}}}{\Delta k}\delta(u_1 - u_2)
\end{aligned} \tag{8.144}$$

In Eq. (8.144), $\delta(ax) = \frac{1}{|a|}\delta(x)$ and $\int_{-\infty}^{+\infty} \delta(x - x_1)\delta(x - x_2)dx = \delta(x_1 - x_2)$ are used [2, 12, 27].

3) Proof of Eq. Eq. (8.33) in **Case 3**), that is, SAI for the basic FT MR image. From Eq. (8.32) and using Eq. (8.29), the correlation of $\bar{x}(u_1)$ and

$\bar{x}(u_2)$ is

$$
\begin{aligned}
&R_{\bar{x}}(u_1, u_2) = E[\bar{x}(u_1)\bar{x}^*(u_2)] \\
&= E[\int_{-\infty}^{\infty} \hat{x}(v_1) W sinc(W(u_1 - v_1)) dv_1 \int_{-\infty}^{\infty} \hat{x}^*(v_2)] W sinc(W(u_2 - v_2)) dv_2] \\
&= \int_{-\infty}^{\infty}\int_{-\infty}^{\infty} E[\hat{x}(v_1)\hat{x}^*(v_2)] W sinc(W(u_1 - v_1)) W sinc(W(u_2 - v_2)) dv_1 dv_2 \\
&= \int_{-\infty}^{\infty}\int_{-\infty}^{\infty} \mu_{\hat{x}(v_1)}\mu^*_{\hat{x}(v_2)} W sinc(W(u_1 - v_1)) W sinc(W(u_2 - v_2)) dv_1 dv_2 \\
&\quad + \frac{\sigma^2_{\mathcal{M}}}{\Delta k}\int_{-\infty}^{\infty}\int_{-\infty}^{\infty} \delta(v_1 - v_2) W sinc(W(u_1 - v_1)) W sinc(W(u_2 - v_2)) dv_1 dv_2 \\
&= \mu_{\bar{x}(u_1)}\mu^*_{\bar{x}(u_2)} + \frac{\sigma^2_{\mathcal{M}}}{\Delta k} W^2 \int_{-\infty}^{\infty} sinc(W(u_1 - v_2)) sinc(W(u_2 - v_2)) dv_2 \\
&= \mu_{\bar{x}(u_1)}\mu^*_{\bar{x}(u_2)} + \frac{\sigma^2_{\mathcal{M}}}{\Delta k} W sinc(W(u_1 - u_2)).
\end{aligned} \tag{8.145}
$$

In Eq. (8.145), $sinc(x) = sinc(-x)$ and $sinc(x) \star sinc(x) = sinc(x)$ are used [2, 12, 27]. Eq. (8.145) is identical to Eq. (8.33).

In Eq. (8.145), when $\frac{\sigma^2_{\mathcal{M}}}{\Delta k} W sinc(W(u_1 - u_2)) = 0$, $R_{\bar{x}} = \mu_{\bar{x}(u_1)}\mu^*_{\bar{x}(u_2)}$, $\bar{x}(u_1)$ and $\bar{x}(u_2)$ become uncorrelated. Thus, the second term on the right side of Eq. (8.145) provides a measure of the correlation of $\bar{x}(u_1)$ and $\bar{x}(u_2)$, which depends on the imaging parameters ($\sigma^2_{\mathcal{M}}$, Δk, W) and is a function of the distance $(u_1 - u_2)$ between two pixels. When $u_1 = u_2$, it gives the variance of the pixel intensity $\sigma^2_{\bar{x}} = Var[\bar{x}(u)] = \frac{\sigma^2_{\mathcal{M}}}{\Delta k} W$.

The following discussion shows the monotonically decreasing magnitude of the second term on the right side of Eq. (8.145). In rectilinear k-space sampling and FT reconstruction, when N k-space samples are acquired and N pixels are reconstructed, a necessary condition for choosing the window width W is $(N-1)\Delta k < W < N\Delta k$, which can be expressed as

$$
W = (N - \epsilon)\Delta k \quad (0 < \epsilon < 1)\,. \tag{8.146}
$$

This condition ensures that the exact (i.e., no more or no less than) N samples are acquired. In this case, the pixel size** is $\delta u = \frac{FOV}{N} = \frac{1}{N\Delta k}$ and the distance between two pixels at u_1 and u_2 is $\Delta u = u_1 - u_2 = m\delta u$, ($|m| \leq (N-1)$). Thus, we have

$$
W\Delta u = \frac{N - \epsilon}{N} m;. \tag{8.147}
$$

**This pixel size is known as the appealing pixel size [7] or the Fourier pixel size [9].

By substituting Eqs. (8.146) and (8.147) into the second term on the right side of Eq. (8.145), its magnitude is

$$c_b(m) = \frac{\sigma^2_{\mathcal{M}}}{\Delta k} W |sinc(\frac{N-\epsilon}{N} m)| = \begin{cases} \frac{\sigma^2_{\mathcal{M}}}{\Delta k} W & (m = 0) \\ \\ \epsilon \sigma^2_{\mathcal{M}} |sinc(\frac{\epsilon}{N} m)| & (m \neq 0). \end{cases} \tag{8.148}$$

For simplicity, the normalized $c_b(m)$ is used, which is given by

$$c_{bn}(m) = (\frac{\sigma^2_{\mathcal{M}}}{\Delta k} W)^{-1} c_b(m) = \begin{cases} 1 & (m = 0) \\ \\ \frac{\epsilon}{N-\epsilon} |sinc(\frac{\epsilon}{N} m)| & (m \neq 0). \end{cases} \tag{8.149}$$

In Eq. (8.149), when $|m|$ varies from 0 to $(N-1)$, due to $|\frac{\epsilon}{N} m| < 1$, $sinc(\frac{\epsilon}{N} m)$ is confined in the main lobe of $sinc(\frac{\epsilon}{N} m)$, where $|sinc(\frac{\epsilon}{N} m)|$ monotonically decreases from 1 to 0 (approaching but not equalizing) as $|m|$ increases from 0 to $(N-1)$.

An example is used to illustrate $c_{bn}(m)$ in Eq. (8.149). In this example, the number of k-space samples and the number of pixels are the same: $N = 256$, and $\epsilon = 0.9,\ 0.09,\ 0.009$. Thus, the window width is $255.1\Delta k$ (very close to $(N-1)\Delta k$), $255.91\Delta k$ (between $(N-1)\Delta k$ and $N\Delta k$), and $W = 255.991\Delta k$ (very close to $N\Delta k$). With these settings, the curves of $c_{bn}(m)$ are shown in Figure 8.8. The curves in Figure 8.8a show that $c_{bn}(m)$ sharply decreases from 1 to the almost zero as $|m|$ increases, starting from 0. In Figure 8.8b, only the first 16 values of $c_{bn}(m)$ $(0 \leq m < 16)$ are shown to clearly demonstrate the decreasing pattern of $c_{bn}(m)$; there is virtually no correlation for $|m| > 1$. In Figure 8.8c, the $c_{bn}(m)$ axis uses the log scale to clearly show the values of $c_{bn}(m)$ $(0 \leq m < 16)$ for the various window widths W. Correlations remain for $|m| > 1$ even though they are very small. The curves in Figures 8.8a through 8.8c demonstrate SAI in the basic FT MR image. The impact of the window width W on SAI is not essential. From Eqs. (8.145) and (8.148) and noticing that $\frac{\sigma^2_{\mathcal{M}}}{\Delta k} W$ is the variance of $\bar{x}$, $c_{bn}(m)$ of Eq. (8.149) is actually the magnitude of the correlation coefficient of intensities of two pixels separated by m pixels.

4) Proof of Eq. (8.39) in **Case 4**), that is, SAI for the filtered FT MR image. From Eq. (8.37) and using Eq. (8.29), the correlation of $x(u_1)$ and $x(u_2)$ is

$$\begin{aligned} R_x(u_1, u_2) &= E[x(u_1) x^*(u_2)] \\ &= E[\int_{-\infty}^{\infty} \hat{x}(v_1) \phi(W(u_1 - v_1)) dv_1 \int_{-\infty}^{\infty} \hat{x}^*(v_2)] \phi(W(u_2 - v_2)) dv_2] \\ &= \int_{-\infty}^{\infty} \int_{-\infty}^{\infty} E[\hat{x}(v_1) \hat{x}^*(v_2)] \phi(W(u_1 - v_1)) \phi(W(u_2 - v_2)) dv_1 dv_2 \\ &= \int_{-\infty}^{\infty} \int_{-\infty}^{\infty} \mu_{\hat{x}(v_1)} \mu^*_{\hat{x}(v_2)} \phi(W(u_1 - v_1)) \phi(W(u_2 - v_2)) dv_1 dv_2 \end{aligned}$$

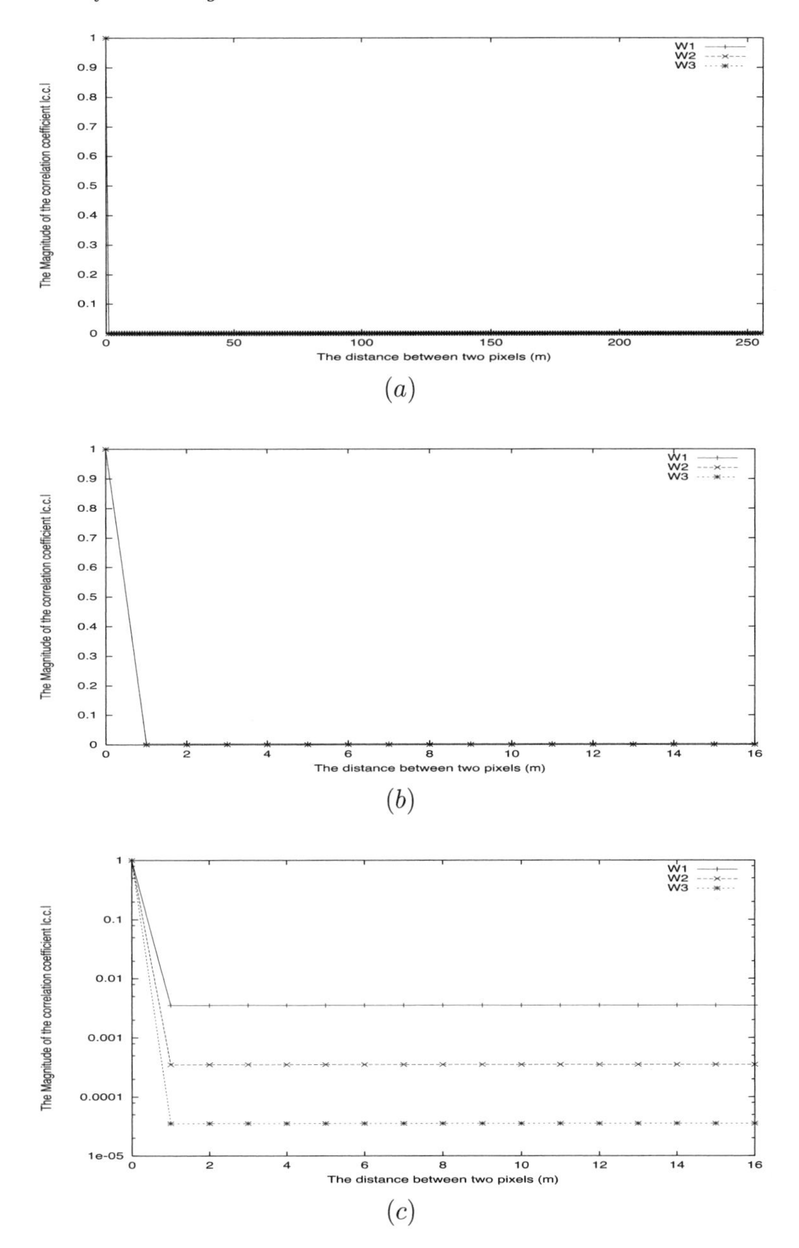

FIGURE 8.8
SAI in the basic FT MR image. $c_{bn}(m)$ of Eq. (8.149) is a measure of the correlation of pixel intensities with respect to the distance between pixels. $N = 256$, $W1 = 255.1\Delta k$, $W2 = 255.91\Delta k$, and $W3 = 255.991\Delta k$. Except for the vertical axis in (c) that uses the log scale, all axes in (a) through (c) are the linear scale. (a) $0 \leq m < 256$, (b) $0 \leq m < 16$, and (c) $0 \leq m < 16$.

$$+\frac{\sigma_{\mathcal{M}}^2}{\Delta k}\int_{-\infty}^{\infty}\int_{-\infty}^{\infty}\delta(v_1-v_2)\phi(W(u_1-v_1))\phi(W(u_2-v_2))dv_1dv_2$$
$$=\mu_{x(u_1)}\mu^*_{x(u_2)}+\frac{\sigma_{\mathcal{M}}^2}{\Delta k}\int_{-\infty}^{\infty}\phi(W(u_1-v_2))\phi(W(u_2-v_2))dv_2$$
$$=\mu_{x(u_1)}\mu^*_{x(u_2)}+\frac{\sigma_{\mathcal{M}}^2}{\Delta k}\phi(W(u_1-u_2))\star\phi(W(u_1-u_2))$$
$$=\mu_{x(u_1)}\mu^*_{x(u_2)}+\frac{\sigma_{\mathcal{M}}^2}{\Delta k}\mathcal{F}^{-1}\{(filt(\frac{k}{W}))^2\}|_{(u_1-u_2)}$$
$$=\mu_{x(u_1)}\mu^*_{x(u_2)}+\frac{\sigma_{\mathcal{M}}^2}{\Delta k}\mathcal{F}^{-1}\{(\Phi(\frac{k}{W})rect(\frac{k}{W}))^2\}|_{(u_1-u_2)}$$
$$=\mu_{x(u_1)}\mu^*_{x(u_2)}+\frac{\sigma_{\mathcal{M}}^2}{\Delta k}W sinc(W(u_1-u_2))f(W(u_1-u_2)). \quad (8.150)$$

where $\Phi(\frac{k}{W})$ and $f(W\Delta u)$ are given by

$$\Phi(\frac{k}{W})=\begin{cases}\frac{1}{2}(1+\cos(\frac{2\pi k}{W})) & \text{(a. Hanning)}\\ \alpha+(1-\alpha)\cos(\frac{2\pi k}{W}) & \text{(b. Hamming)},\end{cases} \quad (8.151)$$

and

$$f(W\Delta u)=\begin{cases}\frac{1.5}{(W\Delta u)^4-5(W\Delta u)^2+4} & \text{(a. Hanning)}\\ 0.0064\frac{(W\Delta u)^4-16.5(W\Delta u)^2+248.375}{(W\Delta u)^4-5(W\Delta u)^2+4} & \text{(b. Hamming}, \alpha=0.54).\end{cases} \quad (8.152)$$

respectively. In the derivation of Eq. (8.150), the property that both $filt$ and ϕ are even functions is used. Eq. (8.150) is identical to Eq. (8.39).

In Eq. (8.150), when $\frac{\sigma_{\mathcal{M}}^2}{\Delta k}W sinc(W(u_1-u_2))f(W(u_1-u_2)=0$, $R_x(u_1,u_2)=\mu_{x(u_1)}\mu^*_{x(u_2)}$, $x(u_1)$ and $x(u_2)$ become uncorrelated. Thus, the second term on the right side of Eq. (8.150) provides a measure of the correlation of $x(u_1)$ and $x(u_2)$, which depends on the imaging parameters ($\sigma_{\mathcal{M}}^2$, Δk, W) and the filter function Φ, and is a function of the distance (u_1-u_2) between two pixels. When $u_1=u_2$, it gives the variance of pixel intensity $\sigma_x^2=Var[x(u)]=\frac{\sigma_{\mathcal{M}}^2}{\Delta k}Wf(0)$ ($f(0)=0.375,\ 0.3974$ for Hanning, Hamming filtering, respectively).

The following shows the monotonically decreasing magnitude of the second term on the right side of Eq. (8.150). By substituting Eqs. (8.146) and (8.147) into the second term on the right side of Eq. (8.150) and using Eq. (8.148), its magnitude is

$$c_f(m)=\frac{\sigma_{\mathcal{M}}^2}{\Delta k}W|sinc(W\Delta u)f(W\Delta u)|$$
$$=\begin{cases}\frac{\sigma_n^2}{\Delta k}Wf(0) & (m=0)\\ \epsilon\sigma_{\mathcal{M}}^2|sinc(\frac{\epsilon}{N}m)f(\frac{N-\epsilon}{N}m)| & (m\neq 0)\,.\end{cases} \quad (8.153)$$

For simplicity, we use the normalized $c_f(m)$ given by

$$c_{fn}(m) = \left(\frac{\sigma^2_{\mathcal{M}}}{\Delta k} W f(0)\right)^{-1} c_f(m)$$
$$= \begin{cases} 1 & (m = 0) \\ \frac{\epsilon}{N-\epsilon}|sinc(\frac{\epsilon}{N}m) f(\frac{N-\epsilon}{N}m)|/f(0) & (m \neq 0). \end{cases} \quad (8.154)$$

When the Hanning filter $\Phi(\frac{k}{W})$ of Eq. (8.151.a) is used, $|f(W\Delta u)|$ of Eq. (8.152.a) monotonically decreases from its maximum to 0 as $|\Delta u|$ increases from δu (a pixel size) to ∞. When the Hamming filter $\Phi(\frac{k}{W})$ of Eq. (8.151.b) is used, using calculus, it has been shown that $|f(W\Delta u)|$ of Eq. (8.152b) monotonically decreases from its maximum at $\Delta u = \delta u$ to 0.0054 at $\Delta u = 6\delta u$, then *slightly* increases to 0.0064 as $|\Delta u|$ increases from $6\delta u$ to ∞. Because this increment is so small and so slow, $|f(W\Delta u)|$ of Eq. (8.152b) can be *considered* constant for $6\delta u < \Delta u < \infty$. Thus, for Hanning and Hamming filters, $|f(W\Delta u)|$ decreases from its maximum to 0 or 0.0064, respectively, as $|\Delta u|$ increases from δu to ∞.

Because (1) $|sinc(\frac{\epsilon}{N}m)|$ *sharply* decreases from 1 to very small values as $|m|$ increases from 0 to $(N-1)$, and (2) $|f(\frac{N-\epsilon}{N}m)|$ decreases as $|m|$ increases from $m = 1$ to $(N-1)$, therefore, when $|m|$ increases from 0 to $(N-1)$, $|sinc(\frac{\epsilon}{N}m) f(\frac{N-\epsilon}{N}m)|$ decreases from 1 to 0 (approaching but not equalizing).

An example is used to illustrate $c_{fn}(m)$ of Eq. (8.154). In this example, the settings are the same as those used in the example for the basic FT: $N = 256$, $\epsilon = 0.9$, 0.09, 0.009. With these settings, the curves of $c_{fn}(m)$ are shown in Figures 8.9 and 8.10, for Hanning and Hamming filters, respectively. The curves in Figures 8.9a and 8.10a show that $c_{fn}(m)$ $(0 \leq m < 256)$ sharply decreases from 1 to almost zero as $|m|$ increases, starting from 0. In Figures 8.9b and 8.10b, only the first 16 values of $c_{fn}(m)$ $(0 \leq m < 16)$ are shown to clearly demonstrate the decreasing patterns of $c_{fn}(m)$; there is virtually no correlation for $|m| > 3$. In Figures 8.9c and 8.10c, $c_{fn}(m)$ axis uses the log scale to clearly show the values of $c_{fn}(m)$ $(0 \leq m < 16)$ for various window widths W. Correlations exist for $|m| > 3$ even though they are very small. The curves in Figures 8.9c and 810.c demonstrate SAI in the filtered FT MR images with Hanning and Hamming filtering, respectively. The impact of the window width W on SAI is not essential. From Eqs. (8.150) and (8.153), and noticing that $\frac{\sigma^2_{\mathcal{M}}}{\Delta k} W f(0)$ is the variance of x, $c_{fn}(m)$ of Eq. (8.154) is actually the magnitude of the correlation coefficient of intensities of two pixels separated by m pixels.

■

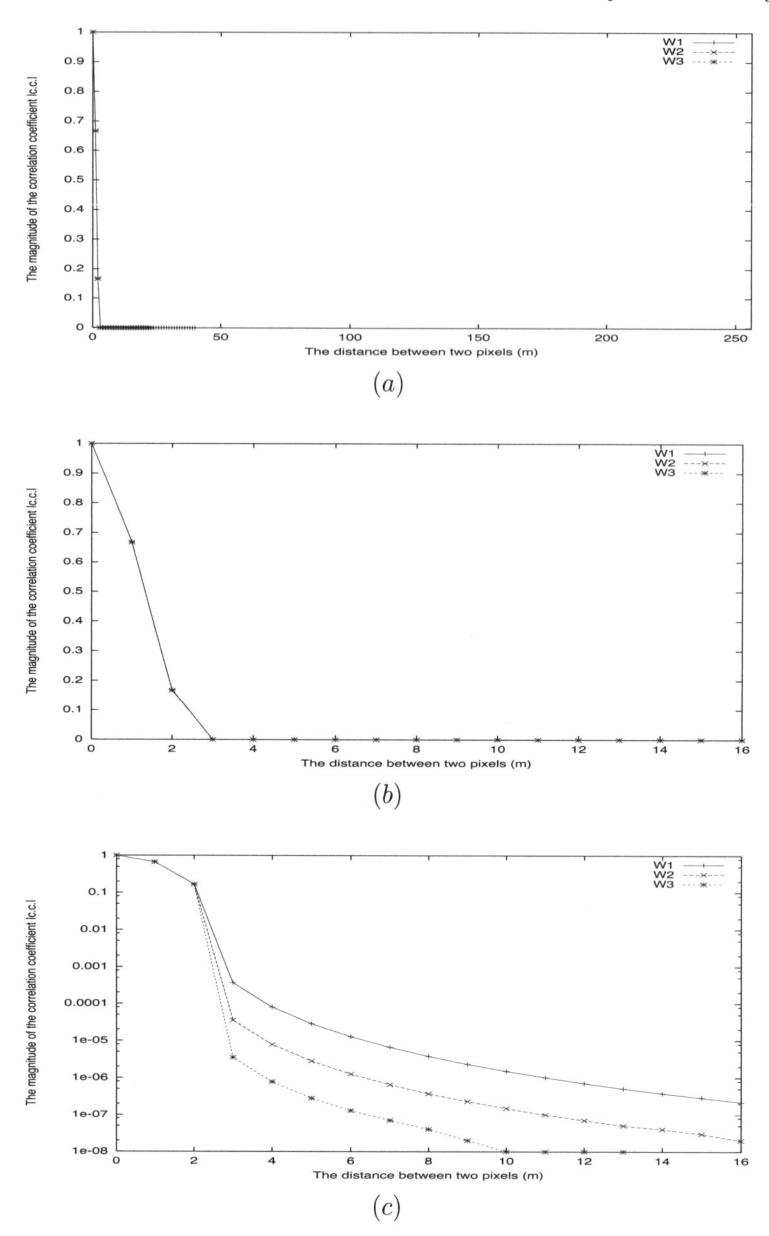

FIGURE 8.9
SAI in the Hanning-filtered FT MR image. $c_{fn}(m)$ of Eq. (8.154) is a measure of the correlation of pixel intensities with respect to the distance between pixels. $N = 256$, $W1 = 255.1\Delta k$, $W2 = 255.91\Delta k$, and $W3 = 255.991\Delta k$. Except for the vertical axis in (c), which uses the log scale, all axes in (a) through (c) use the linear scale. (a) $0 \leq m < 256$, (b) $0 \leq m < 16$, and (c) $0 \leq m < 16$.

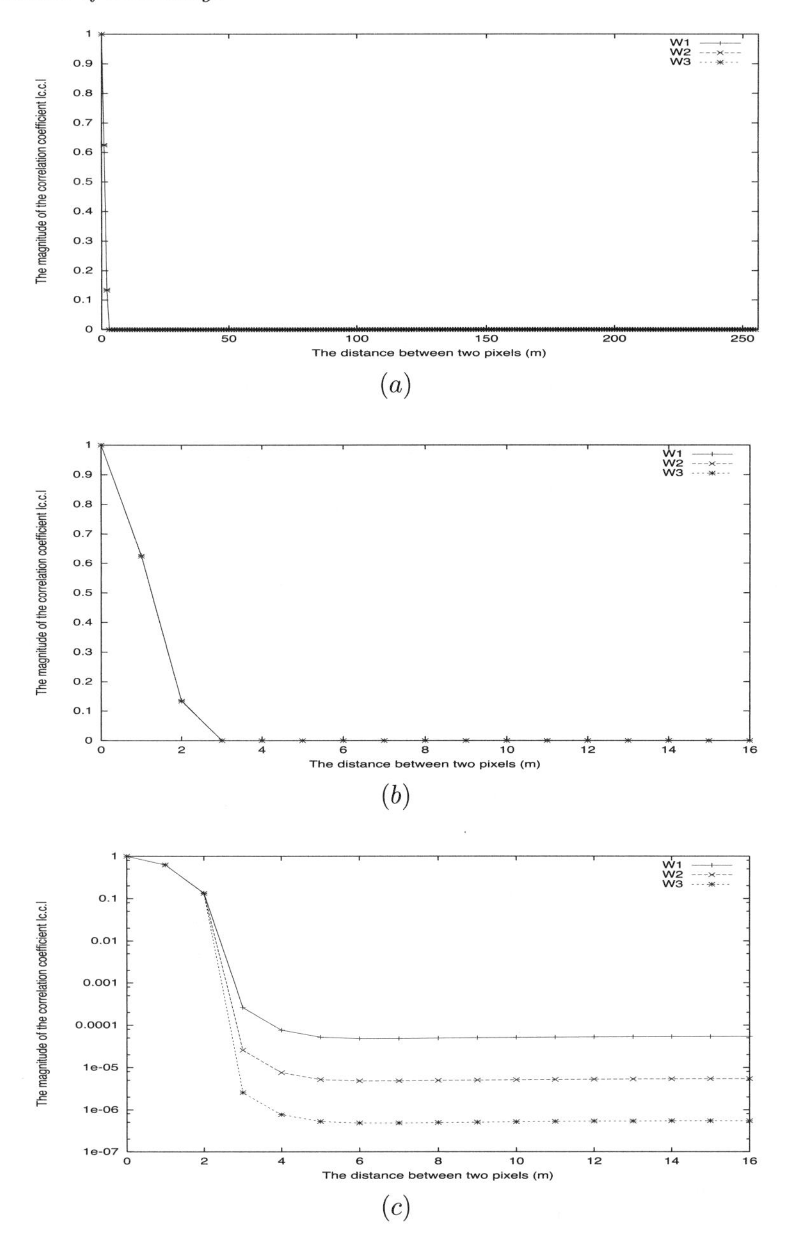

FIGURE 8.10
SAI in the Hamming-filtered FT MR image. $c_{fn}(m)$ of Eq. (8.154) is a measure of the correlation of pixel intensities with respect to the distance between pixels. $N = 256$, $W1 = 255.1\Delta k$, $W2 = 255.91\Delta k$, and $W3 = 255.991\Delta k$. Except for the vertical axis in (c), which uses the log scale, all axes in (a) through (c) use the linear scale. (a) $0 \leq m < 256$, (b) $0 \leq m < 16$, and (c) $0 \leq m < 16$.

8.6.3 Appendix 8C

In this appendix we discuss and summarize the correlations of pixel intensities in FT MR image. We will show (1) the effects of various MR data acquisition schemes and image reconstruction procedures on the correlations of the resultant image, (2) the similarities and differences in the correlations in the basic FT and the filtered FT MR images, and (3) the impacts of different filter functions on the correlations in FT MR image.

(1) The correlations of pixel intensities in the four cases of FT MR image in Section 8.3.1.1 are given by Eqs. (8.25), (8.29), (8.33), and (8.39). Eq. (8.25) in Case 1 and Eq. (8.29) of Case 2 show that when MR data, either continuous or discrete (Nyquist criterion must be satisfied), were acquired in the entire k-space, there was no Gibbs ringing artifact and a true image could be generated. Pixel intensities in the reconstructed image were uncorrelated. This can be interpreted as resulting the independence of k-space samples being preserved by the linear and orthogonal FT over an infinite number of samples.

Eq. (8.33) of the Case 3 shows that when the sampled MR data are acquired in a truncated k-space, Gibbs ringing artifact is introduced and the resultant image is an estimate of the true image. Pixel intensities in this image are correlated (see Eq. (8.33)), which is characterized by SAI. Eq. (8.39) of the case 4 shows that when the filtered, sampled MR data are acquired in a truncated k-space, Gibbs ringing artifact is reduced and the resultant image is a better estimate of the true image. Pixel intensities in this image are correlated (see Eq. (8.39)), which is also characterized by SAI, but in a changed pattern.

Eqs. (8.33) and (8.39) reveal that the finite sampling with a rectangular function in k-space results in the blurring with a *sinc* function in x-space, as shown by $sinc(W\Delta u)$ in (8.33) and (8.39). This is one underlying reason that causes the correlation of pixel intensities in basic and filtered FT MR images, even though the degree of this correlation is very small. In filtered-FT MR imaging, the filtering (i.e., apodization for reducing Gibbs ringing artifact) is another underlying reason for introducing the correlation of pixel intensities, as shown by $f(W\Delta u)$ in Eq. (8.39).

(2) The similarities and differences in SAI in the basic FT and the filtered FT MR images are given below. Figure 9 demonstrates SAI for MR images reconstructed by basic FT. In Figures 8.9a and 8.9b, there is virtually no correlation for $m \geq 1$. The details shown in Figure 8.9c demonstrate that the magnitude of the correlation coefficient is less than 0.01, 0.001, 0.0001, for $m \geq 1$ (for the window width $W = 255.1\Delta k$, $W = 255.91\Delta k$, $W = 255.991\Delta k$). Thus, in practice, pixel intensities can be considered as approximately uncorrelated for $m \geq 1$. Figures 8.10 and 8.11 demonstrate SAI for MR images reconstructed by filtered FT. Similarly, in Figures 8.10a, 8.10b, 8.11a, and 8.11b, there is virtually no correlation for $m \geq 3$. The details shown in Figures 8.10c and 8.11c demonstrate that the magnitude of the correlation coefficient is less than 0.001, 0.0001, 0.00001, for $m \geq 3$ (for the window width $W = 255.1\Delta k$, $W = 255.91\Delta k$, $W = 255.991\Delta k$). Thus, in practice, pixel intensities can be

considered approximately uncorrelated for $m \geq 3$.

By comparing Eq. (8.39) with Eq. (8.33), we notice that a new factor in the correlation, $f(W\Delta u)$, is introduced in the filtered FT. By comparing Figures 8.10c,and 8.11c with Figure 9.c, we find that the filtering operation (which has been used to reduce Gibbs ringing artifact) widens the range of the correlation of pixel intensities from $m = 1$ in the basic FT, to $m = 3$ in the filtered FT; but reduces the degree of the correlation by an order beyond this range ($m \geq 1$ in the basic FT and $m \geq 3$ in the filtered FT).

(3) The impacts of various filter functions (which are used to reduce Gibbs ringing artifact) on the correlations of pixel intensities in the resultant FT MR images are given below. The filter functions used in filtered FT MR imaging are also known as windows in signal analysis. Here, we only discuss Hamming and Hanning filters because the performances of the rectangular and Bartlett filters are poor and the Kaither-Bessel filter is computationally expensive. The comprehensive studies of windows [11–15] show that the spectra of Hamming and Hanning windows have (i) the same mainlobe width $\frac{8\pi}{N}$, (ii) the peak sidelobe levels -41 *db* and -31 *db*, (iii) the sidelobe rolloff rate -6 *db/oct* and -18 *db/oct*, respectively. That is, a Hamming window has the lower peak sidelobe and a Hanning window has the steeper sidelobe rolloff.

In this study, from Eqs. (8.37), (8.38), and (8.23), we have

$$x(u) = \hat{x}(u) \star \mathcal{F}^{-1}\{\Phi(\frac{k}{W})rect(\frac{k}{W})\}, \tag{8.155}$$

where $\Phi(\frac{k}{W})$ is given in Eq. (151), which are Hanning and Hamming filters. The results in Section 8.3.1.1 (Remarks on the variances of FT MR image) indicate that the variance of pixel intensities in Hamming- and Hanning-filtered FT MR images are $0.3974(\frac{1}{N}\sigma_n^2)$ and $0.375(\frac{1}{N}\sigma_n^2)$, respectively. That is about 6% difference. Figures 8.10c and 8.11c show that the magnitude of the correlation coefficient in a Hanning-filtered FT MR image damps out more rapidly than that in Hamming's after six lags. Thus, the Hanning window is a more preferred choice in terms of the smaller variance and the more quickly decreasing correlation.

8.6.4 Appendix 8D

This appendix proves Eqs. (8.49), (8.51), (8.53), (8.56)–(8.60), (8.61)–(8.67), and SAI of the filtered projection $t(u', m\Delta\theta)$ and the pixel intensity $x(u, v)$.

The filtered projection $t(u', \theta)$ is computed by Eq. (8.45), and the measured projection $p(l, \theta)$ is computed by Eq. (7.78). These two equations have the common factors $\mathcal{M}(k, \theta) \cdot \frac{1}{\Delta k}comb(\frac{k}{\Delta k}) \cdot rect(\frac{k}{W_k})$. This fact shows that the filtered projection $t(u', \theta)$ is the filtered (by $|k|$) measured projection $p(l, \theta)$. Thus, statistics of $p(l, \theta)$ can be utilized as the input for the $|k|$ filter to derive statistics for $t(u', \theta)$. That is, three steps (**S.1**, **S.2**, and **S.3**) in the proof of

Property 7.8 can be directly applied to the three steps (A-1, A-2, and A-3) of this proof.

Proof.

A-1) Proof of Eq. (8.49). By using the identical procedures in **S.1** of the Proof of Property 7.8 to derive Eq. (7.90), that is, using u', $R_{\tilde{t}}$, and $\mu_{\tilde{t}}$ to replace l, $R_{\tilde{p}}$, and $\mu_{\tilde{p}}$, respectively, we have

$$R_{\tilde{t}}((u'_1,\theta_1),(u'_2,\theta_2)) = \mu_{\tilde{t}(u'_1,\theta_1)}\mu^*_{\tilde{t}(u'_2,\theta_2)} + \sigma^2_{\mathcal{M}}\delta(u'_1-u'_2)\delta[\theta_1-\theta_2]. \quad (8.156)$$

A-2) Proof of Eq. (8.51). By using the identical procedures in **S.2** of the Proof of Property 7.8 to derive Eq. (7.102), that is, using u', $R_{\hat{t}}$, and $\mu_{\hat{t}}$ to replace l, $R_{\hat{p}}$, and $\mu_{\hat{p}}$, respectively, we have

$$R_{\hat{t}}((u'_1,\theta_1),(u'_2,\theta_2)) = \mu_{\hat{t}(u'_1,\theta_1)}\mu^*_{\hat{t}(u'_2,\theta_2)} + \frac{\sigma^2_{\mathcal{M}}}{\Delta k}\delta(u'_1-u'_2)\delta[\theta_1-\theta_2]. \quad (8.157)$$

A-3) Proof of Eq. (8.53). By using the identical procedures in **S.3** of the Proof of Property 7.8 to derive Eq. (7.112), that is, using u', $R_{\bar{t}}$, and $\mu_{\bar{t}}$ to replace l, $R_{\bar{p}}$, and $\mu_{\bar{p}}$, respectively, we have

$$R_{\bar{t}}((u'_1,\theta_1),(u'_2,\theta_2)) = \mu_{\bar{t}(u'_1,\theta_1)}\mu^*_{\bar{t}(u'_2,\theta_2)} + \frac{\sigma^2_{\mathcal{M}}}{\Delta k}W_k sinc(W_k(u'_1-u'_2))\delta[\theta_1-\theta_2]. \quad (8.158)$$

A-4) Proof of Eq. (8.56), SAI of the filtered projection $t(u',\theta)$, and Eq. (8.60).

a) Proof of Eq. (8.56). From Eq. (8.54) and using Eq. (8.157), the correlation $R_t((u'_1,\theta_1),(u'_2,\theta_2))$ of the filtered projections $t(u'_1,\theta_1)$ and $t(u'_2,\theta_2)$ is

$$\begin{aligned}
&R_t((u'_1,\theta_1),(u'_2,\theta_2)) = E[t(u'_1,\theta_1)t^*(u'_2,\theta_2)] \\
&= E[(\hat{t}(u'_1,\theta_1) \star \phi(W_k u'_1))(\hat{t}^*(u'_2,\theta_2) \star \phi(W_k u'_2))] \\
&= E[\int \hat{t}(v_1,\theta_1)\phi(W_k(u'_1-v_1))dv_1 \cdot \int \hat{t}^*(v_2,\theta_2)\phi(W_k(u'_2-v_2))dv_2] \\
&= \int\int E[\hat{t}(v_1,\theta_1)\hat{t}^*(v_2,\theta_2)]\phi(W_k(u'_1-v_1))\phi(W_k(u'_2-v_2))dv_1dv_2 \\
&= \int\int \mu_{\hat{t}(v'_1,\theta_1)}\mu^*_{\hat{t}(v'_2,\theta_2)}\phi(W_k(u'_1-v_1))\phi(W_k(u'_2-v_2))dv_1dv_2
\end{aligned}$$

$$+\frac{\sigma_{\mathcal{M}}^2}{\Delta k}\int\int\delta(v_1-v_2)\delta[\theta_1-\theta_2]\phi(W_k(u_1'-v_1))\phi(W_k(u_2'-v_2))dv_1dv_2$$

$$= \mu_{t(u_1',\theta_1)}\mu^*_{t(u_2',\theta_2)}$$

$$+\frac{\sigma_{\mathcal{M}}^2}{\Delta k}\delta[\theta_1-\theta_2]\int\phi(W_k(u_1'-v_2))\phi(W_k(u_2'-v_2))dv_2$$

$$= \mu_{t(u_1',\theta_1)}\mu^*_{t(u_2',\theta_2)}$$

$$+\frac{\sigma_{\mathcal{M}}^2}{\Delta k}\delta[\theta_1-\theta_2](\phi(W_k(u_1'-u_2'))\star\phi(W_k(u_1'-u_2')))$$

$$= \mu_{t(u_1',\theta_1)}\mu^*_{t(u_2',\theta_2)}$$

$$+\frac{\sigma_{\mathcal{M}}^2}{\Delta k}\delta[\theta_1-\theta_2]\mathcal{F}_k^{-1}\{(\mathcal{F}_{u_1'-u_2}\{\phi(W_k(u_1'-u_2'))\})^2\}$$

$$= \mu_{t(u_1',\theta_1)}\mu^*_{t(u_2',\theta_2)}$$

$$+\frac{\sigma_{\mathcal{M}}^2}{\Delta k}\delta[\theta_1-\theta_2]\mathcal{F}_k^{-1}\{(k\cdot rect(\frac{k}{W_k}))^2\}|_{(u_1'-u_2')}. \tag{8.159}$$

We have shown that

$$\mathcal{F}_k^{-1}\{(k\cdot rect(\frac{k}{W_k}))^2\}|_{(u_1'-u_2')}$$

$$= \frac{W_k^3}{2}(\frac{1}{2}sinc(W_k(u_1'-u_2'))$$

$$+\frac{\cos(\pi W_k(u_1'-u_2'))-sinc(W_k(u_1'-u_2'))}{(\pi W_k(u_1'-u_2'))^2})$$

$$\triangleq f(W_k(u_1'-u_2')). \tag{8.160}$$

By substituting Eq. (8.160) into Eq. (8.159), we obtain Eq. (8.56).

b) Proof of SAI of $t(u',\theta)$. The second item on the right side of Eq. (8.56) provides a measure of the correlation of the filtered projection $t(u',\theta)$. To prove SAI of $t(u',\theta)$, it is sufficient to analyze $f(W_k(u_1'-u_2'))$ of Eq. (8.57). For simplicity of discussion, let

$$f_1(W_k(u_1'-u_2')) = \tfrac{1}{2}sinc(W_k(u_1'-u_2'))$$

$$f_2(W_k(u_1'-u_2')) = \frac{\cos(\pi W_k(u_1'-u_2'))-sinc(W_k(u_1'-u_2'))}{(\pi W_k(u_1'-u_2'))^2}$$

$$f(W_k(u_1'-u_2')) = \tfrac{W_k^3}{2}(f_1(W_k(u_1'-u_2'))+f_2(W_k(u_1'-u_2'))).$$

Because $f(W_k(u_1'-u_2'))$ is an even function, the proof is shown for $W_k(u_1'-u_2')>0$ only.

Similar to the general sinc function, $f_1(W_k(u_1'-u_2'))$ consists of a main lobe and the side lobes. The peaks of the side lobes occur at $W_k(u_1'-u_2')=(n+\frac{1}{2})\pi$ ($n=1,2\cdots$). Their magnitudes are much smaller than that of the peak of the main lobe (which is 0.5 in this case) and monotonically decrease as $(u_1' - u_2')$ increases. In the limiting case of $(u_1' - u_2') \rightarrow \infty$, $f_1(W_k(u_1' - u_2')) \rightarrow 0$. That is, $f_1(W_k(u_1' - u_2'))$ has an oscillating attenuation pattern for small $(u_1' - u_2')$, overall decreases, and becomes almost zero when $(u_1' - u_2')$ becomes large.

The numerator of $f_2(W_k(u_1' - u_2'))$ is an oscillating function but bounded: $|\cos(\pi W_k(u_1' - u_2')) - sinc(W_k(u_1' - u_2'))| < 2$. Its denominator $(\pi W_k(u_1' - u_2'))^2$ monotonically increases as $(u_1' - u_2')$ increases. As a result, except for some very small $(u_1' - u_2')$ (e.g., as we have shown $f_2(0) = -\frac{1}{3}$—see the paragraph c) below), $f_2(W_k(u_1' - u_2'))$ overall decreases (with the tiny, local fluctuations caused by its numerator) and becomes almost zero when $(u_1' - u_2')$ becomes large. In the limiting case of $(u_1' - u_2') \rightarrow \infty$, $f_2(W_k(u_1' - u_2')) \rightarrow 0$.

Thus, for not very small $(u_1' - u_2')$, $f(W_k(u_1' - u_2'))$ overall decreases and becomes almost zero as $(u_1' - u_2')$ increases. This decrease becomes faster and smoother due to cancellations of the tiny, local fluctuations between $f_1(W_k(u_1' - u_2'))$ and $f_2(W_k(u_1' - u_2'))$. In the limiting case of $(u_1' - u_2') \rightarrow \infty$, $f(W_k(u_1' - u_2')) \rightarrow 0$. Therefore, except for some small $(u_1' - u_2')$, $f(W_k(u_1' - u_2'))$ is *almost* monotonically decreasing as $(u_1' - u_2')$ increases.

These justifications have been confirmed by simulation results. The simulations shown in Figure 8.11 is performed at the view $\theta = \frac{\pi}{4}$. The horizontal axes $(u_1' - u_2')$ in Figure 8.11a–d are in the unit of the pixel. The corresponding vertical axes represent f_1, f_2, f, and $|f|$. Because the magnitudes of f_1, f_2, and f become extremely small for large $(u_1' - u_2')$, the curves are displayed for $(u_1' - u_2') < 64$ only.

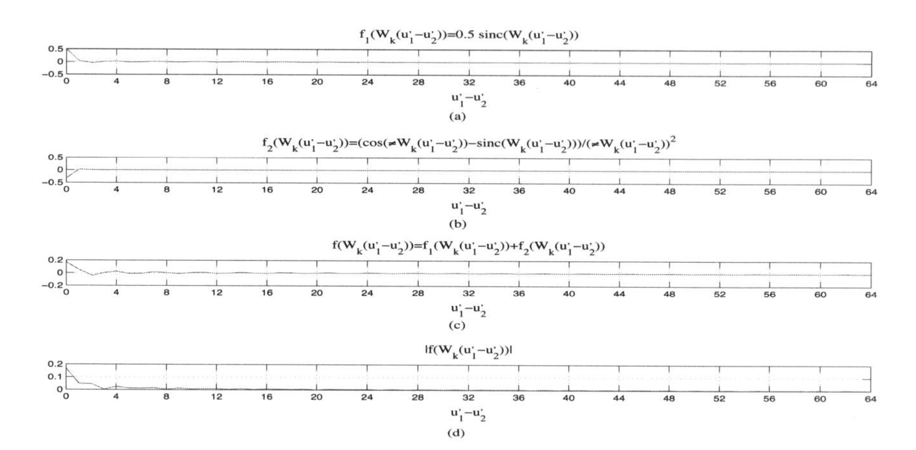

FIGURE 8.11

Simulation results of Eq. (8.160) normalized by $\frac{W_k^3}{2}$, that is, $f(W_k(u_1' - u_2'))$ of Eq. (8.57) normalized by $\frac{W_k^3}{2}$. (a) The first item $f_1(W_k(u_1' - u_2'))$, (b) the second item $f_2(W_k(u_1' - u_2'))$, (c) $f(W_k(u_1' - u_2'))$ itself, and (d) the magnitude $|f(W_k(u_1' - u_2'))|$.

Figure 8.11a shows that $f_1(W_k(u'_1-u'_2))$ has a basic pattern of sinc function. Figure 8.11b shows that $f_2(W_k(u'_1-u'_2))$ is overall decreasing for $(u'_1-u'_2) > 1$. Figure 8.11c shows that $f_1(W_k(u'_1 - u'_2))$ has small fluctuations for small $(u'_1 - u'_2)$ and virtually no fluctuations for $(u'_1 - u'_2) > 32$. Figure 8.11d shows that except for a few isolated values (e.g., $(u'_1-u'_2) = 3$ and 8), $|f(W_k(u'_1-u'_2))|$ *almost* monotonically decreases as $(u'_1 - u'_2)$ increases.

c) Proof of Eq. (8.60). The second item on the right side of Eq. (8.56), that is, Eq. (8.58), is the covariance of $t(u', \theta)$. When $u'_1 = u'_2$, it gives the variance of $t(u', \theta)$, denoted by σ_t^2.

When $u'_1 = u'_2$, the second item of $f(W_k(u'_1 - u'_2))$ in Eq. (8.57), that is, $f_2(W_k(u'_1 - u'_2))$, becomes $\frac{0}{0}$. We have shown that $f_2(0) = -\frac{1}{3}$. Thus, from Eqs. (8.57) and (8.58), the variance of the filtered projection $t(u', \theta)$ in one view is

$$\sigma_t^2 = \sigma_{\mathcal{M}}^2 \frac{W_k}{\Delta k} \frac{W_k^2}{2} \left(\frac{1}{2} - \frac{1}{3}\right) = \sigma_{\mathcal{M}}^2 \frac{W_k}{\Delta k} \left(\frac{1}{12} W_k^2\right). \qquad (8.161)$$

B) Proof of Eq. (8.61), SAI of the pixel intensity $x(u, v)$, and Eq. (8.67).

a) Proof of Eq. (8.61). From Eq. (8.46) and using Eq. (8.56), the correlation $R_x((u_1, v_1), (u_2, v_2))$ of the pixel intensities $x(u_1, v_1)$ and $x(u_2, v_2)$ is given by

$$\begin{aligned}
R_x((u_1, v_1), (u_2, v_2)) &= E[x(u_1, v_1)x^*(u_2, v_2)] \\
&= E[\int_0^{\pi} t(u'_1, \theta_1)d\theta_1 \int_0^{\pi} t^*(u'_2, \theta_2)d\theta_2] \\
&= \int_0^{\pi}\int_0^{\pi} E[t(u'_1, \theta_1)t^*(u'_2, \theta_2)]d\theta_1 d\theta_2 \\
&= \int_0^{\pi}\int_0^{\pi} R_t((u'_1, \theta_1), (u'_2, \theta_2))d\theta_1 d\theta_2 \\
&= \int_0^{\pi}\int_0^{\pi} \mu_{t(u'_1,\theta_1)}\mu^*_{t(u'_2,\theta_2)} d\theta_1 d\theta_2 \\
&\quad + \frac{\sigma_{\mathcal{M}}^2}{\Delta k} \int_0^{\pi}\int_0^{\pi} f(W_k(u'_1 - u'_2))\delta[\theta_1 - \theta_2]d\theta_1 d\theta_2 \\
&= \mu_{x(u_1,v_1)}\mu^*_{x(u_2,v_2)} + \frac{\sigma_{\mathcal{M}}^2}{\Delta k} \int_0^{\pi} f(W_k(u'_1 - u'_2))d\theta \\
&= \mu_{x(u_1,v_1)}\mu^*_{x(u_2,v_2)} + \Delta\theta \frac{\sigma_{\mathcal{M}}^2}{\Delta k} \sum_{m=0}^{M-1} f(W_k(u'_1 - u'_2)).
\end{aligned} \qquad (8.162)$$

Using Eq. (8.47), $(u'_1 - u'_2)$ in the m-th view specified by θ can be expressed as

$$u'_1 - u'_2 = (u_1 - u_2)\cos\theta + (v_1 - v_2)\sin\theta = \Delta r \cos(\theta - \Delta\phi), \qquad (8.163)$$

where Δr is the distance between two pixel centers (u_1, v_1) and (u_2, v_2) in the image, and $\Delta\phi$ is the angle between the line linking these two centers and the

U-axis, given by

$$\Delta r = \sqrt{(u_1 - u_2)^2 + (v_1 - v_2)^2} \quad \text{and} \quad \Delta\phi = \tan^{-1}\left(\frac{v_1 - v_2}{u_1 - u_2}\right). \quad (8.164)$$

By substituting Eq. (8.163) into Eq. (8.162) and using $\theta = m\Delta\theta$ ($m = 0, 1, \cdots, M-1$) and $M\Delta\theta = \pi$, we obtain Eq. (8.61).

Eqs. (8.163) and (8.164) show that $(u_1' - u_2')$ is the projection of the distance Δr (i.e., the vector $(u_1 - u_2)\vec{i} + (v_1 - v_2)\vec{j}$) onto the U'-axis (i.e., the view direction) specified by the view angle $m\Delta\theta$. When m takes values from 0 to $M-1$, the projection $u_1' - u_2' = \Delta r \cos(m\Delta\theta - \Delta\phi)$ may be positive, or negative, or zero, or equal to Δr (its maximum) or $-\Delta r$ (its minimum). That is,

$$-\Delta r \leq \Delta r \cos(m\Delta\theta - \Delta\phi) \leq \Delta r. \quad (8.165)$$

The dimensionality of W_k—the width of the window for $|k|$ filter—is the inverse of the length L^{-1}. The dimensionality of $\Delta r \cos(m\Delta\theta - \Delta\phi)$ is the length L. Thus, $W_k \Delta r \cos(m\Delta\theta - \Delta\phi)$ has no dimensionality, that is, it is just a number, which may be positive, zero, or negative.

Using Eq. (8.163), Eq. (8.162) indicates that the correlation R_x $((u_1, v_1), (u_2, v_2))$ between two pixel intensities $x(u_1, v_1)$ and $x(u_2, v_2)$ in MR image reconstructed via *Filtering by Fourier transform* is determined by the projections $\Delta r \cos(m\Delta\theta - \Delta\phi)$ of the distance $\Delta r = ||(u_1, v_1), (u_2, v_2)||$ on each U'-axis specified by $m\Delta\theta$ ($m = 0, 1, \cdots, M-1$).

b) Proof of SAI of $x(u, v)$. The second item on the right side of Eq. (8.61) is the covariance of $x(u_1, v_1)$ and $x(u_2, v_2)$; hence, Eq. (8.65) provides a measure of the correlation of pixel intensities $x(u_1, v_1)$ and $x(u_2, v_2)$. To prove SAI of $x(u, v)$, it is sufficient to analyze $\sum_{m=0}^{M-1} f(W_k \Delta r \cos(m\Delta\theta - \Delta\phi))$.

SAI of the filtered projection $t(u', \theta)$ of Section 8.3.1.2.A implies that each individual $f(W_k \Delta r \cos(m\Delta\theta - \Delta\phi))$ ($m = 0, 1, \cdots, M-1$) *almost* monotonically decreases as Δr increases. Thus, $\sum_{m=0}^{M-1} f(W_k \Delta r \cos(m\Delta\theta - \Delta\phi))$ *almost* monotonically decreases as Δr increases. This monotonically decreasing pattern becomes more evident due to the cancellation of the tiny, local fluctuations among M individual $f(W_k \Delta r \cos(m\Delta\theta - \Delta\phi))$. In the limiting case of $\Delta r \to \infty$, due to $f(W_k \Delta r \cos(m\Delta\theta - \Delta\phi)) \to 0$ ($m = 0, 1, \cdots, M-1$), we have

$$\lim_{\Delta r \to \infty} \sum_{m=0}^{M-1} f(W_k \Delta r \cos(m\Delta\theta - \phi)) \to 0. \quad (8.166)$$

Eq. (3.126) shows that

$$M \simeq \frac{\pi}{2} N, \quad (8.167)$$

where N is the number of the radial k-space samples in each view. Thus, M is quite large. In Eq. (8.61), we rewrite

$$\Delta\theta \sum_{m=0}^{M-1} f(W_k \Delta r \cos(m\Delta\theta - \phi)) = M\Delta\theta \frac{1}{M} \sum_{m=0}^{M-1} f(W_k \Delta r \cos(m\Delta\theta - \phi)), \tag{8.168}$$

where $\frac{1}{M}\sum_{m=0}^{M-1} f(W_k \Delta r \cos(m\Delta\theta - \phi))$ is an estimate of the mean of $f(W_k \Delta r \cos(\theta - \Delta\phi))$, that is, the sample mean. On the other hand, in Eq. (8.61), θ varies from 0 to π and uniformly takes values $m\Delta\theta$ ($m = 0, 1, \cdots, M-1$). It is reasonable to assume that θ has a uniform distribution over $[0, \pi]$. That is, the pdf of θ, $p(\theta)$, is

$$p(\theta) = \begin{cases} \frac{1}{\pi} & [0, \pi] \\ 0 & \text{elsewhere}. \end{cases} \tag{8.169}$$

Thus, in Eq. (8.61), we rewrite

$$\int_0^{\pi} f(W_k \Delta r \cos(\theta - \phi)) d\theta = \pi \int_0^{\pi} f(W_k \Delta r \cos(\theta - \phi)) p(\theta) d\theta, \tag{8.170}$$

where $\int_0^{\pi} f(W_k \Delta r \cos(\theta - \phi)) p(\theta) d\theta$ is the mean of $f(W_k \Delta r \cos(\theta - \phi))$. Because $M\Delta\theta = \pi$, Eq. (8.168) and Eq. (8.170) are equivalent.

As the mean or the sample mean, its variance is much smaller than those of its individual samples. Thus, $\sum_{m=0}^{M-1} f(W_k \Delta r \cos(m\Delta\theta - \phi))$ is *almost* monotonically decreasing as Δr increases.

These justifications have been confirmed by simulation results. The simulation shown in Figure 8.12 and Figure 13 is for a 256×256 image (i.e., $I = 256$). In its reconstruction, $N \simeq \sqrt{2}I = 362$ and $M \simeq \frac{\pi}{2}N = 569$. In Figures 8.12 and 8.13, the horizontal axis Δr is in the unit of the pixel. The vertical axis shows the normalized magnitude $|\sum_{m=0}^{M-1} f(W_k \Delta r \cos(m\Delta\theta - \phi))|$. Figure 8.12 shows a 2-D plot and Figure 8.13 shows a line profile at the view $\theta = \frac{\pi}{4}$. Because the magnitudes of $|\sum_{m=0}^{M-1} f(W_k \Delta r \cos(m\Delta\theta - \phi))|$ become very small for large Δr, the surface and the curve of $|\sum_{m=0}^{M-1} f(W_k \Delta r \cos(m\Delta\theta - \phi))|$ are only displayed for $\Delta r < 64$ and < 32, respectively.

Figures 8.12 and 8.13 show that (1) except a few points (e.g., $\Delta r = 3$), $|\sum_{m=0}^{M-1} f(W_k \Delta r \cos(\theta - \phi))|$ is monotonically decreases as Δr increases, and (2) there is no virtual correlation for $\Delta r > 8$.

c) Proof of Eq. (8.67). The second item in the right side of Eq. (8.61), i.e., Eq. (8.65), is the covariance of $x(u, v)$. When $(u_1, v_1) = (u_2, v_2)$, it gives variance of $x(u, v)$, denoted by σ_x^2. From Eqs. (8.65) and (8.60), σ_x^2 is given by

$$\sigma_x^2 = \sigma_{\mathcal{M}}^2 \frac{W_k}{\Delta k} \Delta\theta \sum_{m=0}^{M-1} \frac{1}{12} W_k^2 = \sigma_{\mathcal{M}}^2 \frac{W_k}{\Delta k} (\pi \frac{1}{12} W_k^2) = \frac{1}{3} \sigma_{\mathcal{M}}^2 \frac{W_k}{\Delta k} S_{W_k}, \tag{8.171}$$

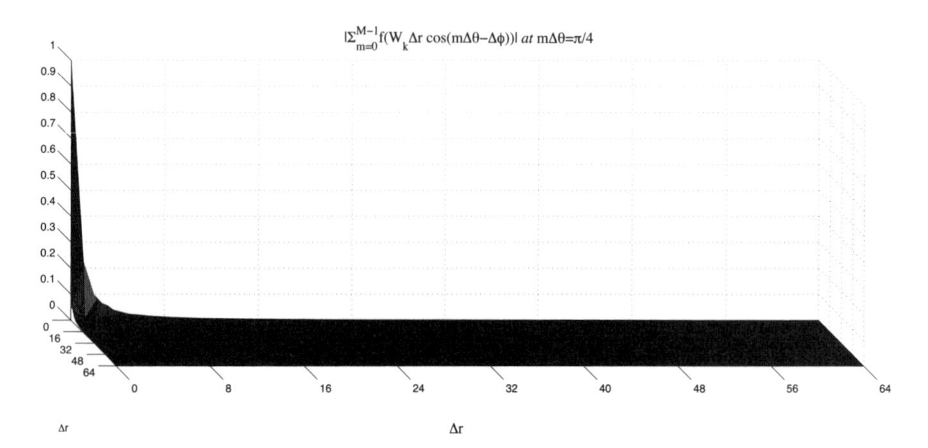

FIGURE 8.12
A 2-D plot of the second item of Eq. (8.162) normalized by $\Delta\theta\frac{\sigma_{\mathcal{M}}^2}{\Delta k}$, that is, Eq. (8.65) normalized by $\Delta\theta\frac{\sigma_{\mathcal{M}}^2}{\Delta k}$, shows that $|\sum_{m=0}^{M-1} f(W_k \Delta r \cos(m\Delta\theta - \phi))|$ monotonically decreases as Δr increases and becomes almost zero when Δr is large. This simulation result is for a 256×256 image. In its reconstruction, $N = 362$, and $M = 569$.

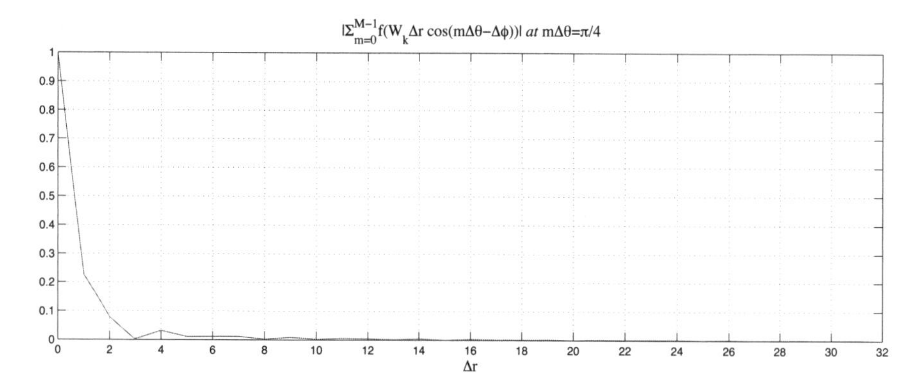

FIGURE 8.13
A line profile in the 2-D plot of Figure 8.12. This line profile is at the view $m\Delta\theta = \frac{\pi}{4}$. It shows that except for a few points, for example, $\Delta r = 3$, $|\sum_{m=0}^{M-1} f(W_k \Delta r \cos(m\Delta\theta - \phi))|$ monotonically decreases as Δr increases and becomes almost zero for $\Delta r > 8$.

where $S_{W_k} = \frac{1}{4}\pi W_k^2$ is the area of a circular region with W_k as its diameter in the k-space of radial sampling.

■

Problems

8.1. Derive Eq. (8.55) and give an intuitive interpretation.

8.2. Derive Eq. (8.133).

8.3. Derive Eq. (8.138).

8.4. Prove Eq. (8.148) and Eq. (8.153).

8.5. Prove Eq. (8.152) and justify the monotonically decreasing patterns.

8.6. Derive Eq. (8.160).

8.7. Derive Eq. (8.161).

8.8. Figure 8.12 shows a 2-D plot of the normalized magnitude $|\sum_{m=0}^{M-1} f(W_k \Delta r \cos(m\Delta\theta - \phi))|$ with $M = \frac{\pi}{2}N = 569$. Make 2-D plots for $M = 284$ and $M = 1136$, respectively, and interpret the results.

References

[1] Abramowitz, M., Stegun, I.: *Handbook of Mathematical Functions.* Dover Publications, New York (1970).

[2] Thomas, J.: *An Introduction to Statistical Communication Theory.* John Wiley & Sons Inc, New York (1969).

[3] Lei, T.: Gibbs ringing artifact, spatial correlation, and spatial resolution in MRI. *SPIE Proceedings* **5368** (2004) 837–847.

[4] Lei, T., Udupa, J.: A new look at Markov random field (mrf) model-based MR image analysis. *SPIE Proceedings* **5747** (2005).

[5] Lei, T., Udupa, J.: Quantification of spatial correlation in x-ray CT and MR images and its applications. *Proc. of the IEEE Nuclear Science Symposium and Medical Imaging Conference* (CD) (2002).

[6] Barrett, H., Swindell, W.: *Radiological Imaging.* Academic Press, Hoboken, New Jersey, (1981).

[7] Nishimura, D.: *Principles of Magnetic Resonance Imaging.* Stanford University, Palo Alto, California, (1996).

[8] Goodman, J.: *Introduction to Fourier Optics.* McGraw-Hill, New York (1996).

[9] Liang, Z.P., Lauterbur, P.: *Principles of Magnetic Resonance Imaging, A Signal Processing Perspective.* IEEE Press, New York (2000).

[10] Haacke, E., Brown, R., Thompson, M., Venkatesan, R.: *Magnetic Resonance Imaging: Physical Principles and Sequence Design.* John Wiley & Sons Inc., New York (1999).

[11] Papoulis, A.: *Signal Analysis.* McGraw-Hill Book Company Inc., New York (1977).

[12] McGillem, C., Cooper, G.: *Continuous and Discrete Signal and System Analysis.* Saunders College Publishing, Philadelphia, Pennsylvania, (1991).

[13] Harris, F.: On the use of windows for harmonic analysis with the discrete fourier transform. *IEEE Proc.* **66**(1) (1978) 51–83.

[14] Tretter, S.: *Introduction to Discrete-Time Signal Processing.* John Wiley & Sons, New York (1976).

[15] Jackson, J.: *Signals, Systems, and Transforms.* Addison-Wesley Publishing Company, Reading, Massachusetts, (1989).

[16] Poularikas, A., Seely, S.: *Signals and Systems.* PWS-KENT Publishing Company, Boston, Massachusetts, (1991).

[17] Lei, T., Udupa, J.: Gibbs ringing artifact and spatial correlation in MRI. *SPIE Proceedings* **5030** (2003) 961–971.

[18] Herman, G.: *Image Reconstruction from Projections.* Academic Press, New York (1980).

[19] Lei, T., Sewchand, W.: Statistical approach to x-ray CT imaging and its applications in image analysis - part 1: Statistical analysis of x-ray CT imaging. *IEEE Trans. Med. Imag.* **11**(1) (1992). 53–61

[20] Papoulis, A.: *Probability, Random Variables and Stochastic Processes.* McGraw-Hill Book Company Inc., New York (1984).

[21] Jenkins, G., Watts, D.: *Spectral Analysis and Its Applications.* Holden-Day, San Francisco, California, (1968).

[22] Fuller, W.: *Introduction to Statistical Time Series.* John Wiley & Sons, New York (1976).

[23] Fuderer, M.: The information content of MR images. *IEEE Trans. Med. Imag.* **7**(4) (1988) 368–380.

[24] Gray, R., Davisson, L.: *Random Processes: A Mathematical Approach for Engineers.* Prentice-Hall, Inc., Englewood Cliffs, New Jersey, (1986).

[25] Press, S.: *Applied Multivariate Analysis.* Robert E. Krieger Publishing Company, Malabar, Florida, (1982).

[26] Bracewell, R.N.: *The Fourier Transform and Its Applications.* McGraw-Hill Book Company, New York (1999).

[27] Papoulis, A.: *The Fourier Integral and Its Applications.* McGraw-Hill Book Company Inc., New York (1962).

[28] Cohen, A.: *Biomedical Signal Processing.* CRC Press, Boca Raton, Florida, (1986)

[29] Anderson, T.W.: *An Introduction to Multivariate Statistical Analysis.* John Wiley & Sons Inc, New York (1984).

[30] Sijbers, J.: *Signal and Noise Estimation from Magnetic Resonance Images,* Ph.D. dissertation. University Antwerp, Antwerp, Belgium (1998).

[31] Kummer, E.: Über die hypergeometrische reihe. *J. Reine Angew. Math.* (15) (1836) 39–83.

[32] Koepf, W.: *Hypergeometric Summation: An Algorithmic Approach to Summation and Special Function Identities.* Vieweg, Braunschweig, Germany: (1998).

[33] Schwartz, M., Bennett, W., Stein, S.: *Communication Systems and Techniques.* McGraw-Hill Book Company Inc., New York (1966).

[34] Trees, H.: *Detection, Estimation, and Modulation Theory, Part I.* John Wiley & Sons Inc., New York (1971).

[35] Lei, T.: *Array Calibration in a Correlated Signal Environment,* Ph.D. dissertation. University of Pennsylvania, Philadelphia, Pennsylvania, (1987).

[36] Gudbjartsson, H., Patz, S.: The Rician distribution of noisy MRI data. *Magn. Reson. Med.* **34** (1995) 910–914.

9

Stochastic Image Models

9.1 Introduction

Chapters 6 and 8 describe statistical properties of X-ray CT imaging and MR imaging at three levels of the image: a single pixel, any two pixels, and a group of pixels (i.e., an image region). When a probabilistic distribution of any pixel intensity with respect to all other pixel intensities in the image is viewed as a stochastic model for the image, then this model can be thought of as the statistical property of an image at its image level. In this way, statistical properties at the three bottom levels of X-ray CT and MR images described in Chapters 6 and 8 can be integrated into those at this top level to build stochastic models. Thus, this chapter is a continuation of Chapters 6 and 8.

Chapters 2 and 3 show that X-ray CT imaging and MR imaging are based on different physical phenomena and their imaging principles are very different. Chapters 5 and 7 show that data acquired in X-ray CT and MR imaging processes represent different physical quantities and their statistical properties are also different. However, Chapters 6 and 8 show that X-ray CT imaging and MR imaging have the very similar statistical properties. For example, in these two types of images, the intensity of a single pixel has a Gaussian distribution; intensities of any two pixels are spatially asymptotically independent; intensities of a group of pixels (i.e., an image region) form a stationary and ergodic random process. These common statistical properties suggest that these two imaging modalities may have some fundamental and intrinsic links.

One possible reason for X-ray CT imaging and MR imaging having very similar statistical properties may be the fact that they both belong to nondiffraction computed tomographic imaging, which is briefly discussed in Chapter 4. In nondiffraction CT imaging, the interaction model and the external measurements (e.g., projections) are characterized by the straight line integrals of some indexes of the medium and the image reconstruction is based on the Fourier slice theorem. The convolution reconstruction method (FBP) for X-ray CT and the projection reconstruction method (PR) for MRI have shown this common feature.

The common statistical properties at the three bottom levels of X-ray CT and MR images also suggest that we can create unified stochastic models for both X-ray CT and MR images. Based on our view of a stochastic image model (i.e., a probabilistic distribution of any pixel intensity with respect to all other pixel intensities in the image), these stochastic image models, in fact, are the probability density functions (pdfs) of the image. This chapter shows two stochastic image models. The first model can be treated as a special case, that is, a simple version of the second model.

The use of stochastic models depends on the application and especially on the image quality. Chapters 5 and 7 indicate that the imaging SNR is one of the fundamental measures of image quality. Chapters 6 and 8 show that the statistical properties of X-ray CT and MR images are related to the image SNR. This chapter will show that SNR plays an important role in the model selection and the model order reduction.

9.2 Stochastic Model I

Let $IMG(J, K)$ denote an image consisting of J pixels and K image regions; $x_{i,j}$ denotes the intensity of the pixel at (i, j) in the image $(1 \leq i \leq J_i,\ 1 \leq j \leq J_j,\ J_iJ_j = J)$. $\mathcal{R}_k$ and $IMG_{\mathcal{R}_k}$ $(k = 1, \cdots, K)$ represent the k-th image region and the k-th region image, respectively. All image regions are mutually exclusive.

Chapter 6 and Chapter 8 show that every pixel intensity has a Gaussian distribution, any two pixel intensities are spatially asymptotically independent, and each image region is a stationary, ergodic Gaussian random process. However, Appendix A of Chapter 6 and Appendix E of Chapter 8 also show that when the image signal-to-noise ratio (SNR) is sufficiently large, pixel intensities of the image can be considered statistically independent. This section describes a stochastic model for the image whose pixel intensities are statistically independent, and gives a general discussion of the relationship between the independence of pixel intensities and the SNR of the image.

9.2.1 Independent Finite Normal Mixture

Property 9.1 The pdf $f(x_{i,j})$ of the pixel intensity $x_{i,j}$ with respect to all other pixel intensities in the image is a sum of the weighted marginal pdfs $g(x_{i,j}|\theta_k)$ $(k = 1, \cdots, K)$ of the pixel intensity $x_{i,j}$,

$$f(x_{i,j}) = \sum_{k=1}^{K} \pi_k g(x_{i,j}|\theta_k), \tag{9.1}$$

where the weight π_k is the probability of the occurrence of the image region $\mathcal{R}_k$ in the image, that is,

$$\pi_k = P(\mathcal{R} = \mathcal{R}_k), \tag{9.2}$$

which is characterized by a multinomial distribution (MN) [1] given by

$$0 < \pi_k < 1 \quad \text{and} \quad \sum_{k=1}^{K} \pi_k = 1, \tag{9.3}$$

each marginal pdf $g(x_{i,j}|\theta_k)$ is a Gaussian characterized by the mean μ_k and the variance σ_k^2 of each image region given by

$$g(x_{i,j}|\theta_k) = \frac{1}{\sqrt{2\pi}\sigma_k} \exp(-\frac{(x_{i,j} - \mu_k)^2}{2\sigma_k^2}), \tag{9.4}$$

with the parameter vector given by

$$\theta_k = (\mu_k, \sigma_k). \tag{9.5}$$

Eq. (9.1) is a new pdf and known as the independent Finite Normal Mixture, abbreviated iFNM.

Proof.

1) For convenience of the derivation, $IMG(J,K)$ is simplified to IMG. Because its K image regions are mutually exclusive, we have

$$IMG = \bigcup_{k=1}^{K} IMG_{\mathcal{R}_k}, \tag{9.6}$$

where $\bigcup$ represents the union. Thus

$$0 < \pi_k = P(\mathcal{R}_k) < 1, \tag{9.7}$$

and

$$\sum_{k=1}^{K} \pi_k = \sum_{k=1}^{K} P(\mathcal{R}_k) = P(\bigcup_{k=1}^{K} \mathcal{R}_k) = 1. \tag{9.8}$$

π_k is characterized by a multinomial distribution.

2) Property 6.1 and Property 8.1 show that the pixel intensity in either the X-ray CT image or the MR image has a Gaussian distribution. Because (a) the intensities of pixels in an image region have the same mean and the variance, and (b) a Gaussian distribution is uniquely determined by its mean and variance, pixel intensities in an image region can be characterized by a (one) Gaussian distribution. Thus, for the pixel $(i,j) \in \mathcal{R}_k$, the pdf of $x_{i,j}$ is given by Eq. (9.4).

3) From 1) and 2), the probability for a pixel (i,j) being in the k-th image region $\mathcal{R}_k$, that is, the probability for a pixel intensity $x_{i,j}$ having $N(\mu_k, \sigma_k)$ distribution, is $\pi_k g(x_{i,j}|\theta_k)$. Thus, the probability for a pixel (i,j) being in the image, that is, the probability of a pixel intensity $x_{i,j}$ with respect to all other pixel intensities in the image, is $\sum_{k=1}^{K} \pi_k g(x_{i,j}|\theta_k)$.

4) It is clear that

$$\sum_{k=1}^{K} \pi_k g(x_{i,j}|\theta_k) > 0, \tag{9.9}$$

and

$$\int_{-\infty}^{\infty} \sum_{k=1}^{K} \pi_k g(x_{i,j}|\theta_k) dx_{i,j} = \sum_{k=1}^{K} \pi_k \int_{-\infty}^{\infty} g(x_{i,j}|\theta_k) dx_{i,j} = \sum_{k=1}^{K} \pi_k = 1. \tag{9.10}$$

Thus, $\sum_{k=1}^{K} \pi_k \, g(x_{i,j}|\theta_k)$ is a pdf and denoted by $f(x_{i,j})$. ■

Note: All pixel intensities x_{ij} are independently identically distributed (i.i.d.) samples drawn from iFNM (Eq. (9.1)).

9.2.2 Independence and Signal-to-Noise Ratio

Property 6.6 shows that for X-ray CT image with high SNR, its pixel intensities are approximately independent. Eq. (6.54) gives its proof. This situation is also observed in MR images. Eq. (8.44) has essentially same form as Eq. (6.54), and will be discussed further in Section 9.4.2. In the following, we elaborate on the relationship between the correlation of pixel intensities and the SNR of the image from a general statistics theory.

Let x_{i_1,j_1} and x_{i_2,j_2} be two pixel intensities, $x_{i_1,j_1} \sim N(\mu_1, \sigma_1)$ and $x_{i_2,j_2} \sim N(\mu_2, \sigma_2)$. Let $C_x((i_1,j_1),(i_2,j_2))$ and $R_x((i_1,j_1),(i_2,j_2))$ be their covariance and correlation, respectively. From the definition of the correlation coefficient $r_x((i_1,j_1),(i_2,j_2))$, we have

$$r_x((i_1,j_1),(i_2,j_2)) = \frac{C_x((i_1,j_1),(i_2,j_2))}{\sigma_1 \sigma_2}, \tag{9.11}$$

which leads to

$$R_x((i_1,j_1),(i_2,j_2)) = \mu_1\mu_2 + r_x((i_1,j_1),(i_2,j_2))\sigma_1\sigma_2. \tag{9.12}$$

Eq. (9.12) can be rewritten as

$$R_x((i_1,j_1),(i_2,j_2)) = \mu_1\mu_2(1 + \frac{1}{\frac{\mu_1}{\sigma_1}\frac{\mu_2}{\sigma_2}} r_x((i_1,j_1),(i_2,j_2))). \tag{9.13}$$

In Eq. (9.13), $\frac{\mu_1}{\sigma_1}$ and $\frac{\mu_2}{\sigma_2}$ represent the signal-to-noise ratio (SNR). Let

$$SNR_1 = \frac{\mu_1}{\sigma_1} \quad \text{and} \quad SNR_2 = \frac{\mu_2}{\sigma_2}; \tag{9.14}$$

we have

$$R_x((i_1,j_1),(i_2,j_2)) = \mu_1\mu_2(1 + \frac{1}{SNR_1 SNR_2} r_x((i_1,j_1),(i_2,j_2))). \tag{9.15}$$

Thus, when SNR is sufficiently large, the second item in parenthesis on the right side of Eq. (9.15) becomes very small and can be ignored. Under this condition (an approximation), Eq. (9.15) becomes

$$R_x((i_1,j_1),(i_2,j_2)) \simeq \mu_1\mu_2 = E[x_{i_1,j_1}]E[x_{i_2,j_2}], \tag{9.16}$$

which implies that x_{i_1,j_1} and x_{i_2,j_2} are approximately uncorrelated, and hence independent.

$\frac{1}{SNR_1 SNR_2} r_x((i_1,j_1),(i_2,j_2))$ of Eq. (9.15) is a measure of the correlation of pixel intensities x_{i_1,j_1} and x_{i_2,j_2}. To illustrate the relation between this measure and the SNR of pixel intensities, numerical examples are shown in Table 9.1, where $|r_x((i_1,j_1),(i_2,j_2))| \leq 1$, for example, 0.5.

TABLE 9.1
Relation between SNR and a Measure of Correlations of Pixel Intensities

$SNR_1 = SNR_2 \geq$	2.24/3.5 db	7.07/8.5 db	22.4/13.5 db
$\frac{1}{SNR_1 SNR_2} r_x((i_1,j_1),(i_2,j_2)) <$	0.100	0.010	0.001

The pixel intensity $x_{i,j}$ consists of two components: signal and noise. The signal component and noise component are viewed as deterministic and random, respectively. The correlation between pixel intensities is mainly determined by their noise components. Eq. (9.14) shows that SNR can be increased by either reducing the noise power or increasing signal strength, or doing both. In each of these three cases, the noise power is reduced either absolutely or relatively (with respect to the signal strength). As a result, the correlation between the noise components of two pixel intensities becomes weaker. Thus, the stronger SNR and the weaker correlation are consistent.

9.3 Stochastic Model II

9.3.1 Markovianity

This section shows that the pixel intensities of X-ray CT and MR images form a Markov process, that is, a Markov random field (MRF) with the proper neighborhood system. This property is called Markovianity. MRF involves multiple pixels. Because pixels in an MRF neighborhood may not be required to be in one group (e.g., in an image region), Markovianity is not addressed in Chapter 6 and Chapter 8, where statistical properties of X-ray CT and MR images are described at the three levels (a single, two, and a group of pixels) of the image.

9.3.1.1 Markov Random Field

SAI (Property 6.2 and Property 8.2) and ECC (Property 8.3) show that when the spatial separation between pixels becomes sufficiently large, the magnitude of the correlation of pixel intensities becomes very small and can be negligible. This property represents a special case of the *local dependence*, which can be characterized by a Markov process or Markov field [1, 2].

Markov process. In this concept, a random process $\mathbf{x}(n)$ is a p-th-order Markov, if the conditional probability of $\mathbf{x}(n)$ given the entire past is equal to the conditional probability of $\mathbf{x}(n)$ given only $\mathbf{x}(n-1), \cdots, \mathbf{x}(n-p)$. Thus, let $\mathbf{x}_i$ $(i = 1, 2, \cdots)$ be discrete random variables, a random process is called p-th-order Markov if

$$P(\mathbf{x}_n | \mathbf{x}_{n-1}, \mathbf{x}_{n-2}, \cdots, \mathbf{x}_1) = P(\mathbf{x}_n | \mathbf{x}_{n-1}, \cdots, \mathbf{x}_{n-p}). \tag{9.17}$$

In a Markov process, the transfer probability is defined by

$$P_{ij}(n, s) = P(\mathbf{x}_n = a_i | \mathbf{x}_s = a_j) \quad (n > s), \tag{9.18}$$

and has the properties

1) Positive

$$P_{ij}(n, s) > 0 \quad \text{and} \quad \sum_j P_{ij}(n, s) = 1; \tag{9.19}$$

2) Reversal

$$P_{ij}(n, s) = P(\mathbf{x}_n = a_i | \mathbf{x}_s = a_j) \quad (n < s); \tag{9.20}$$

3) Transitional

$$P_{ij}(n, s) = \sum_k P_{ik}(n, r) P_{kj}(r, s) \quad (n > r > s), \tag{9.21}$$

which is the discrete version of the Chapman–Kolmogoroff equation.

Markov random field. In this concept, a 2-D random field is a Markov, if at every pixel location we can find a partition $\int^+$ (the future), $\mathcal{N}$ (the present), and $\int^-$ (the past) of the 2-D lattice $\mathcal{L}$ providing support to the sets of random variables $\mathbf{x}^+$, $\mathbf{x}_{\mathcal{N}}$, and $\mathbf{x}^-$ such that

$$P(\mathbf{x}^+ | \mathbf{x}_{\mathcal{N}}, \mathbf{x}^-) = P(\mathbf{x}^+ | \mathbf{x}_{\mathcal{N}}), \tag{9.22}$$

which is illustrated in Figure 9.1.

Lattice. A 2-D lattice $\mathcal{L}$ is defined by a set of pairs of integers

$$\mathcal{L} = \{(i, j), \ 1 \leq i, j \leq M\}. \tag{9.23}$$

● - the future, $*$ - the present, $\cdot$ - the past.

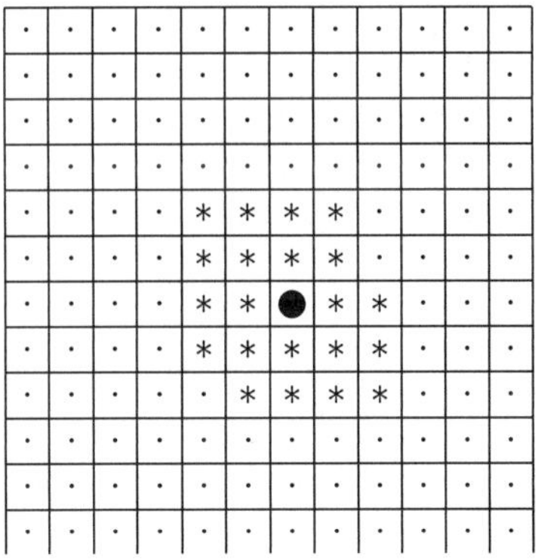

·	·	·	·	·	·	·	·	·	·	·	·
·	·	·	·	·	·	·	·	·	·	·	·
·	·	·	·	·	·	·	·	·	·	·	·
·	·	·	·	·	·	·	·	·	·	·	·
·	·	·	·	*	*	*	*	·	·	·	·
·	·	·	·	*	*	*	*	·	·	·	·
·	·	·	·	*	*	●	*	*	·	·	·
·	·	·	·	*	*	*	*	*	·	·	·
·	·	·	·	·	*	*	*	*	·	·	·
·	·	·	·	·	·	·	·	·	·	·	·
·	·	·	·	·	·	·	·	·	·	·	·
·	·	·	·	·	·	·	·	·	·	·	·

FIGURE 9.1
An illustration of the 2-D lattice and the local dependence $P(\bullet|*, \cdot) = P(\bullet|*)$.

Neighborhood of pixels. The neighborhood $\mathcal{N}_{i,j}$ of the pixel (i,j) $((i,j) \in \mathcal{L})$ is a subset of $\mathcal{L}$, that is, $\mathcal{N}_{i,j} \subset \mathcal{L}$. $\mathcal{N}_{i,j}$ and $\mathcal{N}_{k,l}$ are the neighborhoods of pixels (i,j) and (k,l) if and only if

$$\begin{aligned} &(i,j) \notin \mathcal{N}_{i,j} \quad \text{and} \quad (k,l) \notin \mathcal{N}_{k,l}, \\ &(k,l) \in \mathcal{N}_{i,j} \Longleftrightarrow (i,j) \in \mathcal{N}_{k,l}. \end{aligned} \tag{9.24}$$

Extent of Neighborhood. Let $d = ||(i,j),(k,l)||$ denote the distance between centers of pixels (i,j) and (k,l) in the unit of the pixel; d is an integer. Let p be the order of the neighborhood of pixel (i,j), $\mathcal{N}^p_{i,j}$. If $d \leq \sqrt{2}^{p-1}$, then pixels $(k,l) \in \mathcal{N}^p_{i,j}$. This criterion is valid for $p = 1, 2, 3$.

Clique. Clique c is a subset of $\mathcal{L}$, in which every pair of distinct pixels in c are mutual neighbors, that is, they are in the neighborhoods of each other. Note, by this definition, that individual pixels are cliques.* A function defined on the clique, known as the clique function, is described in Chapter 11.

Neighborhood system. A collection of subsets of $\mathcal{L}$ described by

$$\mathcal{N} = \{\mathcal{N}_{i,j} | (i,j) \in \mathcal{L},\ \mathcal{N}_{i,j} \subset \mathcal{L}\}, \tag{9.25}$$

forms a neighborhood system on the lattice $\mathcal{L}$, which can also be understood as an ordered class of cliques

$$\mathcal{C} = \{c_{i,j} | (i,j) \in \mathcal{L},\ c_{i,j} \subset \mathcal{L}\}. \tag{9.26}$$

*For the pixels located on the edges of an image, their neighborhoods and cliques are specially defined.

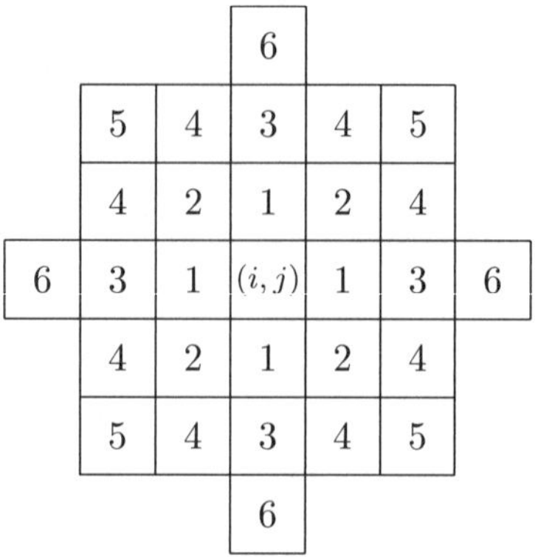

			6			
	5	4	3	4	5	
	4	2	1	2	4	
6	3	1	(i,j)	1	3	6
	4	2	1	2	4	
	5	4	3	4	5	
			6			

$$\mathcal{N}^p = \{\mathcal{N}_{i,j}^p\} \quad \mathcal{N}_{i,j}^p = \{k : k \leq p\}$$

FIGURE 9.2
A structure of a neighborhood system up to the 6th order in which the p-th-order neighborhood system $\mathcal{N}^p$ consists of all lower-order systems $\mathcal{N}^k$ $(k < p)$ and additional pixels marked by p.

Figure 9.2 shows a neighborhood system up to the 6th order, where each square represents a pixel. The integer inside each square represents the lowest order of the neighborhood system to which the pixel belongs. Thus, for an arbitrary pixel, say (i,j), its p-th-order neighborhood system $\mathcal{N}_{i,j}^p$ consists of all its lower-order neighborhood systems: $\mathcal{N}_{i,j}^p = \bigcup_{k=1}^p \mathcal{N}_{i,j}^k = \{k : k \leq p\}$.

Figure 9.3 shows cliques in the 1st and 2nd order neighborhood systems $(\mathcal{N}^p, p = 1,2)$. The clique c depends on the order of the neighborhood systems (p) and the number of pixels in the clique (q). For example, a clique may consist of two pixels (horizontal and vertical pair) in $\mathcal{N}^1$, or two pixels ($45°$ and $135°$ diagonal pair) in $\mathcal{N}^2$, or three pixels (triangle) in $\mathcal{N}^2$, etc.

Markov random field. Let $\mathbf{x}_{i,j}$ $(i,j = 1,2,\cdots)$ be discrete random variables, a discrete 2-D random field is called a Markov [3–8] on $(\mathcal{L}, \mathcal{N})$, if and only if $\forall (i,j) \in \mathcal{L}$, the probability mass function

$$(a) \qquad P(\mathbf{x}_{i,j} = x_{i,j}) > 0,$$

$$(b) \qquad P(\mathbf{x}_{i,j} = x_{i,j} | \mathbf{x}_{k,l} = x_{k,l}, (k,l) \neq (i,j)) \tag{9.27}$$

$$= P(\mathbf{x}_{i,j} = x_{i,j} | \mathbf{x}_{k,l} = x_{k,l}, (k,l) \in \mathcal{N}_{i,j}).$$

Eq. (9.27) can be simplified to

$$P(x_{i,j}) > 0 \quad \text{and} \quad P(x_{i,j} | x_{\mathcal{L}-(i,j)}) = P(x_{i,j} | x_{\mathcal{N}_{i,j}}). \tag{9.28}$$

Gibbs random field. Let $\mathbf{x} = \{\mathbf{x}_{i,j}\}$ $(1 \leq i,j \leq M)$ be defined on $\mathcal{L}$. A probability density function, often known as the Gibbs distribution and

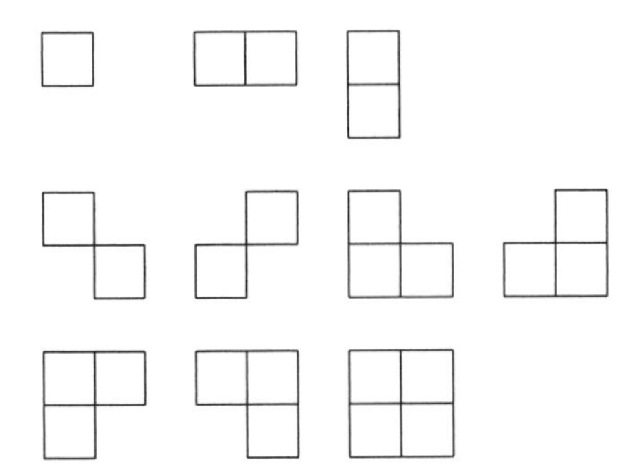

FIGURE 9.3
Clique types in the second-order neighborhood system $\mathcal{N}^2$.

abbreviated GD, is defined by

$$p_G(\mathbf{x}) = Z^{-1} \exp(-\beta^{-1} \sum_c V_c(\mathbf{x})), \tag{9.29}$$

where the partition function $Z = \sum_{\mathbf{x}} \exp(-\beta^{-1} \sum_c V_c(\mathbf{x}))$ is a normalized constant, β is a constant (also known as the temperature), $\sum_c V_c(\mathbf{x})$ is an energy function, and $V_c(\mathbf{x})$ is the clique potential of clique c. $p_G(\mathbf{x})$ specifies a random field called Gibbs random field (GRF), which is defined on $(\mathcal{C}, \mathcal{L})$.

MRF describes the local properties of $\mathbf{x}$, while the Gibbs distribution provides a global description of $\mathbf{x}$. $\mathcal{N}$ of Eq. (9.25) and $\mathcal{C}$ of Eq. (9.26) in the definition of the neighborhood system suggest that there are some intrinsic links between MRF and the Gibbs distribution. The Hammersley–Clifford theorem [3, 9–14] describes the equivalence between MRF and GRF.

For any neighborhood system $\mathcal{N}$, $\mathbf{x}$ is an MRF with respect to $\mathcal{N}$ (Eq. (9.27)) if and only if $p_G(\mathbf{x})$ is a Gibbs distribution with respect to $\mathcal{C}$ (Eq. (9.29)).

It has been shown that this equivalence can be expressed as

$$P(x_{i,j} = x | x_{\mathcal{L}-(i,j)}) = \frac{e^{-\beta^{-1} \sum_{c:(i,j) \in c} V_c(x_{i,j}=x)}}{\sum_{x_{i,j}} e^{-\beta^{-1} \sum_{c:(i,j) \in c} V_c(x_{i,j})}}. \tag{9.30}$$

Eq. (9.30) indicates that the local characteristics of $\mathbf{x}$ described by MRF with respect to $\mathcal{N}$ can be evaluated by a measure given by Gibbs distribution with respect to $\mathcal{C}$ via clique potentials V_c. The ability to move between local and global descriptions is a big advantage of MRF and GRF over other models, and provides the basis for MRF generation, which is described in Appendix 9A.

9.3.1.2 Markovianity

Based on the definition of MRF, a set of *necessary and sufficient conditions* for a random field (RF) to be an MRF is given and proved in the following assertion.

Assertion Let $\mathcal{L}$ be a 2-D finite lattice, $\mathcal{N}$ a neighborhood system on $\mathcal{L}$, $\mathbf{x} = \{\mathbf{x}_{i,j}\}$ $(1 \leq i, j \leq M)$ a 2-D discrete RF on $(\mathcal{L}, \mathcal{N})$. Necessary and sufficient conditions for $\mathbf{x}$ to be an MRF are that $\forall(i,j) \in \mathcal{L}$ and $\forall(k,l) \in \mathcal{L} - \mathcal{N}_{i,j} - (i,j)$; $\mathbf{x}_{i,j}$ and $\mathbf{x}_{k,l}$ are conditionally independent of the neighborhood of (i,j) or (k,l), that is,

$$P(x_{i,j}, x_{k,l}|x_{\mathcal{N}_{i,j}}) = P(x_{i,j}|x_{\mathcal{N}_{i,j}})P(x_{k,l}|x_{\mathcal{N}_{i,j}}) \,, \tag{9.31}$$

where $\mathcal{N}_{i,j}$ denotes the neighborhood of (i,j) and $P(\star|*)$ represents the conditional probability.

Proof

As shown by Eq. (9.28), a discrete 2-D RF $\mathbf{x}$ is called an MRF on $(\mathcal{L}, \mathcal{N})$ if and only if $\forall(i,j) \in \mathcal{L}$,

$$P(x_{i,j}) > 0 \quad \text{and} \quad P(x_{i,j}|x_{\mathcal{L}-(i,j)}) = P(x_{i,j}|x_{\mathcal{N}_{i,j}}). \tag{9.32}$$

Necessary. From the Bayesian theorem, we have

$$\begin{aligned} P(x_{i,j}, x_{k,l}|x_{\mathcal{N}_{i,j}}) &= P(x_{i,j}, x_{k,l}, x_{\mathcal{N}_{i,j}})/P(x_{\mathcal{N}_{i,j}}) \\ &= P(x_{i,j}|x_{k,l}, x_{\mathcal{N}_{i,j}})P(x_{k,l}, x_{\mathcal{N}_{i,j}})/P(x_{\mathcal{N}_{i,j}}) \\ &= P(x_{i,j}|x_{k,l}, x_{\mathcal{N}_{i,j}})P(x_{k,l}|x_{\mathcal{N}_{i,j}}). \end{aligned} \tag{9.33}$$

If $\mathbf{x}$ is an MRF, then $P(x_{i,j}|x_{k,l}, x_{\mathcal{N}_{i,j}}) = P(x_{i,j}|x_{\mathcal{L}-(i,j)}) = P(x_{i,j}|x_{\mathcal{N}_{i,j}})$. Thus, Eq. (9.33) becomes Eq. (9.31), that is, the condition Eq. (9.31) is met.

Sufficient. From the Bayesian theorem, we have

$$\begin{aligned} P(x_{i,j}|x_{\mathcal{L}-(i,j)}) &= P(x_{i,j}|x_{k,l}, x_{\mathcal{N}_{i,j}}) \\ &= P(x_{i,j}, x_{k,l}, x_{\mathcal{N}_{i,j}})/P(x_{k,l}, x_{\mathcal{N}_{i,j}}) \\ &= P(x_{i,j}, x_{k,l}|x_{\mathcal{N}_{i,j}})/P(x_{k,l}|x_{\mathcal{N}_{i,j}}). \end{aligned} \tag{9.34}$$

If the condition Eq. (9.31) is met, then $P(x_{i,j}|x_{\mathcal{L}-(i,j)}) = P(x_{i,j}|x_{\mathcal{N}_{i,j}})$, that is, the condition Eq. (9.32) is met, then $\mathbf{x}$ is an MRF. ■

Applying the above necessary and sufficient conditions to X-ray CT and MR images and using SAI (Property 6.2 and Property 8.2), the Markovianity of X-ray CT and MR images is given and proved below.

Property 9.2 X-ray CT and MR images are embedded in an MRF with a proper neighborhood system in terms of pixel intensities and their correlations.

Proof

Let $||(i,j),(k,l)||$ denote the distance (in the unit of pixel size) between pixels at (i,j) and (k,l). Based on SAI (Property 6.2 and Property 8.2), when $||(i,j),(k,l)||$ increases to a value, say m, the correlation of two pixel intensities becomes very small and negligible. For instance, in an X-ray CT image, $m = 2$ for the parallel projection and $m = 7$ for the divergent projection (Figure 6.3); in an MR image, $m = 1$ for the basic FT image (Figure 8.8) and $m = 3$ for the filtered FT MR image (Figures 8.9 and 8.10). Thus, when $||(i,j),(k,l)|| \geq m$, $\mathbf{x}_{i,j}$ and $\mathbf{x}_{k,l}$ can be considered conditionally uncorrelated, and hence independent due to Gaussianity, that is,

$$P(x_{i,j}, x_{k,l}|(||(i,j),(k,l)|| \geq m))$$

$$= P(x_{i,j}|(||(i,j),(k,l)|| \geq m))P(x_{k,l}|(||(i,j),(k,l)|| \geq m)). \quad (9.35)$$

$\forall(i,j) \in \mathcal{L}$, we can select a neighborhood $\mathcal{N}_{i,j}$ for it such that $\forall(k,l) \in \mathcal{L} - \mathcal{N}_{i,j} - (i,j)$, $||(i,j),(k,l)|| \geq m$. Thus, Eq. (9.35) is equivalent to

$$P(x_{i,j}, x_{k,l}|x_{\mathcal{N}_{i,j}}) = P(x_{i,j}|x_{\mathcal{N}_{i,j}})P(x_{k,l}|x_{\mathcal{N}_{i,j}}). \quad (9.36)$$

Therefore, the sufficient condition Eq. (9.31) is satisfied. Therefore, for such selected neighborhood systems, pixel intensities in X-ray CT and MR images form an MRF. ■

Property 9.2 leads to a procedure for selecting this neighborhood system. Figure 9.2 shows the neighborhood systems up to the 6th order, where each square represents a pixel. The integer number inside each square represents the lowest order of the neighborhood system to which that pixel belongs. The distance between the centers of two adjacent pixels, in either the horizontal or vertical direction, is the pixel, that is, counted by the integer m.

As an example, the numerical values of the curve in Figure 8.9b, $c_{fn}(m_1, m_2)$, which is for the 2-D case[†] are given in Table 9.2. Suppose that under the condition $c_{fn}(m_1, m_2) < 0.1$, pixel intensities can be approximately considered independent; then by mapping Table 9.2 onto Figure 9.2

[†]The curves in Figures 8.8–8.10 are for the 1-D case. Because 2-D FT and all other functions used in FT imaging are separable, the correlations of pixel intensities for the 2-D case are the products of those in two 1-D cases; thus, $c_{fn}(m_1, m_2) = c_{fn}(m_1)c_{fn}(m_2)$.

(1.0000 sits on (i, j) and each number sits on each pixel), the constraint $c_{fn}(m_1, m_2) > 0.1$ will define a region that is exactly the 3rd-order neighborhood of (i, j). This example shows that MR images reconstructed by filtered FT can be considered embedded in an MRF with the 3rd-order neighborhood system under the condition $c_{fn}(m_1, m_2) > 0.1$.

Generally, the threshold t in the condition $c_{fn}(m_1, m_2) > t$ can be different. When t is smaller, the order of the neighborhood system will be higher, which gives a more accurate MRF model, but also requires more computations in the MRF model-based image analysis. Therefore, the choice of t should be dictated by a trade-off between accuracy and efficiency.

TABLE 9.2
$c_{fn}(m_1, m_2)$ of Eq. (8.154) over the Neighborhood Systems

m_2	$m_1 = -3$	-2	-1	0	$+1$	$+2$	$+3$
-3	.00000	.00000	.00001	.00003	.00001	.00000	.00000
-2	.00000	.01778	.08337	**.13333**	.08337	.01778	.00000
-1	.00001	.08337	**.39100**	**.62530**	**.39100**	.08337	.00001
0	.00003	**.13333**	**.62530**	**1.0000**	**.62530**	**.13333**	.00003
$+1$	.00001	.08337	**.39100**	**.62530**	**.39100**	.08337	.00001
$+2$	.00000	.01778	.08337	**.13333**	.08337	.01778	.00000
$+3$	.00000	.00000	.00001	.00003	.00001	.00000	.00000

9.3.2 Correlated Finite Normal Mixture

X-ray CT and MR images consist of piecewise contiguous and mutually exclusive image regions. By using the same notation of Section 9.2, IMG, $\mathcal{R}_k$ and $IMG_{\mathcal{R}_k}$ denote the image, its k-th image region and k-th region image. $IMG = \bigcup_{k=1}^{K} \mathcal{R}_k$. Let $\mathcal{K} = \{1, 2, \cdots, K\}$ be an ordered set of the labels and each image region $\mathcal{R}_k$ $(k = 1, \cdots, K)$ be labeled by a $k \in \mathcal{K}$.

A pixel (i, j) is also given a label $y_{i,j}$. In the k-th image region $\mathcal{R}_k$, because all pixel intensities $x_{i,j}$ have the same mean μ_k and variance σ_k^2 (the definition of the image region) and have the same Gaussian distribution $N(\mu_k, \sigma_k)$ (Property 6.1 and Property 8.1), the pixel label is assigned based on the following role: if $(i, j) \in \mathcal{R}_k$, then $y_{i,j} = k$.

Thus, each pixel (i, j) in an image has two signatures: an intensity $x_{i,j}$ and a label $y_{i,j}$. Hence, the image can be viewed to have two layers: an intensity layer $\mathbf{x} = \{x_{i.j}\}$ that is directly observed and a context layer $\mathbf{y} = \{y_{i.j}\}$ that seems to be hidden.

It is clear that if $x_{i,j} \sim N(\mu_k, \sigma_k)$, then $y_{i,j} = k$; or, if $y_{i,j} = k$, then

$x_{i,j} \sim N(\mu_k, \sigma_k)$. This relationship can be expressed as[‡]

$$\{x_{i,j} \sim N(\mu_k, \sigma_k^2)\} \Longleftrightarrow \{y_{i,j} = k\} .$$

Because $\mathbf{x} = \{x_{i.j}\}$ is embedded in an MRF with respect to a selected neighborhood system (Property 9.2), the one-to-one correspondence between the intensity $x_{i,j}$ and the label $y_{i,j}$ of the pixel (i,j) leads to $\mathbf{y} = \{y_{i.j}\}$ is also embedded in an MRF with respect to the same neighborhood system as $\mathbf{x} = \{x_{i.j}\}$.

A stochastic model for X-ray CT and MR images is described by the following property.

Property 9.3 Let $x_{i,j}$ and $y_{i,j}$ be the intensity and label of the pixel (i,j) in the image, and $\mathcal{N}_{i,j}$ be the neighborhood of (i,j). The conditional pdf $f(x_{i,j}|x_{\neq i,j})$ of a pixel intensity given all other pixel intensities in the image is a sum of the penalized marginal pdfs $g(x_{i,j}|\theta_k)$ $(k = 1, \cdots, K)$ of the pixel intensity $x_{i,j}$,

$$f(x_{i,j}|x_{\neq i,j}) = \sum_{k=1}^{K} P(y_{i,j} = k|y_{\mathcal{N}_{i,j}})g(x_{i,j}|\theta_k), \tag{9.37}$$

where K is the number of image regions in the image, the conditional probability $P(y_{i,j} = k|y_{\mathcal{N}_{i,j}})$ is the penalty of the marginal pdf, given by

$$P(y_{i,j} = k|y_{\mathcal{N}_{i,j}}) = p_G(y_{i,j}) = Z^{-1}\exp(-\beta^{-1}\sum_{c} V_c(y_{i,j})), \tag{9.38}$$

where p_G denotes Gibbs pdf (Eq. (9.29)), Z is the partition function, β is a constant, and $V_c(y_{i,j})$ is the clique potential. Each marginal pdf $g(x_{i,j}|\theta_k)$ is a Gaussian characterized by the mean μ_k and variance σ_k^2 of each region image, given by

$$g(x_{i,j}|\theta_k) = \frac{1}{\sqrt{2\pi}\sigma_k}\exp(-\frac{(x_{i,j} - \mu_k)^2}{2\sigma_k^2}), \tag{9.39}$$

with $\theta_k = (\mu_k, \sigma_k)$ as a parameter vector. Eq. (9.37) is a new pdf and known as the correlated Finite Normal Mixture, abbreviated cFNM.

Proof

Based on Property 9.2 (Markovianity) and Eq. (9.28), we have

$$f(x_{i,j}|x_{\neq i,j}) = P(x_{i,j}|x_{\mathcal{L}-(i,j)}) = P(x_{i,j}|x_{\mathcal{N}_{i,j}}). \tag{9.40}$$

Using the Bayesian theorem, we have

$$P(x_{i,j}|x_{\mathcal{N}_{i,j}}) = \sum_{k=1}^{K} P(x_{i,j}, y_{i,j} = k|x_{\mathcal{N}_{i,j}})$$

[‡]The partial volume effect is not considered here.

$$
\begin{aligned}
&= \sum_{k=1}^{K} P(x_{i,j}, y_{i,j} = k, x_{\mathcal{N}_{i,j}})/P(x_{\mathcal{N}_{i,j}}) \\
&= \sum_{k=1}^{K} P(x_{i,j}|y_{i,j} = k, x_{\mathcal{N}_{i,j}})P(y_{i,j} = k, x_{\mathcal{N}_{i,j}})/P(x_{\mathcal{N}_{i,j}}) \\
&= \sum_{k=1}^{K} P(y_{i,j} = k|x_{\mathcal{N}_{i,j}})P(x_{i,j}|y_{i,j} = k) \\
&= \sum_{k=1}^{K} P(y_{i,j} = k|y_{\mathcal{N}_{i,j}})g(x_{i,j}|\theta_k) \ . \qquad (9.41)
\end{aligned}
$$

Eqs. (9.40) and (9.41) give Eq. (9.37).

With MRF-GD equivalence, the MRF description $P(y_{i,j} = k|y_{\mathcal{N}_{i,j}})$ equals the GD description $p_G(y_{i,j} = k)^{\S}$, which gives Eq. (9.38).

The proof for the marginal pdf $g(x_{i,j}|\theta_k)$ is the same as that in the proof of Property 9.1.

Because (a) $f(x_{i,j}|x_{\neq i,j}) = P(x_{i,j}|x_{\mathcal{N}_{i,j}}) > 0$ and (b) the integration of $f(x_{i,j}|x_{\neq i,j}) =$ is

$$
\begin{aligned}
\int_{-\infty}^{\infty} f(x_{i,j}|x_{\neq i,j})dx_{i,j} &= \int_{-\infty}^{\infty} P(x_{i,j}|x_{\mathcal{N}_{i,j}})dx_{i,j} \\
&= \sum_{k=1}^{K} P(y_{i,j} = k|y_{\mathcal{N}_{i,j}}) \int_{-\infty}^{\infty} g(x_{i,j}|\theta_k)dx_{i,j} \\
&= \sum_{k=1}^{K} P(y_{i,j} = k|y_{\mathcal{N}_{i,j}}) = 1 \ , \qquad (9.42)
\end{aligned}
$$

$f(x_{i,j}|x_{\neq i,j}) =$ of Eq. (9.37) is a new pdf, which is called the correlated Finite Normal Mixture, abbreviated cFNM. ■

9.4 Discussion

9.4.1 Mixture Models and Spatial Regularity

cFNM (Eq. (9.37)) and iFNM (Eq. (9.1)) are derived from different approaches; however, they have a similar function form of pdf. This similarity

§This is based on Besag pseudolikelihood [15], $\mathcal{PL}(\mathbf{y}) = \prod_{(i,j)} P_G(y_{i,j}|y_{\mathcal{N}_{i,j}}) = \prod_{(i,j)}\{\exp(-\frac{1}{\beta}\sum_{c:(i,j)\in c} V_c(y_{i,j} = k))/\sum_{y_{i,j}} \exp(-\frac{1}{\beta}\sum_{c:(i,j)\in c} V_c(y_{i,j}))\}$.

suggests that there are some intrinsic links between them. Indeed, iFNM is a special case, that is, a simple version, of cFNM.

When pixel intensities $x_{i,j}$ are statistically independent, $f(x_{i,j}|x_{\neq i,j})$ of Eq. (9.37) can be expressed as

$$\begin{aligned} f(x_{i,j}|x_{\neq i,j}) &= f(x_{i,j}) \\ &= \sum_{k=1}^{K} P(x_{i,j}, y_{i,j} = k) \\ &= \sum_{k=1}^{K} P(y_{i,j} = k)P(x_{i,j}|y_{i,j} = k) \\ &= \sum_{k=1}^{K} P(y_{i,j} = k)g(x_{i,j}|\theta_k), \end{aligned} \tag{9.43}$$

Section 9.3.2 shows that $P(y_{i,j} = k)$ is equivalent to $P((i,j) \in \mathcal{R}_k)$ - the probability of the occurrence of the k-th image region in the image. From Property 9.1, $P(y_{i,j} = k) = P((i,j) \in \mathcal{R}_k) = \pi_k$. Thus, Eq. (9.43) becomes

$$f(x_{i,j}) = \sum_{k=1}^{K} \pi_k g(x_{i,j}|\theta_k). \tag{9.44}$$

which is Eq. (9.1).

The common items in Eqs. (9.37) and (9.1) are the marginal pdf $g(x_{i,j}|\theta_k)$, which is a Gaussian characterizing the kth image region in the image. The difference between Eq. (9.37) and Eq. (9.1) is the penalty $P(y_{i,j} = k|y_{\mathcal{N}_{i,j}})$ and the weight π_k, which are characterized by a Gibbs distribution (GD) and a multinomial distribution (MN), respectively.

In cFNM, the probability of $y_{i,j} = k$ is conditional: $p(y_{i,j} = k|y_{\mathcal{N}_{i,j}})$. That is, the GD prior represents a regional constraint modulated by the neighborhood on which GD and MRF are based. Neighborhood-based modulation imposes spatial regularity on the regional constraint. In iFNM, the probability of $y_{i,j} = k$ is unconditional: $p(y_{i,j} = k)$. That is, the MN prior represents a regional constraint only, and the spatial regularity is not imposed.

The difference between iFNM and cFNM can be briefly stated as follows. iFNM is a spatially independent model and can be fully characterized by the histogram of the image data. It is clear that images with the same histogram may have very different structures. cFNM imposes certain spatial regularity to overcome this shortcoming. It is a model that can be adapted to structural information or spatial dependence.

9.4.2 Mixture Models and Signal-to-Noise Ratio

Section 9.3.2 gives a general description of the independence and the signal-to-noise ratio of pixel intensities. This section provides a discussion of the interrelationship between iFNM and cFNM and the signal-to-noise ratio of the image. The MR image is used as an example for this discussion.

cFNM is more general but complicated; iFNM is simpler but requires independence. The following discussion provides insights into if iFNM can be utilized as an approximation of cFNM. For the convenience of discussion, the fourth formula of Eq. (8.44) is copied below

$$R_x((u_1,v_1),(u_2,v_2)) = \mu_{x_1}\mu^*_{x_2}$$
$$+ (\frac{W}{\Delta k})^2 \sigma^2_{\mathcal{M}} sinc(W\Delta u, W\Delta v) f(W\Delta u, W\Delta v), \quad (9.45)$$

where $R_x((u_1,v_1),(u_2,v_2))$ is the correlation of two pixel intensities $x(u_1,v_1)$ and $x(u_2,v_2)$, $\sigma^2_{\mathcal{M}}$ is the variance of k-space samples, W and Δk are the window width and the spacing of k-space sampling, $\Delta u = u_1 - u_2$, $\Delta v = v_1 - v_2$, $sinc(\star,*)$ is a 2-D sinc function, and $f(\star,*)$ is a 2-D filter function used in filtered FT MR imaging.

Let $E[x(u_1,v_1)] = \mu_{x_1}e^{i\phi_1}$, $E[x(u_2,v_2)] = \mu_{x_2}e^{i\phi_2}$. From the eighth formula of Eq. (8.44), the signal-to-noise ratio of the pixel intensity can be expressed by $SNR_1 = \mu_{x_1}/(\frac{W}{\Delta k}\sigma_{\mathcal{M}}\sqrt{f(0,0)})$ and $SNR_2 = \mu_{x_2}/(\frac{W}{\Delta k}\sigma_{\mathcal{M}}\sqrt{f(0,0)})$. Thus, Eq. (9.45) can be rewritten as

$$R_x((u_1,v_1),(u_2,v_2)) = \mu_{x_1}\mu_{x_2}e^{i(\phi_1-\phi_2)}$$
$$(1 + \frac{e^{-i(\phi_1-\phi_2)}}{SNR_1 \cdot SNR_2 \cdot f(0,0)} sinc(W\Delta u, W\Delta v) f(W\Delta u, W\Delta v)). \quad (9.46)$$

When SNR is large, for example, $SNR \geq 10$, the second term inside the parentheses on the right side of Eq. (9.46) becomes very small and can be neglected. Thus, Eq. (9.46) can be approximated by

$$R_x((u_1,v_1),(u_2,v_2)) \simeq E[x(u_1,v_1)]E[x^*(u_2,v_2)], \quad (9.47)$$

which implies that when SNR is large, pixel intensities are approximately uncorrelated, and hence independent. Therefore, iFNM can be utilized as an approximation of cFNM.

9.4.3 Mixture Models and Hidden Markov Random Field

Generally, a Hidden Markov random process is derived from the Hidden Markov Model (HMM), that is, a stochastic process generated by a Markov chain (MC) in which its state sequence cannot be observed directly, only through a sequence of the observations. Each observation is assumed to be a stochastic function of the state sequence. We consider a 2-D case, Hidden Markov random field, abbreviated HMRF [16–19].

A HMRF is defined on two random fields: an (observable) intensity field **x** and a (hidden) context field **y**. We have shown that for X-ray CT and

MR imaging, both $\mathbf{x}$ and $\mathbf{y}$ are MRF with respect to the same neighborhold system. In some literature, these two fields are assumed to be conditional independent, that is,

$$P(\mathbf{x}|\mathbf{y}) = \prod_i P(x_i|y_i).$$

The cFNM model in Eq. (9.37) is defined on two MRFs: the intensity field $\mathbf{x}$ and the context field $\mathbf{y}$. This section discusses some relationships between the cFNM model and the HMRF model.

1) In cFNM, the spatial regularity $p(y_{i,j} = k|x_{\mathcal{N}_{i,j}})$ is derived from the correlations of pixel intensities $x_{i,j}$ of the observed image $\mathbf{x}$. As a result, the order of the neighborhood system and the cliques can be selected with the required accuracy and a clearly physical meaning. In HMRF, spatial regularity is assessed by the assumed Markovianity of the hidden contextual field $\mathbf{y}$. It follows that the selections of the neighborhood systems and the cliques are based on some assumptions that are quite arbitrary.

2) cFNM has an explicit expression Eq. (9.37) that directly links spatial regularity with image structure and characteristics, for example, the number of image regions K, the region property (μ_k, σ_k), and the pixel property (Gaussianity). HMM does not directly show these relationships.

3) cFNM has its independent counterpart iFNM. cFNM and iFNM have almost identical formulae. Their interrelationship offers advantages in image analysis. Although iFNM is viewed as degenerated HMRF, iFNM and HMRF have very different expressions.

4) For cFNM, Eq. (9.37) is a pdf that can be directly used to form the conditional expectation in the Expectation-Maximization (EM) algorithm [20, 21], the pseudolikelihood [3, 22], and the mean field-like approximations [19, 23, 24] in parameter estimation. It is difficult to use HMRF to perform these tasks.

5) In some applications, for example, in image analysis, cFNM and HMRF are also different. The standard EM algorithm and its modified version can be directly applied to iFNM and cFNM, respectively. However, it is not so straightforward for HMRF. In cFNM model-based image analysis, pixels are classified into proper image regions with the highest penalized (imposed by spatial regularity) marginal probability. The HMRF model-based approach often adopts Maximum a Posteriori (MAP) operation; its procedure and interpretation are not so direct and clear as in a cFNM model-based approach.

9.5 Appendix

9.5.1 Appendix 9A

This appendix describes MRF image generation and Gibbs sampler, which are based on [6].

9.5.1.1 From Physical Systems to Images

Analogy between Boltzmann factor and Gibbs distribution. In physics, let $\Omega = \{\omega\}$ denote the possible configurations of a system. If the system is in thermal equilibrium with its surroundings, then the probability, or Boltzmann factor, of $\omega \in \Omega$ is given by

$$p(\omega_s) = \frac{e^{-\beta e(\omega_s)}}{\sum_{\omega} e^{-\beta e(\omega)}}, \tag{9.48}$$

where $e(\omega)$ is the energy function of ω and $\beta = \frac{1}{\kappa T}$, κ is Boltzmann's constant and, T is the absolute temperature [25–27]. The physical system tends to be in the low energy state (the most stable) with the higher probability.

The Boltzmann factor (Eq. (9.48)) is identical to the Gibbs distribution (Eq. (9.29)) whose energy function is in terms of the clique potentials. Therefore, minimizing the energy function $e(\omega)$ of Boltzmann factor (i.e., seeking the most stable state of a physical system) is equivalent to the maximizing the Gibbs distribution.

Equivalence between Gibbs distribution and Markov random field. Section 9.3.2 describes MRF-GRF equivalence. Thus, the maximizing Gibbs distribution will lead to generating a most like configuration of MRF. As a benefit of this equivalence, the ability to move between the global and local descriptions provides us with a simple and practical way of creating an MRF via specifying clique potentials, which is easy, instead of local characteristics, which is nearly impossible.

Correspondence between annealing and relaxation. The physical annealing process (gradually reducing temperature of a system) leads to low-energy states, that is, the most probable states under the Gibbs distribution. The result of maximizing the Gibbs distribution, for given initial states, is a highly parallel relaxation algorithm.

When pixel intensities and the image regions are viewed as states of a lattice-based physical system, the above-mentioned analogy, equivalence, and correspondence can be applied to the generation of MRF images. For simplicity of description, we use 1-D notation. For a 2-D lattice $\mathcal{L}$ of pixels (i, j) given by

$$\mathcal{L} = \{(i,j):\ 1 \leq i, j \leq M\}, \tag{9.49}$$

adopting a simple numbering, that is. the row-ordering, of pixels

$$t = j + M(i-1), \tag{9.50}$$

and letting

$$N = M^2, \tag{9.51}$$

we have a 1-D set of pixels

$$S = \{s_1, \cdots, s_t, \cdots, s_N\}. \tag{9.52}$$

Let $\mathbf{x} = \{x_s,\ s \in S\}$ be an MRF over a neighborhood $\{\mathcal{N}_s,\ s \in S\}$ with the state space

$$\Lambda = \{1, \cdots, k \cdots, K\}, \tag{9.53}$$

its configuration space is $\Omega = \prod_s \Lambda$, and the Gibbs distribution is given by $p_G(\mathbf{x}) = \frac{1}{Z} e^{-\sum_{c:s\in c} V_c(\mathbf{x})/T}$ $(\mathbf{x} \in \Omega)$.

9.5.1.2 Statistical Relaxation

Reference [6] introduces a Bayesian paradigm using MRF for the analysis of images. In [6], Theorem A (relaxation), Theorem B (annealing), and Theorem C (ergodicity) establish the convergence properties of the relaxation algorithm. For MRF image generation and Gibbs sampler, we summarize the underlying principles in the following four assertions. Assertion 1 states GRF-MRF equivalence. Assertion 2 shows that the local characteristics can be obtained from the global description. Assertion 3 describes the invariance over the permutations. Assertion 4 states the statistical relaxation rule.

Assertion 1 (Equivalence) Let $\mathcal{N}$ be a neighborhood system. Then $\mathbf{x}$ is an MRF with respect to $\mathcal{N}$ if and only if $p_G(\mathbf{x})$ is a Gibbs distribution with respect to $\mathcal{C}$.

Assertion 2 (Markovianity) Fix $s \in S$, $\mathbf{x} = \{x_1, \cdots, x_s, \cdots, x_N) \in \Omega$. If $P(\mathbf{x})$ is Gibbs, then

$$P(x_s = k | x_r, r \neq s) = \frac{e^{-\sum_{c:s\in c} V_c(x_s=k)/T}}{\sum_{x_s\in\Lambda} e^{-\sum_{c:s\in c} V_c(x_s)/T}}. \tag{9.54}$$

Assertion 3 (Invariance) In a homogeneous MRF, the probability mass function is invariant under permutations of pixels,¶ and for any pixel s

$$P(x_s = k) = \frac{1}{K}. \tag{9.55}$$

Assertion 4 (Statistical relaxation rule) Given the state of the system at time t, say $\mathbf{x}(t)$, one randomly chooses another configuration η and computes the energy change (here we still use the Boltzmann factor)

$$\Delta\varepsilon = \varepsilon(\eta) - \varepsilon(\mathbf{x}(t)) \tag{9.56}$$

¶ $\{s_1, s_2, \cdots, s_N\}$ is one permutation of $\{1, 2, \cdots, N\}$.

and the quantity

$$q = \frac{p(\eta)}{p(\mathbf{x}(t))} = e^{-\beta\Delta\varepsilon}. \tag{9.57}$$

If $q > 1$, the move to η is allowed and $x(t+1) = \eta$; if $q \leq 1$, the transition is made with probability q. Thus, we can choose ξ such that $0 \leq \xi \leq 1$ uniformly and set

$$\begin{array}{ll} \mathbf{x}(t+1) = \eta & (q \geq \xi), \\ \mathbf{x}(t+1) = \mathbf{x}(t) & (q < \xi). \end{array} \tag{9.58}$$

The following example illustrates that the physical system tends to be in the low energy state with the higher probability. Assume the current state has energy ε_0 and two new states have energies ε_1 and ε_2. The energy changes are $\Delta\varepsilon_1 = \varepsilon_1 - \varepsilon_2$ and $\Delta\varepsilon_2 = \varepsilon_2 - \varepsilon_2$.

Case 1. $\Delta\varepsilon_1 < 0$, $\Delta\varepsilon_2 < 0$, and $\Delta\varepsilon_1 > \Delta\varepsilon_2 \quad \Longrightarrow \quad \varepsilon_1 > \varepsilon_2$.
In this case, $q_1 > 1$, $q_2 > 1$, and $q_1 < q_2$. The change will be $\varepsilon_0 \longrightarrow \varepsilon_2$.
Case 2. $\Delta\varepsilon_1 > 0$, $\Delta\varepsilon_2 > 0$, and $\Delta\varepsilon_1 > \Delta\varepsilon_2 \quad \Longrightarrow \quad \varepsilon_1 > \varepsilon_2$.
In this case, $q_1 < 1$, $q_2 < 1$, and $q_1 < q_2$. The change will be $\varepsilon_0 \longrightarrow \varepsilon_2$.
Case 3. $\Delta\varepsilon_1 > 0$, $\Delta\varepsilon_2 < 0$, and $\Delta\varepsilon_1 > \Delta\varepsilon_2 \quad \Longrightarrow \quad \varepsilon_1 > \varepsilon_2$.
In this case, $q_1 < 1$, $q_2 > 1$, and $q_1 < q_2$. The change will be $\varepsilon_0 \longrightarrow \varepsilon_2$.

9.5.1.3 Gibbs Sampler

As shown in Section 7.5.2, an image is a configuration (i.e., a realization) of the underlying random field, and the pixel intensity is a value of the corresponding random variable in the state space of this random field. With this understanding and the four assertions above, MRF generation is organized as follows.

First, generate an initial configuration. Let it be denoted by $\mathbf{x}(0)$. Suppose $\mathbf{x}(0)$ is homogeneous (Assertion 3). The state space is $\Lambda = \{1, \cdots, k, \cdots, K\}$ and the pixel intensities $x_{i,j}(0)$ $(1 \leq i, j \leq M)$ should be uniformly distributed. This can be done in the following way. Generate a random vector $\mathbf{y} = (y_1, \cdots, y_n, \cdots, y_N)$ $(n = j + M(i-1)$ and $N = M^2)$ uniformly distributed on $[0,1]$; assign $x_{i,j}(0)$ $(1 \leq i, j \leq M)$ a value $k \in \Lambda$ based on $y_n \in [0,1]$.

Then change the 2-D index i, j $(1 \leq i, j \leq M)$ to a 1-D index n $(1 \leq n \leq N)$ via $n = j + M(i-1)$ and $N = M^2$. Permute these 1-D indexes and denote them as $(s_1, \cdots, s_n, \cdots, s_N)$. Visit each s_n in natural order, say, the raster scan. During visiting the s_n, for each value $k \in \Lambda$, compute all clique potentials $V_c(x_{s_n})$ based the potential assignment rules and the current pixel intensities $\{x_{s_1}(0), \cdots, x_{s_n}(0), \cdots x_{s_N}(0)\}$ (Assertion 1); further compute the local conditional probabilities of Eq. (9.54) (Assertion 2). Then assign a new value $x_{s_n}(1)$ to replace $x_{s_n}(0)$ (i.e., update) based on the uniformly distributed $y_n \in [0,1]$ (Assertion 4). Note, only one replacement can be done for one visit to $(s_1, \cdots, s_n, \cdots, s_N)$.

After visiting all pixels in $(s_1, \cdots, s_n, \cdots, s_N)$, a new configuration $\mathbf{x}(1)$ is generated. By repeating this type of visit N_{iter} times, we obtain a sequence of configurations $\mathbf{x}(1), \mathbf{x}(2), \cdots, \mathbf{x}(N_{iter})$. Theorems A through C [6] show that they converge to an MRF.

Thus, the corresponding algorithm of the Gibbs sampler consists of the following major steps:

1) Initialize an $M \times M$ lattice by assigning pixel intensities randomly from $\Lambda = \{1,\ 2,\ \cdots,\ K\}$. Call this initial configuration $\mathbf{x}(0)$ (Assertion 3).
2) For s from 1 to $N = M^2$,
   a) Permute N pixels $(s_1,\ \cdots,\ s_N)$ and set an order for visiting each pixel;
   b) Design and compute clique potential functions (Assertion 1);
   c) Compute the local probability for each $k \in \Lambda$ based on Eq. (9.54) (Assertion 2);
   d) Set the intensity of the pixel s based on the statistical relaxation rule (Assertion 4).
3) Repeat 2) N_{iter} times.

Using this algorithm, the various MRF configurations with different resolutions ($M \times M$) are generated. Then the correlated Gaussian noise with different variances σ_0^2 are superimposed on each resolution of each MRF configuration.

Examples of MRF images generated using the above approach are shown in Figure 9.4, where $M = 64$, $N = 4096$, $\Lambda = \{1, 2, 3, 4\}$, and $N_{iter} = 200$. Four types of cliques in the 2nd-order neighborhood system (the horizontal, the vertical, and two diagonal) are used. The region means are set to -45, -15, 15, and 45. $\sigma_0^2 = 20$. Statistical procedures such as random vectors generation, permutations, correlated data matrix, etc., are provided by the International Mathematical and Statistical Library (IMSL) [28]. The algorithm has been implemented in Macintosh computers. Other examples of MRF images are used in Chapters 11 and 12.

Gibbs sampler has been used in many applications, for example, image analysis (segmentation and the restoration), simulated image generation, and partial volume study, etc.[29–35].

Problems

9.1. Prove Eq. (9.30).

9.2. Based on Eq. (9.48), show that the physical system tends to be in the low energy state with the higher probability.

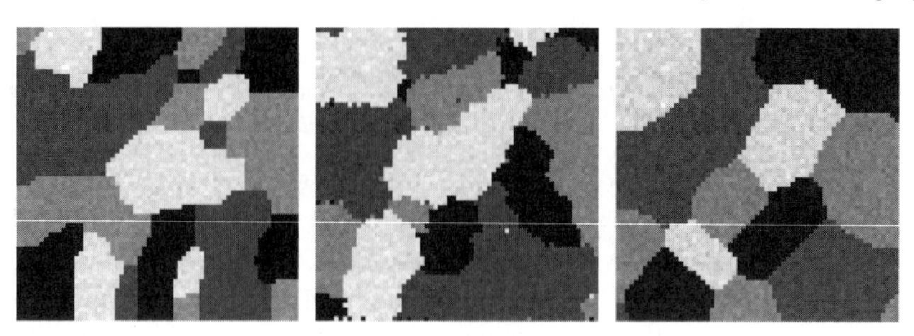

FIGURE 9.4
Examples of MRF configurations generated using Gibbs sampler.

9.3. Following the major steps given in Appendix 9.3, write a program to generate MRF image with different parameters and discuss the results.

References

[1] Papoulis, A.: *Probability, Random Variables and Stochastic Processes.* McGraw-Hill Book Company Inc., New York (1984).

[2] Jain, A.K.: *Fundamentals of Digital Image Processing.* Prentice Hall, Englewood Cliffs, New Jersey, (1989).

[3] Besag, J.: Spatial interaction and the statistical analysis of lattice system (with discussion). *J. Royal Statist. Soc.* **36B**(2) (1974) 192–236.

[4] Kindermann, R., Snell, J.: *Markov Random Field and Their Applications.* American Mathematical Society, Providence, Rhode Island, (1980).

[5] Kashap, R., Chellappa, R.: Estimation and choice of neighbors in spatial interaction models of images. *IEEE Trans. Information Theory* **29** (1983) 60–72.

[6] Geman, S., Geman, D.: Stochastic relaxation, Gibbs distributions, and the Bayesian restoration of images. *IEEE Trans. Pattern Anal. Machine Intell.* **6**(6) (1984) 721–724.

[7] Derin, H., Elliot, H.: Modeling and segmentation of noisy and textured images using Gibbs random fields. *IEEE Trans. Pattern Anal. Machine Intell.* **9**(1) (1987) 39–55.

[8] Li, S.Z.: *Markov Random Field Modeling in Computer Vision.* Springer-Verlag, Berlin, Germany (1995).

[9] Hammersley, J., Clifford, P.: Markov random field in statistics. http://www.statslab.cam.ac.uk/ grg/books/hammfest/ hamm-cliff.pdf (1971).

[10] Clifford, P.: Markov random field in statistics. In G.R. Grimmett and D.J.A. Welsh, (Eds.) *Disorder in Physical Systems*, Oxford University Press, (1990) 19–32.

[11] Grimmett, G.: A theorem about random fields. *Bull. London Math. Soc.* **5**(1) (1973) 81–84.

[12] Preston, C.: Generalized gibbs states and markov random fields. *Adv. Appl. Probability* **5**(2) (1973) 242–261.

[13] Sherman, S.: Markov random fields and gibbs random fields. *srael J. Math.* **14**(1) (1973) 92–103.

[14] Geman, D., Geman, S., Graffigne, C., Dong, P.: Boundary detection by constrained optimization. *IEEE Trans. Pattern Anal. Machine Intell.* **12** (1990) 609–628.

[15] Besag, J.: Statistical analysis of non-lattice data. *The Statistician* **24** (1975) 179–195.

[16] Vlontzos, J., Kung, S.: Hidden Markov models for character recognition. *IEEE Trans. Image Processing* **1** (1992) 539–543.

[17] Rabiner, L.R.: A tutorial on hidden Markov models and selected applications in speech recognition. *Proc. IEEE* **77** (1989) 257–286.

[18] Kundu, A., He, Y., Bahl, P.: Recognition of handwritten word: First and second-order hidden Markov model based approach. *Pattern Recogn* **22**(3) (1989) 283–297.

[19] Forbes, F., Peyrard, N.: Hidden Markov random field model selection criteria based on Mean Field-like approximations. *IEEE Trans. Pattern Anal. Machine Intell.* **25**(9) (2003) 1089–1101.

[20] Dempster, A.P., M.Laird, N., Rubin, D.B.: Maximum likelihood from incomplete data via EM algorithm. *J. R. Statist. Soc.* **39** (1977) 1–38.

[21] Moon, T.: The expectation-maximization algorithm. *IEEE Signal Processing Mag.* **13**(3) (1996) 47–60.

[22] Besag, J.: On the statistical analysis of dirty pictures. *J. Royal Statist. Soc.* **48B** (1986) 259–302.

[23] Zhang, J.: The mean field theory in EM procedures for blind Markov random field image. *IEEE Trans. on Image Proc.* **2**(1) (1993) 27–40.

[24] Celeux, G., Forbes, F., Peyrard, N.: EM procedure using Mean Field-like approximation for Markov model-based image segmentation. *Pattern Recogn.* **36**(1) (2003) 131–144.

[25] Slichter, C.: *Principles of Magnetic Resonance.* Springer-Verlag, Berlin (1980).

[26] Evans, R.: *The Atomic Nucleus.* Robert E. Krieger Publishing Company, Malabar, Florida (1982).

[27] Abragam, A.: *Principles of Nuclear Magnetism.* Oxford University Press, New York (1983).

[28] Developer, I.: *IMSL - Problem-Solving Software Systems.* IMSL, Inc., Houston, Texas (1990).

[29] Choi, H., Haynor, D., Kim, Y.: Multivariate tissue classification of MR images for 3-D volume reconstruction - A statistical approach. *SPIE Proceedings, Med. Imag. III* (1989) 183–193.

[30] Leahy, R., Hebert, T., Lee, R.: Applications of Markov random field models in medical imaging. *Proc. IPMI* **11** (1989) 1–14.

[31] Dubes, R.C., Jain, A.K.: Random field models in image analysis. *J. Appl. Stat.* **16**(2) (1989) 131–164.

[32] Dubes, R.C., Jain, A.K., Nadabar, S., Chen, C.: MRF model-based algorithm for image segmentation. *Proc. IEEE ICPR* (1990) 808–814.

[33] Zhang, Y., Brady, M., Smith, S.: Segmentation of brain MR images through a Hidden Markov Random Field model and the Expectation-Maximization algorithm. *IEEE Trans. Med. Imag.* **20**(1) (2001) 45–57.

[34] Liang, Z., MacFall, J., Harrington, D.: Parameter estimation and tissue segmentation from multispectral MR images. *IEEE Trans. Med. Imag.* **13**(3) (1994) 441–449.

[35] Ji, C., Seymour, L.: A consistent model selection procedure for Markov random field based on penalized pseudolikelihood. *Ann. Appl. Probability* **6** (1996) 423–443.

10

Statistical Image Analysis – I

10.1 Introduction

In most cases, images generated by X-ray CT and MR are all-inclusive. That is, these two imaging modalities cannot directly produce the images of the selected tissue types or organ systems. For example, when the human abdomen is imaged by X-ray CT, the liver, kidney, stomach, pancreas, gallbladder, adrenal glands, spleen, etc., are all shown in the resultant image; when a cross section of the human brain is imaged by MRI, the scalp, bone, gray matter, white matter, and cerebrospinal fluid, etc., are all included in the resultant image. In order to obtain an image of the selected targets of interest, an image processing or image analysis method is often required.

Generally, *imaging* refers to an operation or a process from the data to the picture. X-ray CT represents an operation from photon measurements to a 2-D display of the spatial distribution of the relative linear attenuation coefficient (RLAC); MRI represents a process from free induction decay (FID) signal measurements to a 2-D display of the spatial distribution of the thermal equilibrium macroscopic magnetization (TEMM). *Image processing* refers to an operation or a process from the picture to the picture. The commonly used image processing approaches may include but are not limited to transform, enhancement, and restoration. *Image analysis* refers to an operation or a process from the picture to the "*data.*" Here, *data* may include some image primitives such as edges or regions as well as some quantities and labels related to these primitives. This chapter and the following chapters focus on image analysis.

Various image analysis methods have been developed, and some of them are applied to X-ray CT and MR images. The graph approach [1–5], the classical snakes and active contour approaches [6–10], the Level set methods [11–15], and Active Shape model (ASM) and Active Appearance model (AAM) approaches [16–24] are edge-based approaches. Fuzzy connected object delineation [25–31] and Markov random field (MRF) [32–39] are the region-based approaches. This and the next chapter describe two statistical image analysis methods for X-ray CT and MR images based on the stochastic models I and II given in Chapter 9, respectively.

In analyzing the so-called all-inclusive images as illustrated by the examples given in the beginning of this section, the first step is to determine how many image regions are presented in the image. After this number is detected, the second step is to estimate region parameters, for example, the mean and variance, etc. After these parameters are estimated, the third step is to classify each pixel to the corresponding image regions. By implementing these three steps, an all-inclusive image is partitioned into the separated image regions; each of them represents a tissue type or an organ system. The above detection-estimation-classification approach forms an unsupervised image analysis technique; it is a model-based, data driven approach.

10.2 Detection of Number of Image Regions

Property 9.1 shows that an image whose pixel intensities are statistically independent can be modeled by an independent Finite Normal Mixture (iFNM). Let the image be denoted by $IMG(J, K)$; J and K are the numbers of pixels and image regions, respectively. iFNM pdf is given by

$$f(x) = \textstyle\sum_{k=1}^{K} \pi_k g(x|\theta_k), \tag{10.1}$$

where x denotes the pixel intensity, $\theta_k = (\mu_k, \ \sigma_k^2)$ $(k = 1, \cdots, K)$ is the parameter vector of the k-th image region $\mathcal{R}_k$ (μ_k – the mean, σ_k^2 – the variance), $g(x|\theta_k)$ is a Gaussian pdf of pixel intensities in the k-th image region given by

$$g(x|\theta_k) = \frac{1}{\sqrt{2\pi\sigma_k^2}} \exp(-\frac{(x-\mu_k)^2}{2\sigma_k^2}), \tag{10.2}$$

and π_k $(k = 1, \cdots, K)$ represents the probability of the occurrence of the kth image region in the image and is characterized by a multinomial distribution

$$0 < \pi_k < 1 \quad \text{and} \quad \textstyle\sum_{k=1}^{K} \pi_k = 1. \tag{10.3}$$

The $3K$-dimensional model parameter vector of iFNM (Eq. (10.1)) is defined as

$$\mathbf{r} = (\pi_1, \mu_1, \sigma_1^2, \cdots\cdots, \pi_K, \mu_K, \sigma_K^2)^T. \tag{10.4}$$

In this section and Appendix, for the purpose of derivation, pdf $f(x)$ of Eq. (10.1) is also written as $f(x|\mathbf{r})$ or $f(x, \mathbf{r})$.

Based on Eq. (10.1), detecting the number of image regions is actually the selecting the order of iFNM model. Traditionally, this is implemented by two types of approaches: (1) hypothesis test [40, 41], and (2) power spectrum analysis [42, 43]. The hypothesis test is a general standard method. It first establishes a null hypothesis H_0 : the order is K_0 and an alternative hypothesis H_1 : the order is not K_0, then creates test statistics and derives its distribution;

and finally decides to accept or reject H_0 based on a given confidence level. Power spectrum analysis is particularly useful for time series such as AR, MA, ARMA models. It may include residual flatness and final prediction error (FPE) methods. By checking the flatness of a power spectrum or minimizing prediction error, it determines the order of the models. This section describes a new type of approach for selecting the order of the model.

10.2.1 Information Theoretic Criteria

In Information Theory [44, 45], the relative entropy, also known as Kullback-Leibler divergence (KLD),* provides a way to compare the true pdf and the estimated pdfs. Thus, in statistical identification, the minimization of KLD defines a reasonable criterion for choosing an estimated pdf to best fit the true one [46, 47]. As shown in Appendix 10A, the maximization of the mean log-likelihood provides a practical means to minimize KLD, and the use of the maximum likelihood estimate of parameters of pdf leads to various Information Theoretic Criteria (ITC) [48–54]. For a parametric family of pdfs, the model that best fits the observed data is one that gives a minimum value of ITC.

Thus, selecting the order of the iFNM model is formulated as a model fitting problem, and ITCs provide criteria to evaluate the goodness-of-fit. Several commonly used ITCs are An Information Criterion (AIC) [48, 55, 56] and Minimum Description Length (MDL) [49, 57, 58]. They are defined by

$$\begin{aligned} \text{AIC}(K) &= -2\log(\mathcal{L}(\hat{\mathbf{r}}_{ML})) + 2K_a \quad \textit{and} \\ \text{MDL}(K) &= -\log(\mathcal{L}(\hat{\mathbf{r}}_{ML})) + \tfrac{1}{2}K_a \log J, \end{aligned} \tag{10.5}$$

respectively, where J is the number of independent observations, K is the order of the model, K_a is the number of free adjustable model parameters, $\hat{\mathbf{r}}_{ML}$ is the ML estimate of the model parameter vector, and $\mathcal{L}(\hat{\mathbf{r}}_{ML})$ is the likelihood of the ML estimate of the model parameters.

AIC and MDL address the following general problem. Given a set of independent observed data and a family of models, that is, a parametric family of pdfs, the model that best fits the observed data is one that gives the minimum value of AIC or MDL. Let K_0 be the correct order of the model; it should satisfy

$$\begin{aligned} K_0 &= Arg\{\min_{1\leq K\leq K_{max}} \text{AIC}(K)\} \\ K_0 &= Arg\{\min_{1\leq K\leq K_{max}} \text{MDL}(K)\}, \end{aligned} \tag{10.6}$$

where K_{max} is an up-limit of all possible K.

* "Divergence" rather than "Distance" is used because Kullback-Leibler measure is not symmetric.

In Eq. (10.5), for the iFNM model of Eq. (10.1), the likelihood of ML estimate of the model parameter vector $\mathbf{r}$ is

$$\mathcal{L}(\hat{\mathbf{r}}_{ML}) = \prod_{j=1}^{J} f(x_j|\hat{\mathbf{r}}_{ML}) = \prod_{j=1}^{J}\sum_{k=1}^{K} \hat{\pi}_{k_{ML}} g(x_j|\hat{\theta}_{k_{ML}}), \tag{10.7}$$

where $\hat{\pi}_{k_{ML}}$ and $\hat{\theta}_{k_{ML}} = (\hat{\mu}_{k_{ML}}, \hat{\sigma}^2_{k_{ML}})$ are the ML estimates of π_k and $\theta_k = (\mu_k, \sigma_k^2)$, and the number of free adjustable model parameters in $\mathbf{r}$ is

$$K_a = 3K - 1, \tag{10.8}$$

which is due to $\sum_{k=1}^{K} \pi_k = 1$, only $(3K-1)$, parameters in $\mathbf{r}$ of Eq. (10.4) are independent.

[51] indicates that in the large sample limit, MDL may give the correct value of K_0 while AIC may give an overestimated value of K_0.

In addition to these commonly used ITCs, Appendix 10B proposes a new one. It is based on the principle of minimizing the maximum joint differential entropy of the observed data and the model parameters [59–63]. By seeking maximum entropy, it is able to find a family of models that has the most uncertainty compared with other families. That is, the selected family will include most (all possible) models. By minimizing this maximum entropy, it is able to find one model in this family that has the least uncertainty. That is, the chosen model will best fit the observed data. This min-max approach is based on the joint pdf of x and $\mathbf{r}$ $f(x, \mathbf{r})$; the commonly used ITCs are based on the conditional pdf of x on $\mathbf{r}$ $f(x|\mathbf{r})$.

10.3 Estimation of Image Parameters

The prerequisite for using a information theoretic criterion is to find the ML estimate of the parameter vector $\mathbf{r}$. The ordinary way to obtain ML parameter estimate is to take the derivative of the likelihood function $\mathcal{L}(\mathbf{r})$ with respect to $\mathbf{r}$ and let it equal zero, that is,

$$\mathbf{D}\mathcal{L}(\mathbf{r}) = \frac{\partial \mathcal{L}(\mathbf{r})}{\partial \mathbf{r}} = 0. \tag{10.9}$$

It is readily demonstrated that ML estimates of the K_a $(= 3K-1)$ parameters of the model Eq. (10.1) satisfy the equations

$$\begin{cases} \sum_{j=1}^{J}(g(x_j|\theta_k) - g(x_j|\theta_K)) = 0 \\ \sum_{j=1}^{J} \frac{\pi_k g(x_j|\theta_k)}{f(x_j|\mathbf{r})}(x_j - \mu_k) = 0 \\ \sum_{j=1}^{J} \frac{\pi_k g(x_j|\theta_k)}{f(x_j|\mathbf{r})}(\frac{1}{\sigma_k^2 2} - 1) = 0. \end{cases} \tag{10.10}$$

It has been verified that the likelihood equation Eq. (10.10) cannot be solved explicitly. This implies that iterative methods must be used. The most successful iterative algorithms—Newton–Raphson method and Method of Scoring—require the inversion of the Hessian matrix ($\mathbf{D}^2\mathcal{L}$) and of the Fisher information matrix ($\mathbf{D}\mathcal{L}\mathbf{D}\mathcal{L}^T$) of the high rank $((3K-1)\times(3K-1))$ and always suffer from the numerical singularity. In order to avoid these problems, two algorithms, Expectation-Maximization (EM) [64] and Classification-Maximization (CM) [65], are utilized for ML parameter estimation of iFNM Eq. (10.1).

10.3.1 Expectation-Maximization Method

EM algorithm first estimates the likelihood of the complete (not observed directly) data through the incomplete (observed) data and the current parameter estimates (E-step) and then maximizes the estimated likelihood to generate the updated parameter estimates (M-step). The procedure cycles back and forth between these two steps. The successive iterations increase the likelihood of the estimated parameters and converge (under regularity conditions) to a stationary point of the likelihood [64]. [66] applies the EM algorithm to the ML parameter estimation of iFNM in the case $K = 2$. Appendix 10C gives a derivation of the EM algorithm in the case $K > 2$. The algorithm at the m-th iteration is shown below.

E-step: computing the probability membership $z_{jk}^{(m)}$,

$$z_{jk}^{(m)} = \frac{\pi_k^{(m)} g(x_j|\theta_k^{(m)})}{f(x_j|\mathbf{r}^{(m)})} \qquad (m = 0, 1, 2, \cdots). \tag{10.11}$$

M-step: computing the updated parameter estimates,

$$\begin{cases} \pi_k^{(m+1)} = \frac{1}{J}\sum_{j=1}^{J} z_{jk}^{(m)} \\ \mu_k^{(m+1)} = \frac{1}{J\pi_k^{(m+1)}}\sum_{j=1}^{J} z_{jk}^{(m)} x_j \qquad (m = 0, 1, 2, \cdots) \\ \sigma_k^{2\,(m+1)} = \frac{1}{J\pi_k^{(m+1)}}\sum_{j=1}^{J} z_{jk}^{(m)}(x_j - \mu_k^{(m+1)})^2. \end{cases} \tag{10.12}$$

The algorithm starts from an initial estimate $\mathbf{r}^{(0)}$ and stops when a stopping criterion

$$|\pi_k^{(m+1)} - \pi_k^{(m)}| < \varepsilon \qquad (1 \leq k \leq K) \tag{10.13}$$

is satisfied, where ε is a prespecified small number.

10.3.2 Classification-Maximization Method

The CM algorithm is also a two-step iterative procedure. It can be viewed as a modified K-mean cluster algorithm [65]. It first classifies the entire observed data into the different groups (i.e., image regions) using ML classification criterion and the current parameter estimates (C-step) and then generates the updated parameter estimates in each group according to the ML estimation procedure (M-step). The algorithm at the m-th iteration is shown below.

C-step: data x_j is classified into group G_{k_0} if the modified Mahalanobis distance d_{jk_0} from x_j to the current mean estimate $\mu_{k_0}^{(m)}$ of the k_0-th class is minimized, that is,

$$x_j \in G_{k_0}, \quad if \quad k_0 = Arg\{\min_{1 \leq k \leq K} d_{jk}\}, \tag{10.14}$$

where

$$d_{jk} = (x_j - \mu_k^{(m)})^T (\sigma_k^{2(m)})^{-1} (x_j - \mu_k^{(m)}) + \log \sigma_k^{2(m)}. \tag{10.15}$$

M-step: the updated parameter estimates are given by

$$\begin{cases} \pi_k^{(m+1)} &= \frac{1}{J} J_k^{(m+1)} \\ \mu_k^{(m+1)} &= \frac{1}{J\pi_k^{(m+1)}} \sum_{x_j \in G_k} x_j \qquad (m = 0, 1, 2, \cdots) \\ \sigma_k^{2(m+1)} &= \frac{1}{J\pi_k^{(m+1)}} \sum_{x_j \in G_k} (x_j - \mu_k^{(m+1)})^2, \end{cases} \tag{10.16}$$

where $J_k^{(m+1)}$ is the number of the data that belong to the group G_k. The algorithm starts from an initial estimate $\theta_k^{(0)}$ $(k = 1, \cdots, K)$ and stops when the condition

$$\theta_k^{(m+1)} = \theta_k^{(m)} \qquad (k = 1, \cdots, K) \tag{10.17}$$

is satisfied.

10.3.3 Discussion

Two algorithms have been applied to the simulated data and X-ray CT and MR images and the satisfied results are obtained. Two algorithms increase the likelihood of the estimated parameters over the iterations. Thus, algorithms guarantee convergence to a local maximum. Computer simulations show that for very well-structured data, algorithms converge to the global maximum. However, a general theoretical convergence analysis has not been done. Some statements regarding the convergence of K-means clustering algorithm and EM algorithm are given in [67, 68]. Even though the final estimates obtained by the two algorithms are quite close, differences between them do exist. First, the EM algorithm requires the initial estimates $\pi_k^{(0)}$ $(k = 1, \cdots, K)$ of the weights and the settings of the stopping threshold ε; the CM algorithm does not. Second, computations showed that the EM algorithm needs many more iterations than CM. Third, the CM algorithm seems to have more evident meaning than EM for image analysis (especially for segmentation), but it can only provide an approximate ML estimate. For to these reasons, a hybrid algorithm (CM generates the initial parameter estimates and EM produces the final estimates) has been developed and utilized.

10.4 Classification of Pixels

Having determined the number K_0 of image regions and obtained an ML estimate $\hat{\mathbf{r}}_{ML}$ of the model parameter $\mathbf{r}$, the classification of pixels into different image regions actually becomes a statistical decision problem. Several classifiers are available in statistical decision theory.

Bayesian classifier: If both the priori probability and the cost (loss) functions are known, it will minimize the total expected loss.

Minimax classifier: If the priori probability is not available, it will minimize the total expected loss under the worst possible conditions.

Neyman-Pearson classifier: If neither the priori probability nor the loss functions are available, it will maximize the detection probability for the fixed false-alarm probability.

10.4.1 Bayesian Classifier

The statistical image analysis technique described in this chapter uses the Bayesian classifier, which we will show is actually a likelihood ratio classifier. Let $p(k|x)$ denote the probability that x comes from the k-th image region $(1 \leq k \leq K_0)$. If the classifier decides that x comes from the k-th image region when it actually comes from the l-th image region, then this decision produces a loss denoted by L_{kl}. Because K_0 image regions are under consideration, the total expected loss of assigning x to the k-th image region will be

$$r_k(x) = \sum_{l=1}^{K_0} L_{kl} p(l|x). \tag{10.18}$$

$r_k(x)$ is sometimes called the risk. The Bayesian classifier computes all risks $r_k(x)$ $(1 \leq k \leq K_0)$ and chooses the minimum risk among all $r_k(x)$ to make a decision.

Because

$$p(l|x) = \frac{p(x|l)p(l)}{p(x)}, \tag{10.19}$$

Eq. (10.18) will be

$$r_k(x) = \frac{1}{p(x)} \sum_{l=1}^{K_0} L_{kl} p(x|l)p(l). \tag{10.20}$$

For a given x, with respect to all k $(1 \leq k \leq K_0)$, $p(x)$ is constant; thus

$$r_k(x) = \sum_{l=1}^{K_0} L_{kl} p(x|l)p(l). \tag{10.21}$$

When $K_0 = 2$, we have

$$\begin{aligned} r_1(x) &= L_{11}p(x|1)p(1) + L_{12}p(x|2)p(2) \\ r_2(x) &= L_{21}p(x|1)p(1) + L_{22}p(x|2)p(2). \end{aligned} \tag{10.22}$$

If $r_1(x) < r_2(x)$, then

$$(L_{11} - L_{21})p(x|1)p(1) < (L_{22} - L_{12})p(x|2)p(2). \tag{10.23}$$

Because the loss in the correct decision is less than the loss in the incorrect decision, that is, $L_{ll} < L_{kl}$, we have

$$\frac{p(x|1)}{p(x|2)} > \frac{(L_{22} - L_{12})p(2)}{(L_{11} - L_{21})p(1)}. \tag{10.24}$$

Normally, the zero loss is assigned to the correct decision and the equal losses are assigned to the incorrect decisions, that is, $L_{ll} = 0$ and $L_{lk} = L_{kl}$; thus Eq. (10.24) becomes

$$\frac{p(x|1)}{p(x|2)} > \frac{p(2)}{p(1)}. \tag{10.25}$$

The above equation represents a likelihood ratio criterion.

Applying Eq. (10.25) to iFNM (Eq. (10.1)), we have

$$p(x|k) = g(x|\hat{\theta}_{k_{ML}}) \quad \text{and} \quad p(k) = \hat{\pi}_{k_{ML}}. \tag{10.26}$$

According to the classification criterion Eq. (10.25), the pixel x_j $(1 \leq j \leq J)$ will be classified into the k-th $(1 \leq k \leq K_0)$ image region if

$$\hat{\pi}_{k_{ML}} g(x_j|\hat{\theta}_{k_{ML}}) > \hat{\pi}_{l_{ML}} g(x_j|\hat{\theta}_{l_{ML}}) \qquad (l = 1, \cdots, K_0,\ l \neq k). \tag{10.27}$$

That is,

$$x_j \in \mathcal{R}_{k_0} \qquad \text{if} \qquad k_0 = Arg\{\max_{1 \leq k \leq K_0} \hat{\pi}_{k_{ML}} g(x_j|\hat{\theta}_{k_{ML}})\}, \tag{10.28}$$

where $\mathcal{R}_{k_0}$ denotes the k_0-th image region.

10.5 Statistical Image Analysis

The detection of the number of image regions (Section 10.2), the estimation of image parameters (Section 10.3), and the classification of pixels into image regions (Section 10.4) form an iFNM model-based statistical image analysis method. The method is performed in the following fashion:

1. For a given number K of image regions, the EM or CM algorithm is applied to compute the ML estimate $\hat{\mathbf{r}}_{ML}$ of the model parameters, and then the value of AIC(K) or MDL(K) is computed.

2. By repeating step 1 for all possible K from K_{min} to K_{max}, the number K_0 that minimizes AIC(K) or MDL(K) is chosen as the correct number of image regions.

3. Using the number K_0 and the corresponding ML estimate $\hat{\mathbf{r}}_{ML}$, the Bayesian classifier is applied to classify pixels into K_0 groups so that an image is partitioned into distinctive image regions.

This method is a fully automated, unsupervised data-driven approach. In the following several subsections, we apply it to simulated images, the physical phantom image, and X-ray CT and MR images to demonstrate its operations and performance. The method is applied to both local, region of interest (ROI) images and global, entire CT and MR images. The partitioned region images are displayed either by their mean values in one image or by their original density values in separated images. In order to show the details, some images are displayed in color using an RGB Look-Up table.

10.5.1 Simulated Images

Two 64×64 MRF images were created by Gibbs Sampler (Appendix 9A of Chapter 9) and then superimposed with independent Gaussian samples to

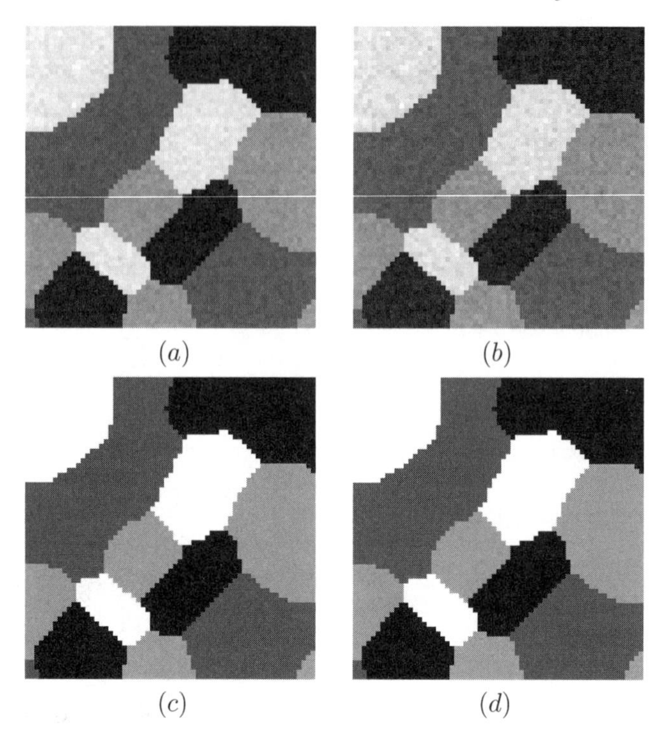

FIGURE 10.1
Two MRF images (a) and (b) with different SNRs. The partitioned region images: the black, dark grey, light grey, and white, (c) and (d).

TABLE 10.1
Settings of Images in Figures 10.1.(a) and (b).

k	1	2	3	4
π_k	0.207764	0.233643	0.310547	0.248047
μ_k	−45.0000	−15.0000	15.0000	45.0000

Image Size - 64 × 64, Number of pixels J = 4096, Number of image regions $K_0 = 4$.

simulate images with iFNM distribution. They are shown in Figures 10.1a and 10.1b.

The settings of these two MRF images are given in Table 10.1.[†] Gaussian samples represent noise from $N(0, \sigma_0^2)$, where σ_0^2 are equal to 5 and 10, for im-

[†]The purpose of setting the means of image regions to both negative and positive values is to simulate pixel intensities of the X-ray CT image where the CT number of the air, water, and bone are negative, zero, and positive.

TABLE 10.2
MDL(K) Values for Simulated MRF images (Figures 10.1.a and 10.1.b) with $K = 2 \sim 6$

	K	2	3	4	5	6
Figure 10.1.a	MDL(K)	19684	17464	**14766**	14779	14791
Figure 10.1.a	MDL(K)	19708	18107	**16184**	16196	16209

ages in Figure 10.1.a and 10.1.b, respectively. The signal-to-noise ratio (SNR) of the image is defined as

$$SNR = \sum_{k=1}^{K} \pi_k SNR_k \quad \text{and} \quad SNR_k = \frac{\mu_k^2}{\sigma_0^2}.$$

More detailed discussion on SNR is given in Section 12.2.1.2. Thus, for images in Figure 10.1.a and 10.1.b, the SNR are equal to 23.2 *db* and 20.2 *db*, respectively.

The detection results are shown in Table 10.2. The minimum value of MDL computed using Eq. (10.6) occurs at $K_0 = 4$, which is correct. The partitioned four region images are shown in Figures 10.1.c and 10.1.d, displayed by mean values of pixel intensities in each image region (in the black, dark grey, light grey, and white). Visual examinations indicate that there is not any classification error. A detailed performance analysis on these two images is given in Section 12.2.

10.5.2 Physical Phantom Image

A physical phantom is made of six cylinders that are inserted in parallel in a base and arranged in a circle. The cross-section of these cylinders is perpendicular to the axis of the cylinder, which is along the direction of the movement of the scanner bed. Figure 10.2.a shows the image of a cross-section of this phantom. The size of this image is 81×81 pixels, which is truncated from a default circular X-ray CT image.

These six cylinders are made of four types of homogeneous materials: Bone, Teflon, Poly (Eth. and Prop.), and 013A (A and B).[‡] In Figure 10.1.a, from the top-left, and clockwise, they are Teflon, Poly (Eth.), 013A (B), 013A (A), Poly (Prop), and Bone. The image of the background (air) is excluded in the image analysis. That is, only the images of cylinders are analyzed.

By assuming $K_{min} = 2$ and $K_{max} = 6$, MDL(K) values computed using Eq. (10.6) are shown in Table 10.3. Results in Table 10.3 suggest that this phantom image has $K_0 = 5$ region images. Each individual region image is shown in Figures 10.2.b–10.2.f. They represent the transition zones (*b*), Poly (Eth and Prop) (c), 013A(A and B) (d), Teflon (e), and Bone (f).

[‡]These cylinders are the test rods used in quality assurance (QA) of the X-ray CT scanner.

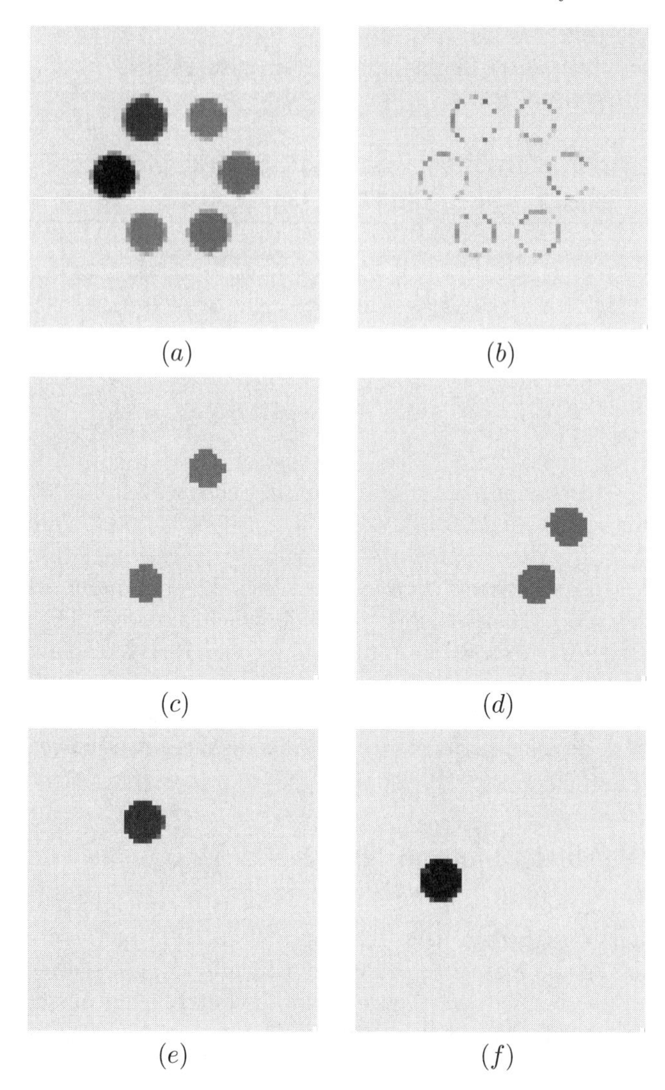

(a) (b) (c) (d) (e) (f)

FIGURE 10.2
X-ray CT image of a physical phantom (a). The partitioned region images: (b) Transition zones, (c) Poly (Eth and Prop), (d) 013A(A and B), (e) Teflon, and (f) Bone.

The number of types of materials in the phantom is 4 and the number of the region images in the phantom image (suggested by the MDL information theoretic criterion) is 5. These results seem contradictory. But, it is believed to be reasonable. The explanation is as follows.

In the image reconstruction of X-ray CT imaging (Chapter 2), an interpola-

TABLE 10.3
MDL(K) Values for the Physical Phantom Image (Figure 10.2.a) with $K = 2 \sim 6$

K	2	3	4	5	6
MDL(K)	3452	3355	3279	**3116**	3246

tion procedure is applied. As a result, in the reconstructed images, a transition zone between two adjacent but distinctive image regions always occurs. Pixel intensities in this zone are different from those in the two adjacent image regions and are usually between them. In other words, there is not an ideal sharp edge between adjacent image regions even though they are homogeneous.

The transition zone normally does not represent the true type of tissue or organ. It is a product of the image reconstruction and often occurs at the edges of image regions. The width of this transition zone is about $1 \sim 2$ pixels. In some cases, it can be interpreted as the partial volume effect.

In Figure 10.3.(a), the X-Y plane defines an 81×81 image; the Z axis defines pixel intensity. The almost flat top surfaces of six vertical "cylinders" represent the pixel intensities at a cross-section of six cylinders of four types of materials. The side surfaces of six vertical "cylinders" represent the pixel intensities in these transition zones that appear as rings at the edges of image regions. Figure 10.3.b shows a line profile at $y = 40$ that passes through two vertical "cylinders," It is clear that several pixels are on the slopes of the side surfaces of vertical "cylinders," and their intensities are between the background and the physical materials of cylinders. As mentioned earlier, the background (air, CT number –1000) is excluded in the image analysis. For display purposes, in Figure 10.3, CT number of the background is set to -200.

10.5.3 X-Ray CT Image

Figure 10.4 shows a cross-sectional X-ray CT image of the chest. For the purpose of image analysis, a region of interest (ROI) that corresponds to the lower part of the right lung is truncated from the image in Figure 10.4 and shown in Figure 10.5 (a) - grayscale, (b) - color). The size of this ROI image is 81×81 pixels. Images in Figures 10.5, 10.6, and 10.7 are displayed in colors that are designed using RGB Look-Up table with CT number -1000 as the blue and $+1000$ as the red.

The number of image regions in this ROI image is unknown. Using an information theoretic criterion to this ROI image, MDL values computed by Eq. (10.6) are given in Table 10.4 that indicate that this number is $K_0 = 6$. The partitioned region images with $K = 2, \cdots, 8$ are shown in Figures 10.6.a–10.6.h, that are displayed by the mean values of pixel intensities in each image region. From the images in Figure 10.6, we observed that when $K < 6$, some major image regions are lumped into one image region, but the results are still meaningful; when $K > 6$, no essential differences in the partitioned region

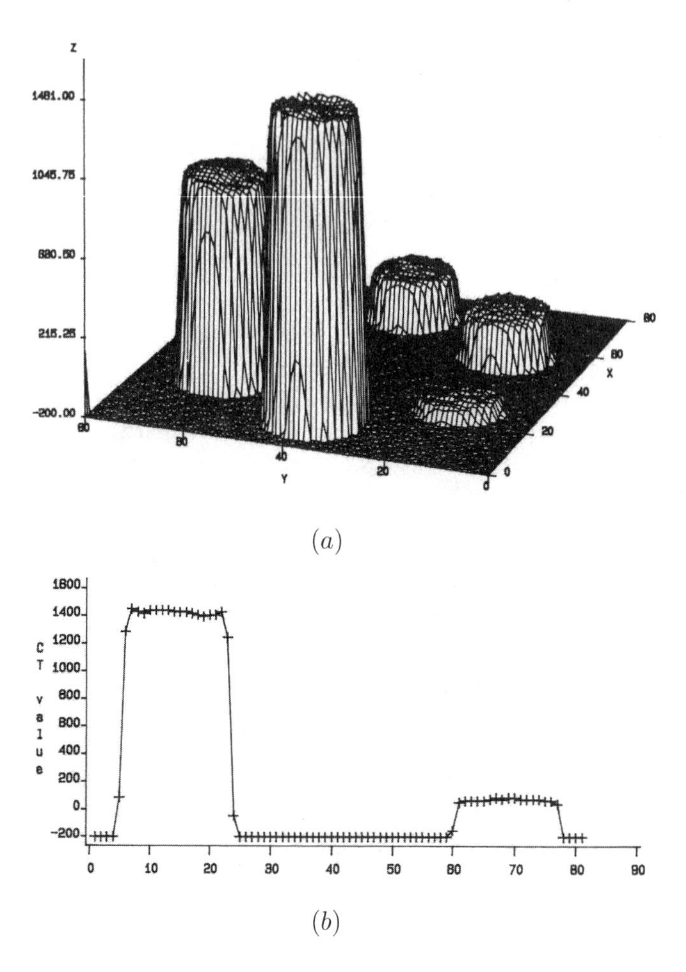

FIGURE 10.3
(a) A 2-D histogram of the phantom image in Figure 10.2.a. The flat plateaus represent the homogeneous image regions, and the inclined surfaces represent the transition zones between image regions and the background. (b) A line profile at $y = 40$. Several pixels are on the sloping surfaces and form the transition zones.

images will be generated.

For $K_0 = 6$, images of each individual image region are shown in Figure 10.7, which represent six types of tissues: (*a*) the lung air; (*b*) the lung parenchyma and fibrosis; (*c*) the pleura, blood vessels, and bronchi; (*d*) a tumor and the tissues outside lung; (*e*) the dense bone (with a spot of tumor); and (*f*) the sponge bone. Quantitative feature information is summarized in Table 10.5.

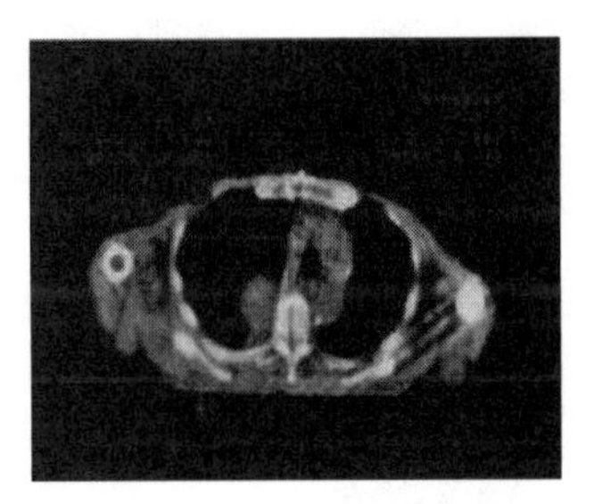

FIGURE 10.4
A cross-sectional X-ray CT image of the chest.

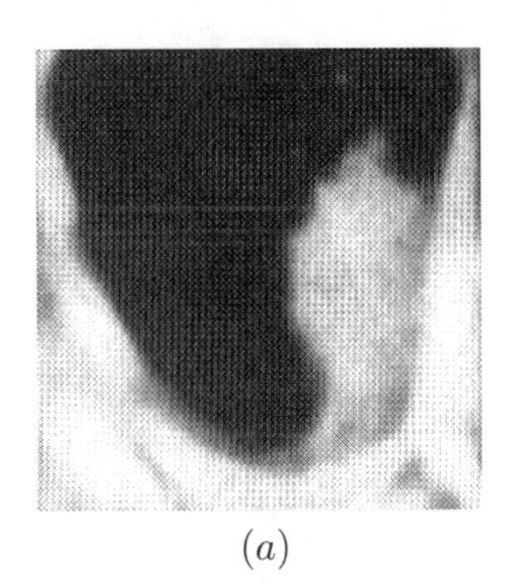

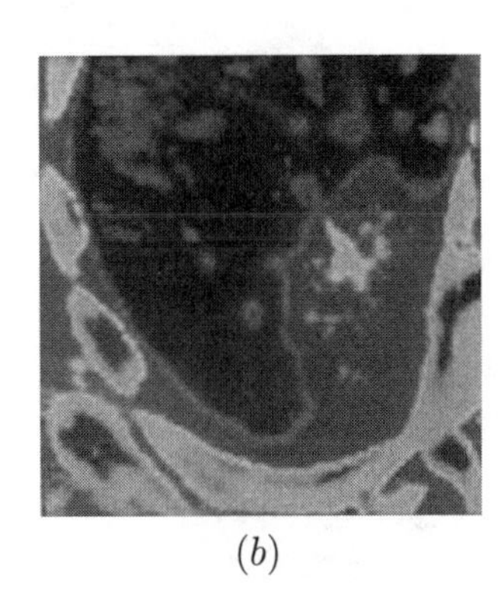

(a) (b)

FIGURE 10.5
An image of the region of interest (ROI) outlined by the red line box in Figure 10.4.

TABLE 10.4
MDL(K) Values for the ROI Image (Figure 10.5) with $K = 2 \sim 9$

K	2	3	4	5	6	7	8	9
MDL(K)	45794	45896	45073	44985	**44914**	44956	45004	44922

TABLE 10.5
Characteristics of the Region Images in Figure 10.7

	Name of Tissue	Size cm^2	Feature
Figure 7.a	Lung air	37	−857
Figure 7.b	Parenchyma, fibrosis	23	−691
Figure 7.c	Pleura, blood vessel, bronchi	14	−301
Figure 7.d	Tumor, tissues outside lung	45	−12
Figure 7.e	Dense bone	18	107
Figure 7.f	Sponge bone	10	266

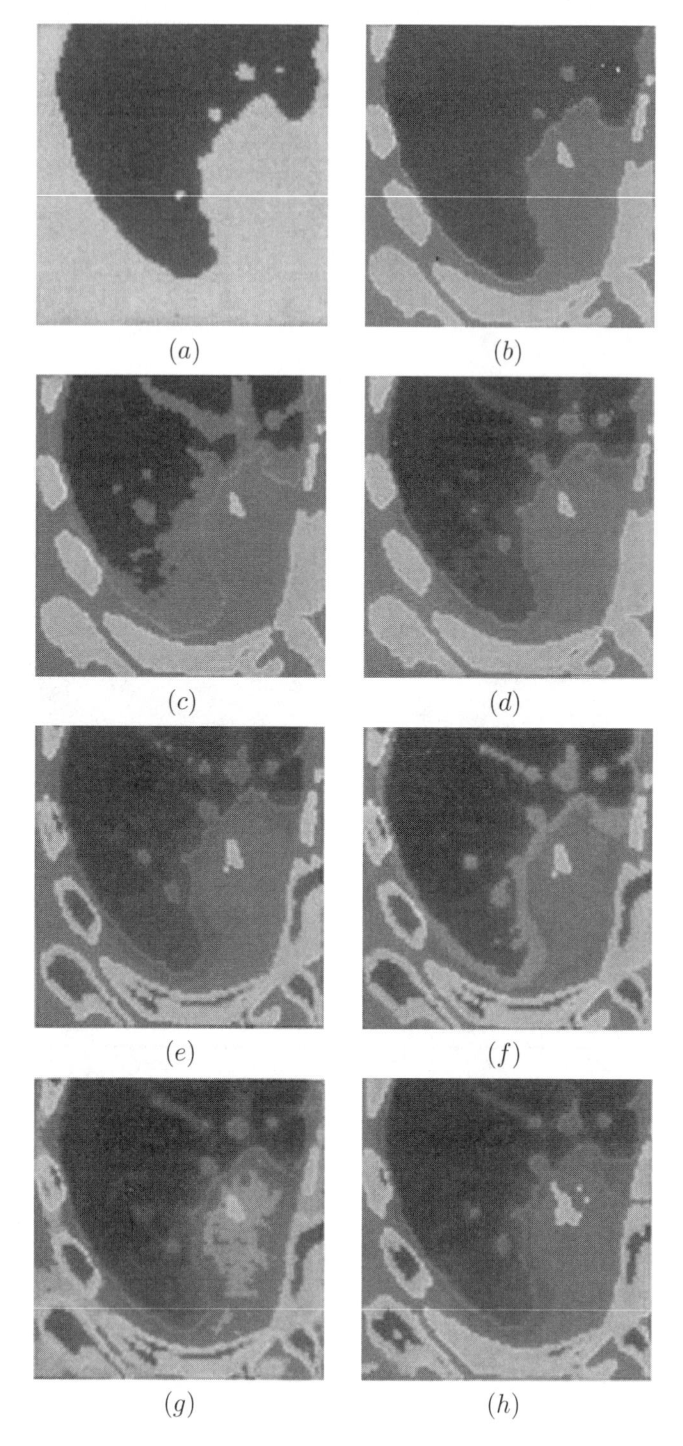

FIGURE 10.6
ROI image of Figure 10.5 is partitioned into $2, \cdots, 8$ region images shown in (a) through (h).

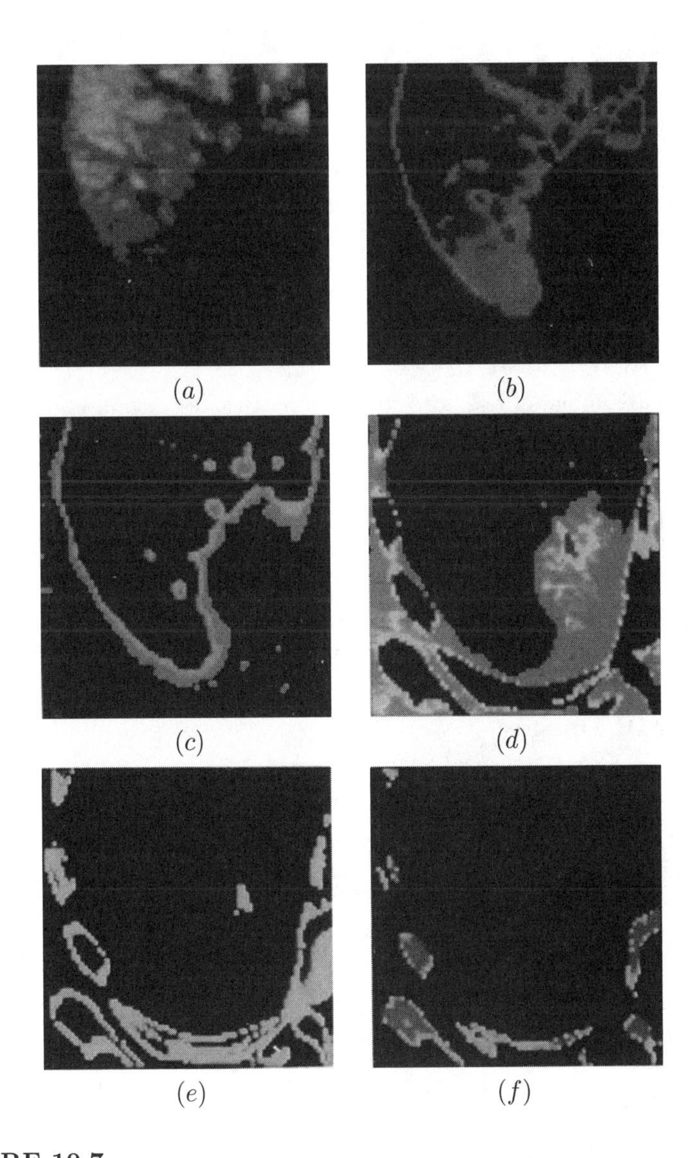

FIGURE 10.7
Images of the 1st, 2nd, 3rd, 4th, 5th, and 6th regions. They represent (a) the lung air; (b) the lung parenchyma and fibrosis; (c) the pleura, blood vessels, and bronchi; (d) a tumor and the tissues outside lung; (e) the dense bone (with a spot of tumor); and (f) the sponge bone.

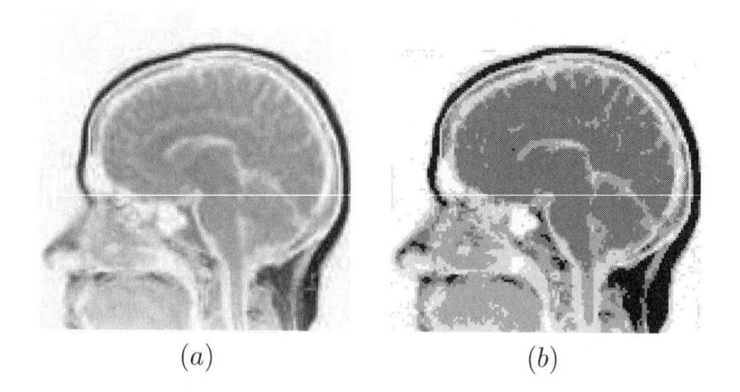

FIGURE 10.8
A sagittal MR image of the head (a). The partitioned region images are displayed in five grey levels—white, light grey, grey, dark grey, and black—in (b).

TABLE 10.6
MDL(K) Values for MRI Image (Figure 10.8.(a)) with $K = 2 \sim 9$

K	2	3	4	5	6	7	8	9
MDL(K)	41165	41155	41140	**41049**	41065	41080	41097	41144

10.5.4 MR Image

A 64 × 64 sagittal MR image of the head is shown in Figure 10.8(a). It is generated by using a quadtree method from the original 256 × 256 image and quantized at 256 grey levels. By assuming $K_{min} = 2$ and $K_{max} = 9$, MDL values computed by using Eq. (10.6) are given in Table 10.6. It shows that this image has five image regions. The partitioned five region images are obtained using iFNM model-based image analysis method and shown in Figure 10.8.(*b*) displayed in five grey levels: black, dark grey, grey, light grey, and white. Information on these five image regions are shown in Table 10.7. These results are meaningful, as expected.

TABLE 10.7
Characteristics of MR Region Images in Figure 10.8.(*b*).

Greylevel	Tissue Type	Pixel #	Mean	Data Range
White	Air	1098	27	24–30
Light grey	Scalp	795	48	31–63
Grey	CSF pathway, cerebrum	827	75	64–88
Dark grey	Grey and white matters	1083	100	89–106
Black	Skin	293	171	107–247

10.6 Appendices

10.6.1 Appendix 10A

Starting from the relative entropy and using the notation $f(x|\mathbf{r})$ for pdf $f(x)$ ($\mathbf{r}$ is the model parameter vector), this appendix outlines the derivation of ITC. The fundamental work on the definition of the information criterion was done by Akaike. The following description is basically based on [48, 55, 56].

1) *The minimization of the Kullback-Leibler divergence (KLD) defines a reasonable criterion for choosing an estimated pdf to best fit the true pdf.*

Let x be a discrete random variable (rv) with the probability mass function (pmf) $f(x|\mathbf{r})$, where $\mathbf{r} = (r_1, \cdots, r_K)^T$ is a parameter vector associated with $f(x|\mathbf{r})$.[§] Let x_i ($i = 1, \cdots, J$) be the outcomes of x. In Information Theory, the quantity defined by

$$H(f(x|\mathbf{r})) = -\sum_{i=1}^{J} f(x_i|\mathbf{r}) \log f(x_i|\mathbf{r}) = E[-\log f(x|\mathbf{r})] \tag{10.29}$$

is known as the entropy of $f(x|\mathbf{r})$ [69]. It is a measure of the uncertainty.[¶]

Let $f(x|\mathbf{r})$ and $\hat{f}(x|\mathbf{r})$ be the true and estimated pmfs of rv x, respectively. The quantity defined by

$$D_{KL}(f(x|\mathbf{r}) \parallel \hat{f}(x|\mathbf{r})) = \sum_{i=1}^{J} f(x_i|\mathbf{r}) \log \frac{f(x_i|\mathbf{r})}{\hat{f}(x_i|\mathbf{r})} = E[\log \frac{f(x|\mathbf{r})}{\hat{f}(x|\mathbf{r})}] \tag{10.30}$$

is known as the relative entropy or Kullback-Leibler divergence (KLD) between $f(x|\mathbf{r})$ and $\hat{f}(x|\mathbf{r})$. It is a noncommutative measure of the difference between the true and estimated pmfs. Eq. (10.30) can be expressed as

$$D_{KL}(f(x|\mathbf{r}) \parallel \hat{f}(x|\mathbf{r})) = \sum_{i=1}^{J} f(x_i|\mathbf{r}) \log f(x_i|\mathbf{r}) - \sum_{i=1}^{J} f(x_i|\mathbf{r}) \log \hat{f}(x_i|\mathbf{r}). \tag{10.31}$$

Based on Eq. (10.29), in Eq. (10.31), $-\sum_{i=1}^{J} f(x_i|\mathbf{r}) \log f(x_i|\mathbf{r}) = H(f(x|\mathbf{r}))$ is the entropy of $f(x|\mathbf{r})$, $-\sum_{i=1}^{J} f(x_i|\mathbf{r}) \log \hat{f}(x_i|\mathbf{r}) = H(f(x|\mathbf{r}), \hat{f}(x|\mathbf{r}))$ is the cross-entropy of $f(x|\mathbf{r})$ and $\hat{f}(x|\mathbf{r})$.

KLD is nonnegative, $D_{KL}(f(x|\mathbf{r}) \parallel \hat{f}(x|\mathbf{r})) \geq 0$, and equal to the zero if and only if $f(x|\mathbf{r}) = \hat{f}(x|\mathbf{r})$.[‖] The smaller the value of $D_{KL}(f(x|\mathbf{r}) \parallel \hat{f}(x|\mathbf{r}))$, the more similar $f(x|\mathbf{r})$ and $\hat{f}(x|\mathbf{r})$ are. Thus, minimizing $D_{KL}(f(x|\mathbf{r}) \parallel \hat{f}(x|\mathbf{r}))$

[§] Compared with the parameter vector $\mathbf{r}$ of Eq. (10.4), each component r_k here represents $(\pi_k, \mu_k, \sigma_k^2)$ in Eq. (10.4).

[¶] The base of the logarithm in Eq. (10.29) can be 2, e, or 10. Thus, the unit of entropy may be in bit, nat, or dit. In this manuscript, log denotes the natural logarithms, that is, the base e and the nat are used.

[‖] This nonnegative property is a result known as Gibbs' inequality.

defines a reasonable criterion for choosing an estimate $\hat{f}(x|\mathbf{r})$ to best fit the true $f(x|\mathbf{r})$.

2) *The maximization of the mean log-likelihood provides a practical way to minimize KLD.*

For the continuous rv x, pmf in Eqs. (10.29) through (10.31) is replaced by pdf and the summation $\sum_{i=1}^{J}$ in Eq. (10.31) is replaced by the integration $\int_x$ accordingly. In this appendix, for simplicity, the notation $f(x|\mathbf{r})$ is used to represent both pmf and pdf. In practice, rv x may be used to characterize the outputs of a system, observations of a process, or measurements of a procedure, etc. The pmf or pdf $f(x|\mathbf{r})$ is viewed as a statistical model for these data. For a $K \times 1$ vector $\mathbf{r}$, $f(x|\mathbf{r})$ is a K-th-order model. In statistical identification, there is a typical case in which (a) the parametric families of the pmf or pdf have the same functional form f but with a different parameter vector $\mathbf{r}$, and (b) f is known and $\mathbf{r}$ is to be estimated. In this case, the estimated pmf or pdf $\hat{f}(x|\mathbf{r})$ is determined by the parameter estimate $\hat{\mathbf{r}}$. For this reason, in the following, $\hat{f}(x|\mathbf{r})$ is replaced by $f(x|\hat{\mathbf{r}})$, and KLD $D_{KL}(f(x|\mathbf{r}) \parallel \hat{f}(x|\mathbf{r}))$ of Eq. (10.31) is replaced by $D_{KL}(f(x|\mathbf{r}) \parallel f(x|\hat{\mathbf{r}}))$.

Let $x_1, \cdots, x_J$ be n independent observations of rv x with pmf or pdf $f(x|\mathbf{r})$. The sample mean of the log-likelihood $\log f(x_i|\hat{\mathbf{r}})$ is given by

$$\frac{1}{J}\sum_{i=1}^{J} \log f(x_i|\hat{\mathbf{r}}) = \frac{1}{J} \log \prod_{i=1}^{J} f(x_i|\hat{\mathbf{r}}), \tag{10.32}$$

which is an estimate of the mean of $\log f(x|\hat{\mathbf{r}})$

$$E[\log f(x|\hat{\mathbf{r}})] = \sum_{i=1}^{J} f(x_i|\mathbf{r}) \log f(x_i|\hat{\mathbf{r}}). \tag{10.33}$$

As n increases indefinitely, Eq. (10.32) approaches $\sum_{i=1}^{J} f(x_i|\mathbf{r}) \log f(x_i|\hat{\mathbf{r}})$ (in the discrete case) or $\int_x f(x|\mathbf{r}) \log f(x|\hat{\mathbf{r}})dx$ (in the continuous case) with Probability one. In this appendix, for simplicity, Eq. (10.32) is called the mean one log-likelihood. The Maximum Likelihood Estimate (MLE) $\hat{\mathbf{r}}_{ML}$ of the model parameter $\mathbf{r}$ maximizes the mean log-likelihood of Eq. (10.32), that is,

$$\hat{\mathbf{r}}_{ML} = Arg\{\max_{\hat{\mathbf{r}}} \frac{1}{J}\sum_{i=1}^{J} \log f(x_i|\hat{\mathbf{r}})\}. \tag{10.34}$$

In Eq. (10.31), the first item on the right side of $\sum_{i=1}^{J} f(x_i|\mathbf{r}) \log\ f(x_i|\mathbf{r})$ is common to all estimate $\hat{\mathbf{r}}$, thus minimizing $D_{KL}(f(x|\mathbf{r}) \parallel f(x|\hat{\mathbf{r}}))$ is equivalent to maximizing $\sum_{i=1}^{J} f(x_i|\mathbf{r}) \log f(x_i|\hat{\mathbf{r}})$. Therefore, the maximization of the mean log-likelihood, that is, the use if MLE $\hat{\mathbf{r}}_{ML}$ of the parameter vector, provides a practical way to minimize KLD.

3) *An Information Criterion.*

When the estimate $\hat{\mathbf{r}}$ is sufficiently close to the true value $\mathbf{r}$, $D_{KL}(f(x|\mathbf{r}) \parallel f(x|\hat{\mathbf{r}})$ of Eq. (10.31) is measured by $\frac{1}{2}(\hat{\mathbf{r}} - \mathbf{r})^T \mathbf{J}(\hat{\mathbf{r}} - \mathbf{r})$ [46, 47], that is,

$$2D_{KL}(f(x|\mathbf{r}) \parallel f(x|\hat{\mathbf{r}}) = (\hat{\mathbf{r}} - \mathbf{r})^T \mathbf{J}(\hat{\mathbf{r}} - \mathbf{r}), \tag{10.35}$$

where $\mathbf{J}$ is the Fisher information matrix whose (ij)-th element is given by $J_{ij} = E[(\frac{\partial \log f(x|\mathbf{r})}{\partial \mathbf{r}_i})(\frac{\partial \log f(x|\mathbf{r})}{\partial \mathbf{r}_j})]$, and T denotes the transpose.

Because MLE $\hat{\mathbf{r}}_{ML}$ is asymptotically efficient [70] under regularity conditions, the mean log-likelihood of Eq. (10.32) is most sensitive to the deviation of the estimated $f(x|\hat{\mathbf{r}}_{ML})$ from the true $f(x|\mathbf{r})$. Thus, in statistical identification, the desirable model should minimize KLD $D_{KL}(f(x|\mathbf{r}) \parallel f(x|\hat{\mathbf{r}}_{ML}))$. However, there may be a difference between $D_{KL}(f(x|\mathbf{r}) \parallel f(x|\hat{\mathbf{r}}_{ML}))$ and $D_{KL}(f(x|\mathbf{r}) \parallel f(x|\hat{\mathbf{r}}))$ of Eq. (10.13).

Let MLE $\hat{\mathbf{r}}_{ML}$ be restricted in a lower k-dimensional subspace ϑ of $\mathbf{r}$ ($k < K$) that does not include the true value $\mathbf{r}$. Assume that $\hat{\mathbf{r}}$ is also restricted in ϑ and yields the maximum of $E[\log f(x|\hat{\mathbf{r}})]$. Because $\sqrt{n}(\hat{\mathbf{r}}_{ML} - \hat{\mathbf{r}})$ is asymptotically distributed as a Gaussian with zero mean and variance matrix $\mathbf{J}^{-1}$ [47, 71, 72], if $\hat{\mathbf{r}}$ is sufficiently close to $\mathbf{r}$, the distribution $n(\hat{\mathbf{r}}_{ML} - \hat{\mathbf{r}})^T \mathbf{J}(\hat{\mathbf{r}}_{ML} - \hat{\mathbf{r}}) = 2nD_{KL}(f(x|\hat{\mathbf{r}}) \parallel f(x|\hat{\mathbf{r}}_{ML}))$ can be approximated by a χ^2-distribution with the degree of freedom K_a under certain regularity conditions and for sufficiently large n [46]. Thus, we have

$$E[2nD_{KL}(f(x|\mathbf{r}) \parallel f(x|\hat{\mathbf{r}}_{ML}))] = 2nD_{KL}(f(x|\mathbf{r}) \parallel f(x|\hat{\mathbf{r}})) + K_a, \tag{10.36}$$

where E denotes the mean of the approximated distribution, and K_a is the number of free adjustable parameters in $\mathbf{r}$ for maximizing the mean log-likelihood of Eq. (10.32). Therefore, when $\hat{\mathbf{r}}$ is replaced by $\hat{\mathbf{r}}_{ML}$ and the quantity $2n(\frac{1}{J}\sum_{i=1}^{J} \log f(x_i|\mathbf{r}) - \frac{1}{J}\sum_{i=1}^{J} \log f(x_i|\hat{\mathbf{r}}_{ML}))$ is used as an estimate of $2nD_{KL}(f(x|\mathbf{r}) \parallel f(x|\hat{\mathbf{r}}))$, a value K_a should be added to it to correct the bias introduced by replacing $\hat{\mathbf{r}}$ with $\hat{\mathbf{r}}_{ML}$.

Thus, $E[2nD_{KL}(f(x|\mathbf{r}) \parallel f(x|\hat{\mathbf{r}}_{ML}))]$ can be estimated by

$$2n(\frac{1}{J}\sum_{i=1}^{J} \log f(x_i|\mathbf{r}) - \frac{1}{J}\sum_{i=1}^{J} \log f(x_i|\hat{\mathbf{r}}_{ML})) + 2K_a. \tag{10.37}$$

For a given data set x, the first term in Eq. (10.37), that is, $2\sum_{i=1}^{J} \log f(x_i|\mathbf{r})$, is common to all competing models. This term can be dropped for the purpose of comparing the values of Eq. (10.37). As a result, an information criterion of the parameter $\mathbf{r}$ is given by

$$-2\sum_{i=1}^{J} \log f(x_i|\hat{\mathbf{r}}_{ML}) + 2K_a \triangleq \mathrm{AIC}(\hat{\mathbf{r}}_{ML}), \tag{10.38}$$

where AIC stands for An Information Criterion. $\mathrm{AIC}(\hat{\mathbf{r}}_{ML})$ can be interpreted as an estimate of $-2nE[\log f(x|\hat{\mathbf{r}}_{ML})]$. It is clear that when there are several specifications of $f(x|\hat{\mathbf{r}}_{ML})$ corresponding to several models, the minimum value of $\text{AIC}(\hat{\mathbf{r}}_{ML})$ leads to the best fitting model.

4) *Other Information Theoretic Criteria (ITC).*

The above analysis directly links the mean of KLD evaluated at MLE of model parameters with the log-likelihood computed at MLE of model parameters and the number of free adjustable parameters. The logarithm uses the base e, that is, $\log = \log_e = \ln$, the natural logarithm. When the base 2 is used, $\log = \log_2$. As indicated in the footnote, the unit of entropy is the bit.

Let $f(x|\mathbf{r})$ represent the true distribution of data, observations, or a precise calculated theoretical distribution, and let $f(x|\hat{\mathbf{r}})$ be a model, description, or approximation of $f(x|\mathbf{r})$. When the logarithm with the base 2 is adopted, KLD measures the mean number of extra bits required to encode samples from $f(x|\mathbf{r})$ using a code based on $f(x|\hat{\mathbf{r}})$, rather than using a code based on $f(x|\mathbf{r})$.

When several models are used to encode the observed data, [49, 57, 58] developed a criterion that selects a model that yields the minimum code length. For a sufficiently large number of samples, the criterion is given by

$$-\sum_{i=1}^{J} \log f(x_i|\hat{\mathbf{r}}_{ML}) + \frac{1}{2} K_a \log n \triangleq \text{MDL}(\hat{\mathbf{r}}_{ML}), \tag{10.39}$$

where MDL stands for Minimum Description Length. The first term is half of the corresponding one in AIC; the second term has an extra factor $\frac{1}{2}\log n$.

[50] developed a Bayesian approach. In this approach, each competing model is assigned a prior probability; the model that yields the maximum posterior probability is selected as the best fitting one. This criterion is given by

$$-2\sum_{i=1}^{J} \log f(x_i|\hat{\mathbf{r}}_{ML}) + K_a \log n \triangleq \text{BIC}(\hat{\mathbf{r}}_{ML}), \tag{10.40}$$

where BIC stands for Bayesian Information Criterion. It is the same as MDL (by a factor 2). A prior probability actually introduces a penalty term for the additional parameters that is stronger than that of the AIC.

10.6.2 Appendix 10B

Starting from the joint differential entropy and using the notation $f(x, \mathbf{r})$ for pdf $f(x)$ ($\mathbf{r}$ is the model parameter vector), this appendix gives the derivation of a new information theoretic criterion, called minimizing maximum entropy (MME).

In Appendix 10A, An Information Criterion (AIC) [48, 55, 56, 73, 74] represents an unbiased estimate of the mean Kulback–Liebler divergence between the pdf of the selected model and the estimated pdf. The Minimum description length (MDL) [49, 57, 58] is based on minimizing code length. The Bayesian Information Criterion (BIC) [50] maximizes posterior probability.

The derivation of this new information theoretic criterion is based on the principle of minimizing the maximum joint differential entropy of the observed data and the model parameters [59–63]. Entropy is a measure of the average uncertainty in random variables [44]. By seeking maximum entropy, it is able to find a family of models (i.e., a group of pdfs) that has the most uncertainty compared with other families. That is, the selected family will include most models. By minimizing this maximum entropy, it is able to find one model in this family (i.e., a specific pdf) that has the least uncertainty. That is, the chosen model will best fit the observations. This minimizing the maximum entropy (MME) procedure can be viewed as a two-stage operation: "*max*" is for a family selection, and "*min*" is for a member determination.

Let x be a random variable, $\mathbf{r} = (r_1, \cdots, r_K)^T$ a $K \times 1$ parameter vector of pdf of x, and $\hat{\mathbf{r}}_K = (\hat{r}_1, \cdots, \hat{r}_K)^T$ the estimate of $\mathbf{r}$.** In this case, the model order is the dimension K of the model parameter vector $\mathbf{r}$. Furthermore, let $f(x, \hat{\mathbf{r}})$ be the joint pdf of x and $\hat{\mathbf{r}}$, $f(x|\hat{\mathbf{r}})$ the conditioned pdf of x on $\hat{\mathbf{r}}$, and $f(\hat{\mathbf{r}})$ the pdf of $\hat{\mathbf{r}}$. The relationship of these three pdfs is

$$f(x, \hat{\mathbf{r}}) = f(x|\hat{\mathbf{r}})f(\hat{\mathbf{r}}), \tag{10.41}$$

and the corresponding entropy relationship [44, 76, 77] is

$$H(f(x, \hat{\mathbf{r}})) = H(f(x|\hat{\mathbf{r}})) + H(f(\hat{\mathbf{r}})). \tag{10.42}$$

Based on Jaynes' Principle, "*Parameters in a model which determine the value of the maximum entropy should be assigned the values which minimizes the maximum entropy,*" [59, 60], if $\hat{\mathbf{r}}_{K_0}$ is a such estimate, then it should minimize

$$\max_{\hat{\mathbf{r}}_K} H(f(x, \hat{\mathbf{r}})) = \max_{\hat{\mathbf{r}}_K} H(f(x|\hat{\mathbf{r}})) + \max_{\hat{\mathbf{r}}_K} H(f(\hat{\mathbf{r}})). \tag{10.43}$$

Entropy can be interpreted as the negative of the expected value of the logarithm of the pdf. The negative of the logarithm of the likelihood of maximum likelihood estimates of parameters is a simple and natural estimate of the maximum conditional entropy [48, 58]. Thus, for a given data set $\{x_1, \cdots, x_J\}$, if x_i $(i = 1, \cdots, J)$ are identically independently distributed, then

$$\max_{\hat{\mathbf{r}}_K} H(f(x|\hat{\mathbf{r}})) = -\log \mathcal{L}(x_1, \cdots, x_J|\hat{\mathbf{r}}_{ML}) = -\sum_{i=1}^{J} \log f(x_i|\hat{\mathbf{r}}_{ML}), \tag{10.44}$$

where $\hat{\mathbf{r}}_{ML} = (\hat{r}_{1_{ML}}, \cdots, \hat{r}_{K_{ML}})^T$ is the ML estimate of $\hat{\mathbf{r}}$, and $\mathcal{L}(x_1, \cdots, x_n|\hat{\mathbf{r}}_{K_{ML}})$ is the likelihood of ML estimates of the parameters.

For the fixed variance σ^2, the Gaussian distribution $N(0, \sigma^2)$ has the largest entropy $\log \sqrt{2\pi e}\sigma$ [44, 76–78]. Under certain general conditions, MLE $\hat{\mathbf{r}}_{ML}$

**Compared with the parameter vector $\mathbf{r}$ of Eq. (10.4), each component r_k here represents $(\pi_k, \mu_k, \sigma_k^2)$ in Eq. (10.4).

has an asymptotic Gaussian distribution; thus, if $\hat{r}_{k_{ML}}$ $(k = 1, \cdots, K)$ are independent, then

$$\max_{\hat{\mathbf{r}}_K} H(f(\hat{\mathbf{r}})) = H(f(\hat{\mathbf{r}}_{ML})) = \sum_{k=1}^{K} H(f(\hat{r}_{k_{ML}})) = \sum_{k=1}^{K} \log \sqrt{2\pi e}\sigma_{\hat{r}_{k_{ML}}}, \tag{10.45}$$

where $\sigma^2_{\hat{r}_{k_{ML}}}$ is the variance of ML estimate $\hat{r}_{k_{ML}}$ of the parameter r_k.

Substituting Eq. (10.44) and Eq. (10.45) into Eq. (10.43), $\hat{\mathbf{r}}_{K_0}$ should minimize

$$\max_{\hat{\mathbf{r}}_K} H(f(x, \hat{\mathbf{r}})) = -\sum_{i=1}^{n} \log f(x_i|\hat{\mathbf{r}}_{ML}) + \sum_{k=1}^{K} \log \sqrt{2\pi e}\sigma_{\hat{r}_{k_{ML}}}. \tag{10.46}$$

Define

$$-\sum_{i=1}^{n} \log f(x_i|\hat{\mathbf{r}}_{ML}) + \sum_{k=1}^{K} \log \sqrt{2\pi e}\sigma_{\hat{r}_{k_{ML}}} \triangleq \text{MME}(\hat{\mathbf{r}}_{ML}), \tag{10.47}$$

where MME stands for minimizing the maximum entropy; the correct order of the model, K_0, should satisfy

$$K_0 = Arg\{\min_{K} \text{MME}(\hat{\mathbf{r}}_{ML})\}. \tag{10.48}$$

10.6.3 Appendix 10C

This appendix describes the derivation of EM algorithm for estimating the model parameters of iFNM (Eq. (10.1)) for the image $IMG(J, K)$ (J – the number of pixels and K – the number of image regions).

Define the $K \times 1$ vector $\mathbf{e}_k$ $(1 \leq k \leq K)$ whose components are all zero except its k-th component equal to unity, that is,

$$\mathbf{e}_k = (e_{kk} = 1, e_{kl} = 0, l = 1, \cdots, k-1, k+1, \cdots, K)^T. \tag{10.49}$$

Assume that a $K \times 1$ discrete random vector (rv) $\mathbf{e}$ takes the values $\mathbf{e}_k$ $(1 \leq k \leq K)$ and its probability mass function (pmf) is defined by

$$v(\mathbf{e} = \mathbf{e}_k) = \pi_k. \tag{10.50}$$

Also assume that the $K \times 1$ vectors $\mathbf{z}_j$ $(1 \leq j \leq J)$ are independently identically distributed (i.i.d.) samples of rv $\mathbf{e}$. The unity component in $\mathbf{z}_j$ $(1 \leq j \leq J)$ indicates an unknown k-th image region $(1 \leq k \leq K)$ associated with the pixel intensity x_j $(1 \leq j \leq J)$.

Because the observed (incomplete) data x_j $(1 \leq j \leq J)$ are i.i.d. samples from iFNM pdf (Eq. (10.1)), x_j—given $\mathbf{z}_j$—are independent of the conditional pdf $g(x_j|\mathbf{z}_j, \theta_k)$. Let y_j $(1 \leq j \leq J)$ be the complete data of x_j, that is, y_j is characterized by both x_j and $\mathbf{z}_j$

$$y_j = (x_j, \mathbf{z}_j)^T, \tag{10.51}$$

the pdf of y_j, denoted by $u(y_j)$, is given by

$$u(y_j) = p(x_j, \mathbf{z}_j) = g(x_j|\mathbf{z}_j, \theta_k)v(\mathbf{z}_j|\pi_k). \tag{10.52}$$

Using the vector-matrix form, let

$$\begin{aligned} \mathbf{x} &= (x_1, \cdots\cdots, x_J)^T, \\ \mathbf{y} &= (y_1, \cdots\cdots, y_J)^T, \\ \mathbf{Z} &= (\mathbf{z}_1, \cdots\cdots, \mathbf{z}_J)^T, \end{aligned} \tag{10.53}$$

where $\mathbf{Z}$ is a $J \times K$ matrix, the complete data of $\mathbf{x}$ are defined by

$$\mathbf{y} = (\mathbf{x}, \mathbf{Z})^T. \tag{10.54}$$

Based on Eq. (10.52), the pdf of complete data $\mathbf{y}$ are given by

$$u(\mathbf{y}|\mathbf{r}) = g(\mathbf{x}|\mathbf{Z}, \theta)v(\mathbf{Z}|\pi), \tag{10.55}$$

where $\theta = (\theta_1, \cdots, \theta_K)^K$ and $\pi = (\pi_1, \cdots, \pi_K)^T$ are $K \times 1$ vectors. Thus, the log-likelihood of i.i.d. data $x_1, \cdots, x_J$ is

$$\log u(\mathbf{y}|\mathbf{r}) = \sum_{j=1}^{J} (\log g(x_j|\mathbf{z}_j, \theta_k) + \log v(\mathbf{z}_j|\pi)). \tag{10.56}$$

Let

$$\mathbf{g}(x_j|\theta) = (\log g(x_j|\mathbf{e}_1, \theta), \cdots, \log g(x_j|\mathbf{e}_K, \theta))^T, \tag{10.57}$$

and

$$\mathbf{v}(\pi) = (\log v(\mathbf{e}_1|\pi), \cdots, \log v(\mathbf{e}_K|\pi))^T, \tag{10.58}$$

Eq. (10.56) can be rewritten as

$$\log u(\mathbf{y}|\mathbf{r}) = \sum_{j=1}^{J} \mathbf{z}_j^T(\mathbf{g}(x_j|\theta) + \mathbf{v}(\pi)). \tag{10.59}$$

The first step in the EM algorithm, the E-step, is to evaluate the expectation of the log-likelihood Eq. (10.59) by using the incomplete data $\mathbf{x}$ and the current parameter estimate $\mathbf{r}^{(m)}$,

$$Q(\mathbf{r}, \mathbf{r}^{(m)}) = \mathbf{E}(\log u(\mathbf{y}|\mathbf{r})|\mathbf{x}, \mathbf{r}^{(m)}), \tag{10.60}$$

where $(m = 1, 2, \cdots)$ denotes the iteration. From Eq. (10.59),

$$Q(\mathbf{r}, \mathbf{r}^{(m)}) = \sum_{j=1}^{J} \left(\mathbf{E}(\mathbf{z}_j|\mathbf{x}, \mathbf{r}^{(m)})\right)^T (\mathbf{g}(x_j|\theta^{(m)}) + \mathbf{v}(\pi^{(m)})). \tag{10.61}$$

Thus, based on the definitions of $\mathbf{z}_j$, $g(x_j|\theta_k)$ and π_k, we have

$$\mathbf{E}(\mathbf{z}_j|\mathbf{x}, \mathbf{r}^{(m)}) = (z_{j1}^{(m)}, \cdots, z_{jK}^{(m)})^T \triangleq \mathbf{z}_j^{(m)}, \tag{10.62}$$

where

$$z_{jk}^{(m)} = \frac{\pi_k^{(m)} g(x_j|\theta_k^{(m)})}{f(x_j|\mathbf{r}^{(m)})} \tag{10.63}$$

is called probability membership.

The second step in the EM algorithm, the M-step, is to compute the updated estimate $\mathbf{r}^{(m+1)}$ by maximizing the estimated likelihood,

$$\mathbf{r}^{(m+1)} = Arg \max_{\mathbf{r}}\{Q(\mathbf{r}, \mathbf{r}^{(m)})\} . \tag{10.64}$$

From Eqs. (10.61) and (10.62),

$$\mathbf{r}^{(m+1)} = Arg \max_{\mathbf{r}}\{\sum_{j=1}^{J}(\mathbf{z}_j^{(m)})^T(\mathbf{g}(x_j|\theta) + \mathbf{v}(\pi))\}. \tag{10.65}$$

Because the parameter vectors θ (in $\mathbf{g}$) and π (in $\mathbf{v}$) are distinct, the first and second terms on the right side of Eq. (10.65) can be maximized separately. For π, noting the constraint $\sum_{j=1}^{k} \pi_k = 1$ and using the Lagrangian multiplier, $\max_\pi \sum_{j=1}^{J} (\mathbf{z}_j^{(m)})^T\mathbf{v}(\pi)$ leads to

$$\pi_k^{(m+1)} = \frac{1}{J}\sum_{j=1}^{J} z_{jk}^{(m)}. \tag{10.66}$$

For θ, using $\pi_k^{(m+1)}$ and noting that $g(x_j|\theta_k)$ is distributed as $N(\mu_k, \sigma_k^2)$, $\max_\theta \sum_{j=1}^{J} (\mathbf{z}_j^{(m)})^T\mathbf{g}(x_j|\theta)$ leads to

$$\mu_k^{(m+1)} = \frac{1}{J\pi_k^{(m+1)}}\sum_{j=1}^{J} z_{jk}^{(m)} x_j, \tag{10.67}$$

$$\sigma_k^{2(m+1)} = \frac{1}{J\pi_k^{(m+1)}}\sum_{j=1}^{J} z_{jk}^{(m)} (x_j - \mu_j^{(m+1)})^2. \tag{10.68}$$

Thus, the EM algorithm for ML parameter estimation of iFNM becomes iteratively computing, starting from an initial estimate $\mathbf{r}^{(0)}$, Eq. (10.63) (E-step) and Eq. (10.66) through Eq. (10.68) (M-step), until a specified stopping criterion

$$|\pi_k^{(m+1)} - \pi_k^{(m)}| < \varepsilon \qquad (1 \leq k \leq K) \tag{10.69}$$

is satisfied, where ε is a prespecified small number.

Rigorously speaking, $\mu_k^{(m+1)}$ of Eq. (10.67) and $\sigma_k^{2(m+1)}$ of Eq. (10.68) are not the true ML estimates of μ_k and σ_k^2. This is because their $(m+1)$-th estimates depend on the $(m+1)$-th estimate of π_k and μ_k, not on (m)-th estimates.

Problems

10.1. What conditions and assumptions are used in the derivation of the minimizing maximum entropy (MME) of Appendix 10B?

10.2. In the derivation of An Information Criterion (AIC), $f(x|\mathbf{r})$ is used, in the derivation of minimizing maximum entropy (MME), $f(x,\mathbf{r})$ is used. Explain why these two probabilities lead two different information theoretic criteria.

10.3. Chapter 12 gives the Cramer-Rao Low Bounds (CRLB) of the variance of the ML estimate $\hat{r}_{k_{ML}}$. Can these CRLBs be used to replace $\sigma^2_{\hat{r}_{k_{ML}}}$ in Eq. (10.47)?

10.4. Derive Eq. (10.66).

10.5. Derive Eqs. (10.67) and (10.68).

10.6. In Eq. (10.69), if we set $\varepsilon = \frac{1}{J}$ (J – the number of pixels in the image), what is the physical meaning of the stopping criterion of Eq. (10.69)?

References

[1] Moore, E.: The shortest path through a maze. *Proc. Intl. Symp. on the Theory of Switching* (Harvard Univ., Apr. 1959) 285–292.

[2] Dijkstra, E.: A note on two problems in connexion with graphs. *Numerische Mathematik* **1** (1959) 269–271.

[3] Martelli, A.: An application of heuristic search methods to edge and contour detection. *Comm. an ACM* **19**(2) (1976) 73–83.

[4] Mortensen, E., Barrett, W.: Interactive segmentation with intelligent scissors. *Graphical Models and Image Processing* **60** (1998) 349–384.

[5] Falcao, A., Udupa, J., Miyazawa, F.: An ultra-fast user-steered image segmentation paradigm: Live-wire-on-the-fly. *IEEE Trans. Med. Imag.* **19**(1) (2000) 55–62.

[6] Kass, M., Witkin, A., Terzopoulos, D.: Snakes: Active contour models. *Intl. J. Comput. Vis.* **1** (1987) 321–331.

[7] Cohen, L.: On active contour models and balloons. *CVGIP: Image Understanding* **53** (1991) 211–218.

[8] Caselles, V., Kimmel R., Sapiro, G.: Geodesic active contours. *Proc. 5th Intl. Conf. Computer Vision* (1995) 694–699.

[9] Leroy, B., Herlin, I., Cohen, L.: Multi-resolution algorithms for active contour models. *Proc. 12th Intl. Conf. Analysis and Optimization of Systems* (1996) 58–65.

[10] Shen, D., Davatzikos, C.: Hammer: Hierarchical attribute matching mechanism for elastic registration. *IEEE Trans. Med. Imag.* **22**(11) (2002) 1421–1439.

[11] Malladi, R., Sethian, J., Vemuri, B.: Shape model with front propagation: A level set approach. *IEEE Trans. Pattern Anal. Machine Intell.* **17** (1995) 158–175.

[12] Teboul, S., Blanc-Feraud, L., Aubert, G., Barlaud, M.: Variational approach for edge-preserving regulariztion using PDE's. *IEEE Trans. Image Processing* **7**(3) (1998) 387–397.

[13] Faugeras, O., Keriven, R.: Variational principles, surface evolution, PDE's, level set methods, and the stereo problem. *IEEE Trans. Image Processing* **7**(3) (1998) 336–344.

[14] Avants, B., Gee, J.: Formulation and evaluation of variational curve matching with prior constraints. *Lect. Notes Comput. Sci.* **2717** (2003) 21–30.

[15] Xu, M., Thompson, P., Toga, A.: An adaptive level set segmentation on a triangulated mesh. *IEEE Trans. Med. Imag.* **32**(2) (2004) 191–201.

[16] Hill, A., Taylor, C.: Automatic landmark generation for point distribution models. *Proc. Br. Machine Vis. Conf.* (1994) 429–438.

[17] Cootes, T., Taylor, C.: Active shape model search using local grey-level models: A quantitative evaluation. *Proc. Br. Machine Vision Conf. (BMVA)* (1994) 639–648.

[18] Hill, A., Cootes, T., Taylor, C.: Active shape model and shape approximation problem. *Image Vision Comput.* **14**(8) (1996) 601–607.

[19] Duta, N., Sonka, M.: Segmentation and interpretation of MR brain images: An improved active shape model. *IEEE Trans. Med. Imag.* **17**(6) (1998) 1049–1062.

[20] Shen, D., Herskovits, E., Davatzikos, C.: An adaptive-focus statistical shape model for segmentation and shape modeling of 3-D brain structures. *IEEE Trans. Med. Imag.* **20**(4) (2001) 257–270.

[21] Cootes, T., Taylor, C.: Modeling object appearance using the grey-level surface. In E. Hancock, Ed., *5th British Machine Vision Conf.* York, United Kingdom (1994) 479–488.

[22] Edwards, G., Taylor, C., Cootes, T.: Face recognition using the Active Appearance Model. *Proc. 5th European Conf. Comput. Vis.* **2** (1998) 581–695.

[23] Cootes, T., Edwards, G., Taylor, C.: Active Appearance Models. *IEEE Trans. Pattern Anal. Machine Intell.* **23**(6) (2001) 681–685.

[24] Davatzikos, C., Tao, X., Shen, D.: Hierarchical active shape models, using the wavelet transform. *IEEE Trans. Med. Imag.* **22**(3) (2003) 414–423.

[25] Kaufmann, A.: *Introduction to the Theory of Fuzzy Subsets.* Academic Press, New York (1975).

[26] Rosenfeld, A.: The fuzzy geometry of image sebsets. *Pattern Recogn. Lett.* **2** (1991) 311–317.

[27] Udupa, J.K., Samarasekera, S.: Fuzzy connectedness and object delineation: Theory, algorithms, and applications in image segmentation. *Graphical Models and Image Processing* **58**(3) (1996) 246–261.

[28] Carvalho, B., Gau, C., Herman, G., Kong, T.: Algorithms for fuzzy segmentation. *Patt. Anal. Appl.* **2** (1999) 73–81.

[29] Saha, P.K., Udupa, J.K.: Iterative relative fuzzy connectedness and object definition: Theory, algorithms, and applications in image segmentation. *Proceedings of IEEE Workshop on Mathematical Methods in Biomedical Image Analysis* (2000) 28–35.

[30] Herman, G., Carvalho, B.: Multiseeded segmentation using fuzzy connectedness. *IEEE Trans. Pattern Anal. Machine Intell.* **23** (2001) 460–474.

[31] Nyul, L., Falcao, A., Udupa, J.: Fuzzy-connected 3D image segmentationat interactive speeds. *Graph. Models* **64** (2002) 259–281.

[32] Besag, J.: Spatial interaction and the statistical analysis of lattice system (with discussion). *J. Royal Statist. Soc.* **36B**(2) (1974) 192–236.

[33] Cross, G., Jain, A.: Markov random field texture models. *IEEE Trans. Pattern Anal. Machine Intell.* **5** (1983) 25–39.

[34] Geman, S., Geman, D.: Stochastic relaxation, Gibbs distributions, and the Bayesian restoration of images. *IEEE Trans. Pattern Anal. Machine Intell.* **6**(6) (1984) 721–724.

[35] Besag, J.: On the statistical analysis of dirty pictures. *J. Royal Statist. Soc.* **48B** (1986) 259–302.

[36] Leahy, R., Hebert, T., Lee, R.: Applications of Markov random field models in medical imaging. *Proc. IPMI* **11** (1989) 1–14.

[37] Bouman, C., Liu, B.: Multiple resolution segmentation of texture images. *IEEE Trans. Pattern Anal. Machine Intell.* **13**(2) (1991) 99–113.

[38] Liang, Z., MacFall, J., Harrington, D.: Parameter estimation and tissue segmentation from multispectral MR images. *IEEE Trans. Med. Imag.* **13**(3) (1994) 441–449.

[39] Krishnamachari, S., Chellappa, R.: Multiresolution Gauss-Markov random field models for texture segmentation. *IEEE Trans. Image Processing* **6**(2) (1997) 251–267.

[40] Bartlett, M.: A note on the multiplying factors for various χ^2 approximations. *J. Roy. Stat. Soc.,* Ser. B. **16** (1954) 296–298.

[41] Lawley, D.: Tests of significance of the latent roots of the covariance and correlations matrices. *Biometrica* **43** (1956) 128–136.

[42] Makhoul, J.: Linear prediction: A tutorial review. *IEEE Proc.* **63**(4) (1975) 561–580.

[43] Cohen, A.: *Biomedical Signal Processing.* CRC Press, Boca Raton, FL (1986).

[44] Cover, T.: *Elements of Information Theory.* John & Sons, New York (1991).

[45] MacKay, D.: *Information Theory, Inference, and Learning Algorithms.* Cambridge University Press, Cambridge, United Kingdom, (2003).

[46] Kullback, S.: *Information Theory and Statistics.* Wiley, New York (1959).

[47] Kullback, S.: *Information Theory and Statistics.* Dover, New York (1968 (reprint)).

[48] Akaike, H.: On entropy maximization principle. P.R. Krishnaiah, Ed., *Applications of Statistics* North-Holland Publishing Company, (1977) 27–41.

[49] Rissanen, J.: Modeling by shortest data description. *J. Automatica* **14** (1978) 465–471.

[50] Schwarz, G.: Estimating the dimension of a model. *Ann. Stat.* **6**(2) (1978) 461–464.

[51] Wax, M., Kailath, T.: Detection of signals by information theoretic criteria. *IEEE Trans. Acoust., Speech, and Signal Proc.* **33**(2) (1985) 387–392.

[52] Yin, Y., Krishnaiah, P.: On some nonparametric methods for detection of the number of signals. *IEEE Trans. Acoust., Speech, and Signal Processing* **35**(11) (1987) 1533–1538.

[53] Wong, K., Zhang, Q., Reilly, J., Yip, P.: On information theoretic criteria for determining the number of signals in high resolution array processing. *IEEE Trans. Acoust., Speech, and Signal Processing* **38**(11) (1990) 1959–1971.

[54] Zhang, J.: The mean field theory in EM procedures for blind Markov random field image. *IEEE Trans. Image Proc.* **2**(1) (1993) 27–40.

[55] Akaike, H.: A new look at the Bayes procedure. *Biometrika* **65**(1) (1978) 53–59.

[56] Akaike, H.: A Bayesian analysis of the minimum AIC procedure. *Ann. Inst. Statist. Math.* **30, Part A** (1978) 9–14.

[57] Rissanen, J.: Consistent order estimation of autoregressive processes by shortest description of data. *Analysis and Optimization of Stochastic Systems* ,Jacobs et al., Eds. Academic Press, New York, (1980) 452–461.

[58] Rissanen, J.: A universal prior for integers and estimation by minimum description length. *Ann. Stat.* **11**(2) (1983) 416–431.

[59] Jaynes, E.: Information theory and statistical mechanics I. *Phys. Rev.* **106**(4) (1957) 620–630.

[60] Jaynes, E.: Information theory and statistical mechanics II. *Phys. Rev.* **108**(2) (1957) 171–190.

[61] Jaynes, E.: *Probability Theory in Science and Engineering.* McGraw-Hill, New York (1961).

[62] Jaynes, E.: The minimum entropy production principle. *Annu. Rev. Phys. Chem.* **31** (1980) 579–601.

[63] Jaynes, E.: On the rationale of maximum-entropy methods. *IEEE Proc.* **70**(9) (1982) 939–952.

[64] Dempster, A.P., Laird, N., Rubin, D.B.: Maximum likelihood from incomplete data via EM algorithm. *J. Roy. Statist. Soc.* **39** (1977) 1–38.

[65] Tou, J., Gonzalaz, R.: *Pattern Recognition Principles.* Addison-Wesley Publication Company, Reading, Massachusetts, (1974).

[66] Titterington, D., Smith, A., Makov, U.: *Statistical Analysis of Finite Mixture Distribution.* John Wiley & Sons Inc., Chichester (1985).

[67] Zhang, J., Modestino, J.: A model-fitting approach to cluster validation with application to stochastic model-based image segmentation. *IEEE Trans. Pattern Analysis and Machine Intelligence* **12**(10) (1990) 1009–1017.

[68] Kelly, P., Derin, H., Hartt, K.: Adaptive segmentation of speckled images using hierarchical random field model. *IEEE Trans. Acoust. Speech. Signal Processing* **36**(10) (1988) 1628–1641.

[69] Shannon, C.E.: A Mathematical Theory of Communication. *Technical report, Bell Syst. Tech. J.* (1948).

[70] Cramer, H.: *Mathematical Methods of Statistics.* Princeton University Press, New Jersey, (1946).

[71] Cramer, H.: *The Elements of Probability Theory.* John Wiley & Sons, New York (1955).

[72] Philippou, A., Roussas, G.: Asymptotic normality of the maximum likelihood estimate in the independent not identically distributed case. *Ann. Inst. Stat. Math.* **27**(1) (1975) 45–55.

[73] Akaike, H.: Information theory and an extension of the maximum likelihood principle. *Proc. 2nd Int. Symp. Inform. Theory, Suppl. Problems of Control and Inform. Theory* (1973) 273–281.

[74] Akaike, H.: A new look at the statistical model identification. *IEEE Trans. Automatic Control* **19**(6) (1974) 716–723.

[75] Rissanen, J.: MDL denoising. *IEEE Trans. Inform. Theory* **46**(7) (2000) 2537–2543.

[76] Thomas, J.: *An Introduction to Statistical Communication Theory.* John Wiley & Sons Inc, New York (1969)

[77] Papoulis, A.: *Probability, Random Variables and Stochastic Processes.* McGraw-Hill Book Company Inc., New York (1984).

[78] Campenhout, J.V., Cover, T.: Maximum entropy and conditional probability. *IEEE Trans. Inform. Theory* **27** (1981) 483–489.

11

Statistical Image Analysis – II

11.1 Introduction

Advanced image analysis techniques, including those applied to medical images, utilize the spatial relationship of pixels in the images. This relationship is mainly determined by the properties of the intensities and geometries of pixels, either local or global. Among various frameworks of image analysis techniques such as listed in Section 10.1, descriptors such as "*connectivity*," "*continuity*," "*smoothness*," and "*hanging togetherness*," etc., are often employed to characterize spatial relationships of pixels.

In the graph approach [1–4], edge contour detection is formulated as a minimum cost (the weighted shortest) path problem on a graph. The path cost function is determined by local image properties: pixel location, intensity, and gradient. It is application specific. Criteria for the *continuity* of arcs and/or segments of edges are different and user specified.

In the classical snakes and active contour approaches [5–8], by minimizing the total energy defined by the models, the edge curve at the points of maximal magnitude of the gradients are located via the external energy while the *smoothness* of the edge curve is kept via the internal energy.

Level set methods [9–12]—a variational approach—seek a minimizer of a functional by solving the associated partial differential equations (PDEs) [13, 14]. These PDEs guide the interface—the evolution of the zero-level curve—toward the boundary of the optimal partitions.

In the Active Shape model (ASM) and Active Appearance model (AAM) approaches [15–23], the corresponding points on each sample of a training set of annotated images are marked and aligned. Eigen-analysis is then applied to build a statistical shape model. Given enough samples, such a model is able to synthesize any image of normal anatomy. By adjusting the parameters that minimize the difference between the synthesized model image and a target image, all structures, represented and modeled in the image, are segmented.

In Fuzzy Connected object delineation [24–27], the strength of *Fuzzy Connectedness* (FC) assigned to a pair of pixels is the strength of the strongest of all paths between this pair, and the strength of a path is the weakest affinity between pixels along the path. The degree of affinity between two pixels is determined by the spatial nearness, region homogeneity, and the expected object feature. A Fuzzy connected object with a given strength, containing a seed pixel, consists of a pool of pixels such that for any pixel inside the pool, the strength of FC between it and the seed is greater than the given strength and otherwise less than the given strength.

Markov random field (MRF) [28–43] is a widely used stochastic model for image segmentation (to yield *homogeneous* regions) and restoration (to smooth regions), because of its ability to impose regularity properties using contextual information. The advantages of MRF-Gibbs equivalence allow us to characterize the local spatial relationships between neighboring pixels via clique potentials and the global spatial relationships among all pixels via Gibbs distribution.

Chapter 9 of this book gives two stochastic models: iFNM and cFNM, to X-ray CT and MR images. Chapter 10 provides an iFNM model-based image analysis method. It has been used to the the images whose pixel intensities are statistically independent. For images with high SNR, pixel intensities can be considered approximately independent. This chapter describes a cFNM model-based image analysis method. It has been used for images whose pixel intensities are statistically correlated. That is a general case. Similar to other image analysis techniques listed in the above brief survey, a cFNM model-based image analysis method utilizes the spatial relationship of pixels; however, unlike those image analysis technique, the spatial relationship of pixels used in this method is evaluated by correlations of pixel intensities, explicitly and quantitatively. Statistical properties of pixel intensities described in Chapters 6 and 8 are integrated in the cFNM model.*

The cFNM model-based image analysis method consists of three steps: the detection of the number of image regions, the estimation of the model parameters, and the classification of pixels into distinctive image regions. A sensor array signal processing method is developed for detection; it is an eigenstructure approach and is independent of the cFNM model. An extended Expectation-Maximization (EM) algorithm is proposed for the estimation; it is an EM-MAP operation with the newly developed design of clique potentials. Classification still uses the Bayesian classifier. The entire image analysis method is a data-driven approach.

*In statistics, second-order statistics such as the correlation and covariance are the simplest and the most commonly used measures for characterizing the relationship of random variables. In visual pattern discrimination, it has been shown that second-order statistics are the important characteristics of human vision [44].

11.2 Detection of the Number of Image Regions

The iFNM model-based image analysis method described in Chapter 10 utilizes Information Theoretic Criteria (ITC) to detect the number of image regions in an image. The underlying reason for this approach is as follows. When pixel intensities are statistically independent, they are i.i.d. samples of the iFNM pdf. Thus, ML estimates of parameters of the iFNM pdf can be easily computed via the EM or CM algorithm, and the mean log-likelihood of ML estimates gives a practical way to minimize Kullback-Leibler divergence (KLD), which leads to a procedure to seek an estimated pdf to best fit the true cFNM pdf.

When pixel intensities are not statistically independent, they are not i.i.d. samples of cFNM pdf. As an impact, it is very difficult to establish the log-likelihood of ML estimates of parameters of the cFNM pdf. Even though this log-likelihood could be formed by ignoring the correlations among pixel intensities, computing ML estimates of parameters of the cFNM pdf, as shown in Section 11.3, is a complicated task. This section describes a new approach. It translates the image region detection problem into a sensor array processing framework. The samples from all *sensors* are utilized to estimate the covariance matrix which is a description of the overall signal environment of this converted sensor array system. By applying eigenstructure analysis to this covariance matrix, the number of distinctive image regions can be detected by counting the *sources* in the converted sensor array system via information theoretic criteria. This new detection scheme is a non-model-based, data-driven approach.

11.2.1 Array Signal Processing Approach

The conversion of an image to an analog sensor array system is performed in the following three steps. As shown in Figure 11.1, (1) By sampling the observed image in two dimensions with a certain step size, a sequence of sub-images is created; (2) by sampling each sub-image in two-dimensions with a certain step size, a sequence of sub-sub-images is created for each sub-image; and (3) by averaging the pixel intensities in each sub-sub-image, a sequence of data is created for each sub-image. In the proposed approach, as shown in Figure 11.2, the distinctive region images are referred to as *sources*, the sampled sub-images are referred to as *sensors*, and the averaged data of each sub-sub-image are referred to as *samples* or *observations* of that sensor. The sampled images and the averaged data have the following statistical properties.

A) Based on the statistical properties—*Spatially asymptotical independence* (SAI) (Property 6.2 for X-ray CT and Property 8.2 for MRI), we have the following observations. (i) Pixel intensities in each sensor (i.e., each sub-image)

Y

$\Downarrow$

$Y_1 \; Y_2 \; Y_3 \; Y_4 \; Y_5 \; Y_6 \; Y_7 \; Y_8 \; Y_9 \; Y_{10} \; Y_{11} \; Y_{12} \; Y_{13} \; Y_{14} \; Y_{15} \; Y_{16}$

$\Downarrow \; \Downarrow \; \Downarrow \; \Downarrow \; \Downarrow \; \Downarrow \; \Downarrow \; \Downarrow \; \Downarrow \; \Downarrow \; \Downarrow \; \Downarrow \; \Downarrow \; \Downarrow \; \Downarrow \; \Downarrow$

$Y_{4,1} \rightarrow x_4(1)$ $\quad$ $Y_{12,1} \rightarrow x_{12}(1)$

$Y_{4,2} \rightarrow x_4(2)$ $\quad$ $Y_{12,2} \rightarrow x_{12}(2)$

$Y_{4,3} \rightarrow x_4(3)$ $\quad$ $Y_{12,3} \rightarrow x_{12}(3)$

$Y_{4,4} \rightarrow x_4(4)$ $\quad$ $Y_{12,4} \rightarrow x_{12}(4)$

FIGURE 11.1
The first sampling (Eq. (11.3)) creates the sensors; the second sampling (Eq. (11.5)) and the averaging (Eq. (11.6)) form the samples. $x_4(k)$ and $x_{12}(k)$ are shown as examples.

are statistically independent, while the corresponding pixel intensities in different sensors are statistically correlated; and (2) pixel intensities in the different sub-sub-images of the same sensor are statistically independent; as a result, the samples of the same sensor are statistically independent, while the corresponding samples in different sensors are statistically correlated.

B) Two sampling processes used in the above conversion comprise an operation called image skipping, which has a minimum loss of statistical precision. This can be intuitively verified from the fact that all sub-images and sub-sub-images of Figure 11.1 have a very similar pattern. As a consequence, all samples of sensors have the same distribution. Based on statistical properties—*Gaussianity* (Property 6.1 for X-ray CT and Property 8.1 for MRI) and the Central-Limit theorem—all samples of sensors have a Gaussian distribution.

Thus, samples of each sensor of this converted sensor array system are i.i.d. and samples across sensors are correlated. The following sections describe the details of this conversion and the problem formulation.

11.2.1.1 Sensor Array System Conversion

For a given image denoted by

$$IMG: \;\; \{Y(i,j),\; 1 \leq i,j \leq I\}, \tag{11.1}$$

sampling it with a step L in both dimensions forms L^2 sub-images denoted by

$$IMG_l : \ \{Y_l(i,j),\ 1 \leq i,j \leq J\} \quad (l = 1,\cdots,L^2), \tag{11.2}$$

where $J = I/L$. This sampling procedure is mathematically described by

$$Y_l(i,j) = Y(iL - L + 1 + l_L,\ jL - L + l - l_L L), \tag{11.3}$$

where l_L denotes the integer part of $\frac{l-1}{L}$. An example is given in Figure 11.1, where for a 64×64 MRF image (IMG), its sixteen 16×16 sub-images (IMG_l) are generated by sampling with step $L = 4$.

For each sub-image $IMG_l, (l = 1,\cdots,L^2)$, sampling it with step N in both dimensions forms N^2 sub-sub-images denoted by

$$IMG_{lk} : \ \{Y_{lk}(i,j),\ 1 \leq i,j \leq M\} \quad (k = 1,\cdots,N^2), \tag{11.4}$$

where $M = J/N$. This sampling procedure is mathematically described by

$$Y_{lk}(i,j) = Y_l(iN - N + 1 + k_N,\ jN - N + k - k_N N), \tag{11.5}$$

where k_N denotes the integer part of $\frac{k-1}{N}$. An example is given in Figure 11.1, where for two sub-images image IMG_l ($l = 4,\ 12$), their corresponding four 8×8 sub-sub-images (IMG_{lk}) are generated by sampling with step $N = 2$.

For each sub-sub-image, $IMG_{lk}, (l = 1,\cdots,L^2, k = 1,\cdots,N^2)$ averaging its $M \times M$ pixel intensities forms a sample. The k-th sample in the l-th sub-image IMG_l, denoted by $x_l(k)$, is given by

$$x_l(k) = \frac{1}{M^2}\sum_{i=1}^{M}\sum_{j=1}^{M} Y_{lk}(i,j). \tag{11.6}$$

An example is given in Figure 11.1, where for two sub-images IMG_l ($l = 4,\ 12$), their four samples ($x_l(k)$) are generated by averaging 8×8 pixel intensities in the corresponding sub-sub-images IMG_{lk}.

The above sampling and averaging procedures generate L^2 sub-images IMG_l ($l = 1,\cdots,L^2$) and N^2 samples $x_l(k)$ ($k = 1,\cdots,N^2$) for each sub-image IMG_l. As defined earlier, the distinctive region images are referred to as *sources*, and the sampled sub-images are referred to as *sensors*. When IMG is composed of q distinctive region images and is sampled into $L^2 = p$ sub-images (e.g., in Figure 11.1, $q = 4$ and $p = 16$), an analog sensor array system with q *sources* and p *sensors* is formed. It is shown in Figure 11.2, where $IMG_{\mathcal{R}_m}$ represents the m-th region image and $s_m(k)$ is the m-th *source signal.* Figures 11.1 and 11.2 provide a way to conceptually imagine this analog sensor array system: (a) *sources* transmit *signals* (i.e., pixel intensities) and *sensors* receive these *signals*, (b) an output (a sample) of a *sensor* is the mixture of the *source signals.*

In the first sampling, when the sampling step L is moderate or large (e.g., $L \geq 3$), the correlations of pixel intensities in each sub-image IMG_l are very

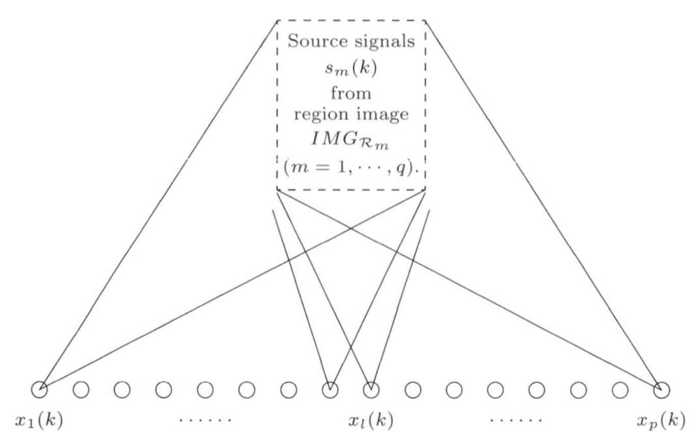

FIGURE 11.2
An analog sensor array system with q sources and p sensors. Source signals $s_m(k)$ and sensor outputs $x_l(k)$ are related via Eq. (11.9). $x_l(k)$ are computed via Eq. (11.6) and an example is shown in Figure 11.1.

small and can be ignored (SAI—Property 6.2 for X-ray CT and Property 8.2 for MRI). That is, pixel intensities in each sub-image IMG_l are statistically independent. Thus, pixel intensities in each sub-sub-image IMG_{lk} and pixel intensities over sub-sub-images of the same sub-image IMG_{lk} (for the same l) are statistically independent. As a result, the samples (an average of $M \times M$ pixel intensities in a sub-sub-image) of the same sensor are statistically independent.

Because the samplings are uniform and the image regions are randomly distributed over the image, the resulting L^2 sub-images IMG_l $(l = 1, \cdots, L^2)$ and N^2 sub-sub-images IMG_{lk} $(k = 1, \cdots, N^2)$ (for the same l) have essentially identical patterns.† Thus, the samples (an average of $M \times M$ pixel intensities in a sub-sub-image) of the same sensor have essentially identical distribution. Because of the Gaussianity of pixel intensity (Property 6.1 for X-ray CT and Property 8.1 for MRI) and the Central Limit theorem, this distribution is Gaussian.

Therefore, for a given l, the samples (Eq. (11.6)) are independently identically distributed (with a Gaussian). That is, N^2 samples $x_l(k)$ from the l-th sub-image IMG_l (i.e., N^2 outputs from the l-th *sensor* of this analog sensor array system) are i.i.d.

†This implies that the corresponding pixel intensities—most but not every one—in the sampled images are similar to each other.

11.2.1.2 Problem Formulation

Let $s(i,j)$ and $n(i,j)$ be the signal and noise components of pixel intensity $Y(i,j)$ $(1 \leq i,j \leq I)$. We have

$$Y(i,j) = s(i,j) + n(i,j). \tag{11.7}$$

Let $\mathcal{R}_m$ and $IMG_{\mathcal{R}_m}$ $(m = 1, \cdots, q)$ denote the m-th *image region* and the m-th *region image*, respectively; then

$$IMG_{\mathcal{R}_m}(i,j) = \begin{cases} Y(i,j) & (i,j) \in \mathcal{R}_m \\ \\ 0 & (i,j) \bar{\in} \mathcal{R}_m. \end{cases} \tag{11.8}$$

By this definition, we have $IMG = \bigcup_{i=1}^{q} IMG_{\mathcal{R}_m}$. Here, $\bigcup$ denotes the union of sets.

Based on Eqs. (11.6) through (11.8), the kth *sample* of the l-th *sensor* of this sensor array system, $x_l(k)$ of Eq. (11.6), can be expressed as

$$x_l(k) = \sum_{m=1}^{q} a_{lm} s_m(k) + n_l(k), \tag{11.9}$$

where $n_l(k)$ is the average of the noise components of pixel intensities in IMG_{lk}, $s_m(k)$ is the average of the signal components of pixel intensities in $IMG_{\mathcal{R}_m} \bigcap IMG_{lk}$, and a_{lm} is the ratio of the number of pixels in $IMG_{\mathcal{R}_m} \bigcap IMG_{lk}$ and the number of pixels in IMG_{lk}. Here, $\bigcap$ denotes the intersection of sets.

Eq. (11.9) indicates that the average of all pixel intensities in a sub-sub-image equals the weighted summation of the mean estimates of all region images in that sub-sub-image plus the average of the noise in the same sub-sub-image. Using vector-matrix notations

$$\begin{aligned} \mathbf{x}(k) &= (x_1(k), \cdots\cdots, x_p(k))^T, \\ \mathbf{s}(k) &= (s_1(k), \cdots\cdots, s_q(k))^T, \\ \mathbf{n}(k) &= (n_1(k), \cdots\cdots, n_p(k))^T, \\ \mathbf{A} &= \begin{pmatrix} a_{11} & \cdots\cdots & a_{1q} \\ \vdots & & \vdots \\ a_{p1} & \cdots\cdots & a_{pq} \end{pmatrix}, \end{aligned} \tag{11.10}$$

Eq. (11.9) can be written as

$$\mathbf{x}(k) = \mathbf{A}\mathbf{s}(k) + \mathbf{n}(k). \tag{11.11}$$

It is important to understand the meanings of the quantities $s_m(k)$ of Eq. (11.9) and $(s_1(k), \cdots\cdots, s_q(k))$ of Eq. (11.10). For a given k, $s_m(k)$—the average of the signal components of pixel intensities in $IMG_{\mathcal{R}_m} \bigcap IMG_{lk}$—is the mean estimate of the m-th region image in the sub-sub-image IMG_{lk}. Over all ks, $s_m(k)$ gives the mean estimate of the m-th region image in the sub-image IMG_l. $s_1(k), \cdots\cdots, s_q(k)$ gives the mean estimate of the 1st, 2nd, $\cdots$, q-th region images at the k-th sample. A mixture of $s_1(k), \cdots\cdots, s_q(k)$—the weighted sum of them—and the noise component $n_l(k)$ forms the outputs of the l-th sensor—the k-th sample of the l-th sensor—of this analog sensor array system. In short, $s_1(k), \cdots\cdots, s_q(k)$ are the samples of the source signals.

Sections 6.4 and 8.4 show that pixel intensities are stationary and ergodic within the image region (Property 6.3 and 6.4 for X-ray CT and Property 8.4 and 8.5 for MRI). In order to make pixel intensities stationary and ergodic over the entire image, we convert the real-valued image to the complex-valued image using the following procedure.

For a real-valued image $IMG: \ \{Y(i,j),\ 1 \leq i,j \leq I\}$, its complex-valued image is $IMG: \ \{Y(i,j)e^{i\phi(i,j)},\ 1 \leq i,j \leq I\}$, where $i = \sqrt{-1}$, $\phi(i,j)$ are the samples from a random variable with uniform distribution on $[-\pi, \pi]$ and independent of $Y(i,j)$. Clearly, the mean of the complex data $Y(i,j)e^{i\phi(i,j)}$ equals zero. As a result of this data conversion, $\mathbf{x}$ of Eq. (11.10) is a Gaussian random vector with zero mean, stationary, and ergodic.

Thus, the signal component $\mathbf{s}$ of $\mathbf{x}$ of Eq. (11.10) is a complex, stationary, ergodic, Gaussian vector with a zero-mean vector and a $p \times p$ (positive definite) covariance matrix

$$\mathbf{C_s} = E[\mathbf{s}(k)\mathbf{s}^{\dagger}(k)], \tag{11.12}$$

where $\dagger$ denotes the conjugate and transpose; the noise component $\mathbf{n}$ of $\mathbf{x}$ of Eq. (11.10) is a complex, stationary, ergodic, Gaussian vector, additive and independent of $\mathbf{s}$, with a zero mean vector and a $p \times p$ covariance matrix

$$\mathbf{C_n} = E[\mathbf{n}(k)\mathbf{n}^{\dagger}(k)] = \sigma^2\mathbf{I}, \tag{11.13}$$

where σ^2 is the variance of the image (a constant), and $\mathbf{I}$ is an identity matrix.

Therefore, the covariance matrix $\mathbf{C_x}$ of $\mathbf{x}(k)$ is

$$\mathbf{C_x} = \mathbf{A}\mathbf{C_s}\mathbf{A}^{\dagger} + \sigma^2\mathbf{I}. \tag{11.14}$$

Thus, the signal environment of this converted sensor array system can be characterized by a complex Gaussian pdf given by [45, 46]

$$f(\mathbf{x}) = \pi^{-p}|\mathbf{C_x}|^{-1}\exp(-\mathbf{x}^{\dagger}\mathbf{C_x}^{-1}\mathbf{x}). \tag{11.15}$$

Under the condition that $p > q$, for the full column rank matrix $\mathbf{A}$ of Eq. (11.10) and the non-singular matrix $\mathbf{C_s}$ of Eq. (11.12), the rank of $\mathbf{A}\mathbf{C_s}\mathbf{A}^{\dagger}$ is q. Thus, the problem of the detection of the number of image regions is converted to a problem of determining the rank of matrix $\mathbf{A}\mathbf{C_s}\mathbf{A}^{\dagger}$. This situation is equivalent to the $(p-q)$ smallest eigenvalues of $\mathbf{A}\mathbf{C_s}\mathbf{A}^{\dagger}$ being equal to zero. Let the eigenvalues of $\mathbf{C_x}$ be

$$\lambda_1 \geq \lambda_2 \geq \cdots \geq \lambda_p; \tag{11.16}$$

then the smallest $(p-q)$ eigenvalues of $\mathbf{C_x}$ are all equal to σ^2:

$$\lambda_{q+1} = \lambda_{q+2} = \cdots = \lambda_p = \sigma^2. \tag{11.17}$$

This implies that the number of image regions q can be determined from the multiplicity of the smallest eigenvalues of the covariance matrix $\mathbf{C_s}$, which can be estimated from the outputs of the converted sensor array system, $\mathbf{x}(k)$ $(k = 1, \cdots, N^2)$.

When the covariances $\mathbf{C_x}$ with different multiplicities of the smallest eigenvalues are viewed as different models, the problem of detecting of the number of image regions is translated as a model fitting problem.

11.2.1.3 Information Theoretic Criterion

As shown in Section 10.2, the model fitting problem can be resolved by applying Information Theoretic Criteria (ITC) [47–49]. Similar to the iFNM model-based image analysis method, we adopt the Minimum Description Length (MDL) criterion [48, 50, 51], which is defined by

$$\mathrm{MDL}(K) = -\log(\mathcal{L}(\hat{\mathbf{r}}_{ML})) + \frac{1}{2}K_a \log J, \tag{11.18}$$

where J is the number of independent observations, K is the order of the model, K_a is the number of free adjustable model parameters, $\hat{\mathbf{r}}_{ML}$ is ML estimate of the model parameter vector, and $\mathcal{L}(\hat{\mathbf{r}}_{ML})$ is the likelihood of an ML estimate of the model parameters.

For this converted sensor array system (p sensors, q sources, $J = N^2$ i.i.d. samples), assume that the number of sources is k $(0 \leq k \leq p-1)$; then the ranks of the covariance matrix $\mathbf{A}\mathbf{C_s}\mathbf{A}^\dagger$ and $\mathbf{C_x}$ are k. Let $\mathbf{C}_\mathbf{x}^{(k)}$ be the covariance matrix associated with the assumed number k, and λ_i and $\mathbf{v}_i$ $(i = 1, \cdots, k)$ be the eigenvalues and eigenvectors of $\mathbf{C}_\mathbf{x}^{(k)}$, respectively. It is known that

$$\mathbf{C}_\mathbf{x}^{(k)} = \sum_{i=1}^{k}(\lambda_i - \sigma^2)\mathbf{v}_i\mathbf{v}_i{}^\dagger + \sigma^2\mathbf{I}, \tag{11.19}$$

where $\mathbf{I}$ is a $p \times p$ identity matrix. Thus, pdf $f(\mathbf{x})$ of Eq. (11.15) is parameterized with a vector given by

$$\theta^{(k)} = (\lambda_1, \cdots, \lambda_k, \sigma^2, \mathbf{v}_1^T, \cdots, \mathbf{v}_k^T). \tag{11.20}$$

and is denoted $f(\mathbf{x}|\theta^{(k)})$.

Using $\mathbf{C}_\mathbf{x}^{(k)}$, the log-likelihood of $f(\mathbf{x}|\theta^{(k)})$ is given by

$$\sum_{i=1}^{N^2} \log f(\mathbf{x}|\theta^{(k)})$$

$$= \sum_{i=1}^{N^2} \log(\pi^{-p}|\mathbf{C}_{\mathbf{x}}^{(k)}|^{-1} \exp(-\mathbf{x}^{\dagger}(i)\mathbf{C}_{\mathbf{x}}^{(k)^{-1}}\mathbf{x}(i)))$$

$$= -N^2 \log |\mathbf{C}_{\mathbf{x}}^{(k)}| - N^2 \mathrm{Tr}\{\mathbf{C}_{\mathbf{x}}^{(k)^{-1}}\hat{\mathbf{C}}_{\mathbf{x}}\}, \tag{11.21}$$

where a constant term $-pN^2 \log \pi$ is omitted and $\hat{\mathbf{C}}_{\mathbf{x}}$ is the sample covariance matrix given by

$$\hat{\mathbf{C}}_{\mathbf{x}} = \frac{1}{N^2} \sum_{i=1}^{N^2} \mathbf{x}(i)\mathbf{x}^{\dagger}(i). \tag{11.22}$$

[52] shows that the ML estimate of $\theta^{(k)}$ is given by

$$\theta_{ML}^{(k)} = (\hat{\lambda}_1, \cdots, \hat{\lambda}_k, \hat{\sigma}^2, \hat{\mathbf{v}}_1^T, \cdots, \hat{\mathbf{v}}_k^T), \tag{11.23}$$

where $\hat{\lambda}_i$ and $\hat{\mathbf{v}}_i$ $(i = 1, \cdots, k)$ are the eigenvalues and eigenvectors of the sample covariance matrix $\hat{\mathbf{C}}_{\mathbf{x}}$, respectively; $\hat{\sigma}^2 = \frac{1}{p-k}\sum_{i=k+1}^{p} \hat{\lambda}_i$; and $\hat{\lambda}_1 > \hat{\lambda}_2 > \cdots \hat{\lambda}_k$. Using Eq. (11.23), Eq. (11.21) becomes

$$\sum_{i=1}^{N^2} \log f(\mathbf{x}|\theta^{(k)}) = N^2(p-k)\log\left(\frac{\prod_{i=k+1}^{p} \hat{\lambda}_i^{1//(p-k)}}{\sum_{i=k+1}^{p} \hat{\lambda}_i/(p-k)}\right). \tag{11.24}$$

The item inside the parentheses represents a ratio of the geometric mean and the arithmetic mean of the $(p-k)$ smallest eigenvalues of $\hat{\mathbf{C}}_{\mathbf{x}}$.

For the number K_a of independently adjustable parameters of Eq. (11.18), Eq. (11.20) or (11.23) indicates

$$K_a(k) = k(2p-k). \tag{11.25}$$

This number is obtained by observing the following facts. $\mathbf{C}_{\mathbf{x}}^{(k)}$ is spanned by k real eigenvalues and k complex p-dimensional eigenvectors. The total number of parameters for spanning $\mathbf{C}_{\mathbf{x}}^{(k)}$ is $2pk + k$. Due to the constraints of mutual orthogonality and normality of the eigenvectors, $k(k-1)$ and $2k$ degrees of freedom are lost. Thus, the number of free adjusted parameters is $k(2p-k)$.

By applying Eqs. (11.24) and (11.25), Eq. (11.18) becomes

$$\mathrm{MDL}(\hat{\theta}_{ML}(k)) = -N^2(p-k)\log\left(\frac{\prod_{i=k+1}^{p} \hat{\lambda}_i^{1//(p-k)}}{\sum_{i=k+1}^{p} \hat{\lambda}_i/(p-k)}\right) + \frac{1}{2}k(2p-k)\log N^2, \tag{11.26}$$

where k $(k = 1, \cdots, p-1)$ is an assumed source number. In Eq. (11.26), the first term is the negative of the log-likelihood of Eq. (11.24), computed at the ML estimate $\hat{\theta}_{ML}(k)$. This term measures how well the model fits the data. The second term is a measure of model complexity. Thus, the best estimate $\hat{q}$ of the source number q is selected such that $\mathrm{MDL}(\hat{\theta}_{ML}(k))$ is minimized:

$$\hat{q} = Arg\{\min_{0<k\leq p-1} \mathrm{MDL}(\hat{\theta}_{ML}(k))\}. \tag{11.27}$$

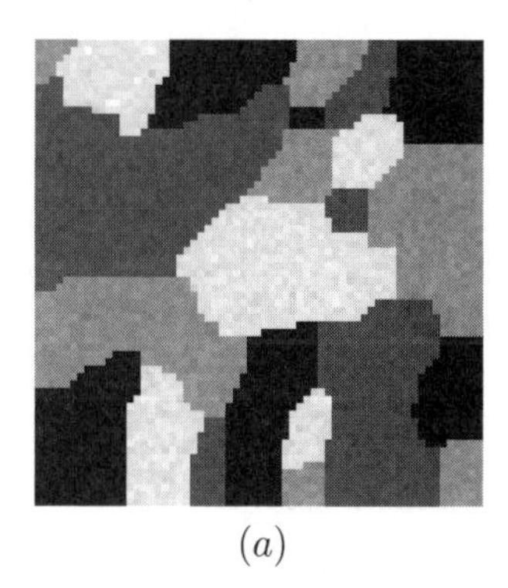

(*a*)

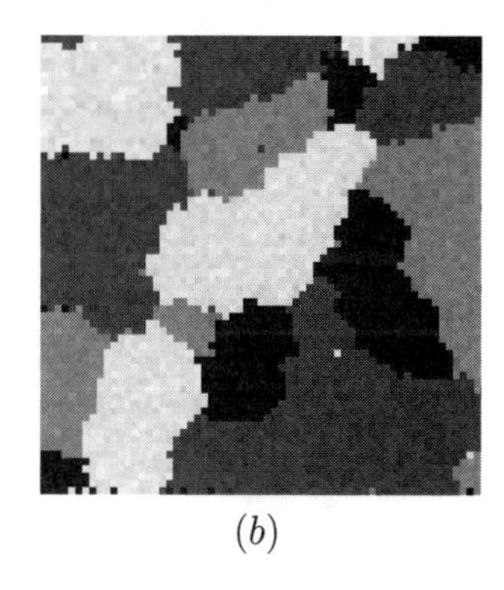

(*b*)

FIGURE 11.3
Two simulated 64 × 64 MRF images.

11.2.2 Examples of Detection Results

This eigenstructure method has been validated for several types of images: simulated images, X-ray CT, and MR images.

11.2.2.1 Simulated Images

TABLE 11.1
Eigen / MDL Values of Two Simulated Images in Figure 11.3

	IMG (*a*)		*IMG* (*b*)	
k	$\hat{\lambda}_k$	MDL(k)	$\hat{\lambda}_k$	MDL(k)
0	3060	1970	1292	1281
1	357	881	255	641
2	240	698	159	528
3	84	507	87	460
4	59	**493**	68	**458**
5	49	503	54	465
6	43	514	39	477
7	35	518	32	505
8	22	523	29	535
9	18	550	25	561
10	13	578	16	585
11	11	613	16	622
12	9	647	13	652
13	8	683	9	683
14	6	716	8	717
15	4	749	5	749

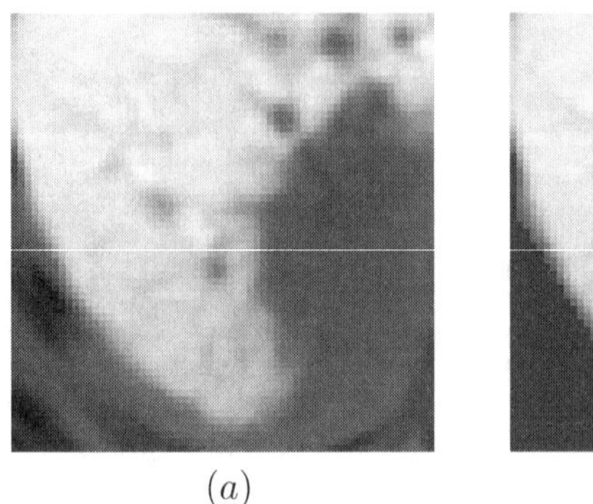
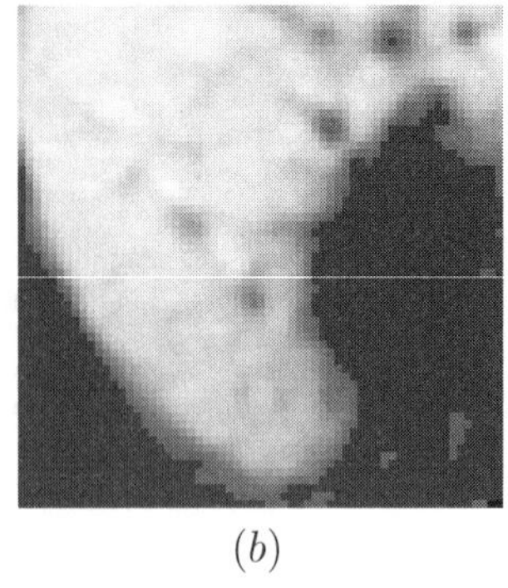

(a) (b)

FIGURE 11.4
X-ray CT image. Two 64×64 ROI images of the lung. The image (a) is original and the image (b) is an image obtained by merging four region images with the higher values of pixel intensities into one region image.

Two 64×64 MRF images were created by Gibbs Sampler (Appendix 9A of Chapter 9) and then were superimposed with correlated Gaussian samples to simulate images with cFNM distribution. Each image has four image regions (black, dark gray, light gray, and white) as shown in Figure 11.3. Applying the eigenstructure method to these two images, results are given in Table 11.1, where one column shows the eigenvalues of Eq. (11.23) and another column shows MDL values of Eq. (11.26). The minimum value of MDL occurs at index $k = 4$, which indicates that the method gives the correct solution. These results are consistent with the results in Figure 10.1 and Table 10.2, where the iFNM model-based image analysis method is utilized the similar MRF images. The parameter settings used here are: $I = 64$, $L = 4$, $J = 16$, $N = 8$, and $M = 2$ (I, L, J, N, and M are defined by Eqs. (11.1) through (11.6)).

Extensive simulation studies have been done. More details are given in Appendix 9A of Chapter 9 and Section 12.2. In these studies, for given images with resolution $I \times I$, various combinations of L and N are used to obtain different settings (I, L, J, N, M). The detection results obtained using the different settings are all correct.

11.2.2.2 X-Ray CT Image

Due to the lack of ground truth or gold standard, the detected number of image regions (e.g., the number of tissue types) in the real medical images are always questioned. To validate this eigenstructure approach in real medical images, the following strategy is applied.

Figure 11.4a shows a 64×64 X-ray CT image of a region of interest (ROI), which corresponds to the lower part of the right lung of a subject. Applying this eigenstructure method to this ROI, results are given in the leftmost columns of Table 11.2. It shows that this ROI has nine image regions. These image regions represent lung air, parenchyma, fibrosis, pleura, blood vessel, bronchi, tumor/tissues outside lung, dense bone, and trabecular bone. By merging four image regions with the higher pixel intensities (that is, CT numbers) into one region, that is, replacing the pixel intensities in these four regions by their averaged mean value, a new image of the same ROI with six image regions is created, which is shown in Figure 11.4b. Applying this eigenstructure method to this new image, the results are given in the rightmost columns of Table 11.2: it has six image regions. This is a correct solution. The parameter settings used in this example are: $I = 64$, $L = 4$, $J = 16$, $N = 8$, and $M = 2$.

Remarks. Images in Figure 11.4 and Figure 10.5 are similar, but the detection results in Table 11.2 and Table 10.4 are different. These facts are not surprising. The main reason is that the image sizes in Figure 11.4 and Figure 10.5 are 64×64 and 81×81, respectively, the former is truncated from the latter. Thus, information of these two images are different. Other reasons are (1) $q = 9$ in Table 11.2 is computed by this eigenstructure method; it is a non-model based, data-driven approach; $q = 6$ in Table 10.4 is computed by iFNM model based, data-driven approach. (2) In this eigenstructure method, pixel intensities of the image are statistically correlated; in the iFNM model-based method, pixel intensities are assumed to be statistically independent.

TABLE 11.2
Eigen / MDL Values of Two X-ray CT ROI Images in Figure 11.4.

	IMG (a)		*IMG (b)*	
k	$\hat{\lambda}_k$	MDL(k)	$\hat{\lambda}_k$	MDL(k)
0	351059	4698	474455	3726
1	13708	2347	16710	1387
2	5670	1834	7655	993
3	1888	1389	2581	682
4	944	1176	1810	597
5	722	1043	1031	520
6	254	833	599	**496**
7	236	783	523	513
8	87	653	399	526
9	55	**639**	283	544
10	41	646	213	574
11	30	654	178	609
12	19	661	148	644
13	9	679	128	680
14	8	714	89	713
15	6	749	81	749

11.2.2.3 MR Image

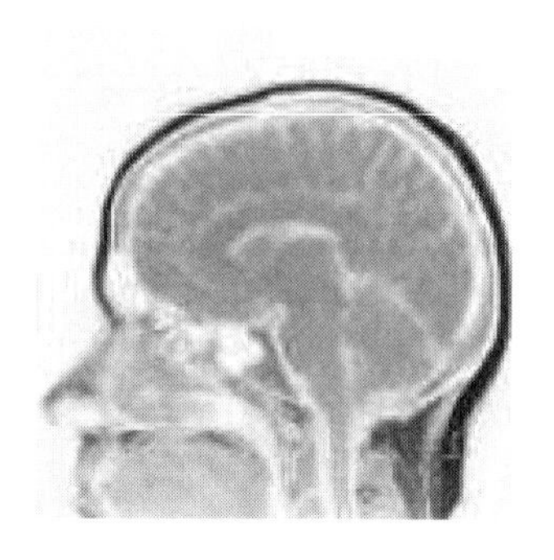

FIGURE 11.5
MR image. A 64 × 64 sagittal image of the head.

TABLE 11.3
Eigen / MDL Values of MR Image in Figure 11.5

	IMG	
k	$\hat{\lambda}_k$	MDL(k)
0	3243	1978
1	298	806
2	244	674
3	78	479
4	75	472
5	49	**457**
6	31	463
7	26	494
8	24	527
9	20	558
10	17	588
11	13	619
12	11	653
13	9	685
14	6	714
15	3	749

A 64×64 sagittal MR image of the head is shown in Figure 11.5, which is the same as the image in Figure 10.8a. Applying the eigenstructure method to this image, the obtained results are given in Table 11.3. It shows that this image has five image regions. This result is consistent with the results in Figure 10.8 and Table 10.6, where the iFNM model-based image analysis method is utilized for this MR image. Five region images are shown in Figure 10.8b and described by the information shown in Table 10.7, which is meaningful as expected. The parameter settings used in this example are $I = 64$, $L = 4$, $J = 16$, $N = 8$ and $M = 2$,

11.2.3 Relationships to Other Related Methods

The earlier version of the eigenstructure method described in Section 11.2.1 was called Block processing. As shown in Section 11.2.1, the settings (I, L, J, N, M) in the sensor array system conversion lead to L^2 $J \times J$ data blocks and $L^2 \times N^2$ $M \times M$ data blocks. This section describes the relationship between this method and other image (signal) analysis methods.

11.2.3.1 Independent Component Analysis

[53, 54] apply second-order independent component analysis (ICA) methods to optical and fMRI image analysis, respectively. The prototype patterns in [53] and the component maps in [54], $s_m(i)$ $(m = 1, \cdots, q,\ i = 1, \cdots, I)$, are defined as source signals, where m denotes the m-th prototype pattern or component map and i denotes the i-th pixel in the images. The observed images $x_l(i)$ $(l = 1, \cdots, L,\ i = 1, \cdots, I)$ are defined as observations, where l denotes the l-th frame or time point and i denotes the i-th pixel in the images. The observations $x_l(i)$ and the sources $s_m(i)$ are linked by a linear mixture model

$$x_l(i) = \sum_{m=1}^{q} a_{lm} s_m(i), \tag{11.28}$$

which can be written in matrix form as

$$\mathbf{x}(i) = \mathbf{A}\mathbf{s}(i). \tag{11.29}$$

[53, 54] show that the joint diagonalization of two symmetric matrices (whose elements are the measurable correlations of observations) leads to a solution for the mixing matrix $\mathbf{A}$, and therefore for the estimates of the source signals $\mathbf{s}(i)$.

Ignoring the noise term of Eq. (11.11), Eqs. (11.29) and (11.11) are identical. Although the present approach and ICA method demonstrate some similarities, they are quite different. The major differences are:

- The aims of [53, 54] are to separate the sources, and second-order ICA methods are applied to each pixel in the image. The present method is to detect the number of sources (not to separate them), and an eigenstructure method is used on the covariance matrix (not on each pixel), which is a description of the entire signal environment of the converted sensor array system and which is computed from the observations.

• In [53, 54], the linear mixture model represents a pixel in the observed image (i.e., an observation) as a weighted summation of the pixels in the different prototype patterns or component maps (i.e., sources). In the present approach, the linear mixture model represents the average of all pixels in a sub-sub-image (i.e., an observation) as the weighted summation of the mean estimates of all object region images (i.e.,the sources) in that sub-sub-image. Thus, the mixing matrices **A** of Eqs. (11.11) and (11.29) have different contents.

• The ICA method requires that the source signals be independent and the number of sources be known. Assessing this independence and detecting the source number become a common concern in many applications of ICA. As demonstrated in Section 11.2.1, this eigenstructure method validates the independence of source signals and detects the source number. Therefore, for some image analysis applications, integrating this eigenstructure method with ICA procedures is good choice.

11.2.3.2 Time Series Model

[38] utilized a p-th-order Autoregressive (AR) model for the forward prediction error. Because the prediction errors are independent, Gaussian likelihood can be easily formed. Carefully studying the covariance matrix of the present method shows that the correlation represented by the matrix of Eq. (11.22) is a p-th-order forward correlation. The patterns of the forward correlation and the forward prediction error are very similar. However, the difference is that the present method does not require models such as AR, SAR, GMRF, etc., to form these statistics [38, 55].

11.2.3.3 Multispectral Image Analysis

[40] proposed a multispectral image processing method for MR brain image segmentation. It assumed that the corresponding pixels in the different spectral images (T_1, T_2, and PD) are correlated and the pixels in the same spectral image are independent. If the sub-images generated by the first sampling process of the present method are treated as different spectral images, then [40] and the present method are essentially the same in nature, and the results obtained by the present method can also be used in multispectral image processing.

11.2.3.4 Multivariate Image Analysis

The procedures developed in the present method are for 2-D image processing. They can be easily applied to 3-D and 4-D image processing with minor modifications. For a single volume image (i.e., 3-D), when the sampling step is L, L^3 sub-images will be generated. For multiple volume images (e.g., 4-D with time as the fourth independent variable), the sampling procedure is no longer needed. In this case, the four-dimensional data array will be reformatted as a *reorganized three-way* array [56], and the resulting covariance matrix will represent the spatio-temporal multivariate Gaussian environment.

11.2.3.5 Some Remarks

A sensor array signal processing approach is developed for detecting the number of image regions. For the structural image, this number provides the necessary information for unsupervised image analysis. For the functional image, this number and its change may indicate the occurrence of the functional activities (*Note*: fMRI here is under a single stimulus condition, that is, not orthogonal.)

This eigenstructure approach is a non-model based, data-led operation. Except for the basic image statistics such as Gaussianity and SAI, it does not refer to any model or hypothesis. It detects the number of image regions in a correlated image environment. The theoretical and simulated/experimental results obtained using this approach were shown to be in good agreement.

11.3 Estimation of Image Parameters

In iFNM model-based image analysis (Chapter 10), both detection and classification processes require estimates of the model parameters. In cFNM model-based image analysis, although the detection process (Section 11.2) does not require estimates of the model parameters, the classification process (Section 11.4), does require these estimates.

To find the estimates of parameters of cFNM pdf (Eq. (9.37)), for the convenience of derivations, some modifications of the notations are made as follows. In Eq. (9.37), let

$$\pi_{k|\mathcal{N}_{i,j}} = P(y_{i,j} = k|y_{\mathcal{N}_{i,j}}), \tag{11.30}$$

and define

$$\mathbf{r} = (\mathbf{r}_1, \cdots, \mathbf{r}_k, \cdots, \mathbf{r}_K)^T \quad \text{and} \quad \mathbf{r}_k = (\pi_{k|\mathcal{N}_{i,j}}, \mu_k, \sigma_k^2). \tag{11.31}$$

Eq. (9.37) can be written as

$$f(x_{i,j}|x_{\neq i,j}, \mathbf{r}) = \sum_{k=1}^{K} \pi_{k|\mathcal{N}_{i,j}} \cdot g(y_{i,j}|\mu_k, \sigma_k^2). \tag{11.32}$$

As a result of these modifications, by comparing Eq. (11.32) with Eq. (9.1), cFNM and iFNM models have a unified form. $\mathbf{r}$, the parameter vector of cFNM pdf, is inserted in f of Eq. (11.32) for the purpose of parameter estimation only. $\pi_{k|\mathcal{N}_{i,j}}$ may vary for each pixel (i, j). It is temporally treated in the same way as μ_k and σ_k^2, that is, as an index k only dependent parameter. The justification of this treatment is given in Appendix 11A. $\pi_{k|\mathcal{N}_{i,j}}$ and (μ_k, σ_k^2) are computed separately.

As shown in Chapter 9, a pixel has two descriptors: a value and a label. An image is characterized by two layers: an intensity field and a context field. In the iFNM model, these two fields are independent Gaussian random fields, and no spatial regularity is imposed. In the cFNM model, they are MRF fields with the same neighborhood system, and the spatial regularity is characterized by the correlations of pixel intensities in the intensity field and imposed by the relations of labels in the context field. The implementation of parameter estimation is regulated by these relations.

11.3.1 Extended Expectation-Maximization Method

Some commonly used MRF model-based image analysis methods adopt the following approach for the parameter estimation. Parameters μ_k and σ_k are estimated first, and the estimates $\hat{\mu}_k$ and $\hat{\sigma}_k$ are obtained by various means. The assignments of the labels $\{y_{i,j}\}$ over the pixel array are then performed by a Maximum a posteriori (MAP) operation based on $p_G(y_{i,j})$ of Eq. (9.38). The main problem with this approach is that the parameters μ_k and σ_k are estimated only once at the beginning of the process. Consequently, they may not fit the updated labels $\{y_{i,j}\}$, that is, the complete data.

To overcome this drawback, parameter estimates $\hat{\mu}_k$, $\hat{\sigma}_k$ and label assignments $\{y_{i,j}\}$ (hence, the parameter estimate $\hat{\pi}_{k|\mathcal{N}_{i,j}}$) should be updated in an interplay fashion based on the observed image $\{x_{i,j}\}$. The mixture of the Expectation-Maximization (EM) algorithm and the Maximum a posteriori (MAP) algorithm is ideally suited to problems of this sort, in which it produces maximum-likelihood (ML) estimates of parameters when there is a mapping from the observed image$\{x_{i,j}\}$ (the incomplete data) to the underlying labels $\{y_{i,j}\}$ (the complete data).

This EM-MAP algorithm consists of two connected components. Component 1 is a standard EM algorithm. For given labels $\mathbf{y}^{(n)} = \{y_{i,j}^{(n)}\}$ ((n) denotes the n-th iteration in updating $\mathbf{y}$ estimate), $\{\pi_{k|\mathcal{N}_{i,j}}^{(n)}\}$ is estimated based on MRF-GD equivalence; an EM algorithm is then applied to the image $\mathbf{x} = \{x_{i,j}\}$ based on the cFNM pdf of Eq. (11.32) to generate the ML parameter estimates $\{\mu_k^{(m+1)},\ \sigma_k^{2(m+1)}\}$ ((m) denotes the m-th iteration in estimating $\{\mu_k,\ \sigma_k^2\}$). Component 2 is a MAP algorithm. $\mathbf{y}^{(n)}$ is updated to $\mathbf{y}^{(n+1)}$ by an operation similar to the Gibbs sampler and then followed by a new estimate $\pi_{k|\mathcal{N}_{i,j}}^{(n+1)}$ and another E-M cycle. This EM-MAP algorithm, starting from $(n) = 0$ and $(m) = 0$, is repeated until the selected stopping criteria are met.

We use a new notation in the derivation of the ML and MAP estimates.

Let $f(z,\theta)$ be the likelihood function of the parameter θ, z the observations. Conventionally, the ML estimate of θ is expressed by $\hat{\theta} = arg\max_{\theta}\{f(z,\theta)\}$. In order to link this ML estimate with an iteration algorithm that numerically generates this ML estimate, a notation $\theta^{(m+1)} = arg\max_{\theta^{(m)}}\{f(z,\theta^{(m)})\}$ is adopted; here (m) denotes the m-th iteration.

Unlike conventional EM algorithm development, instead of maximizing the expectation of the likelihood of the parameters that is shown in Appendix 10C of Chapter 10, we directly maximize the likelihood itself. The justification of this approach is given in Appendix 11A. It shows that the likelihood of the estimated parameters of cFNM Eq. (11.32) $\mathcal{L}$, given $\mathbf{y}^{(m)}$, is‡

$$\begin{aligned}\mathcal{L}(\mathbf{r}^{(m)}) &= \prod_{(i,j)} f(x_{i,j}|x_{\neq i,j},\mathbf{r}^{(m)}) \\ &= \prod_{(i,j)}\sum_{k=1}^{K} \pi_{k|\mathcal{N}_{i,j}}^{(n)} \cdot g(x_{i,j}|\mu_k^{(m)},\sigma_k^{2(m)}).\end{aligned} \tag{11.33}$$

Because $\sum_{k=1}^{K}\pi_{k|\mathcal{N}_{i,j}}^{(n)} = 1$, ML estimates of parameters of the cFNM Eq. (11.32), $\mathbf{r}^{(m+1)}$, can be obtained by

$$\mathbf{r}^{(m+1)} = arg\ max_{\mathbf{r}^{(m)}}\{\ln\mathcal{L}(\mathbf{r}^{(m)}) + \lambda(\sum_{k=1}^{K}\pi_{k|\mathcal{N}_{i,j}}^{(n)} - 1)\}, \tag{11.34}$$

where λ is a Lagrangian multiplier. Appendix 11A shows that the solutions of Eq. (11.34) are given by the following EM-MAP algorithm, which consists of two components.

- Component 1 - EM algorithm

  E-step: computing Bayesian probability $z_{(i,j),k}^{(m)}$ for every (i,j) and k.

  $\pi_{k|\mathcal{N}_{i,j}}^{(n)}$ is determined through MRF-GD equivalence Eq. (9.38)

$$\begin{aligned}\pi_{k|\mathcal{N}_{i,j}}^{(n)} &= P(y_{i,j}^{(n)} = k|y_{\mathcal{N}_{i,j}}^{(n)}) = p_G(y_{i,j}^{(n)}|y_{\mathcal{N}_{i,j}}^{(n)}) \\ &= \frac{1}{Z}\exp(-\frac{1}{\beta}\sum_{c:(i,j)\in c} V_c(y_{i,j}^{(n)})),\end{aligned} \tag{11.35}$$

which remains unchanged in one E-M cycle. $V_c(y_{i,j}^{(n)})$ is the potential of the clique c, and its assignment is described in Section 11.3.2;

$$z_{(i,j),k}^{(m)} = \frac{\pi_{k|\mathcal{N}_{i,j}}^{(n)} g(x_{i,j}|\mu_k^{(m)},\sigma_k^{(m)})}{f(x_{i,j}|x_{\neq i,j},\mathbf{r}^{(m)})}; \tag{11.36}$$

‡The underlying pdf of the likelihood $\mathcal{L}(\mathbf{r}^{(m)})$ is $f(x_{i,j}|x_{\neq i,j},\mathbf{r}^{(m)})$, a cFNM pdf Eq. (11.32); it is not the marginal pdf $g(x_{i,j}|\mu_k^{(m)},\sigma_k^{2(m)})$, a Normal pdf.

M-step: computing the updated estimates of parameters $\mu_k^{(m+1)}$ and $\sigma_k^{2(m+1)}$:

$$\mu_k^{(m+1)} = \frac{\sum_{i,j} z_{(i,j),k}^{(m)} x_{i,j}}{\sum_{i,j} z_{(i,j),k}^{(m)}}$$

$$\sigma_k^{2(m+1)} = \frac{\sum_{i,j} z_{(i,j),k}^{(m)} (x_{i,j} - \mu_k^{(m+1)})^2}{\sum_{i,j} z_{(i,j),k}^{(m)}}. \tag{11.37}$$

E- and M-steps start from an initial estimate (for a given $\mathbf{y}^{(n)}$) and stop by checking if the likelihood quits changing,

$$|\mathcal{L}(\mathbf{r}^{(m+1)}) - \mathcal{L}(\mathbf{r}^{(m)})| < \epsilon, \tag{11.38}$$

where ϵ is a prespecified small number.

The conditional probability of the underlying label $\mathbf{y} = \{y_{i,j}\}$, given image $\mathbf{x} = \{x_{i,j}\}$, is $P(\mathbf{y}|\mathbf{x}) = P(\mathbf{x}, \mathbf{y})/P(\mathbf{x})$. Because $P(\mathbf{x})$ is fixed and its likelihood for a given $\mathbf{x}$ is constant, $P(\mathbf{y}|\mathbf{x}) \propto P(\mathbf{x}|\mathbf{y})P(\mathbf{y})$. Thus, a configuration of $\mathbf{y}$, $\mathbf{y}^{(n+1)}$, can be obtained by a MAP operation over all $\mathbf{y}^{(n)}$,

$$\mathbf{y}^{(n+1)} = arg\ max_{\mathbf{y}^{(n)}} \{P(\mathbf{x}|\mathbf{y}^{(n)}) p_G(\mathbf{y}^{(n)})\}. \tag{11.39}$$

Appendix 11B shows that the conditional field $\mathbf{y}$ given $\mathbf{x}$, $\mathbf{y}|\mathbf{x}$, is also an MRF and its energy function $U(\mathbf{y}^{(n)}|\mathbf{x})$ is

$$U(\mathbf{y}^{(n)}|\mathbf{x}) = \frac{1}{2} \sum_{i,j} ((\frac{x_{i,j} - \mu_k^{(m)}}{\sigma_k^{(m)}})^2 + \ln \sigma_k^{2(n)} + \frac{1}{\beta} \sum_{c:(i,j) \in c} V_c(y_{i,j}^{(n)})). \tag{11.40}$$

Thus, the MAP operation (Eq. (11.39)) is equivalent to an energy minimization (minimizing Eq. (11.40)) and is performed as follows:

- Component 2 - MAP algorithm
  MAP-step: computing the updated underlying labels

$$\mathbf{y}^{(n+1)} = arg\ min_{\mathbf{y}^{(n)}} \{U(\mathbf{y}^{(n)}|\mathbf{x})\}, \tag{11.41}$$

that can be implemented by the Iterated Conditional Modes (ICM) algorithm [34]. We utilize a method that is similar to Gibbs Sampler [32] for computing $\pi_{k|\mathcal{N}_{i,j}}^{(n)}$ of Eq. (11.35) and updating $\mathbf{y}^{(n+1)}$ of Eq. (11.41). This method is described in Section 11.3.3. The MAP algorithm stops when the proper iterations have been performed.

This EM-MAP algorithm is summarized below:

(0) Modeling the image $\{x_{i,j}\}$ as an iFNM (Eq. (10.1)) and applying the standard EM algorithm (Eqs. (10.11)–(10.13)) to generate initial parameter estimates;

(1) MAP algorithm, estimating the pixel labels Eq. (11.41);

(2) EM algorithm:
(2.a) E-step, computing the Bayesian probability Eq. (11.36),
(2.b) M-step, computing the updated parameter estimates Eq. (11.37),
(2.c) Back to (2.a) and repeating until Eq. (11.38) is met,

(3) Back to (1) and repeating until the selected stopping criterion is met.

11.3.2 Clique Potential Design

11.3.2.1 New Formulations of Clique and Its Potential

The clique c in an MRF is a set of sites in which the distinct sites are in the neighborhood of each other. Cliques in the first-and second-order neighborhood systems ($\mathcal{N}^p, p = 1, 2$) are shown in Figure 9.3. The clique c depends on the order p of neighborhood systems and the number q of pixels in the clique. For example, a clique may consist of two pixels (horizontal and vertical pair) in $\mathcal{N}^1$, or two pixels (45° and 135° diagonal pair) in $\mathcal{N}^2$, or three pixels (triangle) in $\mathcal{N}^2$, etc. We use a new expression $c_{p,q}$ to denote a clique consisting of q pixels in the p-th-order neighborhood system. Thus, the third term in the parentheses on the right side of Eq. (11.40) becomes

$$\frac{1}{\beta}\sum_{c:(i,j)\in c} V_c(y_{i,j}^{(n)}) = \frac{1}{\beta}\sum_{c_{p,q}:\ (i,j)\in c_{p,q}}\sum_{p}\sum_{q} V_{c_{p,q}}(y_{i,j}^{(n)}). \tag{11.42}$$

An example in Section 9.3.1.2 shows that the MR image is embedded in a third-order neighborhood system $\mathcal{N}^3$. In a 2-D image, for the order $p = 1, 2, 3$, the number of pixels in the cliques is $q = 2, 3, 4, 5$.

Most MRF model-based image analysis methods use pairwise cliques ($c_{1,2}$ and $c_{2,2}$) only. The lack of explanation why using these cliques only is sufficient becomes a common concern. Assignments of clique potentials $V_{c_{p,q}}$ are quite arbitrary. For instance, for the pairwise cliques $c = \{(i,j),(k,l)\}$, its potential may be

$$\begin{cases} c_{1,2}(y_{i,j}) = -1 \text{ or } 0 & (y_{k,l} = y_{i,j}) \\ c_{1,2}(y_{i,j}) = +1 & (y_{k,l} \neq y_{i,j}). \end{cases}$$

The absence of justifications (either in the physical meaning or in the quantitative values) of why these assignments causes questions. After reviewing the commonly used methods for assigning clique potentials that are given in Appendix 11C, a new and general approach has been developed. Using this method, each clique potential has a unique, meaningful, reasonable assignment and is consistent with the energy minimization in the MAP operation.

The Markovianity of X-ray CT and MR imaging (Property 9.2) is derived from SAI, which describes correlations of the pixel intensities. Thus, the clique potential that characterizes the relations of pixel labels may be related to the correlations of the pixel intensities. Let $r_{(i,j),(k,l)}$ be the correlation coefficient (abbreviated c.c. in this section) of pixel intensity $x_{i,j}$ and $x_{k,l}$. The potential of a clique consisting of multiple pixels is defined as the magnitude of c.c. of the intensities of a pair of pixels that has the longest distance in the clique, that is,

$$V_{c_{p,q}}(y_{i,j}^{(n)}) = |r_{(i',j'),(k',l')}|, \tag{11.43}$$

where $||(i',j'),(k',l')|| = \max\{||(i,j),(k,l)||\}$; here $||(i,j),(k,l)||$ denotes the distance between pixels (i,j) and (k,l).[§] This new definition implies that the potential of the clique in the label field $\mathbf{y} = \{y_{i,j}\}$ is defined by the second-order statistics of pixel intensity in the image field $\mathbf{x} = \{x_{i,j}\}$.

Because $|r_{(i,j),(k,l)}| = |E[\left(\frac{x_{i,j}-\mu_{i,j}}{\sigma_{i,j}}\right)\left(\frac{x_{k,l}-\mu_{k,l}}{\sigma_{k,l}}^*\right)]|$ (* denotes the conjugate), it represents a normalized (by the standard deviation) power at a pixel. Thus, $V_{c_{p,q}}(y_{i,j}^{(n)})$; hence, the third term in the parentheses on the right side of Eq. (11.40), represents the normalized cross-power of pixels in the cliques of the pixel (i,j). The first and second terms in the parentheses the right side of Eq. (11.40), that is, $(\frac{x_{i,j}-\mu_k^{(n)}}{\sigma_k^{(n)}})^2 + \ln \sigma_k^{2(n)}$, represent the normalized self-power of pixel (i,j) and its noise power. Thus, the sum of the self-, cross-, and noise power at each pixel over all pixels, that is,

$$\sum_{i,j}((\frac{x_{i,j}-\mu_k^{(n)}}{\sigma_k^{(n)}})^2 + \ln \sigma_k^{2(n)} + \frac{1}{\beta}\sum_{c:(i,j)\in c} V_c(y_{i,j}^{(n)}))$$

gives the real normalized energy of pixel intensities in the image. Therefore, the clique potential defined by Eq. (11.43) has clear physical meaning and makes the potential (the cross-power), the self-power, and the noise power consistent.

11.3.2.2 Clique Potential Assignment

Based on Figures 9.2 and 9.3 and the new expression of the clique, it is clear that a clique $c_{p,q}$ contains the cliques $c_{s,t}$ that are in the same or lower order neighborhood systems $s \leq p$ with the less number of pixels $t < q$. For instance, in Figure 9.3, a 2×2 square-clique $c_{2,4}$ in $\mathcal{N}^2$ contains four triangle-cliques

[§]The notation $||(i,j),(k,l)||$ for representing the distance between pixels $||(i,j)$ and $(k,l)||$ has been used in the proof of Property 9.2.

$c_{2,3}$ and two pair-cliques $c_{2,2}$ in $\mathcal{N}^2$, as well as four pair-cliques $c_{1,2}$ in $\mathcal{N}^1$. Thus, a new procedure for assigning the clique potential is formed as follows. For a clique $c_{p,q}$, (a) the potentials $V_{c_{s,t}}$ $(s \leq p, t < q)$ are assigned first and the potential $V_{c_{p,q}}$ is then assigned; (b) the potentials are assigned to a value of Eq. (11.43) when all $y_{k,l}^{(n)}$ $((k,l) \in c_{p,q})$ are the same or to an another value otherwise. This procedure makes potential assignments for all cliques unique.

ECC (Property 6.3 and Property 8.3) shows that the magnitude of the c.c. of pixel intensities is given by

$$|r_{(i,j),(k,l)}| = |r_{m,n}| = \exp(-\alpha\sqrt{m^2+n^2}), \tag{11.44}$$

where $m = ||i-k||$ and $n = ||j-l||$ are the distances in the unit of pixel. $r_{(i,j),(k,l)}$ only depends on the distances between pixels, and does not depend on pixel locations. Substituting Eq. (11.44) into Eq. (11.43), we have

$$V_{c_{p,q}}(y_{i,j}^{(n)}) = \frac{1}{\beta}\exp(-\alpha\cdot\max\sqrt{m^2+n^2}). \tag{11.45}$$

We noticed that for a given neighborhood system $\mathcal{N}^p$, all its cliques have the same $max\sqrt{m^2+n^2}$. For instance, for $\mathcal{N}^2$, $max\sqrt{m^2+n^2}$ in cliques $c_{2,2}$, $c_{2,3}$, and $c_{2,4}$ are the same¶: $\sqrt{2}$. It is easy to verify that for $p = 1, 2, 3$, $max\sqrt{m^2+n^2} = \sqrt{2}^{p-1}$. Using $\sqrt{2}^{p-1}$ to replace $max\sqrt{m^2+n^2}$ in Eq. (11.45), which corresponds to the smallest clique potential (hence, the smallest normalized cross-power), we have

$$V_{c_{p,q}}(y_{i,j}^{(n)}) = \frac{1}{\beta}\exp(-\alpha\sqrt{2}^{p-1}). \tag{11.46}$$

This new clique potential assignment is closely related to the energy minimization that is used in cFNM model-based image analysis.

11.3.3 Energy Minimization

11.3.3.1 Energy Minimization

In order to minimize $U(\mathbf{y}^{(n)}|\mathbf{x})$ of Eq. (11.40) using the procedure described above and the potential assignment given by Eq. (11.46), we adopt the following approach. For a given clique $c_{p,q} : (i,j) \in c_{p,q}$, if all $y_{k,l}^{(n)}$ $((k,l) \in c_{p,q})$ in the clique are the same, the potential $V_{c_{p,q}}(y_{i,j}^{(n)})$ should be "rewarded" such that $U(\mathbf{y}^{(n)}|\mathbf{x})$ is toward its ideal minimum energy $\sum_{i,j}((\frac{x_{i,j}-\mu_k^{(n)}}{\sigma_k^{(n)}})^2 + \ln\sigma_k^{2(n)})$ as if the cross-power vanished; if at least one $y_{k,l}^{(n)}$ $((k,l) \in c_{p,q})$ in the clique is

¶This may be an interesting result relating the maximum distance between pixels in cliques and the order of the neighborhood systems that the cliques belong to.

different, the potential $V_{c_{p,q}}(y_{i,j}^{(n)})$ should be "penalized" such that $U(\mathbf{y}^{(n)}|\mathbf{x})$ is toward a non-ideal minimum energy $\sum_{i,j}((\frac{x_{i,j}-\mu_k^{(m)}}{\sigma_k^{(m)}})^2+\ln\sigma_k^{2(n)})+\frac{1}{\beta}V_{c_{p,q}}(y_{i,j}^{(n)})$ as if the cross-power remains with a reasonable value Eq. (11.46). For each (i,j), $V_{c_{p,q}}(y_{i,j}^{(n)})$ are updated by updating $\mathbf{y}^{(n)} = \{y_{i,j}^{(n)}\}$ via $(n+1) \longrightarrow (n)$, which leads to $U(\mathbf{y}^{(n)}|\mathbf{x})$ being eventually minimized.

That is, given a pixel (i,j) in the image $\mathbf{x}$ and other labels $y_{\neq(i,j)}^{(n)}$, the energy minimization sequentially updates each $y_{(i,j)}^{(n)}$ to $y_{(i,j)}^{(n+1)}$ by minimizing $U(y_{(i,j)}^{(n)}|\mathbf{x}, y_{\neq(i,j)}^{(n)})$ with respect to $y_{(i,j)}$.

11.3.3.2 Final Expression of Clique Potential

Thus, with the goal of minimizing energy Eq. (11.40), the clique potential is finally expressed by

$$V_{c_{p,q}}(y_{i,j}^{(n)}) = \frac{1}{\beta}\exp(-\alpha\sqrt{2}^{p-1})(1-\gamma \prod_{c_{p,q}:\ (k,l),(i,j)\in c_{p,q}} \delta[y_{k,l}^{(n)} - y_{i,j}^{(n)}]), \quad (11.47)$$

where γ is a constant and $\delta[z]$ is a Kronecker delta function.

The constants α, β, and γ in Eqs. (11.46) and (11.47) are given as follows. For $\gamma = 1$,

$$V_{c_{p,q}}(y_{i,j}^{(n)}) = \begin{cases} 0 & (\text{all } y_{k,l}^{(n)} = y_{i,j}^{(n)},\ (k,l),(i,j) \in c_{p,q}) \\ \frac{1}{\beta}\exp(-\alpha\sqrt{2}^{p-1}) & (\text{otherwise}); \end{cases} \quad (11.48)$$

and for $\gamma = 2$,

$$V_{c_{p,q}}(y_{i,j}^{(n)}) = \begin{cases} -\frac{1}{\beta}\exp(-\alpha\sqrt{2}^{p-1}) & (\text{all } y_{k,l}^{(n)} = y_{i,j}^{(n)},\ (k,l),(i,j) \in c_{p,q}) \\ +\frac{1}{\beta}\exp(-\alpha\sqrt{2}^{p-1}) & (\text{otherwise}), \end{cases} \quad (11.49)$$

Using Eq. (11.48) or Eq. (11.49) is a choice of applications, but Eq. (11.48) seems to be more meaningful. The constant β in currently used MRF model-based image analysis methods is selected empirically [36, 37, 40, 57, 58]. We adopt $\beta = 1$ [40]. Experimental results of the ECC of Section 8.3.2 shows that $\alpha = 0.8 \sim 1.0$.

Remarks. (1) In most currently used MRF model-based image analysis methods, the single pixel clique [35] is not discussed. Revisiting Eq. (11.40), $(\frac{x_{i,j}-\mu_k^{(m)}}{\sigma_k^{(m)}})^2$ actually represents the potential of the single pixel clique potential. In this view, all clique potentials are assigned by the magnitude of c.c. of corresponding pixel intensities and this assignment is unique—neither omission nor duplication. (2) A common view regarding the advantages of using the MRF model in image analysis is that it can impose spatial regularity in processing using the contextual information of the image. This new clique potential assignment method shows that the contextual information of an image can be expressed by the correlation of pixel intensities and the spatial regularity can be imposed eventually based on the correlation of pixel intensities.

11.4 Classification of Pixels

11.4.1 Bayesian Classifier

In some sense, image analysis can be viewed as an operation to classify pixels in the image into the underlying image regions. For the cFNM model of Eq. (11.32), after the number K_0 of image regions is detected using the sensor array signal processing method (Section 11.2) and the model parameters $\mathbf{r}$ are estimated using the extended EM method (Section 11.3), the classification is performed using the Bayesian classifier. The pixel (i,j) is classified into the k_0 $(k_0 = 1, \cdots, K)$ image region $\mathcal{R}_{k_0}$, if and only if

$$k_0 = arg \max_{1 \leq k \leq K_0} \{\pi_{k|\mathcal{N}_{i,j}} \cdot g(y_{i,j}|\mu_k, \sigma_k^2)\}. \tag{11.50}$$

That is, the pixel (i,j) is classified into the k_0 image region with the highest probability over all other image regions. It has been shown that Bayesian classification and the energy minimization are consistent.

11.5 Statistical Image Analysis

The detection of the number of image regions (Section 11.2), the estimation of image parameters (Section 11.3), and the classification pixels into image regions (Section 11.4) form an cFNM model-based statistical image analysis method. The method is performed in the following fashion:

1. By converting an image to an analog sensor array system, the number K_0 of image regions in the image is detected by an eigenstructure method, which is a non-model-based approach and does not require parameter estimates.

2. With the detected number K_0 of image regions, parameters of the cFNM model are estimated by an extended EM algorithm, that is, a MAP-EM operation with new design of clique potentials.

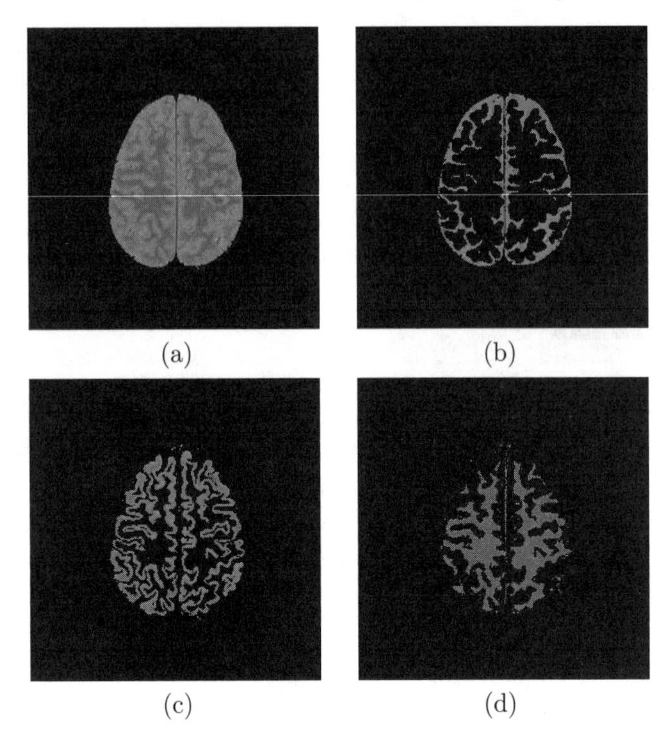

FIGURE 11.6
A Proton Density (PD) weighted MR image of the intracranial (IC) (a), the partitioned region images of the cerebrospinal fluid (CSF) (b), the gray matter (GM) (c), and the white matter (WM) (d).

3. Using the number K_0 and the corresponding ML-MAP estimate $\hat{\mathbf{r}}$, the Bayesian classifier is applied to classify pixels into K_0 groups so that an image is partitioned into distinctive image regions.

This method is a fully automated, unsupervised data-driven approach. The following examples demonstrate its applications in image analysis.

11.5.1 MR Images

Three proton density (PD) weighted MR images of the intracranial (IC) are shown in Figures 11.6a, 11.7a, and 11.8a. By applying the eigenstructure detection method, the number of image regions in these three images are all $K_0 = 3$. Using the cFNM model-based image analysis method described in this chapter, each of these image is partitioned into three region images. They are images of the cerebrospinal fluid (CSF) (Figures 11.6b, 11.7b, and 11.8b), the gray matter (GM) (Figures 11.6c, 11.7c, and 11.8c), and the white matter

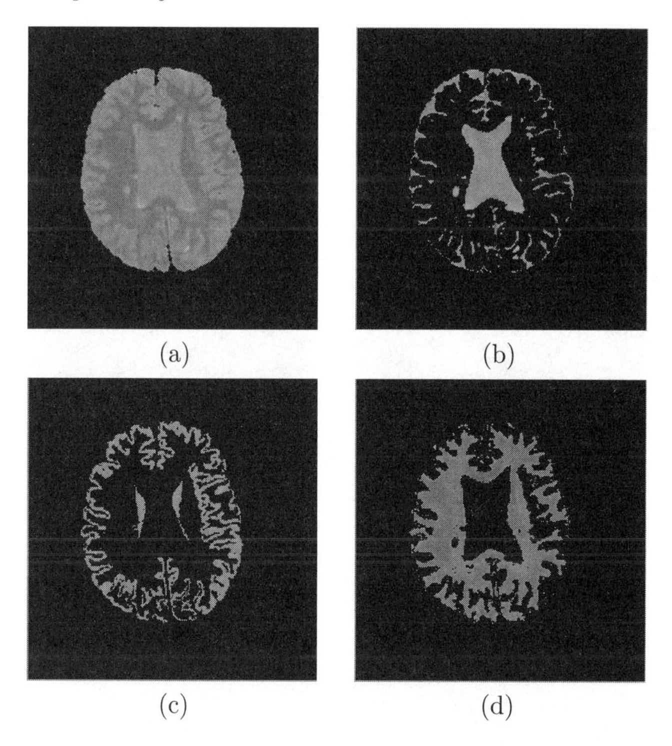

(a) (b)

(c) (d)

FIGURE 11.7
A Proton Density (PD) weighted MR image of the intracranial (IC) (a), the partitioned region images of the cerebrospinal fluid (CSF) (b), the gray matter (GM) (c), and the white matter (WM) (d).

(WM) (Figures 11.6d, 11.7d, and 11.8d). In the processing, the cliques $c_{2,2}$ are used and the iteration number n=3.

11.6 Appendices

11.6.1 Appendix 11A

This appendix proves Eqs. (11.36) and (11.37).

Proof

In parameter estimation, the likelihood of the estimated parameters requires that the observed data must be independently, identically distributed (i.i.d.). In an image $\mathbf{x} = \{x_{i,j}\}$ characterized by cFNM, although $x_{i,j}$ are identically

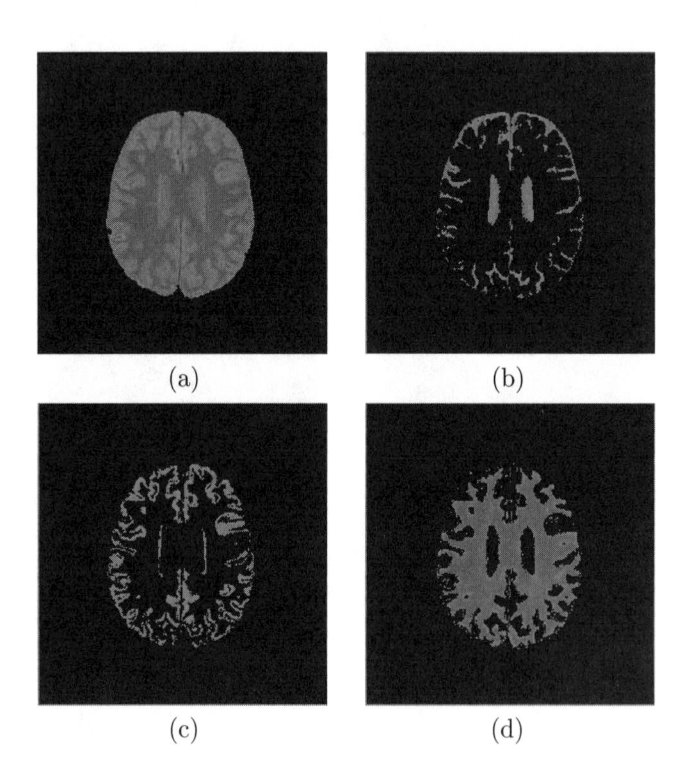

FIGURE 11.8
A Proton Density (PD) weighted MR image of the intracranial (IC) (a), the partitioned region images of the cerebrospinal fluid (CSF) (b), the gray matter (GM) (c), and the white matter (WM) (d).

distributed with a cFNM pdf, they are not independent. As a result, the likelihood of the estimated parameters of the cFNM model cannot be expressed as a product of values of the cFNM pdf at each pixel.

There has been intensive discussion recently regarding the likelihood $\mathcal{L}(\mathbf{r}^{(m)})$ of Eq. (11.33). [59] proposed a consistent procedure based on a penalized Besag pseudo-likelihood [60]; the likelihood is approximated as the product of the values of the pdf at each observation. [61, 62] use a mean field-like approximation based on the mean field principle of statistical physics [63]; an observation does not depend on other observations that are all set to constants independent of that observation. In both cases, the likelihood is factorized as a product of values of the underlying pdf at each observation.

These results lead to the cFNM pdf being approximated by an iFNM pdf. Hence, pixel intensities $\{x_{i,j}\}$ characterized by cFNM are i.i.d. It follows that $\pi_{k|\mathcal{N}_{i,j}}$, which may vary for each pixel (i,j), is temporally treated in the same way as μ_k and σ_k, that is, as an index k only dependent parameter. However, $\pi_{k|\mathcal{N}_{i,j}}$ and (μ_k, σ_k) are computed by Eq. (11.35) and Eq. (11.37), separately. Consequently, because $\sum_{k=1}^{K} \pi_{k|\mathcal{N}_{i,j}}^{(n)} = 1$, from Eq. (11.33) we have a Lagrangian equation

$$L = \ln \mathcal{L}(\mathbf{r}^{(m)}) + \lambda(\sum_{k=1}^{K} \pi_{k|\mathcal{N}_{i,j}}^{(n)} - 1), \tag{11.51}$$

where λ is a Lagrangian multiplier.

By taking partial derivatives with respect to λ, $\pi_{k|\mathcal{N}_{i,j}}^{(n)}$, $\mu_k^{(m)}$ and $\sigma_k^{2(m)}$ and set them to equal zero,

$$\begin{cases} \frac{\partial L}{\partial \lambda} = 0 \\ \frac{\partial L}{\partial \pi_{k|\mathcal{N}_{i,j}}^{(n)}} = \frac{\partial \ln \mathcal{L}}{\partial \pi_{k|\mathcal{N}_{i,j}}^{(n)}} + \lambda = 0 \\ \frac{\partial L}{\partial \mu_k^{(m)}} = \frac{\partial \ln \mathcal{L}}{\partial \mu_k^{(m)}} = 0 \\ \frac{\partial L}{\partial \sigma_k^{2(m)}} = \frac{\partial \ln \mathcal{L}}{\partial \sigma_k^{2(m)}} = 0, \end{cases} \tag{11.52}$$

we have

$$\begin{cases} \sum_{k=1}^{K} \pi_{k|\mathcal{N}_{i,j}}^{(n)} - 1 = 0 \\ \\ \sum_{i,j} \frac{g(x_{i,j}|\mu_k^{(m)},\sigma_k^{(m)})}{f(x_{i,j}|x_{\neq i,j},\mathbf{r}^{(m)})} + \lambda = 0 \\ \\ \sum_{i,j} \frac{\pi_{k|\mathcal{N}_{i,j}^{(n)}} g(x_{i,j}|\mu_k^{(m)},\sigma_k^{(m)})(x_{i,j}-\mu_k^{(m+1)})}{f(x_{i,j}|x_{\neq i,j},\mathbf{r}^{(m)})} = 0 \\ \\ \sum_{i,j} \frac{\pi_{k|\mathcal{N}_{i,j}^{(n)}} g(x_{i,j}|\mu_k^{(m)},\sigma_k^{(m)})((x_{i,j}-\mu_k^{(m+1)})^2-\sigma_k^{2(m)})}{f(x_{i,j}|x_{\neq i,j},\mathbf{r}^{(m)})} = 0. \end{cases} \tag{11.53}$$

The second equation of Eq. (11.53) leads to

$$\lambda = -I \cdot J. \tag{11.54}$$

Let

$$\frac{\pi_{k|\mathcal{N}_{i,j}}^{(n)} g(x_{i,j}|\mu_k^{(m)}, \sigma_k^{(m)})}{f(x_{i,j}|x_{\neq i,j}, \mathbf{r}^{(m)})} = z_{(i,j),k}^{(m)}; \tag{11.55}$$

we obtain, from the third equation of Eq. (11.53),

$$\mu_k^{(m+1)} = \frac{\sum_{i,j} z_{(i,j),k}^{(m)} x_{i,j}}{\sum_{i,j} z_{(i,j),k}^{(m)}}, \tag{11.56}$$

and subsequently from the fourth equation of Eq. (11.53)

$$\sigma_k^{2(m+1)} = \frac{\sum_{i,j} z_{(i,j),k}^{(m)} (x_{i,j} - \mu_k^{(m+1)})^2}{\sum_{i,j} z_{(i,j),k}^{(m)}}. \tag{11.57}$$

Note: $\sigma_k^{2(m+1)}$ is determined by $\mu_k^{(m+1)}$, not by $\mu_k^{(m)}$.

From Eq. (11.30) (the definition of $\pi_{k|\mathcal{N}_{i,j}}^{(n)}$) and Eq. (9.38) (MRF-GD equivalence), $\pi_{k|\mathcal{N}_{i,j}}^{(n)}$ is determined by

$$\begin{aligned} \pi_{k|\mathcal{N}_{i,j}}^{(n)} &= P(y_{i,j}^{(n)} = k|y_{\mathcal{N}_{i,j}}^{(n)}) \\ &= p_G(y_{i,j}^{(n)}|y_{\mathcal{N}_{i,j}}^{(n)}) \\ &= \frac{1}{Z} \exp(-\frac{1}{\beta} \sum_{c:(i,j)\in c} V_c(y_{i,j}^{(n)})), \end{aligned} \tag{11.58}$$

where $c : (i, j)$ is the clique at the pixel (i, j) and $V_c(y_{i,j}^{(n)})$ is the potential of the clique $c : (i, j)$. $\pi_{k|\mathcal{N}_{i,j}}^{(n)}$ remains unchanged in one E-M cycle. ■

11.6.2 Appendix 11B

This appendix proves that $\mathbf{y}|\mathbf{x}$ is an MRF and its energy function is given by Eq. (11.40).

Proof

As shown in Section 9.4, an HMRF, that is, given the context field $\mathbf{y}$, the intensity field $\mathbf{x}$ is conditionally independent[‖] [36, 40, 62], that is,

$$P(\mathbf{x}|\mathbf{y}^{(n)}) = \prod_{i,j} P(x_{i,j}|y_{i,j}^{(n)}) = \prod_{i,j} \frac{1}{\sqrt{2\pi\sigma_k^{2(n)}}} \exp(-\frac{(x_{i,j} - \mu_k^{(n)})^2}{2\sigma_k^{2(n)}}). \quad (11.59)$$

An approximation of $p_G(\mathbf{y}^{(n)})$ in Eq. (9.29) is the Besag pseudo-likelihood [60], i.e.,

$$p_G(\mathbf{y}^{(n)}) = \prod_{i,j} p_G(y_{i,j}^{(n)}|y_{\mathcal{N}_{i,j}}^{(n)}) = \prod_{i,j}(\frac{1}{Z} \exp(-\frac{1}{\eta} \sum_{c:(i,j)\in c} V_c(y_{i,j}^{(n)}))). \quad (11.60)$$

Thus, we have

$$P(\mathbf{x}|\mathbf{y}^{(n)})p_G(\mathbf{y}^{(n)}) = \frac{1}{(\sqrt{2\pi}Z)^{I\cdot J}}$$

$$\exp(-\frac{1}{2}\sum_{i,j}(\frac{x_{i,j} - \mu_k^{(m)}}{\sigma_k^{(m)}})^2 + \ln \sigma_k^{2(m)} + \frac{2}{\eta} \sum_{c:(i,j)\in c} V_c(y_{i,j}^{(n)}))). \quad (11.61)$$

Let

$$U(\mathbf{y}^{(n)}|\mathbf{x}) \triangleq \frac{1}{2}\sum_{i,j}((\frac{x_{i,j} - \mu_k^{(m)}}{\sigma_k^{(m)}})^2 + \ln \sigma_k^{2(m)} + \frac{2}{\eta} \sum_{c:(i,j)\in c} V_c(y_{i,j}^{(n)})), \quad (11.62)$$

from Eqs. (11.61) and (11.62), we can verify

$$\begin{aligned} &\sum_{\mathbf{y}^{(n)}} \left(\int_{\mathbf{x}} P(\mathbf{x}|\mathbf{y}^{(n)})p_G(\mathbf{y}^{(n)})d\mathbf{x} \right) \\ &= \frac{1}{(\sqrt{2\pi}Z)^{I\cdot J}} \sum_{\mathbf{y}^{(n)}} \left(\int_{\mathbf{x}} \exp(-U(\mathbf{y}^{(n)}|\mathbf{x}))d\mathbf{x} \right) \\ &= 1\,, \end{aligned} \quad (11.63)$$

and

$$(\sqrt{2\pi}Z)^{I\cdot J} = \sum_{\mathbf{y}^{(n)}} \left(\int_{\mathbf{x}} \exp(-U(\mathbf{y}^{(n)}|\mathbf{x}))d\mathbf{x} \right) \triangleq Z_{\mathbf{y}|\mathbf{x}}. \quad (11.64)$$

[‖]In some sense, this conditional independence can be understood as a result of the one-to-one correspondence $\{y_{i,j} = k\} \Longleftrightarrow \{x_{i,j} \sim N(\mu_k, \sigma_k^2)\}$.

Thus, Eq. (11.61) can be written as

$$P(\mathbf{x}|\mathbf{y}^{(n)})p_G(\mathbf{y}^{(n)}) = \frac{1}{Z_{\mathbf{y}|\mathbf{x}}} \exp(-U(\mathbf{y}^{(n)}|\mathbf{x})). \tag{11.65}$$

Therefore, from Eq. (11.65) and the relation $P(\mathbf{y}|\mathbf{x}) \propto P(\mathbf{x}|\mathbf{y})p_G(\mathbf{y})$, we have that the conditional field $\mathbf{y}$ given $\mathbf{x}$, that is, $\mathbf{y}|\mathbf{x}$, is an MRF and the energy function of the corresponding GD is $U(\mathbf{y}^{(n)}|\mathbf{x})$ Eq. (11.62). Let $\eta = 2\beta$; Eq. (11.62) gives Eq. (11.40). ■

11.6.3 Appendix 11C

This appendix reviews clique potential assignment approaches used in other MRF model-based image analysis methods and compares them with the approach described in Section 11.3.

(1) The strategy of using the correlation coefficient for clique potential assignment (Section 11.3) has been echoed by an approach reported in [55], which proposed a spatial correlation-based method for neighbor set selection in the simultaneous autoregressive (SAR) and Gauss-Markov (GMRF) random field image models.** In that approach, the additive noise was assumed to be uncorrelated for the SAR model

$$E[e(\mathbf{s})e(\mathbf{s} \oplus \mathbf{r})] = 0 \quad (\forall \mathbf{s}), \tag{11.66}$$

and correlated for the GMRF model

$$E[e(\mathbf{s})e(\mathbf{s} \oplus \mathbf{r})] = \begin{cases} -\theta_{\mathbf{r}}\rho & (\mathbf{r} \in \mathcal{N}) \\ \rho & (\mathbf{r} = 0) \\ 0 & \text{otherwise}, \end{cases} \tag{11.67}$$

where $\mathbf{s} \in \Omega$, $\Omega = \{\mathbf{s} = (i,j) : \ i,j \in J\}$, $J = \{0, 1, \cdots, M-1\}$; the neighbor sets are defined as $\mathcal{N} = \{\mathbf{r} = (k,l) : \ k,l \in \{-(M-1), \cdots, -1, 0, 1 \cdots, (M-1)\};\ k^2 + l^2 > 0\}$, and $\oplus$ denotes modulo M addition in each index. The most likely neighbor sets for the given model are those that satisfy

$$\max\{SCF(\mathbf{r}) - \frac{1}{\rho}E[e(\mathbf{s})e(\mathbf{s} \oplus \mathbf{r})]\} < \xi_{max}, \tag{11.68}$$

where ξ_{max} is the correlation error desired for the model, $SCF(\mathbf{r})$ is the sample correlation function estimated by

$$SCF(\mathbf{r}) = \sum_{\mathbf{s} \in \Omega} \eta(\mathbf{s})\eta(\mathbf{s} \oplus \mathbf{r}) / \sum_{\mathbf{s} \in \Omega} \eta^2(\mathbf{s}) \quad (\forall \mathbf{r} \in \Omega), \tag{11.69}$$

**Property 9.2 provides a procedure for selecting a neighborhood system to an MRF field. See the example in Section 9.3.1.2 given by Table 9.2.

where $\eta(\mathbf{s})$ is the residual image given by

$$\eta(\mathbf{s}) = y(\mathbf{s}) - \sum_{\mathbf{r}\in\mathcal{N}} \hat{\theta} y(\mathbf{s} \oplus \mathbf{r}) \quad (\forall \mathbf{s} \in \Omega), \tag{11.70}$$

where $y(\mathbf{s}) = \sum_{\mathbf{r}\in\mathcal{N}} \theta_{\mathbf{r}} y(\mathbf{s} \oplus \mathbf{r}) + e(\mathbf{s})$ is the model equation; and $\hat{\theta} = \{\hat{\theta}_{\mathbf{r}} : \mathbf{r} \in \mathcal{N}\}$ is an estimate of the interaction coefficient vector that can be obtained by $\min \sum_{\mathbf{r}\in\mathcal{N}} (SCF(\mathbf{r}) - \frac{1}{\rho} E[e(\mathbf{s})e(\mathbf{s} \oplus \mathbf{r})])^2$.

Reviewing this approach, we find that $\frac{1}{\rho} E[e(\mathbf{s})e(\mathbf{s} \oplus \mathbf{r})]$ in Eq. (11.68) is the correlation coefficient of $e(\mathbf{s})$, $SCF(\mathbf{r})$ in Eq. (11.68), and Eq. (11.69) is an estimate of the correlation coefficient of $e(\mathbf{s})$. Thus, this method, in fact, is to minimize the maximum error between the estimated correlation coefficients and the desired correlation coefficients. It is also clear that the interaction coefficient vector $\hat{\theta} = \{\hat{\theta}_{\mathbf{r}} : \mathbf{r} \in \mathcal{N}\}$ is a set of correlation coefficients. Due to the lack of knowledge of these correlation coefficients, $\hat{\theta}$ must be estimated by minimizing $\sum_{\mathbf{r}\in\mathcal{N}} (SCF(\mathbf{r}) - \frac{1}{\rho} E[e(\mathbf{s})e(\mathbf{s} \oplus \mathbf{r})])^2$, which is very intensive computationally. Therefore, in neighborhood system selection, the strategy of Section 11.3 and this approach are the same with regard to nature (both are based on correlation coefficient), but the former will be straightforward and much simpler.

(2) The strategy of using the correlation coefficient for clique potential assignment (Section 11.3) has also been mirrored by currently used MRF modeling methods.

[36] chooses a prior in which pixels of the same type as their neighbors are rewarded and pixels that differ from their neighbors are penalized. Thus, the potentials of two pixel cliques in a second-order neighborhood system are

$$V_c(y) = \begin{cases} 1 - 2\delta[y_i - y_j] & (y_j : 4 \text{ first-order neighbors}) \\ (1 - 2\delta[y_i - y_j])/\sqrt{2} & (y_j : 4 \text{ second-order neighbors}), \end{cases} \tag{11.71}$$

where $\delta[x]$ denotes the Kronecker Delta function.

In [40], the potentials of two pixel cliques in a second-order neighborhood system in the 3-D case are

$$V_c(y) = \begin{cases} 1 - \delta[y_i - y_r] & (y_r : 6 \text{ first-order neighbors}) \\ (1 - \delta[y_i - y_r])/\sqrt{\gamma} & (y_r : 12 \text{ second-order neighbors}), \end{cases} \tag{11.72}$$

where $\gamma = 2$; and for 8 third-order neighbors, $\gamma = 3$.

In [35], the associated clique potentials in a second-order neighborhood system are defined as

$$V_c(y) = \begin{cases} -\xi & (\text{all } y_{i,j} \text{ in clique are equal}) \\ +\xi & (\text{otherwise}), \end{cases} \tag{11.73}$$

where ξ is the specified parameter. For the single pixel clique, its potential is

$$V_c(y) = \alpha_k \quad (y_{i,j} = q_k), \tag{11.74}$$

where α_k controls the percentage of pixels in each region type.

Comparing the above listed methods and the strategy of Section 11.3, we find that (1) parameter settings of clique potentials in the former are "hard" (e.g., $+\xi$ versus $-\xi$, or $+1$ versus -1, or $+1$ versus 0), and those in the latter are "adaptive," (2) the settings in the former are somewhat heuristic and those in the latter are reasonable and consistent with the statistical properties of X-ray CT and MR images.

Problems

11.1. Prove Eq. (11.9).

11.2. Give an interpretation of the source signals $s_m(k)$ $(m = 1, \cdots, q)$ of Eq. (11.9).

11.3. Prove that the complex data $Y(i,j)e^{i\phi(i,j)}$ have zero mean.

11.4. Prove that $\mathbf{x}$ of Eq. (11.10) is a Gaussian random vector with zero mean.

11.5. Prove that $\mathbf{x}$ of Eq. (11.10) is stationary and ergodic.

11.6. Verify that the covariance matrix $\mathbf{S}$ of Eq. (11.12) is positive definite.

11.7. Prove Eq. (11.24).

11.8. Verify that for a given neighborhood system $\mathcal{N}^p$, all its cliques have the same $max\sqrt{m^2+n^2}$.

11.9. Prove that for $p = 1, 2, 3$, $max\sqrt{m^2+n^2} = \sqrt{2}^{p-1}$.

11.10. Prove Eq. (11.63).

11.11. Justify that the clique potential assignment (Eq. (11.47)), the energy minimization (Section 11.3.3), and the Gibbs sampler (Appendix 9A of Chapter 9) are consistent.

11.12. Elaborate on Bayesian classification (Eq. (11.50)) and the energy minimization are consistent.

References

[1] Bellman, R.: On a routing problem. *Q. Appl. Math.* **16** (1958) 87–90.

[2] Dial, R.: Shortest-path forest with topological ordering. *Comm. ACM* **12**(11) (1969) 632–633.

[3] Frieze, A.: Minimum paths in directed graphs. *Operational Res. Quart.* **28**(2,i) (1977) 339–346.

[4] Falcao, A., Udupa, J., Samarasekera, S., Sharma, S., Hirsh, B., Lotufo, R.: User-steered image segmentation paradigms: Live-wire and Live-lane. *Graphical Models and Image Processing* **60**(4) (1998) 233–260.

[5] Terzopoulos, D., Fleischer, K.: Deformable models. *Vis. Comput.* **4** (1988) 306–331.

[6] Davatzikos, C., Prince, J.: Convexity analysis of active contour models. *Proc. Conf. Information Science and Systems* (1994) 581–587.

[7] Davatzikos, C., Prince, J.: An active contour model for mapping the cortex. *IEEE Trans. Med. Imag.* **14**(3) (1995) 65–80.

[8] Xu, C., Prince, J.: Snakes, shapes, and gradient flow. *IEEE Trans. Image Process.* **7**(3) (1998) 359–369.

[9] Yezzi, A., Jr.: Modified curvature motion for image smoothing and enhancement. *IEEE Trans. Image Process.* **7**(3) (1998) 345–352.

[10] Chan, F., Lam, F., Zhu, H.: Adaptice thresholding by variational method. *IEEE Trans. Image Process.* **7**(3) (1998) 468–473.

[11] Chan, T., Vese, L.: Active contour without edges. *IEEE Trans. Image Process.* **10**(2) (2001) 266–277.

[12] Aujol, J.F., Aubert, G., Blanc-Feraud, L.: Wavelet-based Level set evolution for classification of textured images. *IEEE Trans. Image Process.* **12**(12) (2003) 1634–1641.

[13] Charles, A., Porsching, T.: *Numerical Analysis of Parial Differntial Equations.* Prentice-Hall, Englewood Cliffs, New Jersey, (1990).

[14] Ames, W.: *Numerical Methods of Parial Differntial Equations.* Academic, New York (1992).

[15] Cootes, T., Hill, A., Taylor, C.: Use of active shape models for locating structure in medical images. *Image Vision Comput.* **12**(6) (1994) 355–365.

[16] Cootes, T., Taylor, C., Cooper, D.: Active shape models - their training and application. *Computer Vision and Image Understanding* **61**(1) (1995) 38–59.

[17] Jain, A., Zhong, Y., Lakshamanan, S.: Object matching using deformable templates. *IEEE Trans. Pattern Anal. Machine Intell.* **18**(3) (1996) 267–278.

[18] Wang, Y., Staib, L.: Boundary finding with prior shape and smoothness models. *IEEE Trans. Pattern Anal. Machine Intell.* **22**(7) (2000) 738–743.

[19] Grenander, U., Miller, M.: Representations of knowledge in complex systems. *J. Roy. Statist. Soc. B* **56** (1993) 249–603.

[20] Lanitis, A., Taylor, C., Cootes, T.: Automatic interpretation and coding of face images using flexible models. *IEEE Trans. Pattern Anal. Machine Intell.* **19**(7) (1997) 743–756.

[21] Cootes, T., Beeston, C., Edwards, G.: A unified framework for atlas matching using Active Appearance Models. *Lect. Notes Comput. Sci.* **1613** (1999) 322–333.

[22] Davies, R., Twining, C., Cootes, T., Waterton, J., Taylor, C.: A minimum description length approach to statistical shape model. *IEEE Trans. Med. Imag.* **21**(5) (2002) 525–537

[23] Shen, D., Zhan, Y., Davatzikos, C.: Segmentation of prostate boundaries from ultrasound images using statistical shape model. *IEEE Trans. Med. Imag.* **22**(4) (2003) 539–551.

[24] Rosenfeld, A.: Fuzzy digital topology. *Inform. Contr.* **40**(1) (1979) 76–87.

[25] Udupa, J.K., Saha, P., Lotufo, R.: Fuzzy connected object definition in images with respect to co-objects. *Proc. SPIE: Med. Imag.* **3661** (1999) 236–245.

[26] Saha, P.K., Udupa, J.K., Odhner, D.: Scale-based fuzzy connected image segmentation: Theory, algorithm, and validation. *Computer Vision and Image Understanding* **77** (2000) 145–174.

[27] Herman, G., Carvalho, B.: Multiseeded segmentation using fuzzy connectedness. *IEEE Trans. Pattern Anal. Machine Intell.* **23** (2001) 460–474.

[28] Besag, J.: Spatial interaction and the statistical analysis of lattice system (with discussion). *J. Roy. Statist. Soc.* **36B**(2) (1974) 192–236.

[29] Kindermann, R., Snell, J.: *Markov Random Field and Their Applications.* American Mathematical Society, Providence, Rhode Island, (1980).

[30] Cross, G., Jain, A.: Markov random field texture models. *IEEE Trans. Pattern Anal. Machine Intell.* **5** (1983) 25–39.

[31] Kashap, R., Chellappa, R.: Estimation and choice of neighbors in spatial interaction models of images. *IEEE Trans. Information Theory* **29** (1983) 60–72.

[32] Geman, S., Geman, D.: Stochastic relaxation, Gibbs distributions, and the Bayesian restoration of images. *IEEE Trans. Pattern Anal. Machine Intell.* **6**(6) (1984) 721–724.

[33] Chellappa, R., Lashyap, R.L.: Texture synthesis using 2-D noncausal autoregressive models. *IEEE Trans. Acoustics, Speech, Signal Processing* **33** (1985) 194–203.

[34] Besag, J.: On the statistical analysis of dirty pictures. *J. Royal Statist. Soc.* **48B** (1986) 259–302.

[35] Derin, H., Elliot, H.: Modeling and segmentation of noisy and textured images using Gibbs random fields. IEEE Trans. Pattern Anal. Machine Intell. **9**(1) (1987) 39–55.

[36] Leahy, R., Hebert, T., Lee, R.: Applications of Markov random field models in medical imaging. *Proc. IPMI* **11** (1989) 1–14.

[37] Dubes, R.C., Jain, A.K.: Random field models in image analysis. *J. Appl. Stat.* **16**(2) (1989) 131–164.

[38] Bouman, C., Liu, B.: Multiple resolution segmentation of texture images. *IEEE Trans. Pattern Anal. Machine Intell.* **13**(2) (1991) 99–113.

[39] Manjunath, B., Chellappa, R.: Unsupervised texture segmentation using Markov random field models. *IEEE Trans. Pattern Anal. Machine Intell.* **13** (1991) 478–482.

[40] Liang, Z., MacFall, J., Harrington, D.: Parameter estimation and tissue segmentation from multispectral MR images. *IEEE Trans. Med. Imag.* **13**(3) (1994) 441–449.

[41] Bouman, C., Shapiro, M.: A multiscale random field model for Bayesian image segmentation. *IEEE Trans. Image Process.* **3** (1994) 162–176.

[42] Krishnamachari, S., Chellappa, R.: Multiresolution Gauss-Markov random field models for texture segmentation. *IEEE Trans. Image Process.* **6**(2) (1997) 251–267.

[43] Willsky, A.: Multiresolution Markov model for signal and image processing. *Proc. IEEE* **90**(8) (2002) 1396–1458.

[44] Julesz, B.: Visual pattern discrimination. *IRE Trans. Inform. Theory* **8**(2) (1962) 84–92.

[45] Anderson, T.W.: *An Introduction to Multivariate Statistical Analysis.* John Wiley & Sons Inc., New York (1984).

[46] Wax, M., Kailath, T.: Detection of signals by information theoretic criteria. IEEE Trans. Acoust., Speech, and Signal Process. **33**(2) (1985) 387–392

[47] Akaike, H.: On entropy maximization principle. P.R. Krishnaiah, Ed., *Applications of Statistics*, North-Holland Publishing Company, Amsterdam, The Netherlands, 1977, 27–41.

[48] Rissanen, J.: Modeling by shortest data description. *J. Automatica* **14** (1978) 465–471.

[49] Schwarz, G.: Estimating the dimension of a model. *Ann. Stat.* **6**(2) (1978) 461–464.

[50] Rissanen, J.: Consistent order estimation of autoregressive processes by shortest description of data. *Analysis and Optimization of Stochastic Systems*, Jacobs et al., Eds., Academic, New York, 1980, 452–461.

[51] Rissanen, J.: A universal prior for integers and estimation by minimum description length. *Ann. Stat.* **11**(2) (1983) 416–431.

[52] Andersen, A.: Asymptotic theory for principal component analysis. *Ann. J. Math. Stat.* **34** (1963) 122–148.

[53] Schiebl, I., stetter, M., Mayhew, J., McLoughlin, N., Lund, J.S., Obermayer, K.: Blind signal separation from optical imaging recordings with extended spatial decorrelation. IEEE Trans. Biomed. Eng. **47**(5) (2000) 573–577.

[54] McKeown, M.J., Makeig, S., Brown, G.G., Jung, T.P., Kindermann, S.S., Bell, A.J., Sejnowski, T.J.: Analysis of fMRI data by blind separation into independent spatial components. *Human Brain Mapping* **6** (1998) 160–188.

[55] Khotanzad, A., Bennett, J.: A spatial correlation based method for neighbor set selection in random field image models. *IEEE Transactions on Image Process.* **8**(5) (1999) 734–740.

[56] Geladi, P., Grahn, H.: *Multivariate Image Analysis.* John Wiley & Sons Inc., New York, 1996.

[57] Choi, H., Haynor, D., Kim, Y.: Multivariate tissue classification of MR images for 3-D volume reconstruction—A statistical approach. *SPIE Proc. Med. Imag. III* (1989) 183–193.

[58] Dubes, R.C., Jain, A.K., Nadabar, S., Chen, C.: MRF model-based algorithm for image segmentation. *Proc. IEEE ICPR* (1990) 808–814.

[59] Ji, C., Seymour, L.: A consistent model selection procedure for Markov random field based on penalized pseudolikelihood. *Ann. Appl. Probability* **6** (1996) 423–443.

[60] Besag, J.: Statistical analysis of non-lattice data. *The Statistician* **24** (1975) 179–195.

[61] Celeux, G., Forbes, F., Peyrard, N.: EM procedure using Mean Field-like approximation for Markov model-based image segmentation. *Pattern Recog.* **36**(1) (2003) 131–144.

[62] Forbes, F., Peyrard, N.: Hidden Markov random field model selection criteria based on Mean Field-lik approximations. *IEEE Trans. Pattern Analy. Machine Intell.* **25**(9) (2003) 1089–1101.

[63] Chandler, D.: *Introduction to Modern Statistical Mechanics.* Oxford University Press (1987).

12

Performance Evaluation of Image Analysis Methods

12.1 Introduction

With the rapid development of image analysis techniques [1–39], an increasing interest has been directed toward the performance evaluation of these techniques. Commonly used evaluation criteria may include accuracy, precision, efficiency, consistency, reproducibility, robustness, etc. In order to make assessments based on these criteria (e.g., the accuracy), observers often compare the results obtained using these techniques with the corresponding ground truth or the gold standard.

The ground truth may be seen as a conceptual term relative to the knowledge of the truth concerning a specific question. The gold standard may be seen as the concrete realization of the ground truth or an accepted surrogate of truth [40]. Due to the complexity of the structures of living objects and the irregularity of the anomalies, the ground truth or the gold standard of these structures and anomalies is unknown, inaccurate, or even difficult to establish. As a result, subjective criteria and procedures are often used in the performance evaluation, which can lead to inaccurate or biased assessments.

This chapter describes two approaches for the precise and quantitative evaluation of the performance of image analysis techniques. Instead of comparing the results obtained by the image analysis techniques with the ground truth or the gold standard, or using some statistical measures, these two approaches directly assess the image analysis technique itself. The first approach gives analytical assessments of the performance of each step of the image analysis technique. The second approach is focused on the validity of the image analysis technique with its fundamental imaging principles.

The first approach is applied to the iFNM model-based image analysis method (Chapter 10), which consists of three steps: detection, estimation, and classification. (1) For detection performance, probabilities of over- and under-detection of the number of image regions are defined, and the corresponding formulas in terms of model parameters and image quality are derived. (2) For estimation performance, both EM and CM algorithms are showed to produce asymptotically unbiased ML estimates of model parameters in the case of no-overlap. Cramer-Rao bounds of the variances of these estimates are derived. (3) For classification performance, a misclassification probability for the Bayesian classifier is defined, and a simple formula, based on parameter estimates and classified data, is derived to evaluate classification errors.

The results obtained by applying this method to a set of simulated images show that, for images with a moderate quality ($SNR > 14.2$ db, i.e., $\mu/\sigma \geq 5.13$), (1) the number of image regions suggested by the detection criterion is correct and the error-detection probabilities are almost zero; (2) the relative errors of the weight and the mean are less than 0.6%, and all parameter estimates are in the Cramer-Rao estimation intervals; and (3) the misclassification probabilities are less than 0.5%. These results demonstrate that for this class of image analysis methods, the detection procedure is robust, the parameter estimates are accurate, and the classification errors are small.

A strength of this approach is that it not only provides the theoretically approachable accuracy limits of image analysis techniques, but also shows the practically achievable performance for the given images.

The second approach is applied to the cFNM model-based image analysis method (Chapter 11), which also consists of three steps: detection, estimation, and classification. (1) For detection performance, although the cFNM model-based image analysis method uses a sensor array eigenstructure-based approach (which is different from the information criterion-based approach used in the iFNM model-based image analysis method), the probabilities of over- and under-detection of the number of image regions are defined in a similar way. The error-detection probabilities are shown to be functions of image quality, resolution, and complexity. (2) For estimation performance, when the EM algorithm is used, the performances of iFNM and cFNM model-based image analysis methods are similar. (3) For classification performance, the cFNM model-based image analysis method uses the MAP criterion to assess its validity with the underlying imaging principles and shows that the results obtained by MAP are toward the physical ground truth that is to be imaged.

12.2 Performance of the iFNM Model-Based Image Analysis Method

This section analyzes the iFNM model-based image analysis method of Chapter 10. It evaluates its performance at three steps: detection, estimation, and classification.

12.2.1 Detection Performance

This subsection describes the detection performance of the iFNM model-based image analysis method. It first defines the probabilities of over- and under-detection of the number of image regions, then derives their expressions in terms of the signal-to-noise ratio, resolution, and complexity of the image, and finally shows some results.

12.2.1.1 Probabilities of Over-Detection and Under-Detection of the Number of Image Regions

Let $IMG(J, K)$ denote an image of J pixels ($x_j, j = 1, \cdots, J$) and K image regions ($\mathcal{R}_k, k = 1, \cdots, K$). The iFNM model-based image analysis method (Chapter 10) uses an information criterion, Minimum Description Length (MDL), to detect the number of image regions [41–43]. For the simplicity of the derivation, Eq. (10.5) is rewritten in the following form:

$$I_K = -\ln \mathcal{L}_K(\hat{\mathbf{r}}) + p_K(J), \tag{12.1}$$

where I_K represents MDL(K), and $\hat{\mathbf{r}}$ is the maximum likelihood (ML) estimate of the model parameter vector $\mathbf{r}$ of Eq. (10.4). The first term of Eq. (12.1) is the log-likelihood of the ML estimate of the model parameters and is given by

$$\mathcal{L}_K(\hat{\mathbf{r}}) = \prod_{j=1}^{J} f(x_j|\hat{\mathbf{r}}), \tag{12.2}$$

and the second term of Eq. (12.1) is a chosen penalty function of K and J and equals

$$p_K(J) = (0.5 \ln J)(3K - 1), \tag{12.3}$$

where $(3K - 1)$ is the number of free-adjustable parameters of iFNM.

The information theoretic criterion says that, given a set of J independent observations $(x_1, \cdots, x_J)$ and a family of models (i.e., a parameterized family of pdfs $f(x|\mathbf{r})$), the model that best fits the observed data is one that gives the minimum I_K. Thus, let K_0 and K_1 denote the correct and the suggested (by the information criterion) number of image regions, respectively, then the probabilities of over-detection and under-detection of the number of image regions can be defined by

$$\begin{aligned} P_{ov}(K_1 - K_0 > 0) &= P(I_{K_1} < I_{K_0}), \\ P_{ud}(K_1 - K_0 < 0) &= P(I_{K_1} < I_{K_0}), \end{aligned} \tag{12.4}$$

respectively. Let $P_e(K_1 - K_0 \neq 0)$ denote the error-detection probability, which represents either P_{ov} or P_{ud} under the different relations between K_1 and K_0. Eq. (12.4) can be written as

$$P_{ov} = P_e \quad (K_1 > K_0) \quad \text{and} \quad P_{ud} = P_e \quad (K_1 < K_0). \tag{12.5}$$

Substituting Eq. (12.3) into Eq. (12.1) and then applying Eq. (12.1) to I_{K_0} and I_{K_1} in Eq. (12.4), we have

$$P_e = P(\ln \mathcal{L}_{K_0}(\hat{\mathbf{r}}) - \ln \mathcal{L}_{K_1}(\hat{\mathbf{r}}) < (1.5 \ln J)(K_0 - K_1)). \tag{12.6}$$

In Eq. (12.2), when $\hat{\sigma}_k^2 > \frac{1}{2\pi}$, $(k = 1, \cdots, K)$ (this condition is always satisfied in medical images), $0 < f(x_j|\hat{\mathbf{r}}) < 1$. [44] shows that $\ln w = \sum_{k=1}^{\infty}(-1)^{k+1}\frac{(w-1)^k}{k}$ $(0 < w \leq 2)$. Using the linear term $(w - 1)$ of this series representation as an approximation of $\ln w$ (Section A.5 of Appendix 12A gives a detailed discussion on this approximation) and applying this simplification to Eq. (12.4), we have

$$\ln \mathcal{L}_K(\hat{\mathbf{r}}) = \sum_{j=1}^{J} \ln f(x_j|\hat{\mathbf{r}}) \simeq \sum_{j=1}^{J} f(x_j|\hat{\mathbf{r}}) - J. \tag{12.7}$$

Assuming that $\hat{\pi}_k = 1/K$ and $\hat{\sigma}_k^2 = \hat{\sigma}^2$ $(k = 1, \cdots, K)^*$, Eq. (12.7) becomes

$$\ln \mathcal{L}_K(\hat{\mathbf{r}}) = \frac{1}{\sqrt{2\pi}K\hat{\sigma}} \sum_{j=1}^{J}\sum_{k=1}^{K} \exp(-\frac{(x_j - \hat{\mu}_k)^2}{2\hat{\sigma}^2}) - J. \tag{12.8}$$

Pixel x_j is a sample from the iFNM $f(x|\hat{\mathbf{r}})$. After classification, however, x_j belongs to one and only one component of iFNM $f(x|\hat{\mathbf{r}})$, say $g(x|\theta_k)$; that is, to one and only one image region, say $\mathcal{R}_k$. Let $n_0 = 0$, $n_K = J$. Without loss of generality, assume $\{x_{n_{k-1}+1}, \cdots\cdots, x_{n_k}\} \in \mathcal{R}_k$ $(k = 1, \cdots, K)$. [44] also shows that $e^{-\frac{1}{2}w} = \sum_{k=0}^{\infty}(-1)^k\frac{(w/2)^k}{k!}$ $(-\infty < w < \infty)$. Using the linear term $(1 - \frac{1}{2}w)$ of this series representation as an approximation of $e^{-\frac{1}{2}w}$ (Section A.5 of Appendix 12A gives detailed discussion on this approximation) and applying this simplification to Eq. (12.8), we have

$$\ln \mathcal{L}_K(\hat{\mathbf{r}}) \simeq \frac{1}{\sqrt{2\pi}K\hat{\sigma}} \sum_{k=1}^{K} \sum_{j=n_{k-1}+1}^{n_k} \exp(-\frac{(x_j - \hat{\mu}_k)^2}{2\hat{\sigma}^2}) - J \tag{12.9}$$

---

*(1) When π_ks are very different, the minor regions (i.e., regions with smaller π_k) may be ignored in the image analysis. The performance evaluation of this section is confined to the case where all π_k are similar. As a result, the error-detection probability will truthfully represent the detection ability of the image analysis technique itself and will not be affected by the sizes of regions which are difficult to be assumed generally. (2) σ_k^2s are determined by two factors: the inhomogeneity in each object and the total noise. Inhomogeneity in each object causes a slow variation of gray levels in each image region, which can be taken into account in the region mean estimation. In this way, σ_k^2 will be solely affected by noise. When σ_k^2s are different, Eq. (12.21) is used to define $\hat{\sigma}^2$. (3) The above two assumptions (with respect to π_k and σ_k^2) can be satisfied by some simulation methods. For example, when the Gibbs sampler [28, 33] is used to generate a Markov random field image and Gaussian noise is superimposed on the image, then the sizes of the image regions can be controlled (to be similar), the means of the image regions will be different, and the variances of the image regions will be the same. Examples are given in the following sections.

$$\simeq \frac{1}{\sqrt{2\pi}K\hat{\sigma}}[J - \frac{1}{2}\sum_{k=1}^{K}\sum_{j=n_{k-1}+1}^{n_k}\frac{(x_j - \hat{\mu}_k)^2}{\hat{\sigma}^2}] - J. \quad (12.10)$$

Define

$$\sum_{k=1}^{K}\sum_{j=n_{k-1}+1}^{n_k}\frac{(x_j - \hat{\mu}_k)^2}{\hat{\sigma}^2} = Y_K, \quad (12.11)$$

Eq. (12.10) becomes

$$\ln \mathcal{L}_K(\hat{\mathbf{r}}) \simeq \frac{1}{\sqrt{2\pi}K\hat{\sigma}}[J - \frac{1}{2}Y_K] - J. \quad (12.12)$$

Because $\sum_{j=n_{k-1}+1}^{n_k}\frac{(x_j-\hat{\mu}_k)^2}{\hat{\sigma}^2}$ has a χ^2 distribution with the degree of freedom $(n_k - n_{k-1} - 1)$; Y_K has a χ^2 distribution with the degree of freedom $\sum_{k=1}^{K}(n_k - n_{k-1} - 1) = J - K$ [45].

Applying Eq. (12.12) to $\mathcal{L}_{K_0}(\hat{\mathbf{r}})$ and $\mathcal{L}_{K_1}(\hat{\mathbf{r}})$ in Eq. (12.6) with $\hat{\sigma} = \hat{\sigma}_0$ (when $K = K_0$) and $\hat{\sigma} = \hat{\sigma}_1$ (when $K = K_1$), we have

$$\begin{aligned} P_e = P(&\frac{1}{K_1\hat{\sigma}_1}Y_{K_1} - \frac{1}{K_0\hat{\sigma}_0}Y_{K_0} \\ &< (3\sqrt{2\pi}\ln J)(K_0 - K_1) + 2J(\frac{1}{K_1\hat{\sigma}_1} - \frac{1}{K_0\hat{\sigma}_0})), \end{aligned} \quad (12.13)$$

In the following sections and the appendices, σ_0 is the true value and $\hat{\sigma}_0$ is its estimate. Let

$$\begin{cases} Z = \frac{1}{K_1\hat{\sigma}_1}Y_{K_1} - \frac{1}{K_0\hat{\sigma}_0}Y_{K_0} \\ \Delta_1 = (3\sqrt{2\pi}\ln J)(K_0 - K_1) \\ \Delta_2 = 2J(\frac{1}{K_1\hat{\sigma}_1} - \frac{1}{K_0\hat{\sigma}_0}) \\ \Delta = \Delta_1 + \Delta_2. \end{cases} \quad (12.14)$$

Eq. (12.13) becomes

$$P_e = P(Z < \Delta). \quad (12.15)$$

Appendix 12A shows that the pdf of Z is

$$h(z) = \begin{cases} Ce^{\frac{K_0\hat{\sigma}_0}{2}z}[\sum_{l=0}^{m}\binom{l}{m}\frac{(2m-l)!}{a^{2m-l+1}}(-z)^l] \ (z < 0) \\ Ce^{-\frac{K_1\hat{\sigma}_1}{2}z}[\sum_{l=0}^{m}\binom{l}{m}\frac{(2m-l)!}{a^{2m-l+1}}(z)^l] \ (z \geq 0), \end{cases} \quad (12.16)$$

where

$$\begin{cases} \binom{l}{m} = \frac{m!}{l!(m-l)!} \\ m = \frac{J-K_0}{2} - 1 \quad \text{or} \quad m = \frac{J-K_1}{2} - 1 \\ a = \frac{K_0\hat{\sigma}_0 + K_1\hat{\sigma}_1}{2} \\ C = \frac{1}{(m!)^2}(\frac{K_0\hat{\sigma}_0 K_1\hat{\sigma}_1}{4})^{m+1}. \end{cases} \tag{12.17}$$

P_{ov} and P_{ud} are given by

$$P_{ov} = P_{ud} = \sum_{l=0}^{m} \binom{m}{2m-l} \frac{\gamma^{l+1} e_0}{(\gamma + \gamma^{-1})^{2m-l+1}} \quad (\Delta < 0), \tag{12.18}$$

$$P_{ud} = \sum_{l=0}^{m} \binom{m}{2m-l} \frac{(\gamma^{l+1} + \gamma^{-l-1}(1 - e_1))}{(\gamma + \gamma^{-1})^{2m-l+1}} \quad (\Delta \geq 0), \tag{12.19}$$

where

$$\begin{cases} \gamma = \sqrt{\frac{K_1\hat{\sigma}_1}{K_0\hat{\sigma}_0}} \\ e_0 = \sum_{j=0}^{l} \frac{(-\frac{K_0\hat{\sigma}_0}{2}\Delta)^j}{j!} e^{\frac{K_0\hat{\sigma}_0}{2}\Delta} \\ e_1 = \sum_{j=0}^{l} \frac{(\frac{K_1\hat{\sigma}_1}{2}\Delta)^j}{j!} e^{-\frac{K_1\hat{\sigma}_1}{2}\Delta}. \end{cases} \tag{12.20}$$

Although P_{ov} and P_{ud} have the same functional form in the case that $\Delta < 0$ (see Eq. (12.18)), their values are different. This is because the parameters used in Eq. (12.18), for example, K_1 and $\hat{\sigma}_1$, have different values in the over- and under-detection cases, respectively. $\Delta > 0$ occurs only in the case of under-detection: either $\hat{\sigma}$ is very large or $\gamma = 1$, which leads to $\Delta_2 = 0$. From Eqs. (12.18) through (12.20), it is clear that the error-detection probabilities of the number of image regions, P_{ov} and P_{ud}, are functions of the number of pixels (J), the number of image regions (K_0, K_1), and the variances of the image ($\hat{\sigma}_0$, $\hat{\sigma}_1$).

12.2.1.2 Error-Detection Probabilities and Image Quality

When the variance of the entire image is defined by

$$\hat{\sigma}^2 = \sum_{k=1}^{K} \hat{\pi}_k \hat{\sigma}_k^2, \tag{12.21}$$

TABLE 12.1
Settings of the Images in Figures 12.1a–12.1l.

k	1	2	3	4
π_k	0.207764	0.233643	0.310547	0.248047
μ_k	-45.0000	-15.0000	15.0000	45.0000
σ_k^2	σ_0^2	σ_0^2	σ_0^2	σ_0^2

Image size - 64×64, Number of pixels $J = 4096$, Number of image regions $K_0 = 4$.

TABLE 12.2
Variance σ_0^2 and Signal-to-Noise Ratio SNR(db) of Images in Figures 12.1a–l.

	a	b	c	d	e	f	g	h	i	j	k	l
σ_0^2	1	5	10	20	30	40	50	60	70	80	90	100
db	30.2	23.2	20.2	17.2	15.4	14.2	13.2	12.4	11.7	11.2	10.7	10.2

TABLE 12.3
Detection Results.

	MDL(K)							
	$K=2$	$K=3$	$K=4$	$K=5$	$K=6$	K_1	$P_{ov}(+1)$	$P_{ud}(-1)$
a	19662	15964	11471	11484	11496	4	$8.1 \cdot 10^{-20}$	$\simeq 0$
b	19684	17464	14766	14779	14791	4	$2.8 \cdot 10^{-32}$	$\simeq 0$
c	19708	18107	16184	16196	16209	4	$\simeq 0$	$\simeq 0$
d	19754	18741	17595	17608	17621	4	$\simeq 0$	$2.9 \cdot 10^{-33}$
e	19793	19095	18385	18398	18411	4	$\simeq 0$	$3.3 \cdot 10^{-16}$
f	19829	19329	18884	18897	18910	4	$\simeq 0$	$1.6 \cdot 10^{-7}$
g	19863	19496	19212	19225	19238	4	$\simeq 0$	$8.4 \cdot 10^{-3}$
h	19895	19622	19435	19448	19460	4	$\simeq 0$	$6.2 \cdot 10^{-1}$
i	19926	19719	19593	19606	19618	4	$\simeq 0$	$9.9 \cdot 10^{-1}$
j	19955	19797	19709	19722	19735	4	$\simeq 0$	$\simeq 1$
k	19983	19861	19798	19811	19824	4	$\simeq 0$	$\simeq 1$
l	20011	19915	19870	19882	19895	4	$\simeq 0$	$\simeq 1$

Images in Figures 12.1a–l ($K_0 = 4$)

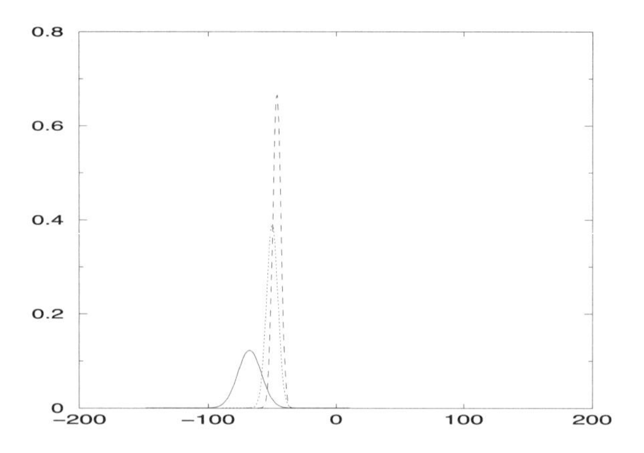

FIGURE 12.2
The pdfs of over-detection in the case where $\sigma_0^2 = 10,\ 40,\ 70$ (curves from left to right).

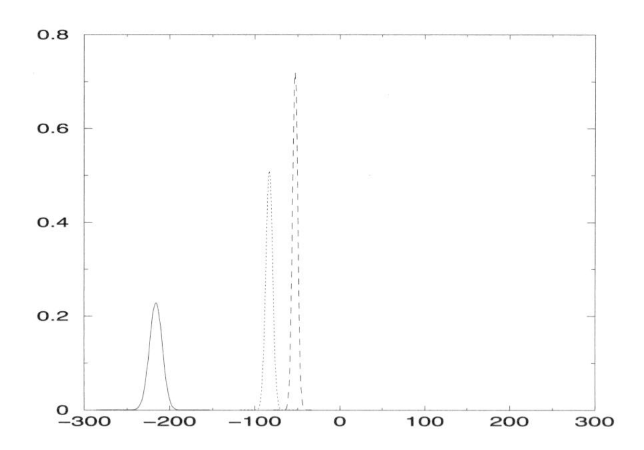

FIGURE 12.3
The pdfs of under-detection in the case $\sigma_0^2 = 10,\ 40,\ 70$ (curves from left to right).

Z' is $\Delta' = \frac{\Delta - \mu}{\sigma}$. Thus, the error-detection probability, $P_e = P(Z < \Delta)$ of Eq. (12.15), can be approximated by the cumulative distribution function (cdf) of Z' at $Z' = \Delta'$, that is, $\Phi(\Delta')$ of Eq. (12.51). For images in Figures 12.1. $a - l$, $\Phi(\Delta')$ are given in Table 12.4. Because $\Phi(\Delta')$ (cdf of the standard Gaussian) monotonically increases with Δ', Table 12.4 shows that the over-detection probability decreases and the under-detection probability increases as the variance of the image increases.

TABLE 12.4
Approximated Probabilities of Over- and Under-Detection of the Number of Image Regions for Images in Figures 12.1a–l.

	μ	σ	Δ	$P_{ov}(+1)$	μ	σ	Δ	$P_{ud}(-1)$
a	−205	18.4	−473	$\Phi(-14.5)$	−906	5.57	−1752	$\Phi(-152)$
b	−94	5.49	−251	$\Phi(-28.6)$	−344	2.49	−627	$\Phi(-114)$
c	−68	3.26	−198	$\Phi(-39.9)$	−216	1.75	−371	$\Phi(-88.6)$
d	−53	1.90	−168	$\Phi(-60.5)$	−133	1.21	−203	$\Phi(-57.9)$
e	−48	1.35	−159	$\Phi(-82.2)$	−101	0.94	−139	$\Phi(-40.4)$
f	−50	1.02	−163	$\Phi(-111.)$	−83.	0.78	−104	$\Phi(-26.9)$
g	−50	0.82	−163	$\Phi(-138.)$	−71.	0.68	−80.	$\Phi(-13.2)$
h	−48	0.69	−159	$\Phi(-161.)$	−61.	0.61	−61.	$\Phi(+0.00)$
i	−46	0.60	−155	$\Phi(-182.)$	−53.	0.55	−44.	$\Phi(+16.4)$
j	−43	0.54	−148	$\Phi(-194.)$	−45.	0.51	−28.	$\Phi(+33.3)$
k	−37	0.48	−137	$\Phi(-208.)$	−37.	0.49	−13.	$\Phi(+49.0)$
l	−32	0.45	−126	$\Phi(-209.)$	−30.	0.46	3.0	$\Phi(+71.7)$

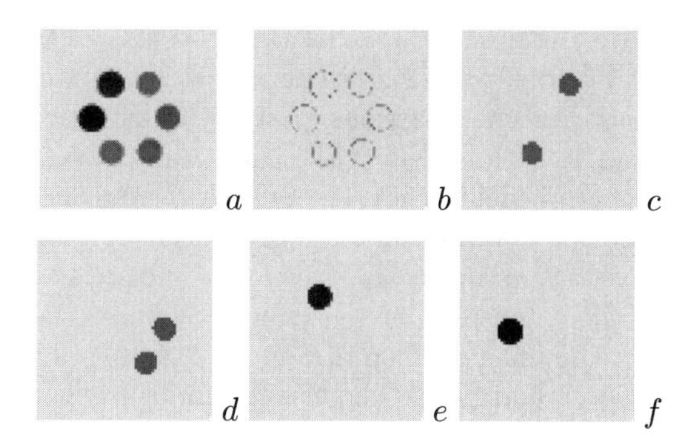

FIGURE 12.4
X-ray CT image of a physical phantom (a) and its five components (b $\sim$ f).

2) Results from a real image

Figure 12.4a shows a real X-ray CT phantom image that is shown in Section 10.5.1. For convenience of description, some contents of Section 10.5.1 are used in this section. This physical phantom consists of six cylinders (made of four types of materials: Poly, 013A, Teflon, and Bone) with nearly the same diameters. These cylinders are mounted on a base in such a way that their axes are parallel to each other and equiangularly located on a circle. The resolution of the X-ray CT scanner that was used to generate this image is $1.5 \times 1.5 \times 5.0$ mm^3. While imaging, the axes of the cylinders are set parallel to the moving direction of the scanner bed. Thus, the image shown in Figure 12.4.a includes images of the cross-sections of all cylinders.[‡] In Figure 12.4.a, starting from the top-left location, in clockwise order, are Teflon, Poly, 013A, 013A, Poly, and Bone. The partitioned image regions by the iFNM model-based image analysis method are shown in Figures 12.4b-f.

TABLE 12.5
Ground Truth of the Images in Figures 12.4a.

k	1	2	3	4	5
π_k	0.2664	0.2167	0.2370	0.1354	0.1445
μ_k	−319.0	−120.0	60.00	800.0	1300.0
σ_k^2	18926	635.0	158.8	28224	74529

Image size - 64×64, Number of pixels $J = 4096$, Number of image regions $K_0 = 5$.

First, we elaborate the ground truth of this real image. Qualitatively, the image in Figure 12.4.a consists of six circular disks and six circular rings. six disk images represent the cross-sections of six cylinders. six ring images are the results of (1) partial volume effect (i.e., part of a pixel is in the air and the other part of this pixel is inside the cylinder) and (2) interpolation in image reconstruction (i.e., even though the pixel is entirely inside either air or a cylinder, its intensity is an interpolated value of the air's and the cylinder's intensities). Thus, the intensities of the ring pixels and the disk pixels are different. Therefore, the image shown in Figure 12.4a has five image regions: ring, Poly, 013A, Teflon, and Bone. The last four image regions represent four types of materials and the first one does not represent a specific material. However, they are all components of this real X-ray CT image.

Quantitative information, that is, the ground truth, of the image in Figure 12.4.a is summarized in Table 12.5. π_k $(k = 1, \cdots, 5)$ are determined by the size of the cylinders. Here, we assume that each ring is closed (i.e., no break). μ_k $(k = 2, \cdots, 5)$ are the given CT numbers of those materials. μ_1 is an averaged value of the CT numbers of air and μ_k $(k = 2, \cdots, 5)$, weighted by π_k $(k = 2, \cdots, 5)$. As stated in the footnote, σ_k^2 $(k = 2, \cdots, 5)$ are determined by two factors: the inhomogeneity in each object (characterized by $\sigma_{k_i}^2$) and the total noise (characterized by $\sigma_{k_n}^2$): $\sigma_k^2 = \sigma_{k_i}^2 + \sigma_{k_n}^2$. In this real image, $\sigma_{k_i} \simeq 0.05\mu_k$ and $\sigma_{k_n} \simeq 0.20\mu_k$. σ_1^2 is the average of σ_k^2 $(k = 2, \cdots, 5)$, weighted by π_k $(k = 2, \cdots, 5)$. The signal-to-noise ratio of each component is given by $SNR_k = \frac{\mu_k^2}{\sigma_k^2}$. The SNR of this ground truth image is 12.6 db.

Then we describe the detection results, which are given in Table 12.6. They shows that the number of image regions indicated by the MDL criterion is

[‡]The background (air) is excluded in the image analysis.

TABLE 12.6
Detection Results.

K	2	3	4	5	6	K_1	$P_{ov}(+1)$	$P_{ud}(-1)$
MDL(K)	3452	3355	3279	3116	3246	5	$\simeq 0$	5.1×10^{-1}

Images in Figures 12.4a ($K_0 = 5$).

correct: $K_1 = Arg\{\min_{1<K<7} I_K\} = 5$ is equal to the correct number of image regions $K_0 = 5$. Probabilities of over- and under-detection of the number of image regions are $P_{ov}(+1) \simeq 0$ and $P_{ud}(-1) \simeq 51\%$, which are similar to $P_{ov}(+1) \simeq 0$ and $P_{ud}(-1) \simeq 62\%$ of the image in Figure 12.1h (Table 12.3), where $SNR = 12.4$ *db* (Table 12.2).

12.2.2 Estimation Performance

This subsection describes the estimation performance of the iFNM model-based image analysis method. It first shows that both the EM and CM algorithms produce asymptotically unbiased ML estimates of model parameters in the case of no-overlap, then derives Cramer-Rao bounds of variances of these estimates, and finally shows some results.

12.2.2.1 Asymptotically Unbiased ML Estimates of FNM Model Parameters

1) ML Estimates of iFNM Model Parameters

Let $IMG(J, K)$ denote an image of J pixels $(x_j, j = 1, \cdots, J)$ and K image regions $(\mathcal{R}_k, k = 1, \cdots, K)$. The iFNM model-based image analysis method (Chapter 10) uses the EM algorithm to estimate the model parameters. Eqs. (10.11)-(10.13) are the EM solution derived by maximizing the *expectation* of iFNM likelihood function (Appendix 10C of Chapter 10). By maximizing the iFNM likelihood function *itself*, the ML estimates of iFNM model parameters can be obtained. The joint likelihood of J independent pixel intensities x_j $(j = 1, \cdots, J)$ is given by Eq. (12.2). The ML estimate $\hat{\mathbf{r}} = \{(\hat{\pi}_k, \hat{\mu}_k, \hat{\sigma}_k^2), k = 1, \cdots, K\}$ of iFNM model parameters of Eq. (10.4) can be obtained by

$$\text{maximizing} \quad L = \ln \mathcal{L}(\hat{\mathbf{r}}) + \lambda(\sum_{k=1}^{K} \pi_k - 1) \tag{12.26}$$

over the parameter set $\{(\pi_k, \mu_k, \sigma_k^2), k = 1, \cdots, K\}$ and a Lagrangian multiplier λ.

The solutions of Eq. (12.26) are

$$\begin{cases} \hat{\lambda} = -J \\ \hat{\pi}_k = \frac{1}{J}\sum_{j=1}^{J} z_{jk} \\ \hat{\mu}_k = \frac{1}{J\hat{\pi}_k}\sum_{j=1}^{J} z_{jk}x_j \\ \hat{\sigma}_k^2 = \frac{1}{J\hat{\pi}_k}\sum_{j=1}^{J} z_{jk}(x_j - \hat{\mu}_k)^2, \end{cases} \tag{12.27}$$

where $k = 1, \cdots, K$ and the quantities

$$z_{jk} = \frac{\hat{\pi}_k g(x_j|\hat{\theta}_k)}{f(x_j|\hat{\mathbf{r}})} \tag{12.28}$$

are known as Bayesian probabilities of the j-th pixel to be classified into the k-th image region for every $j = 1, \cdots, J$ and $k = 1, \cdots, K$.

2) Bayesian Probability z_{jk} and Probability Membership $z_{jk}^{(m)}$

Theoretically, any Gaussian random variable X takes the values from the interval $(-\infty, \infty)$. In medical imaging, due to the physical limitations, any X that represents a measured quantity (such as pixel intensity) takes values from the interval $(-M, M)$, where $M > 0$. That is, X has a truncated Gaussian distribution. In the iFNM model, each Gaussian random variable X_k ($k = 1, \cdots, K$) that represents pixel intensity x_k takes values from the interval $(x_k^{'}, x_k^{''})$.[§] If all these intervals are mutually exclusive, then it is said that the random variables X_k ($k = 1, \cdots, K$) are no-overlapping.

The Bayesian probability z_{jk} of Eq. (12.28) and the probability membership $z_{jk}^{(m)}$ of Eq. (10.11) are identical in their functional forms. However, they are different. z_{jk} cannot be really computed, because it requires the parameter estimates $(\hat{\pi}_k, \hat{\mu}_k, \hat{\sigma}_k^2)$ that are unknown and to be determined. While $z_{jk}^{(m)}$ can be computed using the incomplete data $(\{x_j,\ j = 1, \cdots, J\})$ and the current parameter estimates $(\pi_k^{(m)}, \mu_k^{(m)}, \sigma_k^{2\,(m)})$, and will be used to compute the updated parameter estimates $(\pi_k^{(m+1)}, \mu_k^{(m+1)}, \sigma_k^{2\,(m+1)})$.

It is easy to prove that Bayesian probabilities z_{jk} of Eq. (12.28) have two properties

$$\sum_{k=1}^{K} z_{jk} = 1 \quad (j = 1, \cdots, J) \quad \text{and} \quad E\{z_{jk}\} = \pi_k \quad (k = 1, \cdots, K). \tag{12.29}$$

[§]This notation and $\{x_{n_{k-1}+1}, \cdots, x_{n_k}\} \in \mathcal{R}_k$ of Section 12.2.1.1 are equivalent, that is, $x_k^{'} < \{x_{n_{k-1}+1}, \cdots, x_{n_k}\} \leq x_k^{''}$.

Suppose that an image consists of K image regions, $\mathcal{R}_k$ $(k = 1, \cdots, K)$. In the case of no-overlap, after classification, $\{x_{n_{k-1}+1}, \cdots\cdots, x_{n_k}\} \in \mathcal{R}_k$, $(k = 1, \cdots, K)$. This is equivalent to the condition $g(x_j|\theta_k) = 0$ if $x_j \notin \mathcal{R}_k$. Thus, from Eq. (12.28)

$$z_{jk} = \begin{cases} 1 & (x_j \in \mathcal{R}_k) \\ & \\ 0 & (x_j \notin \mathcal{R}_k). \end{cases} \tag{12.30}$$

Eq. (12.30) is also true for the probability membership, $z_{jk}^{(m)}$. It can be verified by numerical simulations. Two examples are used here to show the evolution of $z_{jk}^{(m)}$ in the EM algorithm. Example 1 is for the no-overlap case and example 2 is for the overlap case. In these examples, two pixels x_{2048} and x_{2049} are randomly selected from two 64×64 images shown in Figures 12.1a and 12.1h. After classification, each of these two pixels is classified into one of the four image regions. The results are summarized in Tables 12.7 and 12.8. In the case of no-overlap (example 1, as shown in Table 12.7), for each j, when k is correct, $z_{jk}^{(m)}$ increases with m until it reaches 1; when k is incorrect, $z_{jk}^{(m)}$ decreases with m until it reaches 0. When $m = 20$: $z_{2048,2}^{(20)} = 1$, $z_{2048,k}^{(20)} = 0$ $(k \neq 2)$; $z_{2049,3}^{(20)} = 1$, $z_{2049,k}^{(20)} = 0$ $(k \neq 3)$. In the case of overlap (example 2, as shown in Table 12.8), for each j, when k is correct, $z_{jk}^{(m)}$ increases (toward 1); when k is incorrect, $z_{jk}^{(m)}$ decreases (toward 0); $z_{jk}^{(m)}$ may not reach 1 or 0. When $m = 55$: $z_{2048,2}^{(55)} = 0.994$, $z_{2048,k}^{(55)} \simeq 0$ $(k \neq 2)$; $z_{2049,3}^{(55)} = 0.999$, $z_{2049,k}^{(55)} \simeq 0$ $(k \neq 3)$. Tables 12.7 and 12.8 also show that the likelihood monotonically increases with m until the stopping criterion is satisfied.

TABLE 12.7
$z_{j,k}^{(m)}$ of EM Algorithm – Example #1: Pixels from Image in Figure 12.1a

	$z_{2048,k}^{(m)}$				$z_{2049,k}^{(m)}$				
	$k=1$	$k=2$	$k=3$	$k=4$	$k=1$	$k=2$	$k=3$	$k=4$	$\ln \mathcal{L}$
m=1	0.288	0.452	0.224	0.035	0.033	0.218	0.452	0.296	−19934
5	0.306	0.485	0.205	0.004	0.001	0.190	0.479	0.330	−19646
10	0.171	0.613	0.215	0.001	0.000	0.203	0.543	0.254	−19606
15	0.000	0.791	0.209	0.000	0.000	0.222	0.748	0.030	−17759
18	0.000	0.989	0.011	0.000	0.000	0.069	0.931	0.000	−15492
19	0.000	1.000	0.000	0.000	0.000	0.002	0.998	0.000	−13905
20	0.000	1.000	0.000	0.000	0.000	0.000	1.000	0.000	−11762
21	0.000	1.000	0.000	0.000	0.000	0.000	1.000	0.000	−11762

3) Asymptotically Unbiased ML Estimates of iFNM Model Parameters

EM, CM, and ML solutions for estimating iFNM model parameters are

TABLE 12.8
$z_{j,k}^{(m)}$ of EM Algorithm – Example #2: Pixels from Image in Figure 12.1h

	$z_{2048,k}^{(m)}$				$z_{2049,k}^{(m)}$				
	$k=1$	$k=2$	$k=3$	$k=4$	$k=1$	$k=2$	$k=3$	$k=4$	$\ln \mathcal{L}$
m=1	0.112	0.594	0.282	0.012	0.005	0.180	0.623	0.192	−20485
11	0.000	0.714	0.286	0.000	0.000	0.182	0.813	0.005	−19603
21	0.000	0.943	0.057	0.000	0.000	0.059	0.941	0.000	−19502
31	0.000	0.996	0.004	0.000	0.000	0.004	0.995	0.000	−19413
41	0.000	0.996	0.004	0.000	0.000	0.000	0.999	0.001	−19391
51	0.000	0.994	0.005	0.000	0.000	0.000	0.999	0.001	−19389
53	0.000	0.994	0.006	0.000	0.000	0.000	0.999	0.001	−19389
55	0.000	0.994	0.006	0.000	0.000	0.000	0.999	0.001	−19389

shown in subsections 10.3.1, 10.3.2, and 12.2.2.1.1, respectively. This subsection first shows that EM and CM solutions provide ML estimates of iFNM model parameters in the case of no-overlap. Eq. (12.27) is the ML solution, but it cannot be really used to compute the parameter estimates because z_{jk} depends on the parameter estimates $(\hat{\pi}_k, \hat{\mu}_k, \hat{\sigma}_k^2)$ $(k = 1, \cdots, K)$, which are unknown and to be determined. Eq. (12.27), however, is an intuitively appealing form for the solution of Eq. (12.26), and is also analogous to the corresponding EM solution Eq. (10.12). Section 12.2.2.1.2 established that, in the case of no-overlap, $z_{jk} = z_{jk}^{(m)}$. Thus, Eq. (10.12) (EM solution) will be exactly the same as Eq. (12.27) (ML solution). Moreover, Eq. (10.16) (CM solution) is a special case of Eq. (10.12) when $z_{jk}^{(m)} = 1$. Therefore, in the case of no-overlap, Eqs. (10.12) and (10.16) produce ML estimates of the iFNM model parameters.

Section 12.2.2.1.2 also showed that in the case of overlap, $z_{jk} \simeq z_{jk}^{(m)}$. That is, parameter estimates by EM algorithm may not be exactly the same as that by the ML procedure. The reason for this difference is that the EM solution is obtained by maximizing the *expectation* of the iFNM likelihood function, while the ML solution is obtained by maximizing the iFNM likelihood function *itself*.

Next, this subsection shows that the EM and CM solutions provide asymptotically unbiased ML estimates of iFNM model parameters in the case of no-overlap. In the case of no-overlap, the expectation of $\hat{\pi}_k$ of Eq. (12.27), by using Eq. (12.29), is

$$E\{\hat{\pi}_k\} = \frac{1}{J}\sum_{j=1}^{J} E\{z_{jk}\} = \pi_k. \tag{12.31}$$

The expectation of $\hat{\mu}_k$ of Eq. (12.27), using Eq. (12.30), is

$$E\{\hat{\mu}_k\} = \frac{1}{J\hat{\pi}_k}\sum_{j=1}^{J} E\{z_{jk}x_j\} = \frac{1}{J\hat{\pi}_k}\sum_{x_j \in G_k} E\{x_j\} = \mu_k. \tag{12.32}$$

The expectation of $\hat{\sigma}_k^2$ of Eq. (12.27), using Eq. (12.30), is

$$E\{\hat{\sigma}_k^2\} = \frac{1}{J\hat{\pi}_k}\sum_{j=1}^{J} E\{z_{jk}(x_j - \hat{\mu}_k)^2\}$$

$$= \frac{1}{J\hat{\pi}_k}\sum_{x_j \in G_k} E\{(x_j - \hat{\mu}_k)^2\} = \sigma_k^2(1 - \frac{1}{J\hat{\pi}_k}), \tag{12.33}$$

which finally leads to

$$\lim_{J\to\infty} E\{\hat{\sigma}_k^2\} = \sigma_k^2. \tag{12.34}$$

Thus, $(\hat{\pi}_k, \hat{\mu}_k, \hat{\sigma}_k^2)$ are the asymptotically unbiased ML estimates of $(\pi_k, \mu_k, \sigma_k^2)$ in the case of no-overlap. Therefore, $(\pi_k^{(m+1)}, \mu_k^{(m+1)}, \sigma_k^{2(m+1)})$ of Eqs. (10.12) and (10.16) are the asymptotically unbiased ML estimates of $(\pi_k, \mu_k, \sigma_k^2)$ under the same condition. The stopping criterion of the EM algorithm is Eq. (10.13): $|\mathcal{L}^{(m+1)} - \mathcal{L}^{(m)}| < \epsilon$. When this criterion is satisfied, as shown in Tables 12.7 and 12.8, the probability memberships $z_{jk}^{(m)}$ $(j = 1, \cdots, J,\ k = 1, \cdots, K)$ do not change with further iterations. From Eq. (10.12), $\pi_k^{(m)}$, $\mu_k^{(m)}$, $\sigma_k^{2(m)}$ $(k = 1, \cdots, K)$ will not change. The stopping criterion of CM algorithm is Eq. (10.17): $\mu_k^{(m+1)} = \mu_k^{(m)}$ and $\sigma_k^{2(m+1)} = \sigma_k^{2(m)}$ $(k = 1, \cdots, K)$. When this criterion is satisfied, there will be no pixel interchange among the image regions in future iterations. Thus, when the stopping criteria (Eqs. (10.13) and (10.17)) are satisfied, the classification of pixels into image regions will not change. Therefore, in the case of no-overlap, the EM and CM algorithms produce asymptotically unbiased ML estimates of the iFNM model parameters when the stopping criteria are satisfied.

12.2.2.2 Cramer–Rao Low Bounds of Variances of the Parameter Estimates

1) For Weight Estimation

Parameters θ_k and π_k of the iFNM model are linked by $\pi_k = P(\theta = \theta_k)$, which shows that π_k is the probability of occurrence of the k-th component $g(x_j|\theta_k)$ of iFNM. The discrete distribution $P(\theta = \theta_k)$ can be expressed by the matrix

$$\begin{pmatrix} \theta = \theta_1, \cdots, \theta_k, \cdots, \theta_K \\ p_\theta = \pi_1, \cdots, \pi_k, \cdots, \pi_K \end{pmatrix} \quad \text{or} \quad \begin{pmatrix} \theta = \theta_k, & \theta \neq \theta_k \\ p_\theta = \pi_k, & p_\theta = 1 - \pi_k \end{pmatrix}. \tag{12.35}$$

Thus

$$E\left[\left(\frac{\partial \ln p_\theta}{\partial \pi_k}\right)^2\right] = \sum_\theta \left(\frac{\partial \ln p_\theta}{\partial \pi_k}\right)^2 p_\theta = \frac{1}{\pi_k(1-\pi_k)}. \tag{12.36}$$

Because $E(\hat{\pi}_k) = \pi_k$ and x takes J independent values, the Cramer–Rao Low Bound of variance of the estimate $\hat{\pi}_k$ is

$$Var\{\hat{\pi}_k\} \geq \left(JE\left[\left(\frac{\partial \ln p_\theta}{\partial \pi_k}\right)^2\right]\right)^{-1} = J^{-1}\pi_k(1-\pi_k) \triangleq CRLB_{\hat{\pi}_k}. \quad (12.37)$$

Eq. (12.37) indicates that the Cramer–Rao Low Bound of variance of the estimate $\hat{\pi}_k$ is the same as that of p in a binomial distribution $B(n,p)$ [45]. This is because we treat the multinomial distribution $\theta = \theta_k$, $(k = 1, \cdots, K)$ as a binomial distribution: $\theta = \theta_k$ *versus* $\theta \neq \theta_k$. [47, 48] use a fuzzy classification variable and the theory of overlap to prove that in the case of no-overlap among all classes, $Var\{\hat{\pi}_k\} \geq J^{-1}\pi_k(1-\pi_k)$, which is exactly the same as Eq. (12.37).

2) For Mean Estimation

From Eq. (12.26), we have

$$\frac{\partial \ln \mathcal{L}}{\partial \mu_k} = (\sigma_k^2)^{-1}\sum_{j=1}^{J} z_{jk}(x_j - \mu_k). \quad (12.38)$$

Due to independence among x_j,

$$E\left[\left(\frac{\partial \ln \mathcal{L}}{\partial \mu_k}\right)^2\right] = (\sigma_k^2)^{-2}[\sum_{j=1}^{J} E({z_{jk}}^2(x_j - \mu_k)^2) + \sum_{i,j=1, i\neq j}^{J} E(z_{ik}(x_i - \mu_k))E(z_{jk}(x_j - \mu_k))]. \quad (12.39)$$

In the case of no-overlap, Eq. (12.39) becomes

$$E\left[\left(\frac{\partial \ln \mathcal{L}}{\partial \mu_k}\right)^2\right] = (\sigma_k^2)^{-2}\sum_{x_j \in G_k} E(x_j - \mu_k)^2 = J\pi_k(\sigma_k^2)^{-1}. \quad (12.40)$$

Because $E(\hat{\mu}_k) = \mu_k$, the Cramer–Rao Low Bound of variance of $\hat{\mu}_k$ is

$$Var\{\hat{\mu}_k\} \geq \left(E\left[\left(\frac{\partial \ln \mathcal{L}}{\partial \mu_k}\right)^2\right]\right)^{-1} = \sigma_k^2/J\pi_k \triangleq CRLB_{\hat{\mu}_k}. \quad (12.41)$$

[49] shows that the pixels in each image region form a truncated Gaussian stochastic process that is (spatially) stationary and ergodic in mean and variance. Thus, instead of ensemble averaging using many independent images, the mean of an image region can be estimated by spatially averaging over the pixels inside that image region (i.e., by the sample mean) of just one image.

In order to form this sample mean, the Bayesian probabilities, z_{jk}, must be assigned for every $j = 1, \cdots, J$ and $k = 1, \cdots, K$. The above derivation shows that the EM and CM algorithms can solve this assignment and finally lead to the sample mean. The Cramer-Rao Low Bound of variance of the sample mean (with the sample size $J\hat{\pi}_k$) is $\sigma_k^2/J\hat{\pi}_k$ [45], which is exactly the same as Eq. (12.41).

3) For Variance Estimation

From Eq. (12.26), we have

$$\frac{\partial \ln \mathcal{L}}{\partial \sigma_k^2} = \frac{1}{2}(\sigma_k^2)^{-2} \sum_{j=1}^{J} z_{jk}((x_j - \mu_k)^2 - \sigma_k^2). \tag{12.42}$$

Due to independence among x_j,

$$E\left[\left(\frac{\partial \ln \mathcal{L}}{\partial \sigma_k^2}\right)^2\right] = \frac{1}{4}(\sigma_k^2)^{-4}[\sum_{j=1}^{J} E({z_{jk}}^2((x_j - \mu_k)^2 - \sigma_k^2)^2)$$
$$+ \sum_{i,j=1,i\neq j}^{J} E(z_{ik}((x_i - \mu_k)^2 - \sigma_k^2))E(z_{jk}((x_j - \mu_k)^2 - \sigma_k^2))]. \tag{12.43}$$

In the case of no-overlap, Eq. (12.43) becomes

$$\begin{aligned} E\left[\left(\frac{\partial \ln \mathcal{L}}{\partial \sigma_k^2}\right)^2\right] &= \frac{1}{4}(\sigma_k^2)^{-4} \sum_{x_j \in G_k} E((x_j - \mu_k)^2 - \sigma_k^2)^2 \\ &= \frac{1}{4}(\sigma_k^2)^{-4} \sum_{x_j \in G_k} E((x_j - \mu_k)^4 - \sigma_k^4). \end{aligned} \tag{12.44}$$

Because $E(x_j - \mu_k)^4 = 3\sigma_k^4$, Eq. (12.44) is

$$E\left[\left(\frac{\partial \ln \mathcal{L}}{\partial \sigma_k^2}\right)^2\right] = \frac{1}{2} J\pi_k \sigma_k^{-4}. \tag{12.45}$$

The bias of the estimate $\hat{\sigma}_k^2$ is defined by $b(\sigma_k^2) = E(\hat{\sigma}_k^2) - \sigma_k^2$. Thus, from Eq. (12.33), $b(\sigma_k^2) = -\frac{\sigma_k^2}{J\pi_k}$. Therefore, the Cramer-Rao Low Bound of variance of the estimate $\hat{\sigma}_k^2$ is

$$\begin{aligned} Var\{\hat{\sigma}_k^2\} &\geq \left(1 + \frac{\partial b(\sigma_k^2)}{\partial \sigma_k^2}\right) \cdot \left(E\left[\left(\frac{\partial \ln \mathcal{L}}{\partial \sigma_k^2}\right)^2\right]\right)^{-1} \\ &= (2\sigma_k^4/J\pi_k)(1 - \frac{1}{J\pi_k}) \triangleq CRLB_{\hat{\sigma}_k^2}. \end{aligned} \tag{12.46}$$

In the limiting case, because $\lim_{J\to\infty}\left(1+\frac{\partial b(\sigma_k^2)}{\partial \sigma_k^2}\right) = 1$, $CRLB_{\hat{\sigma}_k^2} = 2\sigma_k^4/J\pi_k$. The Cramer-Rao Low Bound of variance of the sample variance (with the sample size $J\pi_k$) is $(2\sigma_k^4/J\pi_k)(1-\frac{1}{J\pi_k})$ [45], which is exactly the same as Eq. (12.46). This result indicates that both the EM and CM algorithms finally lead to the sample variance.

Eqs. (12.37), (12.41), and (12.46) show that the Cramer–Rao Low Bounds of variances of the iFNM model parameter estimates are functions of σ_k^2, J, and π_k. More specifically, these bounds will decrease when (a) image quality becomes better (i.e., higher SNR or lower σ_k^2), (b) the resolution is higher (i.e., larger J), and (c) the complexity is less (i.e., smaller K_0 or bigger π_k). These observations are similar to and consistent with the discussions under detection performance (Section 12.2.1.2).

12.2.2.3 Results of Estimation Performance

The relative error and the estimation interval of parameter estimates are utilized to illustrate the accuracy of estimates. The relative errors of the parameter estimates are defined by

$$\begin{aligned}
\varepsilon_\pi &= \max_{1\le k\le 4}\{\tfrac{\hat{\pi}_k-\pi_k}{\pi_k}\} \\
\varepsilon_\mu &= \max_{1\le k\le 4}\{\tfrac{\hat{\mu}_k-\mu_k}{\mu_k}\} \\
\varepsilon_\sigma &= \max_{1\le k\le 4}\{\tfrac{\hat{\sigma}_k^2-\sigma_k^2}{\sigma_k^2}\},
\end{aligned} \tag{12.47}$$

the estimation intervals of the parameter estimates are defined by

$$\begin{aligned}
\varpi_{\hat{\pi}_k} &= (\pi_k - \sqrt{CRLB_{\hat{\pi}_k}},\ \pi_k + \sqrt{CRLB_{\hat{\pi}_k}}) \\
\varpi_{\hat{\mu}_k} &= (\mu_k - \sqrt{CRLB_{\hat{\mu}_k}},\ \mu_k + \sqrt{CRLB_{\hat{\mu}_k}}) \\
\varpi_{\hat{\sigma}_k^2} &= (\sigma_k^2 - \sqrt{CRLB_{\hat{\sigma}_k^2}},\ \sigma_k^2 + \sqrt{CRLB_{\hat{\sigma}_k^2}}),
\end{aligned} \tag{12.48}$$

where $(\pi_k, \mu_k, \sigma_k^2)$ and $(\hat{\pi}_k, \hat{\mu}_k, \hat{\sigma}_k^2)$ are the true, and the estimated values of the parameters, $(CRLB_{\hat{\pi}_k}, CRLB_{\hat{\mu}_k}, CRLB_{\hat{\sigma}_k^2})$ are given by Eqs. (12.37), (12.41), and (12.46).

1) Results from simulated images

The simulated images shown in Figure 12.1 are also used here for evaluating estimation performance. The parameter settings (i.e., the true values of the parameters) are listed in Table 12.1. The parameter estimates and the Cramer-Rao Low Bounds of the variances of these estimates are summarized in Table

12.9. Eq. (12.37) shows that the Cramer–Rao Low Bounds of the variances of weight estimates do not depend on the variance σ_0^2 of the image. Eqs. (12.41) and (12.46) show that the Cramer–Rao Low Bounds of the variances of mean and variance estimates depend on the variance σ_0^2 of the image and increase as σ_0^2 increases. These facts are also reflected in Table 12.9.

The accuracy of the estimation can be evaluated by judging if the relative errors are small and the estimates $\hat{\pi}_k$, $\hat{\mu}_k$, and $\hat{\sigma}_k^2$ fall into these estimation intervals. Using Tables 12.1 and 12.9, the results are summarized in Table 12.10, which shows that for the images with $SNR > 14.2$ db, the relative errors of the weight and the mean are less than 0.6%; all parameter estimates are in the Cramer-Rao estimation intervals.[¶]

2) Results from the real image

For the real image shown in Figure 12.4a, the ground truth is given in Table 12.5. The parameter estimates and Cramer–Rao Low Bounds of the variances of these estimates are given in Table 12.11. Using Tables 12.5 and 12.11, we note that the relative errors of the weights and the means ($k = 2, \cdots, 5$) are less than 8.4% and 2.1%, respectively, $\hat{\pi}_k \in \varpi_{\hat{\pi}_k}$ ($k = 2, \cdots, 5$), $\hat{\mu}_k \in \varpi_{\hat{\mu}_k}$ ($k = 2, \cdots, 5$), and $\hat{\sigma}_k^2 \in \varpi_{\hat{\sigma}_k^2}$ ($k = 5$).[‖]

For the image shown in Figure 12.1h (its SNR = 12.4 db is similar to SNR = 12.6 db of this real image), $\hat{\pi}_k \in \varpi_{\hat{\pi}_k}$ ($k = 1, 2, 4$), $\hat{\mu}_k \in \varpi_{\hat{\mu}_k}$ ($k = 1, 2, 3, 4$). These results show that the estimation performance for the images in Figures 12.1h and 12.4a are very similar.

12.2.3 Classification Performance

Let $IMG(J, K)$ denote an image of J pixels ($x_j, j = 1, \cdots, J$) and K image regions ($\mathcal{R}_k, k = 1, \cdots, K$). The iFNM model-based image analysis method (Chapter 10) uses a Bayesian classifier to classify pixels into image regions. The decision rule in pixel classification is

$$x_j \in \mathcal{R}_{k_0} \text{ if } \hat{\pi}_{k_0} g(x|\hat{\theta}_{k_0}) > \hat{\pi}_k g(x|\hat{\theta}_k) \ \ (k = 1, \cdots, K_0,\ k \neq k_0). \quad (12.49)$$

Suppose an image is partitioned into K_0 image regions, $\mathcal{R}_1, \cdots, \mathcal{R}_{K_0}$. Then the probability of misclassification, given that the true image region is $\mathcal{R}_{k_0}$, is [50]

$$P_{mis}(\bullet|k_0) = \hat{\pi}_{k_0} \sum_{k=1, k\neq k_0}^{K_0} \int_{\mathcal{R}_k} g(x|\hat{\theta}_{k_0}) dx$$

[¶]Only the variances of the 2nd image region of images in Figures 12.1.a - 12.1.c and 12.1.f are slightly outside $\varpi_{\hat{\sigma}_2^2}$.

[‖]Here we consider only the second through fifth image regions (i.e., Poly, 013A, Teflon, and Bone), because the first image region (rings) does not represent any real physical object, and is a product of image reconstruction.

TABLE 12.9
Estimation Results (Images in Figures 12.1a–l)

	k	$\hat{\pi}_k$	$\hat{\mu}_k$	$\hat{\sigma}_k^2$	$CRLB_{\hat{\pi}_k}$	$CRLB_{\hat{\mu}_k}$	$CRLB_{\hat{\sigma}_k^2}$
a	1	0.207764	−45.0016	1.02169	0.00004	0.00117	0.00235
	2	0.233643	−14.9936	0.94865	0.00004	0.00104	0.00208
	3	0.310547	15.0057	1.00741	0.00005	0.00078	0.00157
	4	0.248047	44.9790	0.98504	0.00005	0.00098	0.00196
b	1	0.207764	−45.0042	5.11387	0.00004	0.00587	0.05875
	2	0.233643	−14.9844	4.72885	0.00004	0.00522	0.05224
	3	0.310547	15.0146	5.03336	0.00005	0.00393	0.03930
	4	0.248047	44.9520	4.92275	0.00005	0.00492	0.04921
c	1	0.207764	−45.0048	10.1955	0.00004	0.01175	0.23502
	2	0.233643	−14.9783	9.45399	0.00004	0.01044	0.20899
	3	0.310547	15.0214	10.0634	0.00005	0.00786	0.15723
	4	0.248047	44.9320	9.85142	0.00005	0.00984	0.19685
d	1	0.207538	−45.0210	20.1751	0.00004	0.02350	0.94007
	2	0.233939	−14.9820	19.1059	0.00004	0.02090	0.83595
	3	0.310581	15.0403	20.0926	0.00005	0.01572	0.62893
	4	0.247943	44.9080	19.6336	0.00005	0.01969	0.78740
e	1	0.206936	−45.0667	29.7995	0.00004	0.03525	2.11516
	2	0.234568	−15.0128	29.0683	0.00004	0.03135	1.88088
	3	0.310768	15.0607	30.3491	0.00005	0.02358	1.41509
	4	0.247749	44.8940	29.4103	0.00005	0.02953	1.77165
f	1	0.207425	−45.0305	40.5117	0.00004	0.04700	3.76028
	2	0.233631	−15.0033	38.0433	0.00004	0.04180	3.34378
	3	0.311782	15.0780	41.3012	0.00005	0.03145	2.51572
	4	0.247160	44.9020	39.0368	0.00005	0.03937	3.14961
g	1	0.209074	−44.9135	52.0223	0.00004	0.05875	5.87544
	2	0.230335	−14.9907	45.0555	0.00004	0.05224	5.22466
	3	0.314779	15.0809	53.7758	0.00005	0.03931	3.93082
	4	0.245811	44.9460	48.4869	0.00005	0.04921	4.92125
h	1	0.211273	−44.7658	63.6950	0.00004	0.07051	8.46063
	2	0.224996	−15.0240	50.5216	0.00004	0.06270	7.52351
	3	0.320597	15.0782	68.4264	0.00005	0.04717	5.66038
	4	0.243132	45.0580	57.3469	0.00005	0.05906	7.08661
i	1	0.214228	−44.5700	75.8200	0.00004	0.08226	11.5159
	2	0.216772	−15.1219	54.0811	0.00004	0.07315	10.2403
	3	0.330609	15.0692	86.3911	0.00005	0.05503	7.70440
	4	0.238390	45.2783	65.2230	0.00005	0.06889	9.64567
j	1	0.217733	−44.3372	88.2984	0.00004	0.09401	15.0411
	2	0.206143	−15.2758	56.4073	0.00004	0.08359	13.3751
	3	0.344258	15.0501	107.650	0.00005	0.06289	10.0629
	4	0.231866	45.5868	72.3280	0.00005	0.07874	12.5984
k	1	0.221723	−44.0702	101.176	0.00004	0.10576	19.0364
	2	0.193388	−15.4692	57.8880	0.00004	0.09404	16.9279
	3	0.360892	15.0107	132.279	0.00005	0.07075	12.7358
	4	0.223996	45.9524	79.0594	0.00005	0.08858	15.9449
l	1	0.226134	−43.7729	114.403	0.00004	0.11751	23.5018
	2	0.178967	−15.6751	58.8092	0.00004	0.10449	20.9986
	3	0.379492	14.9407	159.894	0.00005	0.07862	15.7233
	4	0.215405	46.3330	85.7858	0.00005	0.09843	19.6850

TABLE 12.10
The Relative Errors and Estimation Intervals of the Parameter Estimates of the Images in Figures 12.1a–l

$\hat{\pi}_k$	Figures 12.1a–f	$SNR \geq 14.2$ db	$\varepsilon_\pi < 0.4\%$	$\hat{\pi}_k \in \varpi_{\hat{\pi}_k}$
$\hat{\mu}_k$	Figures 12.1–h	$SNR \geq 12.4$ db	$\varepsilon_\mu < 0.6\%$	$\hat{\mu}_k \in \varpi_{\hat{\mu}_k}$
$\hat{\sigma}_k^2$	Figures 12.1a–f	$SNR \geq 14.2$ db	$\varepsilon_\sigma < 5.0\%$	$\hat{\sigma}_k^2 \in \varpi_{\hat{\sigma}_k^2}$

TABLE 12.11
Estimation Results (Images in Figures 12.4.a)

k	$\hat{\pi}_k$	$\hat{\mu}_k$	$\hat{\sigma}_k^2$	$CRLB_{\hat{\pi}_k}$	$CRLB_{\hat{\mu}_k}$	$CRLB_{\hat{\sigma}_k^2} \times 10^6$
1	0.2348	−634.6	56271	0.00044	160.4	6.020
2	0.2167	−122.5	125	0.00038	6.6	0.008
3	0.2460	58.9	488	0.00041	1.5	0.001
4	0.1467	815.5	307284	0.00026	470.4	26.11
5	0.1558	1290.2	62701	0.00028	1164.5	170.9

$$= \hat{\pi}_{k_0} \sum_{k=1, k \neq k_0}^{K_0} \left(\Phi\left(\frac{x_k^{''} - \hat{\mu}_{k_0}}{\hat{\sigma}_{k_0}}\right) - \Phi\left(\frac{x_k^{'} - \hat{\mu}_{k_0}}{\hat{\sigma}_{k_0}}\right)\right), \tag{12.50}$$

where $\bullet|k_0$ represents the event that x comes from $\mathcal{R}_{k_0}$ but is classified (by the classifier) as coming from $\mathcal{R}_k$ $(k \neq k_0)$: $x_k^{'} < x \leq x_k^{''}$, and $\Phi(y)$ is cdf of the standard Gaussian random variable given by

$$\Phi(y) = \frac{1}{\sqrt{2\pi}} \int_{-\infty}^{y} e^{-\frac{x^2}{2}} dx. \tag{12.51}$$

12.2.3.1 Results of Classification Performance

1) Results from simulated images

The simulated images shown in Figure 12.1 are used here for evaluating classification performance. Image analysis is implemented using the iFNM model-based image analysis method of Chapter 10. The resultant images are shown in Figure 12.5, with the same labelings as for Figure 12.1. Image regions are represented by the mean values of pixels in each image region. In Figure 12.5, four grAy levels (white, light gray, dark grey, and black) are used for the four image regions.

Comparing these images with their counterparts in Figure 12.1 shows that, for the images in Figure 12.5a–12.5d, there is almost no classification error; for the images in Figure 12.5e–12.5f, there is a very small amount of error; for the images in Figure 12.5g–12.5l, there are some errors. These errors are mainly due to isolated pixels inside one image region being misclassified into another image region. The region shapes, however, are all preserved. This type of error occurs due to the fact that the Bayesian criterion (Eq. (12.49)) actually

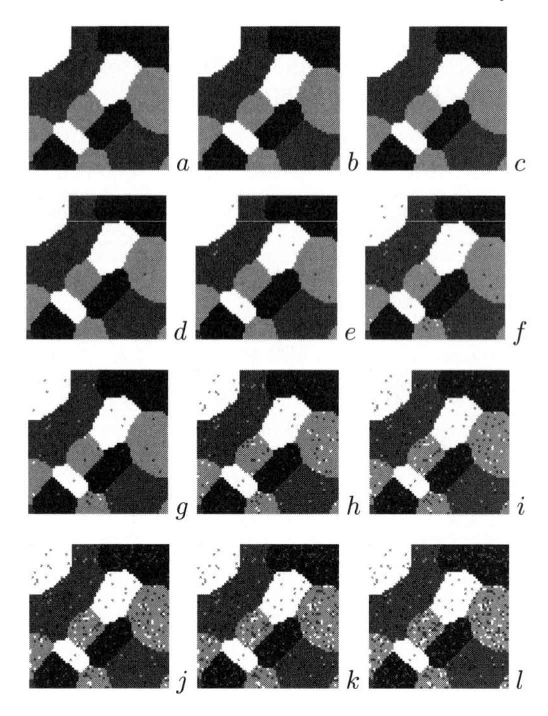

FIGURE 12.5
Partitioned image regions of twelve simulated images shown in Figure 12.1.

is based on Bayesian Probability (Eq. (12.28)) and Probability Membership (Eq. (10.11)), which treat pixels as being statistically independent and do not apply any spatial and statistical relationships among pixels (e.g., connectedness and correlation) in the classification process. This type of error can be removed using smoothing operations [51, 52].

Quantitatively, the classification results are summarized in Table 12.12. The interval $(x_k^{'}, x_k^{''}]$ in Table 12.12 represents the intensity range of the pixels classified in the k-th image region. It shows that for the images in Figures 12.5a–12.5d, the pixel intensities in the different image regions do not overlap; for the images in Figures 12.5e–12.5f, the pixel intensities have a small overlap (in some image regions); for the images in Figures 12.5g–12.5l, the pixel intensities have some overlap. Using Table 12.12 and Eq. (12.50), the misclassification probabilities are given in Table 12.13. Table 12.14 summarizes the classification performance, which shows that for the images with $SNR > 14.2$ db, the misclassification probabilities are less than 0.5%. Misclassification probabilities $< 0.015\%$, $< 0.447\%$, and $< 4.678\%$ are equivalent to the misclassifications of less than 1, 20, and 200 pixels, respectively, in this simulation study.

TABLE 12.12
Classification Results (Images in Figures 12.5a–l)

	$\mathcal{R}_1 : (x_1^{'}, x_1^{''}]$	$\mathcal{R}_2 : (x_2^{'}, x_2^{''}]$	$\mathcal{R}_3 : (x_3^{'}, x_3^{''}]$	$\mathcal{R}_4 : (x_4^{'}, x_4^{''}]$	*overlap*
a	−48, −42	−18, −12	11, 18	42, 48	*No*
b	−52, −37	−21, −7	6, 22	38, 52	*No*
c	−56, −34	−24, −4	2, 25	35, 55	*No*
d	−60, −31	−30, −2	0, 29	30, 58	*No*
e	−63, −30	−30, −1	0, 30	31, 61	*Partial*
f	−66, −30	−30, −1	−1, 30	31, 64	*Partial*
g	−69, −30	−30, −1	−1, 31	31, 66	*Yes*
h	−71, −29	−29, −2	−1, 31	31, 68	*Yes*
i	−73, −29	−29, −2	−2, 32	32, 70	*Yes*
j	−75, −29	−29, −3	−3, 32	32, 72	*Yes*
k	−77, −28	−28, −4	−4, 33	33, 73	*Yes*
l	−78, −28	−28, −5	−5, 34	34, 75	*Yes*

TABLE 12.13
Misclassification Probabilities for the Images in Figures 12.5

	$P_{mis}(\bullet \vert 1)$	$P_{mis}(\bullet \vert 2)$	$P_{mis}(\bullet \vert 3)$	$P_{mis}(\bullet \vert 4)$
a	0.00000	0.00000	0.00000	0.00000
b	0.00000	0.00000	0.00000	0.00000
c	0.00000	0.00000	0.00000	0.00000
d	0.00009	0.00010	0.00015	0.00004
e	0.00060	0.00127	0.00114	0.00075
f	0.00189	0.00447	0.00399	0.00211
g	0.00404	0.00720	0.00917	0.00556
h	0.00509	0.01100	0.01494	0.00771
i	0.00790	0.01447	0.02229	0.01194
j	0.01117	0.01750	0.03171	0.01277
k	0.01221	0.02236	0.03900	0.01626
l	0.01586	0.02433	0.04678	0.01971

TABLE 12.14
Classification Performance for the Images in Figures 12.5

$a - d$	No overlap	17.2 db $\leq SNR \leq$ 30.2 db	$P_{mis} < 0.015\%$
$e - f$	Small overlap	14.2 db $\leq SNR \leq$ 15.5 db	$P_{mis} < 0.447\%$
$g - l$	Some overlap	10.2 db $\leq SNR \leq$ 13.2 db	$P_{mis} < 4.678\%$

TABLE 12.15
Classification Results (Images in Figures 12.4)

$\mathcal{R}_1 : (x_1^{'}, x_1^{''}]$	$\mathcal{R}_2 : (x_2^{'}, x_2^{''}]$	$\mathcal{R}_3 : (x_3^{'}, x_3^{''}]$	$\mathcal{R}_4 : (x_4^{'}, x_4^{''}]$	$\mathcal{R}_5 : (x_5^{'}, x_5^{''}]$
−975, −130	−130, −104	−60, 83	83, 936	936, 1447

TABLE 12.16
Misclassification Probabilities for the Images in Figure 12.4.

$P_{mis}(\bullet\|1)$	$P_{mis}(\bullet\|2)$	$P_{mis}(\bullet\|3)$	$P_{mis}(\bullet\|4)$	$P_{mis}(\bullet\|5)$
0.00042	0.00000	0.03386	0.04801	0.01225

2) Results from the real image

For the real image shown in Figure 12.4.a, the partitioned images are given in Figure 12.4.b - 12.4.f which represent the images of the rings, Poly, 013A, Teflon, and Bone, respectively. Quantitatively, the classification results are summarized in Tables 12.15 and 12.16, which show that the pixel intervals have some overlap and the misclassification probabilities are less than 4.9%. These misclassification probabilities and those for the image in Figure 12.5.h (SNR=12.4 db) are of the same level.

12.3 Performance of the cFNM Model-Based Image Analysis Method

This section analyzes the cFNM model-based image analysis method of Chapter 11. It evaluates its performance at three steps: detection, estimation, and classification. Compared with the performance evaluation of the iFNM model-based image analysis method in Section 12.2, this evaluation is very brief.

1) For detection performance, the cFNM model-based image analysis method translates the image region detection problem into a sensor array processing framework and detects the number of image regions based on the signal eigenstructure of the converted array system. The detection scheme still uses the information theoretic criterion (ITC), but the likelihood function (i.e., the first term of the expression for ITC) is not based on the cFNM pdf. Because the ITC is used, the probabilities of over- and under-detection of the number of image regions are defined in the same way as in the case of the iFNM model-based image analysis method (Section 12.2.1).

It has been shown that error-detection probabilities are functions of the image quality, resolution, and complexity. The simulation results show that for images with moderate quality, the probability of over-detection of the number of image regions is negligible, and the probability of under-detection of the number of image regions is very small. These results are similar to those in the iFNM model-based image analysis method.

2) For estimation performance, the cFNM model-based image analysis method used the EM algorithm (with Gibbs distribution or MRF as a priori). Its performance is essentially the same as that of the iFNM model-based image analysis method given in Section 12.2.2.

3) For classification performance, the cFNM model-based image analysis method uses the MAP criterion. A strategy has been developed to assess the validity of using MAP with the underlying imaging principles. That is, the results obtained using MAP are judged if they are toward the physical ground truth that is to be imaged. To elaborate this strategy, MR images are used as an example.

In statistical physics, let $\Omega = \{\omega\}$ denote the possible configurations of a nuclear spin system. If the system is in thermal equilibrium with its surroundings, then the probability, or Boltzmann's factor, of ω is given by

$$p(\omega) = \frac{1}{\sum_{\omega} \exp(-\gamma\varepsilon(\omega))} \exp(-\gamma\varepsilon(\omega)), \tag{12.52}$$

where $\varepsilon(\omega)$ is the energy function of ω and $\gamma = \frac{1}{\kappa T}$, κ is the Boltzmann's constant, and T is the absolute temperature [53–56]. Eq. (12.52) indicates that the nuclear spin system tends to be in the low energy state, or with the higher probability to be in the low energy state.

In MR imaging, pixel intensities represent thermal equilibrium macroscopic magnetization (TEMM), that is, the vector sum of nuclear spin moments in unit volume. Chapter 9 shows that pixel intensities of MR images are characterized by an MRF $\mathbf{X}$ or a corresponding Gibbs distribution $P(\mathbf{x})$.

$$P(\mathbf{x}) = \frac{1}{\sum_{\mathbf{x}} \exp(-\beta^{-1} \sum_c V_c(\mathbf{x}))} \exp(-\beta^{-1} \sum_c V_c(\mathbf{x})), \tag{12.53}$$

which, in the functional form, is identical to the Boltzmann's factor Eq. (12.52).

Eq. (12.52) and Eq. (12.53) characterize the same nuclear spin system in the *microscopic* and the *macroscopic* states, respectively. Thus, given an image $\mathbf{x}$, maximizing the posterior probability $P(\mathbf{y}|\mathbf{x}) = P(\mathbf{x}|\mathbf{y})P(\mathbf{y})$, that is, seeking a configuration $\hat{\mathbf{y}}$ for the underlying MRF or for the corresponding Gibbs distribution, is actually to seek a thermal equilibrium of the nuclear spin system. This thermal equilibrium is well defined and is the physical ground truth to be imaged.

Given the interplay between the microscopic and macroscopic states of a process and of the analytical modeling of the propagations from one to the other, we can account for the geometric locality of the pixels/spins as well as for their probabilistic dynamics suited to the underlying structure of an image and its perturbations.

12.4 Appendices

12.4.1 Appendix 12A

This appendix derives pdf of Z, Eq. (12.16), and the probabilities of the over and under-detection of the number of image regions, P_{ov} Eq. (12.18) and P_{ud} Eq. (12.19).

A.1 pdf of Z

This section proves Eq. (12.16).

Proof.

Y_K of Eq. (12.11) has a χ^2 distribution with a degree of freedom $(J-K)$. Its pdf is

$$p_{Y_K}(y) = \begin{cases} \frac{1}{2^{\frac{J-K}{2}}\Gamma(\frac{J-K}{2})} e^{-\frac{y}{2}} y^{\frac{J-K}{2}-1} & (y > 0) \\ 0 & (y \leq 0). \end{cases} \tag{12.54}$$

Let $Z_{K_0} = \frac{1}{K_0\hat{\sigma}_0} Y_{K_0}$ and $Z_{K_1} = \frac{1}{K_1\hat{\sigma}_1} Y_{K_1}$. Their pdfs are

$$p_{Z_K}(z) = \begin{cases} \frac{(\frac{K_0\hat{\sigma}_0}{2})^{\frac{J-K_0}{2}}}{\Gamma(\frac{J-K_0}{2})} e^{-\frac{K_0\hat{\sigma}_0}{2}z} z^{\frac{J-K_0}{2}-1} & (z > 0) \\ 0 & (z \leq 0), \end{cases} \tag{12.55}$$

$$p_{Z_K}(z) = \begin{cases} \frac{(\frac{K_1\hat{\sigma}_1}{2})^{\frac{J-K_1}{2}}}{\Gamma(\frac{J-K_1}{2})} e^{-\frac{K_1\hat{\sigma}_1}{2}z} z^{\frac{J-K_1}{2}-1} & (z > 0) \\ 0 & (z \leq 0). \end{cases} \tag{12.56}$$

Assume that Z_{K_0} and Z_{K_1} are independent. The pdf $h(z)$ of $Z = Z_{K_1} - Z_{K_0}$ is

$$\begin{aligned} h(z) &= \int_{-\infty}^{\infty} p_{Z_{K_1}}(u) p_{Z_{K_0}}(u-z) du \\ &= \begin{cases} \int_0^{\infty} p_{Z_{K_1}}(u) p_{Z_{K_0}}(u-z) du & (z < 0) \\ \int_z^{\infty} p_{Z_{K_1}}(u) p_{Z_{K_0}}(u-z) du & (z \geq 0). \end{cases} \end{aligned} \tag{12.57}$$

Substituting Eqs. (12.55), (12.56) into Eq. (12.57), using formula [44]

$$\int_0^{\infty} x^n e^{-ax} dx = \frac{n!}{a^{n+1}} \quad (a > 0)\ , \tag{12.58}$$

for the case $z < 0$, and by using formula [44]

$$\int_u^\infty x^n e^{-bx} dx = e^{-bu} \sum_{k=0}^{n} \frac{n!}{k!} \cdot \frac{u^k}{b^{n-k+1}} \quad (u > 0,\ b > 0), \tag{12.59}$$

and the mathematical induction method for the case $z \geq 0$, we have

$$h(z) = \begin{cases} Ce^{\frac{K_0\hat{\sigma}_0}{2}z} \sum_{l=0}^{m_0} \binom{l}{m_0} \frac{(m_0+m_1-l)!}{a^{m_0+m_1-l+1}} (-z)^l & (z < 0) \\ \\ Ce^{-\frac{K_1\hat{\sigma}_1}{2}z} \sum_{l=0}^{m_0} \binom{l}{m_0} \frac{(m_0+m_1-l)!}{a^{m_0+m_1-l+1}} (z)^l & (z \geq 0), \end{cases} \tag{12.60}$$

where

$$\begin{cases} \binom{l}{m_0} = \frac{m_0!}{l!(m_0-l)!} \\ \\ m_0 = \frac{J-K_0}{2} - 1 \quad \text{and} \quad m_1 = \frac{J-K_1}{2} - 1 \\ \\ a = \frac{K_0\hat{\sigma}_0 + K_1\hat{\sigma}_1}{2} \\ \\ C = \frac{1}{m_0! m_1!} \left(\frac{K_0\hat{\sigma}_0}{2}\right)^{\frac{J-K_0}{2}} \left(\frac{K_1\hat{\sigma}_1}{2}\right)^{\frac{J-K_1}{2}}. \end{cases} \tag{12.61}$$

It has been verified that $\int_{-\infty}^{\infty} h(z)dz = 1$. ■

A.2 Thresholds Δ_1, Δ_2, and Δ

Δ_1, Δ_2, and Δ of Eq. (12.14) are functions of J, K_0, and K_1. Intensive simulations** showed that they have the following properties:

(1) $\Delta_1(K_1 > K_0) < 0$ and $\Delta_1(K_1 < K_0) > 0$,

(2) $\Delta_2(K_1 \neq K_0) \leq 0$ and monotonically increases with σ_0,

(3) $|\Delta_2(K_1 > K_0)| < |\Delta_2(K_1 < K_0)|$,

(4) $\Delta < 0$ for small and moderate σ_0 and $\Delta > 0$ only for very large σ_0.

These findings can be justified as follows.

For (1), it is straightforward to verify using the definition in Eq. (12.14).

For (2), because the frequency of occurrence of $|K_1 - K_0| = 1$ is higher than that of $|K_1 - K_0| > 1$, we set $|K_1 - K_0| = 1$ in the simulations. In the case of over-detection ($K_1 > K_0$), a new region (say, the l-th region,

---

**About the simulation study. A Gibbs Sampler algorithm has been developed (Appendix 9A of Chapter 9) based on the theory in [28]. The algorithm has been implemented on Macintosh computers with IMSL [57]. Using this algorithm, the various MRF configurations with different resolutions (256×256, 128×128, and 64×64) were generated. The MRF shown in Figures 12.1 and 12.5 is one of them. Then Gaussian noise with different variances $\sigma_0^2 = 1, 5, 10i, 100(j+1), 500(k+2)$ ($i = 1, \cdots, 10$, $j = 1, \cdots, 9$, $k = 1, \cdots, 8$) was superimposed on each resolution of each MRF configuration. Finally, the iFNM model-based image analysis method (Chapter 10) was applied to these (over 600) images to perform detection, estimation, and classification.

$1 \leq l \leq K_1$) with a few pixels is created. This newly created region has a large variance (but not very large) and a very small weight ($\hat{\pi}_l \simeq 0$) such that $\hat{\pi}_l\hat{\sigma}_l^2$ is very small. Other regions (say, the k-th region, $1 \leq k \leq K_1$ and $k \neq l$) are essentially the same as the original regions. Thus $\hat{\sigma}_1^2 = \sum_{k=1}^{K_1} \hat{\pi}_k\hat{\sigma}_k^2 \simeq \sum_{k=1,\, k\neq l}^{K_1} \hat{\pi}_k\hat{\sigma}_k^2 \simeq \sum_{k=1}^{K_0} \hat{\pi}_k\hat{\sigma}_k^2 = \hat{\sigma}_0^2$, or $\hat{\sigma}_1$ is slightly larger than $\hat{\sigma}_0$. As a result, $K_1\hat{\sigma}_1 > K_0\hat{\sigma}_0$. In the case of under-detection ($K_1 < K_0$), some regions (say, the i-th and j-th regions, $1 \leq i, j \leq K_0$) are merged into one region (say, the l-th region, $1 \leq l \leq K_1$). This newly merged region has a **very** large variance and a big weight ($\hat{\pi}_l = \hat{\pi}_i + \hat{\pi}_j$) such that $\hat{\pi}_l\hat{\sigma}_l^2 \gg \hat{\pi}_i\hat{\sigma}_i^2 + \hat{\pi}_j\hat{\sigma}_j^2$. Other regions (say, the k-th region, $1 \leq k \leq K_1$ and $k \neq l$) remain almost the same as the original regions. Thus, $\hat{\sigma}_1^2 = \sum_{k=1}^{K_1} \hat{\pi}_k\hat{\sigma}_k^2 \gg \sum_{k=1}^{K_0} \hat{\pi}_k\hat{\sigma}_k^2 = \hat{\sigma}_0^2$. As a result, $K_1\hat{\sigma}_1 > K_0\hat{\sigma}_0$ (even though $K_1 < K_0$). So, in both of the above cases, $\Delta_2 = 2J(\frac{1}{K_1\hat{\sigma}_1} - \frac{1}{K_0\hat{\sigma}_0}) \leq 0$. The equality occurs if and only if $K_1\hat{\sigma}_1 = K_0\hat{\sigma}_0$. Simulations also showed that when σ_0 increases, $\hat{\sigma}_0$ and $\hat{\sigma}_1$ increase. Therefore, Δ_2 will monotonically increase.

For (3), we also set $|K_1 - K_0| = 1$. From the discussion in (2) above, we know that $\Delta_2 = 2J(\frac{1}{K_1\hat{\sigma}_1} - \frac{1}{K_0\hat{\sigma}_0}) \leq 0$, but due to $\hat{\sigma}_1 \simeq \hat{\sigma}_0$ ($K_1 > K_0$) and $\hat{\sigma}_1 \gg \hat{\sigma}_0$ ($K_1 < K_0$), $|\frac{1}{K_1\hat{\sigma}_1} - \frac{1}{K_0\hat{\sigma}_0}|$ is close to zero when $K_1 > K_0$ and $|\frac{1}{K_1\hat{\sigma}_1} - \frac{1}{K_0\hat{\sigma}_0}|$ is quite different from zero when $K_1 < K_0$. Therefore, $|\Delta_2(K_1 > K_0)| < |\Delta_2(K_1 < K_0)|$.

For (4), in the case of over-detection ($K_1 > K_0$), because $\Delta_1 < 0$ and $\Delta_2 < 0$, we have $\Delta < 0$. In the case of under-detection ($K_1 < K_0$), due to $|\Delta_1| \propto \ln J$ when $\Delta_1 \geq 0$ and $|\Delta_2| \propto J$ when $\Delta_2 < 0$; therefore, for the not very large σ_0, $\Delta = \Delta_1 + \Delta_2 < 0$. For very large σ_0, $|\frac{1}{K_1\hat{\sigma}_1} - \frac{1}{K_0\hat{\sigma}_0}|$ is very small so that Δ_2 becomes very small. As a result, $\Delta = \Delta_1 + \Delta_2 > 0$.

A.3 Probabilities P_{ov} and P_{ud}

This section proves Eq. (12.18) and Eq. (12.19).

Proof.

For the probability of over-detection P_{ov}, because $\Delta < 0$, from Eqs. (12.5), (12.15), and (12.14) and using Eq. (12.60), we have

$$P_{ov} = \int_{-\infty}^{\Delta} h(z)dz = \frac{\gamma^{m_1-m_0}}{(\gamma + \gamma^{-1})^{m_0+m_1+1}} \sum_{l=0}^{m_0} \binom{m_1}{m_0+m_1-l}[\gamma(1+\gamma^2)^l e_0], \tag{12.62}$$

where

$$\gamma = \sqrt{\frac{K_1\hat{\sigma}_1}{K_0\hat{\sigma}_0}} \quad \text{and} \quad e_0 = \sum_{j=0}^{l} \frac{(-\frac{K_0\hat{\sigma}_0}{2}\Delta)^j}{j!} e^{\frac{K_0\hat{\sigma}_0}{2}\Delta}. \tag{12.63}$$

For the probability of under-detection P_{ud}, because $\Delta_1 > 0$ and $\Delta_2 < 0$, Δ may be negative (when σ_0^2 is small or moderate) or positive (when σ_0^2 is **very** large). When $\Delta < 0$, P_{ud} has the same functional form as in Eq. (12.62) but with different values of the same parameters. When $\Delta > 0$, from Eqs. (12.5), (12.15), and (12.14) and using Eq. (12.60), we have

$$P_{ud} = \int_{-\infty}^{\Delta} h(z)dz = \int_{-\infty}^{0} h(z)dz + \int_{0}^{\Delta} h(z)dz = \frac{\gamma^{m_1-m_0}}{(\gamma+\gamma^{-1})^{m_0+m_1+1}}$$
$$\sum_{l=0}^{m_0} \binom{m_1}{m_0+m_1-l} [\gamma(1+\gamma^2)^l + \gamma^{-1}(1+\gamma^{-2})^l(1-e_1)], \tag{12.64}$$

where

$$\gamma = \sqrt{\frac{K_1\hat{\sigma}_1}{K_0\hat{\sigma}_0}} \quad \text{and} \quad e_1 = \sum_{j=0}^{l} \frac{(\frac{K_1\hat{\sigma}_1}{2}\Delta)^j}{j!} e^{-\frac{K_1\hat{\sigma}_1}{2}\Delta}. \tag{12.65}$$

Thus, Eq. (12.62) finally becomes

$$P_{ov} = \sum_{l=0}^{m_0} \binom{m_1}{m_0+m_1-l} \frac{\gamma^{m_1-m_0+l+1} e_0}{(\gamma+\gamma^{-1})^{m_1+m_0-l+1}}. \tag{12.66}$$

Eq. (12.64), when $\Delta \geq 0$, becomes

$$P_{ud} = \sum_{l=0}^{m_0} \binom{m_1}{m_0+m_1-l} \frac{\gamma^{m_1-m_0+l+1}(1+\gamma^{-2l-2}(1-e_1))}{(\gamma+\gamma^{-1})^{m_0+m_1-l+1}}, \tag{12.67}$$

and when $\Delta < 0$, becomes

$$P_{ud} = \sum_{l=0}^{m_0} \binom{m_1}{m_0+m_1-l} \frac{\gamma^{m_1-m_0+l+1} e_0}{(\gamma+\gamma^{-1})^{m_1+m_0-l+1}} e_0. \tag{12.68}$$

Because $J >> K_0$ and $J >> K_1$, $\frac{J-K_0}{2} - 1 \simeq \frac{J-K_1}{2} - 1$, that is, $m_0 \simeq m_1$ (Eq. (12.61)). Let $m_0 \simeq m_1 \triangleq m$, Eqs. (12.60) and (12.61) become Eqs. (12.16) and (12.17), and Eqs. (12.66), (12.68), and (12.67) become Eqs. (12.18) and (12.19), respectively. ■

A.4 Special Cases

When $\gamma = 1$, Δ_2 in Eq. (12.14) becomes zero, which leads to $\Delta = \Delta_1$. Thus, Eq. (12.15) becomes

$$P_e = P(Z < \Delta) = P(Z < \Delta_1), \tag{12.69}$$

and Eqs. (12.18) and (12.19) become

$$P_{ov} = \sum_{l=0}^{m} \frac{\binom{m}{2m-l} e_0}{2^{2m-l+1}} \quad \text{and} \quad P_{ud} = \sum_{l=0}^{m} \frac{\binom{m}{2m-l}(2-e_1)}{2^{2m-l+1}}. \tag{12.70}$$

respectively. Also, when $\gamma = 1$, the pdf $h(z)$ of Eq. (12.16) is symmetric, that is, $h(z) = h(-z)$.

One of the possible cases for $\gamma = 1$ is the case of correct detection ($K_1 = K_0$) and accurate estimation ($\hat{\pi}_k = \pi_k$, $\hat{\mu}_k = \mu_k$, $\hat{\sigma}_k^2 = \sigma_k^2$). In this special case, Δ_1 of Eq. (12.14) becomes zero, which leads to $\Delta = 0$. As a result, Eqs. (12.18) and (12.19) become $P_{ov}(0) = \int_{-\infty}^{0} h(z)dz$ and $P_{ud}(0) = \int_{-\infty}^{0} h(z)dz$, respectively. Using $h(z) = h(-z)$, we obtain $P_{ov}(0) = P_{ud}(0) = 0.5$.

The interpretation is as follows. Let K_1 take various values (e.g., $K_1 = 1, \cdots, 100$). K_0 is one of them. $P_{ov}(0) = P_{ud}(0) = 0.5$ indicates that K_1 approaches K_0 with probability 1: 50% from the left side (under-detection side) and 50% from the right side (over-detection side), in this special case.

A.5. About $\ln w \simeq w - 1$ and $e^{-\frac{1}{2}w} \simeq 1 - \frac{1}{2}w$

Here we justify these two simplifications used in Section 12.2.1.1. [44] shows that

$$\ln w = \sum_{k=1}^{\infty} a_k = \sum_{k=1}^{\infty} (-1)^{k+1} \frac{(w-1)^k}{k}$$
$$= (w-1) - \frac{1}{2}(w-1)^2 + \frac{1}{3}(w-1)^3 - \cdots \quad (0 < w \leq 2). \tag{12.71}$$

In our case, $w = f(x_j|\hat{\mathbf{r}})$. As indicated in Section 12.2.1.1, $0 < f(x_j|\hat{\mathbf{r}}) < 1$. Thus, let $a_k = (-1)^{k+1} \frac{(w-1)^k}{k}$; we have $|a_k| < 1$ and $|\frac{a_{k+1}}{a_k}| < 1$ ($k = 1, 2, \cdots\cdots$). [44] also shows that

$$e^{-\frac{1}{2}w} = \sum_{k=0}^{\infty} b_k = \sum_{k=0}^{\infty} (-1)^k \frac{(w/2)^k}{k!}$$
$$= 1 - \frac{w}{2} + \frac{1}{2!}\left(\frac{w}{2}\right)^2 - \frac{1}{3!}\left(\frac{w}{2}\right)^3 + \cdots \quad (-\infty < w < \infty). \tag{12.72}$$

In our case, $w = \left(\frac{x_j - \hat{\mu}_k}{\hat{\sigma}}\right)^2$. As indicated in Section 12.2.1.1, pixel x_j belongs to a (one) Gaussian component of iFNM. This means that with probability 70%, $|x_j - \hat{\mu}_k| < \hat{\sigma}$, that is, $w < 1$. Thus, letting $b_k = (-1)^k \frac{(w/2)^k}{k!}$, we have $|b_k| < 1$ and $|\frac{b_{k+1}}{b_k}| < 1$ ($k = 1, 2, \cdots\cdots$).

For $\ln w$, when more terms in the series representation Eq. (12.71) (say, $a_1 + a_2 + a_3$) are used to approximate $\ln w$, then it will be more accurate than the linear term $a_1 = w - 1$. However, by taking this approximation, the log likelihood $\ln \mathcal{L}_K(\hat{\mathbf{r}})$ of Eq. (12.7) will become too complicated for the derivation of a formula similar to Eq. (12.12). Another linear approximation, $\alpha_1 w - \alpha_2$, may be a better choice than $(w-1)$ for approximating $\ln w$, because it improves the accuracy and still keeps the simplicity of the approximation. Similarly, for $e^{-\frac{1}{2}w}$, $\beta_1 - \beta_2 w$ may be a better choice than $b_0 + b_1 = 1 - \frac{w}{2}$ for approximating $e^{-\frac{1}{2}w}$.

If we use $\ln w \simeq \alpha_1 w - \alpha_2$ and $e^{-\frac{1}{2}w} \simeq \beta_1 - \beta_2 w$, then we can show that Eqs. (12.12) and (12.13) will be

$$\ln \mathcal{L}(\hat{\mathbf{r}}) \simeq \frac{\alpha_1}{\sqrt{2\pi}K\hat{\sigma}}[\beta_1 J - \frac{\beta_2}{2}Y_K] - \alpha_2 J, \tag{12.73}$$

and

$$\begin{aligned} P_e &= P(\frac{1}{K_1\hat{\sigma}_1}Y_{K_1} - \frac{1}{K_0\hat{\sigma}_0}Y_{K_0}) \\ &< \frac{1}{\alpha_1\beta_2}(3\sqrt{2\pi}\ln J)(K_0 - K_1) + 2J\frac{\beta_1}{\beta_2}(\frac{1}{K_1\hat{\sigma}_1} - \frac{1}{K_0\hat{\sigma}_0})), \end{aligned} \tag{12.74}$$

respectively. Using Eq. (12.14), Eq. (12.74) becomes

$$P_e = P(Z < \frac{1}{\alpha_1\beta_2}\Delta_1 + \frac{\beta_1}{\beta_2}\Delta_2) \triangleq P(Z < \Delta_1^{'} + \Delta_2^{'}), \tag{12.75}$$

where $\Delta_1^{'} = \frac{1}{\alpha_1\beta_2}\Delta_1$ and $\Delta_2^{'} = \frac{\beta_1}{\beta_2}\Delta_2$.

Graphically, it is easy to verify that $\alpha_1, \alpha_2 > 1$ and $0 < \beta_1, \beta_2 < 1$. More precisely, these coefficients can be determined by minimizing the errors $|(\alpha_1 w - \alpha_2) - \ln w|$ and $|(\beta_1 - \beta_2 w) - e^{-\frac{1}{2}w}|$. This minimization can be numerically implemented by a least squares error fitting approach: $\min_{\alpha_1,\alpha_2}\sum_i(\alpha_1 w_i - \alpha_2 - \ln w_i)^2$ and $\min_{\beta_1,\beta_2}\sum_i(\beta_1 - \beta_2 w_i - e^{-\frac{1}{2}w_i})^2$, over a set of w_i.

For example, by uniformly taking ten data points (samples) w_i over $(0,1)$ and using the above least squares error approach, the numerical study shows that $\alpha_1\beta_2 = 0.9814$ and $\frac{\beta_1}{\beta_2} = 2.5068$. $\alpha_1\beta_2 \simeq 1$ could be the result of the fact that the opposite effects caused by $\ln w$ and $e^{-\frac{1}{2}w}$ cancel each other. This result leads to $\Delta_1' \simeq \Delta_1$ and $\Delta_2' < \Delta_2$ (because $\Delta_2 < 0$), that is, probability $P_e' = P(Z < \Delta_1' + \Delta_2')$ is less than $P_e = P(Z < \Delta_1 + \Delta_2)$. This observation has also been verified by other data samples w_i. Therefore, the probability P_e of Eq. (12.13) is consistent with P_e' (which are obtained using more accurate linear approximations of $\ln w$ and $e^{-\frac{1}{2}w}$) and also can be used as an upper limit of P_e'.

12.4.2 Appendix 12B

This appendix derives Eq. (12.22).

Proof.

From [46], we have

$$\sum_{k=1}^{K}\sum_{j=1}^{J_k}(x_j - \hat{\mu})^2 = \sum_{k=1}^{K} J_k(\hat{\mu}_k - \hat{\mu})^2 + \sum_{k=1}^{K}\sum_{j=1}^{J_k}(x_j - \hat{\mu}_k)^2, \tag{12.76}$$

where

$$\begin{cases} J_k = \text{Number of } x_j \in \mathcal{R}_k \\ \hat{\mu} = \frac{1}{J}\sum_{j=1}^{J} x_j \\ \hat{\mu}_k = \frac{1}{J_k}\sum_{j=1}^{J_k} x_j \\ J = \sum_{k=1}^{K} J_k. \end{cases} \tag{12.77}$$

Because

$$\sum_{k=1}^{K} J_k(\hat{\mu}_k - \hat{\mu})^2 = \sum_{k=1}^{K} J_k{\hat{\mu}_k}^2 - J\hat{\mu}^2, \tag{12.78}$$

and (from Eq. (12.21)),

$$\sum_{k=1}^{K}\sum_{j=1}^{J_k}(x_j - \hat{\mu}_k)^2 = J\hat{\sigma}^2, \tag{12.79}$$

Eq. (12.76) becomes

$$\sum_{k=1}^{K}\sum_{j=1}^{J_k}(x_j - \hat{\mu})^2 + J\hat{\mu}^2 = \sum_{k=1}^{K} J_k{\hat{\mu}_k}^2 + J\hat{\sigma}^2. \tag{12.80}$$

For a given image, $\sum_{k=1}^{K}\sum_{j=1}^{J_k}(x_j - \hat{\mu})^2 + J\hat{\mu}^2$ is constant. Thus, we have

$$\sum_{k=1}^{K_1} J_k{\hat{\mu}_k}^2 + J\hat{\sigma}_1^2 = \sum_{k=1}^{K_0} J_k{\hat{\mu}_k}^2 + J\hat{\sigma}_0^2. \tag{12.81}$$

Eq. (12.81) leads to Eq. (12.22). ■

Problems

12.1. Prove the properties of Eq. (12.29).

12.2. Derive Eq. (12.37).

12.3. Derive Eq. (12.41).

12.4. Derive Eq. (12.46).

12.5. The mathematical induction method is used in the derivation of Eq. (12.60). Show the details of this mathematical induction.

References

[1] Dijkstra, E.: A note on two problems in connexion with graphs. *Numerische Mathematik* **1** (1959) 269–271.

[2] Dial, R.: Shortest-path forest with topological ordering. *Comm. ACM* **12**(11) (1969) 632–633.

[3] Mortensen, E., Barrett, W.: Interactive segmentation with intelligent scissors. *Graphical Models and Image Process.* **60** (1998) 349–384.

[4] Terzopoulos, D., Fleischer, K.: Deformable models. *Vis. Comput.* **4** (1988) 306–331.

[5] Cohen, L.: On active contour models and balloons. *CVGIP: Image Understanding* **53** (1991) 211–218.

[6] Caselles, V., Kimmel, R., Sapiro, G.: Geodesic active contours. *Proc. 5th Intl. Conf. Computer Vision* (1995) 694–699.

[7] Davatzikos, C., Prince, J.: An active contour model for mapping the cortex. *IEEE Trans. Med. Imag.* **14**(3) (1995) 65–80.

[8] Shen, D., Davatzikos, C.: Hammer: Hierarchical attribute matching mechanism for elastic registration. *IEEE Trans. Med. Imag.* **22**(11) (2002) 1421–1439.

[9] Chan, F., Lam, F., Zhu, H.: Adaptice thresholding by variational method. *IEEE Trans. Image Process.* **7**(3) (1998) 468–473.

[10] Faugeras, O., Keriven, R.: Variational principles, surface evolution, PDE's, level set methods, and the stereo problem. *IEEE Trans. Image Process.* **7**(3) (1998) 336–344.

[11] Chan, T., Vese, L.: Active contour without edges. *IEEE Trans. Image Process.* **10**(2) (2001) 266–277.

[12] Avants, B., Gee, J.: Formulation and evaluation of variational curve matching with prior constraints. *Lect. Notes Comput. Sci.* **2717** (2003) 21–30.

[13] Xu, M., Thompson, P., Toga, A.: An adaptive level set segmentation on a triangulated mesh. *IEEE Trans. Med. Imag.* **32**(2) (2004) 191–201.

[14] Cootes, T., Taylor, C., Cooper, D.: Active shape models—Their training and application. *Computer Vision and Image Understanding* **61**(1) (1995) 38–59.

[15] Duta, N., Sonka, M.: Segmentation and interpretation of MR brain images: An improved active shape model. *IEEE Trans. Med. Imag.* **17**(6) (1998) 1049–1062.

[16] Wang, Y., Staib, L.: Boundary finding with prior shape and smoothness models. *IEEE Trans. Pattern Anal. Machine Intell.* **22**(7) (2000) 738–743.

[17] Cootes, T., Edwards, G., Taylor, C.: Active appearance models. *IEEE Trans. Pattern Anal. Machine Intell.* **23**(6) (2001) 681–685.

[18] Davatzikos, C., Tao, X., Shen, D.: Hierarchical active shape models, using the wavelet transform. *IEEE Trans. Med. Imag.* **22**(3) (2003) 414–423.

[19] Shen, D., Zhan, Y., Davatzikos, C.: Segmentation of prostate boundaries from ultrasound images using statistical shape model. *IEEE Trans. Med. Imag.* **22**(4) (2003) 539–551.

[20] Kaufmann, A.: *Introduction to the Theory of Fuzzy Subsets.* Academic, New York (1975).

[21] Rosenfeld, A.: Fuzzy digital topology. *Inform. Contr.* **40**(1) (1979) 76–87.

[22] Herman, G., Carvalho, B.: Multiseeded segmentation using fuzzy connectedness. *IEEE Trans. Pattern Anal. Machine Intell.* **23** (2001) 460–474.

[23] Nyul, L., Falcao, A., Udupa, J.: Fuzzy-connected 3D image segmentationat interactive speeds. *Graph. Models* **64** (2002) 259–281.

[24] Besag, J.: Spatial interaction and the statistical analysis of lattice system (with discussion). *J. Roy. Statist. Soc.* **36B**(2) (1974) 192–236.

[25] Kindermann, R., Snell, J.: *Markov Random Field and Their Applications.* American Mathematical Society, Providence, Rhode Island, (1980).

[26] Cross, G., Jain, A.: Markov random field texture models. *IEEE Trans. Pattern Anal. Machine Intell.* **5** (1983) 25–39.

[27] Kashap, R., Chellappa, R.: Estimation and choice of neighbors in spatial interaction models of images. *IEEE Trans. Information Theory* **29** (1983) 60–72.

[28] Geman, S., Geman, D.: Stochastic relaxation, Gibbs distributions, and the Bayesian restoration of images. *IEEE Trans. Pattern Anal. Machine Intell.* **6**(6) (1984) 721–724.

[29] Chellappa, R., Lashyap, R.L.: Texture synthesis using 2-D noncausal autoregressive models. *IEEE Trans. Acoustics, Speech, Signal Process.* **33** (1985) 194–203.

[30] Besag, J.: On the statistical analysis of dirty pictures. *J. Roy. Statist. Soc.* **48B** (1986) 259–302.

[31] Derin, H., Elliot, H.: Modeling and segmentation of noisy and textured images using Gibbs random fields. *IEEE Trans. Pattern Anal. Machine Intell.* **9**(1) (1987) 39–55.

[32] Leahy, R., Hebert, T., Lee, R.: Applications of Markov random field models in medical imaging. *Proc. IPMI* **11** (1989) 1–14.

[33] Dubes, R.C., Jain, A.K.: Random field models in image analysis. *J. Appl. Stat.* **16**(2) (1989) 131–164.

[34] Bouman, C., Liu, B.: Multiple resolution segmentation of texture images. *IEEE Trans. Pattern Anal. Machine Intell.* **13**(2) (1991) 99–113.

[35] Manjunath, B., Chellappa, R.: Unsupervised texture segmentation using Markov random field models. *IEEE Trans. Pattern Anal. Machine Intell.* **13** (1991) 478–482.

[36] Liang, Z., MacFall, J., Harrington, D.: Parameter estimation and tissue segmentation from multispectral MR images. *IEEE Trans. Med. Imag.* **13**(3) (1994) 441–449.

[37] Bouman, C., Shapiro, M.: A multiscale random field model for Bayesian image segmentation. *IEEE Trans. Image Process.* **3** (1994) 162–176.

[38] Krishnamachari, S., Chellappa, R.: Multiresolution Gauss-Markov random field models for texture segmentation. *IEEE Trans. Image Process.* **6**(2) (1997) 251–267.

[39] Willsky, A.: Multiresolution Markov model for signal and image processing. *Proc. IEEE* **90**(8) (2002) 1396–1458.

[40] Jannin, P., Fitzpatrick, J., Hawkes, D., Pennec, X., Shahidi, R., Vannier, M.: Validation of medical image processing in image-guided therapy. *IEEE Trans. Med. Imag.* **21**(12) (2002) 1445–1449.

[41] Schwarz, G.: Estimating the dimension of a model. *Ann. Stat.* **6**(2) (1978) 461–464.

[42] Rissanen, J.: A universal prior for integers and estimation by minimum description length. *Ann. Stat.* **11**(2) (1983) 416–431.

[43] Wax, M., Kailath, T.: Detection of signals by information theoretic criteria. *IEEE Trans. Acoust., Speech, and Signal Proc.* **33**(2) (1985) 387–392.

[44] Gradshteyn, I.S., Ryzhik, I.M.: *Table of Integrals, Series, and Products.* Academic Press, New York (1994).

[45] Fisz, M.: *Probability Theory and Mathematical Statistics.* John Wiley & Sons, Inc., New York (1965).

[46] Scheffe, H.: *The Analysis of Variance.* John Wiley & Sons, Inc., New York (1959).

[47] Perlovsky, L.I.: Cramer-Rao bounds for the estimation of means in a clustering problem. *Pattern Recog. Lett.* **8**(1) (1988) 1–3.

[48] Perlovsky, L.I.: Cramer-Rao bounds for the estimation of normal mixtures. *Pattern Recogn. Lett.* **10**(3) (1989) 141–148.

[49] Lei, T., Sewchand, W.: Statistical approach to x-ray CT imaging and its applications in image analysis—Part 1: Statistical analysis of x-ray CT imaging. *IEEE Trans. Med. Imag.* **11**(1) (1992) 53–61.

[50] Anderson, T.W.: *An Introduction to Multivariate Statistical Analysis.* John Wiley & Sons Inc,, New York (1984).

[51] Zhang, J.: The mean field theory in EM procedures for blind Markov random field image. *IEEE Trans. Image Proc.* **2**(1) (1993) 27–40.

[52] Descombes, X., Kruggel, F., von Cramon, D.Y.: Spatio-temporal fMRI analysis using Markov random field. *IEEE Trans. Medical Imaging* **17**(6) (1998) 1028–1039.

[53] Cho, Z.H., Jones, J.P., Singh, M.: *Foundations of Medical Imaging.* John Wiley & Sons, Inc., New York (1993).

[54] Nishimura, D.: *Principles of Magnetic Resonance Imaging.* Stanford University, Palo Alto, California, (1996).

[55] Haacke, E., Brown, R., Thompson, M., Venkatesan, R.: *Magnetic Resonance Imaging: Physical Principles and Sequence Design.* John Wiley & Sons Inc., New York (1999).

[56] Liang, Z.P., Lauterbur, P.: *Principles of Magnetic Resonance Imaging, A Signal Processing Perspective.* IEEE Press, New York (2000).

[57] Developer, I.: IMSL—Problem-Solving Software Systems. IMSL, Inc., Houston, Texas, (1990).

Index